# Processing Metabolomics and Proteomics Data with Open Software

## A Practical Guide

**New Developments in Mass Spectrometry**

*Editor-in-chief:*
Frank Sobott, *University of Leeds, UK*

*Series editors:*
Juan F. Garcia-Reyes, *Universidad de Jaén, Spain*
Marek Domin, *Boston College, USA*

*Titles in the Series:*

1: Quantitative Proteomics
2: Ambient Ionization Mass Spectrometry
3: Sector Field Mass Spectrometry for Elemental and Isotopic Analysis
4: Tandem Mass Spectrometry of Lipids: Molecular Analysis of Complex Lipids
5: Proteome Informatics
6: Capillary Electrophoresis–Mass Spectrometry for Metabolomics
7: Lipidomics: Current and Emerging Techniques
8: Processing Metabolomics and Proteomics Data with Open Software: A Practical Guide

*How to obtain future titles on publication:*
A standing order plan is available for this series. A standing order will bring delivery of each new volume immediately on publication.

*For further information please contact:*
Book Sales Department, Royal Society of Chemistry, Thomas Graham House, Science Park, Milton Road, Cambridge, CB4 0WF, UK
Telephone: +44 (0)1223 420066, Fax: +44 (0)1223 420247
Email: booksales@rsc.org
Visit our website at www.rsc.org/books

# *Processing Metabolomics and Proteomics Data with Open Software*

## *A Practical Guide*

Edited by

**Robert Winkler**
*CINVESTAV Unidad Irapuato, Mexico*
*Email: robert.winkler@cinvestav.mx*

ROYAL SOCIETY OF **CHEMISTRY**

New Developments in Mass Spectrometry No. 8

Print ISBN: 978-1-78801-721-3
PDF ISBN: 978-1-78801-988-0
EPUB ISBN: 978-1-78801-990-3
Print ISSN: 2045-7545
Electronic ISSN: 2045-7553

A catalogue record for this book is available from the British Library



The Royal Society of Chemistry is a charity, registered in England and Wales, Number 207890, and a company incorporated in England by Royal Charter (Registered No. RC000524), registered office: Burlington House, Piccadilly, London W1J 0BA, UK, Telephone: +44 (0) 20 7437 8656.

For further information see our web site at www.rsc.org

Printed in the United Kingdom by CPI Group (UK) Ltd, Croydon, CR0 4YY, UK

# Preface

The development of data analysis software accompanies technological advances in mass spectrometry. Naturally, academic laboratories are often the first to implement new ideas and algorithms into scientific software. The public release of code has led to the community-driven creation, testing, and improvement of applications, and the compilation of complete data-analysis platforms. The definition of open data formats has facilitated the exchange of data between researchers and the analysis of data by a variety of diverse programs.

Open software for the analysis of mass spectrometry data has reached a maturity that makes it suitable for professional use by analytical chemists in academia and industry. Further, exciting new tools, such as data mining by machine learning/artificial intelligence and DevOps—agile software development/testing/delivery—are accelerating developments in all fields of computer science.

However, documentation for the multiple existing programs and their optimal use is scattered in package manuals, read-mes, and online forums. This information fragmentation complicates the efficient use of open-source software, especially for students, scientists, and professionals who analyse mass spectrometry data occasionally or for first-time users.

In my laboratory, we develop analytical platforms for mass spectrometry, including hardware, software, and databases. Consequently, programming is an essential part of our daily work. Furthermore, we regularly provide mass spectrometry primers to new lab members and offer optional courses in our postgraduate programs. Many of our collaborators are more interested in biological questions than technical details, but also want to understand their data or reanalyse their results. As a result, we identified the

New Developments in Mass Spectrometry No. 8
Processing Metabolomics and Proteomics Data with Open Software: A Practical Guide
Edited by Robert Winkler

Published by the Royal Society of Chemistry, www.rsc.org

need for a book that provides the very basics, facilitates an understanding of the technology, and that also covers practical aspects of mass spectrometry data analysis.

Consequently, when Marek Domin asked me to edit a book about metabolomics for the Royal Society of Chemistry, I suggested that the scope be widened to mass spectrometry (MS) data analysis using open-source software. We discussed possible contents with the Editorial Board of the Royal Society of Chemistry and agreed on a book that strongly focuses on MS data analysis with open source software and statistical methods, but also covers the most prominent MS-based omics methods, namely metabolomics and proteomics.

The primary target audience of this book is the students and scientists who are experiencing mass spectrometry for the first time, as well as MS professionals who currently work with commercial data processing software and want to explore new ways of interpreting their results. Furthermore, (bio) informaticians working on omics datasets or MS software development can profit from the book since the book presents diverse modules, toolkits, and strategies for workflow design and data mining.

A central feature of this book is the inclusion of example data and workflows, which provides the reader with the opportunity to reproduce the presented protocols. The demo data are downloadable from public repositories free of charge and serve as useful practice material for hands-on courses or as templates for the development of individual workflows.

The book comprises three parts. **Part A** describes general concepts and holds five main chapters:

1. Chapter one provides the fundamentals, such as distinct research approaches (hypothesis-driven/exploratory/data mining) and mass spectrometry technology. Furthermore, it encourages the use of open software.
2. Chapter two explains typical unit operations applied to raw mass spectrometry data and the design of practical MS data processing workflows.
3. Chapter three presents different "flavours" of metabolomics—targeted metabolomics, untargeted metabolomics, fluxomics and metabolite imaging, technologies and software tools for metabolomics data evaluation.
4. Chapter four acquaints the reader with the complexity of the proteome and presents experimental strategies for qualitative and quantitative analyses.
5. Chapter five demonstrates, using practical examples, statistical methods for sample comparisons, clustering, dimension reduction, biomarker discovery and the creation of predictive modelling employing machine learning.

**Part B** gives an overview of available mass spectrometry software, including general MS data processing programs, metabolomics/proteomics software, and development tools (modules for R, Java, Python, and Docker). **Part C** briefly comments on the dynamics in MS software development, and speculates on future developments.

Of course, I am aware that this book has various limitations. The field of open software development is evolving rapidly. Currently, Docker (presented in chapter 19) and other DevOps concepts are changing the world of software development, testing, and deployment. For emerging and growing areas of mass spectrometry, such as ambient ionisation mass spectrometry (AIMS) methods, ion mobility separation (IMS), and mass spectrometry imaging (MSI), the development of software, data formats, and data banks has just begun. Therefore, this book only provides a snapshot of the current state-of-the-art.

I warmly thank all authors and editors who participated in the book, for the time and energy that they have invested. All of these individuals are also active developers and supporters of open source software and hence contribute extraordinarily to the evolution of science. Also, I greatly acknowledge the editorial support of the Royal Society of Chemistry, represented by Janet Freshwater and Katie Morrey, and the editors, reviewers, and proof-readers for their valuable feedback. Finally, I thank all users of open source software, because only an active community of developers and practitioners motivates further development.

Robert Winkler

# Contents




New Developments in Mass Spectrometry No. 8
Processing Metabolomics and Proteomics Data with Open Software: A Practical Guide
Edited by Robert Winkler



Published by the Royal Society of Chemistry, www.rsc.org

## Part C Conclusion

# Part A

# General Section

CHAPTER 1

# *Introduction*

ROBERT WINKLER*[a,b]

[a]Department of Biochemistry and Biotechnology, Center for Research and Advanced Studies (CINVESTAV) Irapuato, Km. 9.6 Libramiento Norte Carr. Irapuato-León, 36824 Irapuato, Guanajuato, Mexico; [b]Mass Spectrometry Group, Max Planck Institute for Chemical Ecology, Hans-Knöll-Straße 8, 07745 Jena, Germany
*E-mail: robert.winkler@cinvestav.mx

## 1.1 Hypothesis-driven *versus* Exploratory Research

Before diving into the structure of mass spectrometry data, processing workflows, statistical tools, and knowledge representations, we need to have a clear idea of the purpose of a project; in the absence of a well-defined question, any further effort is pointless.

The human mind tends to build empirical knowledge from individual observations and experiences. Knowledge obtained this way might be of practical use, but does not withstand rigorous scientific criteria. Sir Karl Raimund Popper (1959) highlighted such thinking errors by the use of a simple example:

*Now it is far from obvious, from a logical point of view, that we are justified in inferring universal statements from singular ones, no matter how numerous; for any conclusion drawn in this way may always turn out to be false: no matter how many instances of white swans we may have observed, this does not justify the conclusion that all swans are white.*[1]

New Developments in Mass Spectrometry No. 8
Processing Metabolomics and Proteomics Data with Open Software: A Practical Guide
Edited by Robert Winkler

Published by the Royal Society of Chemistry, www.rsc.org

This *problem of induction* was investigated by David Hume (1748) who questioned the causality of *matters of fact* which are not supported by deductive logic.[2]

According to *The Scientific Method* of Popper, the creation of universally valid models requires a hypothesis as a starting point. Deductions and statements are driven by the hypothesis, which, in turn, can be tested theoretically or experimentally. Popper introduced the concept of *falsifiability*. Often, hypotheses or theories cannot be verified by experimental evidence, but individual deduced statements can be falsified by testing to evaluate their limits.[1] The differences between deductive and inductive approaches are shown in Figure 1.1.

Returning from epistemology to practical experimental design, two main strategies are distinguishable, namely hypothesis-driven and the exploratory approaches. Having a well-defined hypothesis-driven analytical question greatly facilitates experimental planning. For example, if the regulation of a certain biosynthetic pathway is proposed, related intermediates and products can be monitored for with a targeted metabolomics

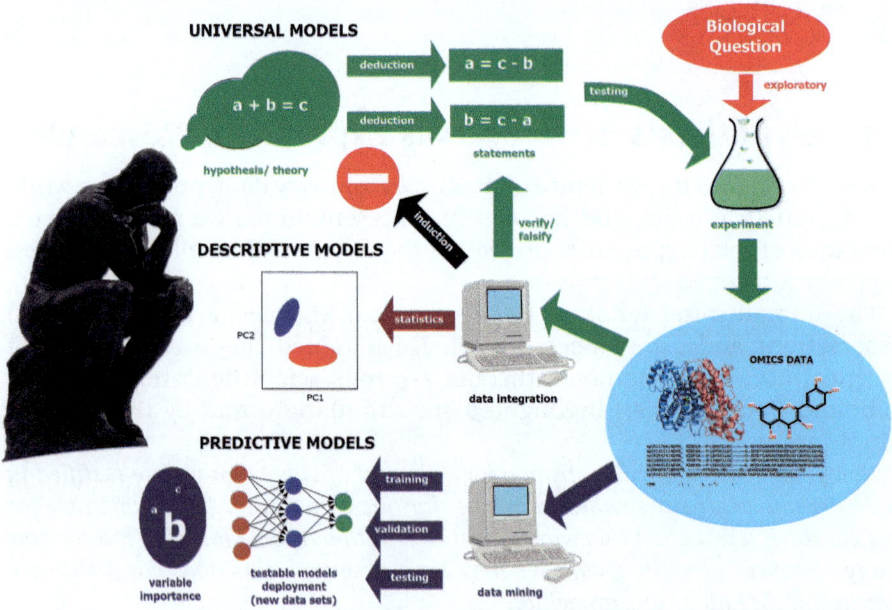


**Figure 1.1** The scientific method starts with a hypothesis and then tests the validity of deductions made either theoretically or experimentally. Most omics studies are exploratory. The induction of universal models based on observation is not valid. Alternatively, data mining methods can be used to create predictive models and to elucidate the importance of variables in a system. Reproduced from ref. 3 [https://doi.org/10.3389/fpls.2016.00195] under the terms of a CC BY 4.0 license [https://creativecommons.org/licenses/by/4.0/].

strategy, which increases the quality of the data for the compounds of interest and reduces the acquisition of less relevant (for this question) data. In addition, alternative analysis methods, such as HPLC, optical methods, ELISA testing, and complementary studies, such as qPCR, are considered to increase the sample number and further strengthen the necessary evidence.

In stark contrast, many omics studies lack a clearly defined hypothesis; rather, they strive to explore possible differences between two or more groups of organisms on a particular level, for example a transcriptome, proteome, or metabolome. Therefore, the expected outcome of comparative omics experiments are patterns that support the formulation of hypotheses.[4]

Data mining (DM) is considered to be a hybrid between deductive and inductive research. Although DM is data driven and lacks human creativity, predictive models can be built within a defined numerical space. DM extends 'classic' statistical tools such as principal components analysis (PCA) by artificial intelligence (AI) and machine learning. The outcomes of a DM procedure are classification models, associations, and the importance of variables. An excellent practical guide to DM using the R statistical software was published by Williams (2011).[5]

The training of a new model starts with a question that should be answered using available data. In case of a metabolomics study, possible questions might include:

- Do the wild types of a genotype and its mutant display distinct metabolic identities? ⇒ Classification model.
- If so, which metabolites are indicative for each group? ⇒ Variable importance analysis.
- Which level of glucose can I expect for parameter set X (which is not part of the training data)? ⇒ Quantitative model.
- Which of the metabolites directly or indirectly interact with each other? ⇒ Association analysis.

Importantly, only part of the dataset is used to construct the actual model; two other parts of the data set serve for validating the model and testing its performance, for example, to estimate a realistic classification error.[6]

For example, the random forest (RF) algorithm deals well with mass-spectrometry-based omics data that are characterized by a large number of variables and relatively few observations. Furthermore, RF analysis is tolerant to noisy data and may even be employed as a filter to discriminate between informative signals and random ones.

It is worth noting that DM methods are not knowledge biased and do not give meaning to signals other than statistical ones; therefore, they can help the scientist to discover unexpected relationships in a dataset. Altogether,

DM methods are very useful for the processing of MS data, and their practical value (*e.g.*, for classification models in biology and medicine) is only now becoming recognized.[7]

## 1.2 Mass Spectrometry Basics

John Bennet Fenn, who shared the Nobel Prize in Chemistry (2002) with Koichi Tanaka and Kurt Wüthrich *for the development of methods for identification and structure analyses of biological macromolecules* (https://www.nobelprize.org/prizes/chemistry/2002/summary/), described the concept of mass spectrometry in a few words:

*Mass spectrometry is the art of measuring atoms and molecules to determine their molecular weight. Such mass or weight information is sometimes sufficient, frequently necessary, and always useful in determining the identity of a species. To practice this art one puts charge on the molecules of interest, i.e., the analyte, then measures how the trajectories of the resulting ions respond in vacuum to various combinations of electric and magnetic fields. Clearly, the sine qua non of such a method is the conversion of neutral analyte molecules into ions. For small and simple species the ionisation is readily carried by gas-phase encounters between the neutral molecules and electrons, photons, or other ions. In recent years, the efforts of many investigators have led to new techniques for producing ions of species too large and complex to be vaporized without substantial, even catastrophic, decomposition.*

Many different solutions that address the technical challenges of mass spectrometry have been found, and only basic MS principles are expounded upon in this section, which should provide the reader with sufficient knowledge to understand data provided by mass spectrometry. A more detailed introduction can be found in the book *The Expanding Role of Mass Spectrometry in Biotechnology* by Gary Siuzdak (2006),[8] for example.

The acronyms used in this book follow the glossary of the Analytical Methods Committee (AMC) of the Royal Society of Chemistry;[9] other terms are used in accordance with IUPAC Recommendations 2013.[10]

As illustrated in Figure 1.2, a mass spectrometric system consists of the components described below.

### 1.2.1 The Sample Introduction Unit

Although this might appear to be trivial, the sample needs to be transferred into the mass spectrometer. Whereas the sample is at ambient temperature and pressure, the mass analyser requires vacuum conditions; hence, the sample stream entering the mass analyser must be free of liquids ('dry') and excess gas needs to be removed by differential pumping. For calibration or the direct flow injection (DLI) of a small number of individual samples, solutions can be introduced using a syringe pump, while an autosampler is usually used for larger sample numbers. To avoid 'batch effects' during data acquisition and analysis, it is strongly recommended that samples be analysed in random order.

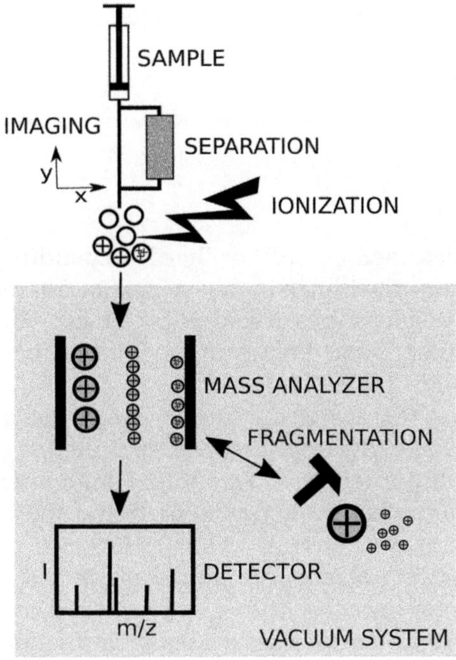


**Figure 1.2** Components of a mass spectrometer.

## 1.2.2 The Separation/Imaging Component

Fractionation of a complex sample improves the detection and quantification of compounds, because matrix effects and ion suppression are reduced. Consequently, mass spectrometry is often augmented with gas or liquid chromatography (GC and LC). The retention times of the analysed compounds provide additional analytical dimensions for identification purposes. During mass spectrometry imaging (MSI), surfaces are scanned and mass spectra are combined with the location data of each sampled spot. MSI generates images that visualize the distribution of mass signals in a planar projection.

## 1.2.3 The Ionization Unit

To activate neutral molecules for mass spectrometry, they need to be converted into positive or negative ions. This process occurs in the **ion source**. A neutral molecule 'M' can be charged by different processes:[8]

**Positively charged ions** are formed by:

- **Protonation**
  $M + H^+ \rightarrow MH^+$
- **Cationization**
  $M + cation^+ \rightarrow [M + cation]^+$
- **Electron ejection**
  $M - e^- \rightarrow M^+$.

**Negatively charged ions** can be generated by:

- **Deprotonation**
  $M - H^+ \rightarrow [M - H]^-$
- **Electron capture**
  $M + e^- \rightarrow M^{-\bullet}$.

Ion sources are classified as 'soft' or 'hard', depending on the ionization energy, with the latter providing a higher degree of fragmentation. The type of ionization defines which types of molecular ion are expected to be observable. The most frequently used types of ion sources in metabolomics and proteomics are listed in Table 1.1.

Electron ionization (EI) is the method of choice for highly volatile molecules with low polarity ('hydrophobic' compounds), *e.g.* in drug metabolomics, lipidomics and forensics.[13] Due to the high energy level of 70 eV,[8] organic molecules are extensively fragmented and the intact $M^{+\bullet}$ signal is generally of low intensity or absent.[12] Usually, EI MS is coupled to gas chromatography (GC). GC-EI MS is highly standardized and has been employed in biochemical labs for decades. Therefore, numerous free and commercial databases with reference spectra are available and facilitate the identification of compounds.[13,14]

Hydrophobic gas phase molecules also can be ionized by chemical ionization (CI). Compared to EI, the energy level is lower. Therefore, the fragmentation is reduced and the detection of molecular ions with sufficient intensity is more likely.[8]

Atmospheric pressure chemical ionization (APCI) and atmospheric pressure photoionization (APPI) ionize vaporized neutral molecules with medium and low polarity. Compared to APCI and ESI, APPI is more robust regarding buffer salts and ion suppression,[15] and ionizes more hydrophobic compounds such as polycyclic aromatic hydrocarbons (PAHs).[16]

Electrospray ionization (ESI) favours the creation of protonated molecules and positively charged adducts in positive mode, and deprotonated molecules in negative mode. The samples enter the ion source with a solvent flow, either by direct liquid introduction (DLI), or from a coupled LC. Mainly polar compounds are ionized, including many organic small molecules, lipids (except for nonpolar lipids, *e.g.* sterols and triacylglycerols), peptides and proteins. ESI is a very soft ionization method with low levels of fragmentation, which allows the detection of intact biomolecule ions. Even complete viruses can be observed by ESI while maintaining their virulence.[17] Large biomolecules present multiple charges ($[M + nH]^{n+}$), resulting in ions of a mass-to-charge ratio ($m/z$) that can be analyzed with most mass analyzers (see below, Table 1.2).[18] ESI with very low flow rates, nanoelectrospray ionization (nanoESI), is especially useful in proteomics, because of its increased tolerance against buffer salts and high sensitivity.[19] Since biological samples often contain alkali metals (*i.e.*, sodium and potassium), $[M + Na]^+$ and $[M + K]^+$ ions are often observed in ESI mass spectra. The addition of ammonium

**Table 1.1** Ionization sources used in metabolomics and proteomics. Assembled from Siuzdak (2006),[8] Awad *et al.* (2015)[11] and Maher *et al.* (2015).[12] APCI, atmospheric pressure chemical ionization; APPI, atmospheric pressure photoionization; CI, chemical ionization; EI, electron ionization; ESI, electrospray ionization; MALDI, matrix-assisted laser desorption/ionization.

| Ion source type | Principle | Sample phase | Energy level | Applications |
|---|---|---|---|---|
| EI | Electron capture and ejection | Gas | High | Suitable for volatile molecules with low polarity; extensive fragmentation; large databases available; widely used in GC-MS based metabolomics |
| CI | Chemical charge transfer | Gas | Medium | Similar range of compounds as EI (volatile and hydrophobic compounds), but less fragmentation |
| APCI | Corona discharge and proton transfer | Gas | Medium | Ionization of small molecules with medium and low polarity; less fragmentation than EI |
| APPI | Photon energy transfer | Gas | Medium | Ionization of small compounds of medium and low polarity; more robust against buffer salts than APCI and ESI |
| ESI/nanoESI | Evaporation of charged droplets/Coulombic explosions | Liquid | Low | Ionization of polar compounds in a wide molecular range; reduced sample flow and high sensitivity in nanoESI; standard method in metabolomics (ESI) and proteomics (nanoESI) |
| MALDI | Photon absorption/proton transfer | Solid | Low | Imaging and high-throughput analysis; ionization requires a suitable matrix; matrix effects can be problematic for method development and data analysis. |

acetate can be used experimentally to dissociate alkali metal adducts, leading to the formation of protonated or $[M + NH_4]^+$ ions instead. In contrast, the addition of lithium can be used to ionize sugars by cationization.[20] ESI/nanoESI the most commonly used ion source in LC-MS metabolomics and proteomics, because of its wide *m*/*z* range and huge diversity of detectable organic molecules.

Matrix-assisted laser desorption/ionization (MALDI) is a soft method for large biomolecules.[21] MALDI requires the co-crystallization of the analyte molecules with a matrix. Depending on the choice of matrix, many different compound classes can be ionized (*e.g.* proteins, lipids, DNA). Since the prepared samples need to be solid, a direct coupling of MALDI to LC-MS systems is not practical. In contrast, MALDI is frequently used in mass spectrometry

**Table 1.2** Mass analyzers for metabolomics and proteomics applications. Modified from Junot *et al.* (2014),[25] and updated with the technical information of major mass spectrometer suppliers. Q, quadrupole analyzer; q, quadrupole collision cell (fragmentation); QqQ, triple quadrupole; IT, ion trap; QIT, quadrupole (cubic) ion trap; LIT, linear ion trap; ToF, time-of-flight; FT ICR, Fourier transform ion cyclotron resonance; T, Tesla.

| Mass analyzer type | Resolution [FWHM] | Mass accuracy [ppm] | Mass range [*m*/*z*] | Dynamic range | Applications |
|---|---|---|---|---|---|
| Triple quadrupole (QqQ) | ≤7500 | 5–500 | ≤3000 | $10^5$–$10^6$ | Routine quality control; trace analysis (*e.g.* pesticide residues) |
| Ion traps (QIT/LIT) | ≤10 000 | 50–500 | ≤4000 | $10^4$ | Routine chemical profiling; fragmentation studies (MS$^n$); metabolomics and proteomics with spectral matching |
| Quadrupole-linear ion trap (Qq-LIT) | ≤10 000 | 50–500 | ≤2000 | $10^5$–$10^6$ | Same as QIT/LIT, but with enhanced dynamic range |
| Time-of-flight (ToF) | ≤20 000 | <1–2 | ≤20 000 | $10_4$–$10_5$ | Comparative metabolomics; no fragmentation possible |
| Quadrupole-time-of-flight (Qq-ToF) | ≤60 000 | <1–2 | ≤40 000 | $10^4$–$10^5$ | Metabolomics and proteomics; *de novo* studies (metabolites and peptides) |
| Orbitrap | ≤1 000 000 | <1 | ≤6 000 | $10^3$–$10^4$ | Metabolomics and proteomics; *de novo* studies (metabolites and peptides); analysis of complex mixtures; top–down proteomics |
| Quadrupole-Fourier transform ion cyclotron resonance (Qq-FT ICR) | >10 000 000 | <1 | ≤10 000 | $10_3$–$10_4$ | Special analytical questions (proteoforms, dissolved organic matter) |

imaging (MSI)[22] and high-throughput screening. The main challenges of MALDI are related to the interference of matrix substances with target molecules, and the lacking reproducibility between experiments. Advanced data analysis methods[23] and novel matrix compounds[24] therefore can enhance the utility of MALDI.

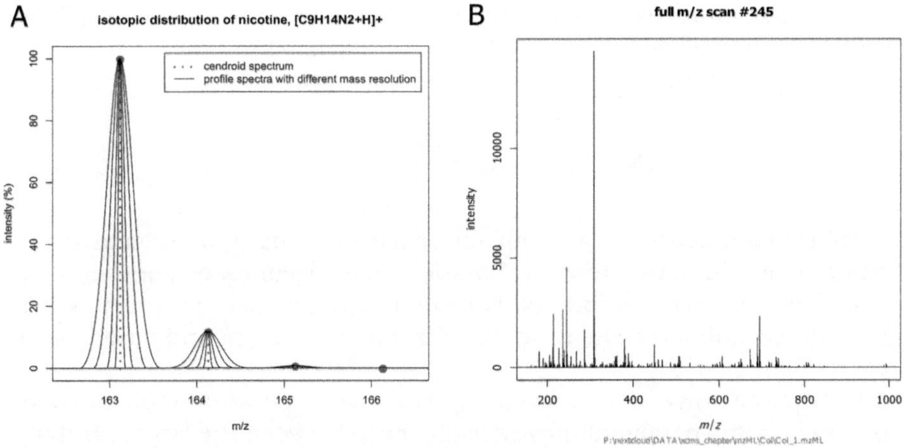


**Figure 1.3** (A) Theoretical mass spectrum of nicotine and its sum formula. (B) Experimental full scan mass spectrum from an LC-MS metabolomics data set of an *Arabidopsis* flower extract. The spectrum is of a single scan (#245) from *Arabidopsis* inflorescence.

Mass analyzers require the analyte in a gaseous state (see Figure 1.2). Therefore, molecules that are already charged in solution, still need to be transferred into the gas phase:

$$M^{+}\text{solution} \rightarrow M^{+}\text{gas}$$

This **vaporization** occurs in various ion sources such as APCI, ESI/nanoESI, MALDI.[8]

### 1.2.4 The Mass Analyzer

Mass analyzers separate ions based on their mass-to-charge ($m/z$) ratios and different types of mass analyzer have inherent differences in their analytical performance. The most advertized parameter of a mass analyzer is its specified **mass resolving power** or **resolution.** The value of this parameter indicates to which degree signals of similar mass-to-charge ratio can be distinguished (see Figure 1.3). Resolution is calculated by:

$$\text{Resolution} = \frac{m}{\Delta m}$$

where $m$ is the mass-to-charge ratio of a signal and $\Delta m$ the mass peak width, with the full-width-at half-maximum (FWHM) of the peak commonly taken as the reference. High-resolution instruments are advantageous for challenging analytical tasks, such as the analysis of complex mixtures, structural elucidation of natural products, and top-down protein studies (multiply charged ions). The resolution depends on the settings of the mass analyzer (method) and the observed signal. Consequently, the maximum mass resolving power of a device is specified together with the $m/z$ of the peak (*e.g.* 240 000 at 200 $m/z$).

The **mass accuracy** expresses the deviation of the measured mass $m_{\mathrm{m}}$ from its exact (theoretical) value $m_{\mathrm{e}}$ and can be expressed in terms of *m*/*z* tolerance (*e.g.* ± 0.15 *m*/*z*) or relatively, in parts per million (ppm):

$$\text{Mass accuracy} = \frac{10^6 \times \left|(m_{\mathrm{m}} - m_{\mathrm{e}})\right|}{m_{\mathrm{e}}}$$

Mass accuracy depends on the operational conditions of the MS and its calibration. In addition to external calibration, mass accuracy can be increased by the use of internal calibrants. For example, high mass accuracy is crucial for determining the chemical sum formula of a compound based on its mass-to-charge ratio.

The **dynamic range** of the mass analyzer is also important for quantitative experiments; this parameter provides the 'ratio between the largest and the smallest detectable signals' (http://mass-spec.lsu.edu/msterms/index.php/Dynamic_range). A high dynamic range facilitates the evaluation of minor constituents in complex mixtures containing multiple molecules.

Another important performance parameter of a mass analyzer is its **scan cycle time**; *i.e.*, the time required to collect a single mass spectrum. For time-critical applications, such as fast metabolomics screening, a mass analyzer or method with low resolution and/or a reduced mass range, among others, is required; however, a shorter scan cycle time might be a preferable option.

Typical specifications of mass analyzers commonly used in metabolomics and proteomics are given in Table 1.2.

Overall, no ideal mass analyzer exists; hence, a compromise is required, which depends on the principle use of the mass analyzer and the available budget.

A triple quadrupole (QqQ) mass spectrometer is an excellent choice for the routine analysis of known compounds, because it provides high sensitivity over a wide dynamic range. QqQs are often used for quantification with multiple reaction monitoring (MRM) methods (see below), *e.g.* in food trace analysis. The resolution and the mass accuracy of quadrupole analyzers are modest. On the other hand, quadruples are reliable and economic. Therefore, quadrupoles are often used in hybrid MS systems, *e.g.* as upstream mass filters, and in collision cells.

Ion traps are constructed in two basic models: quadrupole (cubic) ion traps (QITs) and linear ion traps (LITs). In contrast to quadrupoles, that operate with a continuous ion flow, ion traps are able to collect ions. Thus, a low analyte concentration can be compensated by an increased ion collection time. Consequently, ion trap analyzers can reach high sensitivities. The resolution and the mass accuracy of ITs are limited, but the fragmentation of ions is possible to obtain structural information. Thus, LITs can be used for proteomics and metabolomics, if reference data are available ('spectral matching'). The sequential (manual or automated) selection and fragmentation of ions ('$\mathrm{MS}^n$ capability') allows for in-depth molecule structure studies.

Time-of-flight (ToF) mass analyzers provide high resolution and mass accuracy. At the same time, ToF mass analyzers provide an extended dynamic range, which makes them suitable for chemical profiling/comparative metabolomics studies. Qq-ToF devices additionally provide $MS^2$ capability and are widely used in metabolomics and proteomics.

The introduction of the Orbitrap mass analyzers about two decades ago[26] was game-changing, because devices with extremely high resolution became more accessible for routine laboratories. Orbitraps are suitable for most metabolomic and proteomic studies, including the analysis of complex mixtures. The high resolution and mass accuracy facilitate chemical structure elucidation and *de novo* peptide sequencing. Additionally, Orbitrabs can be used for fragmentation studies of entire proteins ('top-down proteomics').[27]

Fourier transform ion cyclotron resonance (FT ICR) devices represent the high-end mass analyzers and require significant investments in scientific infrastructure. Since the Orbitrap analyzers are catching up in analytical performance, FT ICR mass analyzers are reserved for the most demanding analytical tasks such as the characterization of proteoforms[28] or crude samples of terrestrial dissolved organic matter (DOM).[29]

Ion mobility separation (IMS) determines the ion collision cross section (CCS) of a molecule, which gives information about its geometrical shape.[30,31] This additional analytical dimension permits isobaric compounds (molecules of the same mass) to be discriminated, and the study of (bio)molecular confirmation. IMS forces charged molecules to travel in an electrostatic field against the resistance of atmospheric molecules or a gas steam. An IMS cell operates at an elevated pressure compared to the mass analyzer and, therefore, is usually coupled upstream with the mass analyzer (*e.g.* LC-IMS-Qq-ToF). Currently, eight different IMS instrument types have been described,[32] which are coupled to mass analyzers, with drift tubes and travelling wave separators being the most common ones.[33] An increasing number of ion mobility-mass spectrometry (IM-MS) devices are now commercially available, which has led to a growing number of reported applications and reference data, which has stimulated the active development of IM-MS software.

The analytical performance of a device depends strongly on its operating conditions, which include calibration, maintenance, installation site (vibrations, electricity, temperature, and humidity), and the device method. For example, ToF analyzers have wide mass ranges that can extend beyond 100 000 *m*/*z*, which facilitates the analysis of complete, singly charged proteins. However, sensitivity and resolution will be much below the technical specifications in such an experimental set-up.

To interpret data, practical values of instrumental parameters need to be determined. For example, a mass accuracy of 3–10 ppm is realistic when analysing complex mixtures of metabolites by ToF without internal calibration, compared to the specified value of 1 ppm under optimal conditions. Of course, high mass accuracy facilitates the identification of compounds by database matching and the *de novo* creation of sum formulas; however, even mass accuracies below 1 ppm are insufficient to identify metabolites by mass alone.[34] Taking into account isotope patterns and chemical rules helps to narrow down possible structure candidates, Kind and Fiehn reported 'Seven Golden Rules' for the heuristic filtering of chemical formulas.[35]

A combination of different analytical dimensions ($MS^n$ analysis, isotope pattern, IMS collision cross section and chromatographic retention time) and/or the co-analysis of an authentic reference material is usually necessary to reliably identify a compound by MS.[36] Claiming novel organic structures usually also requires the reporting of X-ray or nuclear magnetic resonance (NMR) data.

### 1.2.5 Fragmentation

A compound ion can be selected and fragmented on a mass analyzer engendered with that capability in order to obtain structural information. With the exception of simple (*i.e.*, without another analyzer coupled) quadruple and ToF MS analyzers, all MS systems have this feature. Depending on the construction of the mass analyzer, fragmentation is either performed in a separate fragmentation cell or in the same cell. Collision-induced dissociation (CID) is provoked if an inert gas is introduced into a chamber containing ions at high vacuum.[37] High-energy collision-induced dissociation (HCID) is based on the same principle, but it provides high MS/MS resolution on Orbitrap instruments;[38] (H)CID fragmentation is widely used in metabolomics and proteomics applications.

Electron capture dissociation (ECD), electron transfer dissociation (ETD), and other ECD-derived methods that are referred to as 'ExD' expose multiply charged 'parent ions' to free electrons. The capture of these electrons by these molecules results in their fragmentation. ExD methods provide complementary fragmentation information and are mainly used in MS-based protein studies.[39]

Often, fragmentation spectra are referred to as '$MS^2$' or 'MS/MS' spectra, suggesting that they are obtained in a second experiment following the acquisition of the MS scan. However, fragmentation can also occur in the ion source ('in-source fragmentation', 'in-source decay', or 'in source collision-induced dissociation', InS), depending on the stability of the structures and the applied ionization energy. For example, while EI ionization leads to a high proportion of fragmented ions, APCI and 'soft' ionization methods also display fragmentation events, for example, by laser-induced photodissociation (PD) during MALDI.

Information about the expected fragmentation behaviour of a molecule greatly supports its structural elucidation or identification.[36,40] Currently, databases of small molecules mainly contain fragmentation spectra following EI ionization and CID, and these reference databases sometimes contain spectra that were collected with different fragmentation energies. Consequently, it might be necessary to collect spectra with different fragmentation settings in order to identify unknown ions.

In proteomics, theoretical peptide fragments are calculated from the amino acid sequence. Depending on where the peptide bond is broken, N-terminal fragment ions are referred to as '*a*, *b*,' and '*c*', while C-terminal ions are labelled '*x*, *y*,' and '*z*'; *b* and *y* ions are expected with CID or HCID, which are the most common fragmentation methods in proteomics, while *c* and *z* ions are mainly expected for ECD/ETD.

The selection and further fragmentation of selected fragment ions is possible in MS systems, resulting in $MS^3$, or even $MS^4$ and $MS^5$ (*etc.*) if this process is continued. Such '$MS^n$' fragmentations provide detailed structural information, but require a rather manual procedure with few reference spectra available in databases. Nevertheless, $MS^n$ provides valuable structural information from devices with low or modest mass resolutions, such as ion-trap analyzers.

### 1.2.6 Detector

The detector is usually not deliberately chosen by the user, but is an integral and fixed component of an MS system. Consequently, not much attention is usually paid to this critical component. This section is based on information from a concise review of the history of ion current detectors.[41]

Most mass analyzers require a detector to sense the intensity of ions in a certain *m*/*z* region and to translate them into an electrical/computational signal. One ion per second corresponds to $1.6 \times 10^{-19}$ A, which can technically be measured; however, the minimum detectable ion intensity of a detector is usually about a thousand times higher. The most frequently used detectors are secondary electron multipliers (SEMs), such as micro-channel plate (MCP) detectors. However, Faraday cup, ion-to-photon, and cryogenic detectors have also been used, but the latter ones are mainly employed for special applications, such as the detection of macroions. Instead of a separate ion detector, FT-ICR, and Orbitrap analyzers record the current induced by resonating ions. This process is non-destructive and leads to high analyzer resolution, accuracy, and sensitivity.[41]

### 1.2.7 Mass Spectra and Mass Chromatograms

Mass analyzers determine the mass-to-charge (*m*/*z*) ratios of introduced ions. Figure 1.3A shows the theoretical mass spectrum of pure protonated nicotine with its sum formula. Based on its chemical composition, its **monoisotopic mass** can be calculated to be 163.123 *m*/*z*. This mass corresponds to the ion that is composed only of the most abundant elemental isotopes. Subsequent less-intense peaks represent ions that contain $_{13}C$, $^2H$, or other isotopes. The pattern of *m*/*z* values and their relative intensities for a particular compound is referred to as the 'isotopic distribution'. The isotopic distribution of protonated nicotine shown in Figure 1.3A was simulated with the `OrgMassSpecR` R package (http://orgmassspec.github.io/).[42,43] The program uses the data tables for *Atomic Weights and Isotopic Compositions with Relative Atomic Masses*, maintained by the National Institute of Standards and Technology (NIST) (http://physics.nist.gov/PhysRefData/Compositions/). These data are based on those published by the Commission on Isotopic Abundances and Atomic Weights (CIAAW) of the International Union of Pure and Applied Chemistry (IUPAC) (http://www.ciaaw.org/atomic-weights.htm),[44] and the new evaluation of atomic masses, Ame2012.[45]

Two different *m*/*z*-signal representations are shown, namely **centroid** spectra that consist only of *m*/*z* values and their respective intensities, whereas in **profile** spectra, peaks are formed by curves defined by various data points. Centroid data are either derived from profile spectra (by peak picking) or collected directly during MS in centroid acquisition mode. Since centroid spectra contain less information, their data files are smaller and allow faster scan-cycle times. Most MS data processing workflows convert profile spectra into centroid spectra. On the other hand, profile spectra contain important information about the quality of the MS data and the performance of the MS analyzer.

Figure 1.3A shows profile spectra with different theoretical **resolutions.** If the resolution is too low, neighbouring peaks cannot be distinguished and isotopic resolution is lost. In this case, the average mass of a molecule will be determined. Nevertheless, even average spectra are suitable for determining the molecular weights of proteins from multiply charged ESI-MS signals[46] (http://www.bioprocess.org/esiprot/esiprot_form.php). To determine accurate masses, the peak shapes should be symmetrical. In the case of ToF analyzers, peak shapes and, therefore, the mass accuracies are affected by analyte concentration.[47]

Comparing theoretical and experimental isotopic distributions is an efficient strategy for testing the validity of possible sum formulas from MS data, and, therefore, has been implemented in a variety of commercial and open source programs. With isotope distribution matching, likely chemical sum formulas can be derived even from spectra with modest mass accuracies and resolutions.[34,35,48]

The **intensity** of a mass signal can be used for quantification purposes; however, we should keep in mind that ions are observed rather than neutral molecules. Therefore, the abundant compounds in a sample may be invisible to MS if no ions are formed with the chosen ionization source. On the other hand, the levels of highly ionizable molecules can be overestimated.

Figure 1.3B displays an MS spectrum generated from experimental data. In short, *Arabidopsis* flowers were extracted with methanol and subjected to LC-MS. More information on this dataset and its evaluation is provided in Section 8.2. The data file contains 800 mass scans that were recorded during the chromatographic run. The presented scan (#245) corresponds to a retention time of 10.64 minutes. The strongest signal in the spectrum, with 309.0977 *m*/*z*, is referred to as the 'base peak'.

This dataset is already centroided, but still contains many signals that later might be considered to be noise and removed. The density of peaks in this spectrum underlines the importance of high mass resolution for complex samples.

Fragmentation spectra are presented in the same format as full-scan MS spectra, but also report the fragmentation level and the precursor mass selected for fragmentation (*e.g.* MSMS 195, $MS^2$ 195, $MS^3$ 195 → 82).

Chromatograms can be constructed from the ion intensities, and a total ion current chromatogram (TICC) plots the total ion current as a function of retention time. Figure 1.4A shows the base peak chromatogram (BPC) of the above-mentioned *Arabidopsis* flower extract analyzed by LC-MS. Only the most intense signal of each scan is considered for a BPC, and an extracted ion chromatogram (XIC) is finally constructed from the ion intensities in a chosen *m*/*z* window. An example is shown in Figure 1.4B, where only signals in the 736.2–736.4 *m*/*z* range were evaluated. LC-MS **features** are defined by their retention times and mass-to-charge ratios, both with their respective tolerances. The integrated ion intensities of such features are used for quantitative LC-MS analyses.

## 1.2.8 LC-MS Analysis and Data Acquisition Strategies

LC-MS analysis and data acquisition strategies are defined by the analytical scope of a project, and one needs to first choose between a **targeted** or an **untargeted** approach. In the first case, the compounds of interest are known and the complete workflow is directed towards the optimized detection of

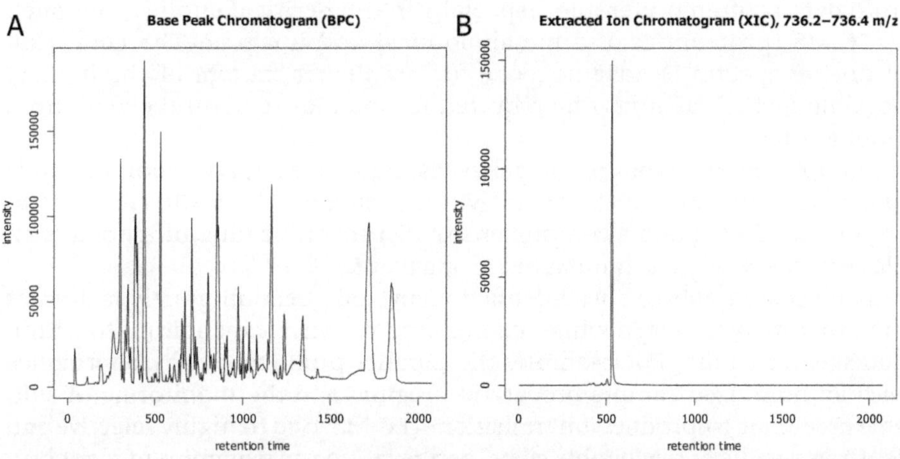


**Figure 1.4** (A) Base peak chromatogram (BPC) and (B) the extracted ion chromatogram (XIC) for a feature of interest.

these analytes. However, the range of molecules of interest must also be restricted to something reasonable in the untargeted strategy (*e.g.*, according to polarity, or *m*/*z* range, *etc.*).

On the basis of the properties of the expected molecules and the complexity of the sample, an **extraction method** (disintegration method, solvents, centrifugation, filtration, *etc.*) and **separation method** need to be defined, after which the **ionization method, polarity,** and a suitable **mass analyzer** need to be chosen appropriately.

An **automated precursor-ion fragmentation** process, also known as data-dependent acquisition (DDA), provides structural information for the identification of compounds. However, conventional DDA methods with MS and alternating $MS^2$ scans suffer from low efficiencies due to scan-cycle time limitations. Therefore, separate target-directed DDA experiments for the acquisition of fragmentation data provide superior results.[49] In addition, the additional data provided by **MS full scan mode** (*i.e.*, without intermittent fragmentation scans) deliver a higher data density for quantitative analyses.

For the same reason, the sensitivities toward ions can be improved by **selecting the ions of interest.** In selected ion monitoring (SIM), only the defined ions of interest are recorded, thereby maximizing the sensitivities for those signals. In addition, only $MS^n$ ions of a defined *m*/*z* precursor can be acquired in selected reaction monitoring (SRM). The strategy is referred to as consecutive reaction monitoring (CRM) when product ions from multiple sequential fragmentations are observed. Finally, SRM can be applied to multiple product ions from one or more precursor ions, resulting in multiple reaction monitoring (MRM).

Data are collected in either **profile** or **centroid mode**. As explained above, centroid spectra contain fewer data points, but are less bulky and data are acquired faster by the mass analyzer. The acquisition of centroid data is often preferable, especially in time-critical applications such as LC-MS (proteomics and metabolomics) and imaging. The collection of profile spectra is advantageous for producing spectra of the highest possible quality, as might be required for manual structural elucidation, among others.

Multiple sample types (organs, tissues, liquids, *etc.*), extraction methods (polar, non-polar, pH, *etc.*), and analytical strategies (MS, NMR, *etc.*) need to be combined to obtain a comprehensive picture of the physiology of a complex organism, such as human being or a plant.[50,51]

As discussed above, a well-defined analytical question greatly facilitates the experimental design while maintaining resource expenditure to within manageable limits. For example, the parallel profiling of phytohormones with an optimized sample preparation regime and the monitoring of suitable precursor-to-product ion transitions (LC-$MS^2$) led to highly selective and sensitive results at reasonable costs, and revealed the dynamics of plant hormones in thermodormant seeds.[52]

## 1.3 Why Open Software for Mass Spectrometry?

Mass spectrometers are built by companies, so why shouldn't we analyze our data with the software they provide with their devices? Indeed, good commercial software exists for MS data analysis in (bio)chemistry and medicine. Integrated hardware–software solutions that are optimized for specific applications, easy to use, and validated, generate results of consistent quality. In addition, if questions arise, professional support is available. However, 'all-in' solutions are expensive, not only when first acquired, but also if additional software modules or software maintenance (updates/upgrades) are required. Frequently, only proprietary file formats are supported, which complicate data analyses with alternative programs (*e.g.*, for data mining).[53] In the worst case, a commercial platform is not supported any more, which may lead to the 'locked-in data syndrome'. The decision on the (dis)continuation of proprietary software is made by the owning companies based on the market situation. Therefore, using community data formats with long-term readability is highly recommended, even when using a commercial data processing platform.

In the worst-case scenario, a commercial platform becomes obsolete and is no longer supported, which may lead to 'locked-in data syndrome'. The decision to (dis)continue proprietary software is made by the associated company based on the market. Therefore, the use of long-term-readable community data formats is highly recommended, even when a commercial data processing platform is used.

The development of open software is not driven by a financial motivation, but by academic interest. Often, scientists become developers because the available commercial software is not suitable for their objectives or is too costly. In the simplest case, a custom script is attached to commercial software to extend its functionality; this is not open software, but an 'in-house solution'. Important requirements for releasing a program as open-source software (OSS) include the obligation to make the source program freely available to others and to allow others to modify and redistribute it. This is in stark contrast to proprietary programs that are usually delivered as compiled programs, and may not be copied, modified, or reverse engineered. The exact terms are defined in the end-user license agreements (EULAs). The Open Source Initiative defines ten OSS criteria and maintains a list of compatible licenses (https://opensource.org/).

OSS authors should release their code with a suitable license to define the terms and conditions of its use and copyright issues, among others. The community-driven development of OSS follows rules different to those used in industrial production. Yochai Benkler (2006) introduced the concept of *peer production* projects as *large-scale, non-hierarchical, decentralized collaborations among multiple contributors*. These are the result of the transition from an industrial economy to a networked information economy and made possible by *1. Diversity of Incentives*, where creators of information goods are motivated by diverse reasons, and are not exclusively interested in financial rewards (copyright), and *2. Technological Shifts*, in which the general availability of computers and internet access means that everyone is a potential author or distributor of information (democratization of publishing).[54]

The advantages of open research have been impressively demonstrated by the production of an off-patent anti-parasitic drug, for which research accelerated after disclosure of the relevant information to the public, because more experts became involved in the project. Furthermore, the entire process is transparent and all project-related information is backed by a larger community and not maintained by only a few individuals.[55]

The frequency of software updates and upgrades is another important distinctive feature. In OSS, bug fixes and function updates are released continuously and instantaneously following their implementation. On the other hand, new versions of commercial software are distributed less frequently. Since results can be affected by changes in the software, the corresponding software version (or snap-shot identifier) should always be documented.

From a socio-economical perspective, the use of OSS presents enormous potential. A current initiative of the Free Software Foundation (https://fsfe.org/) postulates the publication of code funded by public funds (https://publiccode.eu/). OSS may be installed on multiple computers without purchasing (additional) licenses, which leads to substantial cost reductions, especially in academic institution with multiple users. In addition, the program code can be inspected. As an example, Dr Chris Rath used my ESIprot program (https://bitbucket.org/lababi/esiprot) in his class and encountered a bug that I was able to quickly fix. Consequently, all participants profit from the OSS concept.

Table 1.3 compares the typical characteristics of commercial and open source software.

**Table 1.3** Key differences between proprietary and open source software.

| Sales argument | Commercial | Open source |
|---|---|---|
| Integration with hardware | Very good | Usually none |
| Usability | Easy to use | Very diverse |
| Documentation | Good | Variable |
| Support | Commercial service | Community-based |
| Cost | Expensive | Free |
| Data formats | Proprietary formats | Community formats |
| Updates/upgrades | Versions | Continuous releases |
| Continuity | Company decisions | Interests of the community |
| Software structure | Monolithic | Modular |
| Motivation | Monetary | Academic |
| License | Restrictive | Permissive |

The priorities in an industrial or service laboratory are different to those of academia and education. Software stability, result reproducibility, and regulatory compliance are mandatory. Program malfunction or the loss of data following updates or third-party changes (*e.g.*, public web API) are unacceptable; consequently, companies such as LabKey (https://www.labkey.com/) offer different versions of their software platforms. The free community server edition offers core open source features and is supported by a user forum, whereas professional/enterprise versions include proprietary modules and additional support. Such 'multi-licensing' models provide an excellent compromise between the dynamics of peer software production and those of a productive environment and the requirements of a productive environment.

## References


1. K. R. Popper, *The Logic of Scientific Discovery*, Basic Books, Oxford, England, 1959.
2. D. Hume, *Philosophical Essays Concerning Human Understanding*, A. Millar, 1748.
3. R. Winkler, Popper and the Omics, *Front. Plant Sci.*, 2016, **7**, 1–3.
4. W. Weckwerth, Metabolomics in systems biology, *Annu. Rev. Plant Biol.*, 2003, **54**, 669–689.
5. G. Williams, *Data Mining with Rattle and R: The Art of Excavating Data for Knowledge Discovery (Use R!)*, Springer, 1st edn, 2011.
6. G. J. Williams, Rattle: A Data Mining GUI for R, *R J.*, 2009, **1**, 45–55.
7. R. Winkler, An evolving computational platform for biological mass spectrometry. Workflows, statistics and data mining with MASSyPup64, *PeerJ*, 2015, **3**, 1–34.
8. G. Siuzdak, *The Expanding Role of Mass Spectrometry in Biotechnology*, Mcc Press, San Diego, 2nd edn, 2006, vol. 15.
9. Analytical Methods Committee AMCTB No. 81, A 'Periodic Table' of mass spectrometry instrumentation and acronyms, *Anal. Methods*, 2017, **9**, 5086–5090.
10. K. K. Murray, R. K. Boyd, M. N. Eberlin, G. J. Langley, L. Li and Y. Naito, Definitions of terms relating to mass spectrometry (IUPAC Recommendations 2013), *Pure Appl. Chem.*, 2013, **85**, 1515–1609.
11. H. Awad, M. M. Khamis and A. El-Aneed, Mass Spectrometry, Review of the Basics: Ionization, *Appl. Spectrosc. Rev.*, 2015, **50**, 158–175.
12. S. Maher, F. P. Jjunju and S. Taylor, Colloquium: 100 years of mass spectrometry: Perspectives and future trends, *Rev. Mod. Phys.*, 2015, **87**, 113–135.
13. A. Garcia and C. Barbas, Gas chromatography-mass spectrometry (GC-MS)-based metabolomics, *Methods Mol. Biol.*, 2011, **708**, 191–204.
14. T. Tohge and A. R. Fernie, Web-based resources for mass-spectrometry-based metabolomics: A user's guide, *Phytochemistry*, 2009, **70**, 450–456.

15. K. A. Hanold, S. M. Fischer, P. H. Cormia, C. E. Miller and J. A. Syage, Atmospheric Pressure Photoionization. 1. General Properties for LC/MS, *Anal. Chem.*, 2004, **76**, 2842–2851.
16. M. Smoker, K. Tran and R. E. Smith, Determination of Polycyclic Aromatic Hydrocarbons (PAHs) in Shrimp, *J. Agric. Food Chem.*, 2010, **58**, 12101–12104.
17. B. Bothner and G. Siuzdak, Electrospray ionization of a whole virus: Analyzing mass, structure, and viability, *ChemBioChem*, 2004, **5**, 258–260.
18. J. Fenn, M. Mann, C. Meng, S. Wong and C. Whitehouse, Electrospray ionization for mass spectrometry of large biomolecules, *Science*, 1989, **246**, 64–71.
19. M. Wilm and M. Mann, Analytical Properties of the Nanoelectrospray Ion Source, *Anal. Chem.*, 1996, **68**, 1–8.
20. K. P. Madhusudanan, Multiple lithium exchange under lithium cationization of cyclodextrins, *J. Mass Spectrom.*, 2003, **38**, 409–416.
21. M. Karas and F. Hillenkamp, Laser desorption ionization of proteins with molecular masses exceeding 10,000 daltons, *Anal. Chem.*, 1988, **60**, 2299–2301.
22. F. Kaftan, V. Vrkoslav, P. Kynast, P. Kulkarni, S. Böcker, J. Cvačka, M. Knaden and A. Svatoš, Mass spectrometry imaging of surface lipids on intact Drosophila melanogaster flies, *J. Mass Spectrom.*, 2014, **49**, 223–232.
23. J. D. Pazmiño-Arteaga, A. Chagolla, C. Gallardo-Cabrera, A. F. Ruiz-Márquez, A. T. González-Rodríguez, M. O. Camargo-Escalante, A. Tiessen and R. Winkler, Screening for Green Coffee with Sensorial Defects Due to Aging During Storage by MALDI-ToF Mass Fingerprinting, *Food Anal. Meth.*, 2019, **12**, 1571–1576.
24. J. Weißflog and A. Svatoš, 1,8-Di(piperidinyl)-naphthalene – rationally designed MAILD/MALDI matrix for metabolomics and imaging mass spectrometry, *RSC Adv.*, 2016, **6**, 75073–75081.
25. C. Junot, F. Fenaille, B. Colsch and F. Bécher, High resolution mass spectrometry based techniques at the crossroads of metabolic pathways, *Mass Spectrom. Rev.*, 2014, **33**, 471–500.
26. A. Makarov, Electrostatic Axially Harmonic Orbital Trapping: A High-Performance Technique of Mass Analysis, *Anal. Chem.*, 2000, **72**, 1156–1162.
27. K. Scheffler, in *Shotgun Proteomics: Methods and Protocols*, ed. D. Martins-de-Souza, Springer New York, New York, NY, 2014, pp. 465–487.
28. S. M. Patrie, Top-Down Mass Spectrometry: Proteomics to Proteoforms, *Adv. Exp. Med. Biol.*, 2016, **919**, 171–200.
29. C. Simon, V.-N. Roth, T. Dittmar and G. Gleixner, Molecular Signals of Heterogeneous Terrestrial Environments Identified in Dissolved Organic Matter: A Comparative Analysis of Orbitrap and Ion Cyclotron Resonance Mass Spectrometers, *Front. Earth Sci.*, 2018, **6**, 1–16.

30. B. Harper, E. K. Neumann, S. M. Stow, J. C. May, J. A. McLean and T. Solouki, Determination of ion mobility collision cross sections for unresolved isomeric mixtures using tandem mass spectrometry and chemometric deconvolution, *Anal. Chim. Acta*, 2016, **939**, 64–72.
31. R. Cumeras, E. Figueras, C. E. Davis, J. I. Baumbach and I. Gràcia, Review on Ion Mobility Spectrometry. Part 1: Current instrumentation, *Analyst*, 2015, **140**, 1376–1390.
32. R. Cumeras, E. Figueras, C. E. Davis, J. I. Baumbach and I. Gràcia, Review on Ion Mobility Spectrometry. Part 2: Hyphenated methods and effects of experimental parameters, *Analyst*, 2015, **140**, 1391–1410.
33. J. C. May and J. A. McLean, Ion Mobility-Mass Spectrometry: Time-Dispersive Instrumentation, *Anal. Chem.*, 2015, **87**, 1422–1436.
34. T. Kind and O. Fiehn, Metabolomic database annotations *via* query of elemental compositions: Mass accuracy is insufficient even at less than 1 ppm, *BMC Bioinf.*, 2006, **7**, 234.
35. T. Kind and O. Fiehn, Seven Golden Rules for heuristic filtering of molecular formulas obtained by accurate mass spectrometry, *BMC Bioinf.*, 2007, **8**, 105.
36. B. L. Milman, General principles of identification by mass spectrometry, *TrAC, Trends Anal. Chem.*, 2015, **69**, 24–33.
37. E. de Hoffmann, Tandem mass spectrometry: A primer,, *J. Mass Spectrom.*, 1996, **31**, 129–137.
38. C. Tu, J. Li, S. Shen, Q. Sheng, Y. Shyr and J. Qu, Performance Investigation of Proteomic Identification by HCD/CID Fragmentations in Combination with High/Low-Resolution Detectors on a Tribrid, High-Field Orbitrap Instrument, *PLoS One*, 2016, **11**, e0160160.
39. Y. Qi and D. A. Volmer, Electron-based fragmentation methods in mass spectrometry: An overview, *Mass Spectrom. Rev.*, 2017, **36**, 4–15.
40. A. Vaniya and O. Fiehn, Using fragmentation trees and mass spectral trees for identifying unknown compounds in metabolomics, *TrAC, Trends Anal. Chem.*, 2015, **69**, 52–61.
41. J. Roboz, in *The Encyclopedia of Mass Spectrometry*, ed. M. L. Gross and R. M. Caprioli, Elsevier, Boston, 2016, pp. 183–188.
42. T. A. Addona, S. E. Abbatiello, B. Schilling, S. J. Skates, D. R. Mani, D. M. Bunk, C. H. Spiegelman, L. J. Zimmerman, A.-J. L. Ham, H. Keshishian, S. C. Hall, S. Allen, R. K. Blackman, C. H. Borchers, C. Buck, H. L. Cardasis, M. P. Cusack, N. G. Dodder, B. W. Gibson, J. M. Held, T. Hiltke, A. Jackson, E. B. Johansen, C. R. Kinsinger, J. Li, M. Mesri, T. A. Neubert, R. K. Niles, T. C. Pulsipher, D. Ransohoff, H. Rodriguez, P. A. Rudnick, D. Smith, D. L. Tabb, T. J. Tegeler, A. M. Variyath, L. J. Vega-Montoto, Å. Wahlander, S. Waldemarson, M. Wang, J. R. Whiteaker, L. Zhao, N. L. Anderson, S. J. Fisher, D. C. Liebler, A. G. Paulovich, F. E. Regnier, P. Tempst and S. A. Carr, Multi-site assessment of the precision and reproducibility of multiple reaction monitoring–based measurements of proteins in plasma, *Nat. Biotechnol.*, 2009, **27**, 633–641.

43. W.-L. Liao, G.-Y. Heo, N. G. Dodder, I. A. Pikuleva and I. V. Turko, Optimizing the Conditions of a Multiple Reaction Monitoring Assay for Membrane Proteins: Quantification of Cytochrome P450 11A1 and Adrenodoxin Reductase in Bovine Adrenal Cortex and Retina, *Anal. Chem.*, 2010, **82**, 5760–5767.
44. M. Berglund and M. E. Wieser, Isotopic compositions of the elements 2009 (IUPAC Technical Report), *Pure Appl. Chem.*, 2011, **83**, 397–410.
45. M. Wang, G. Audi, A. H. Wapstra, F. G. Kondev, M. MacCormick, X. Xu and B. Pfeiffer, The Ame2012 atomic mass evaluation, *Chin. Phys. C*, 2012, **36**, 1603–2014.
46. R. Winkler, ESIprot: A universal tool for charge state determination and molecular weight calculation of proteins from electrospray ionization mass spectrometry data, *Rapid Commun. Mass Spectrom.*, 2010, **24**, 285–294.
47. H. C. Köfeler and M. L. Gross, Correction of accurate mass measurement for target compound verification by quadrupole time-of-flight mass spectrometry, *J. Am. Soc. Mass Spectrom.*, 2005, **16**, 406–408.
48. R. Winkler, SpiderMass: Semantic database creation and tripartite metabolite identification strategy, *J. Mass Spectrom.*, 2015, **50**, 538–541.
49. Y. Wang, R. Feng, R. Wang, F. Yang, P. Li and J.-B. Wan, Enhanced MS/MS coverage for metabolite identification in LC-MS-based untargeted metabolomics by target-directed data dependent acquisition with time-staggered precursor ion list, *Anal. Chim. Acta*, 2017, **992**, 67–75.
50. M. M. Ulaszewska, C. H. Weinert, A. Trimigno, R. Portmann, C. A. Lacueva, R. Badertscher, L. Brennan, C. Brunius, A. Bub, F. Capozzi, M. C. Rosso, C. E. Cordero, H. Daniel, S. Durand, B. Egert, P. G. Ferrario, E. J. M. Feskens, P. Franceschi, M. Garcia-Aloy, F. Giacomoni, P. Giesbertz, R. González-Domínguez, K. Hanhineva, L. Y. Hemeryck, J. Kopka, S. E. Kulling, R. Llorach, C. Manach, F. Mattivi, C. Migné, L. H. Münger, B. Ott, G. Picone, G. Pimentel, E. Pujos-Guillot, S. Riccadonna, M. J. Rist, C. Rombouts, J. Rubert, T. Skurk, P. S. C. S. Harsha, L. V. Meulebroek, L. Vanhaecke, R. Vázquez-Fresno, D. Wishart and G. Vergères, Nutrimetabolomics: An Integrative Action for Metabolomic Analyses in Human Nutritional Studies, *Mol. Nutr. Food Res.*, 2019, **63**, 1800384.
51. A. D. Hegeman, Plant metabolomics–meeting the analytical challenges of comprehensive metabolite analysis, *Briefings Funct. Genomics*, 2010, **9**, 139–148.
52. S. D. S. Chiwocha, S. R. Abrams, S. J. Ambrose, A. J. Cutler, M. Loewen, A. R. S. Ross and A. R. Kermode, A method for profiling classes of plant hormones and their metabolites using liquid chromatography-electrospray ionization tandem mass spectrometry: An analysis of hormone regulation of thermodormancy of lettuce (Lactuca sativa L.) Seeds, *Plant J.*, 2003, **35**, 405–417.

53. R. Weiskirchen, S. Weiskirchen, P. Kim and R. Winkler, Software solutions for evaluation and visualization of laser ablation inductively coupled plasma mass spectrometry imaging (LA-ICP-MSI) data: A short overview, *J. Cheminf.*, 2019, **11**, 16.
54. Y. Benkler, *The Wealth of Networks: How Social Production Transforms Markets and Freedom*, Yale University Press, New Haven, CT, USA, 2006.
55. M. Woelfle, P. Olliaro and M. H. Todd, Open Science Is a Research Accelerator, *Nat. Chem.*, 2011, **3**, 745–748.

CHAPTER 2

# Mass Spectrometry Data Operations and Workflows

MAGNUS PALMBLAD*

Leiden University Medical Center, Center for Proteomics and Metabolomics, Postzone P1-Q, P.O. Box 9600, Leiden, 2300 RC, The Netherlands
*E-mail: n.m.palmblad@lumc.nl

## 2.1 Operations

Mass spectrometry is an extremely versatile analytical technique. It follows, therefore, that the number of applicable data analysis operations and the number of ways in which they can be combined in workflows are enormous. However, many workflows in proteomics and metabolomics, especially those starting from raw data, share a number of core operational components. Thanks to community efforts within ELIXIR – the European distributed infrastructure for life-science information – we now have the semantic tools, ontologies, to describe scientific domains, bioinformatics analysis operations and mass spectrometry data types and formats. The wording used in this chapter follows the EDAM[1] preferred terms. To give an idea of the number of (free and open source) software packages available, this chapter sometimes refers to the number of entries in the ELIXIR bio.tools registry (http://bio.tools).[2] The bio.tools registry is a community-based effort, collecting information on software in the bioinformatics domain. Recently, several hundred pieces of software for the analysis of mass spectrometry data in proteomics and metabolomics were added from http://ms-utils.org

New Developments in Mass Spectrometry No. 8
Processing Metabolomics and Proteomics Data with Open Software: A Practical Guide
Edited by Robert Winkler

Published by the Royal Society of Chemistry, www.rsc.org

and other resources. In total, bio.tools currently (August 2019) contains systematic, functional annotations of close to 13 000 software packages, making it by far the richest collection of bioinformatics software descriptions of any kind, and a uniquely powerful resource for building bioinformatics data analysis workflows, including the analysis of mass spectrometry data in the omics domains. This chapter will briefly discuss common operations in such workflows, and give a few examples of open software for performing them. The emphasis is on tools performing a single task and performing it well, rather than monolithic software, libraries or toolboxes covering a wide range of functions. However, this book contains chapters dedicated to the open source OpenMS,[3,4] MZmine[5]/MZmine 2,[6] XCMS[7,8] and Trans-Proteomic Pipeline,[9,10] each covering most of the operations described below. Individual modules and operations can, in many cases, be performed independently from the command line, in Windows or under Linux. This makes them well suited for integration in automated workflows, containerization and deploying on the cloud for large-scale analyses. The table below summarizes the operations described here and gives a few examples of free and open software able to perform these operations. Operations specific to mass spectrometry imaging, such as co-registration and region-of-interest determination, are not included, but the operations that are apply to mass spectrometry imaging data as well (Table 2.1).

### 2.1.1 Formatting

Most raw data is produced by mass spectrometry vendor software in a proprietary format. However, all major vendors provide libraries for accessing data in this format. These libraries are used by formatting – format conversion – software such as msconvert,[11] ThermoRawFileReader and ThermoRawFileParser,[12] uses vendor libraries to read and convert the raw data into an open format such as mzXML[13] or mzML.[14] Most third-party, academic software supports at least one of these two formats. As a curiosity, although mzML was intended to supersede mzXML in 2009, the latter is still in almost as frequent use according to a poll at the 2018 ASMS mass spectrometry software workshop. Most workflows combining multiple third-party or academic software from multiple sources, therefore, begin by formatting the raw data into one of the open, community-supported, formats.

### 2.1.2 Alignment

In bioinformatics, *alignment* is often understood to refer to the alignment of two or more molecular sequences. However, in mass spectrometry data analysis workflows, we also align spectra and chromatograms (the time axis in chromatographic separations). Conceptually, alignment is closely related to calibration. The main difference is that for alignment, there is no notion of a 'correct' *m*/*z* or retention time. Rather spectral or chromatographic features are aligned to other features of the same type. Chromatographic alignment

**Table 2.1** Operations, simplified definitions and examples of tools capable of performing the operation. Many of these operations can be performed by software packages, including OpenMS, XCMS, MZmine and the Trans-Proteomic Pipeline. Each of these packages have their own dedicated chapter in this book.

| Operation | Simple definition | Open source examples |
|---|---|---|
| Formatting | Reformatting a data file or equivalent entity in memory | Msconvert, ThermoRawFileParser |
| Alignment | Aligning one or more chromatograms or mass spectra | Msalign, SpecArray, ROIMCR, CPM |
| Peak detection | Finding peaks in a spectrum or chromatogram | DeconTools, cosmiq, mMass, Dinosaur |
| Identification | Assigning molecular structures to peaks or spectra | Comet, X!Tandem, SpiderMass |
| Calibration | Calibrating mass spectra (in the *m*/*z* dimension) | Deltamass, Recal2, VIPER |
| Quantification | Measuring observations into abundance of analytes | Multi-Q, isobariQ, SILVER, MS-Spectre |
| Quality control | Assessing of the quality of data or interpretations | Qcmetrics, Proteinspector, Q4SRM |
| Statistical analysis | Analyzing data to determine its statistical properties | Triqler, MSStats |
| Visualization | Illustrating data graphically using software | Unipept, Mass++, MsViz, jmzML, rMSI |
| Deposition | Submitting data and analysis results to repositories | PRIDE Converter,[a] ProCon[a] |

[a]Strictly speaking, these tools perform formatting operation, but with the explicit purpose of combining and generating files for deposition.

was a key component of the accurate mass and time (AMT) paradigm developed by Smith and co-workers[15] to match identifications by tandem mass spectrometry between runs. A large body of literature on the topic exists, but will not be reviewed here. However, in general, correct pairing of features for alignment typically requires either accurate mass in both datasets to be aligned[16] or the unambiguous identification of at least the elemental composition of a metabolite or peptide by tandem mass spectrometry in at least one of the datasets.[17] If robust, it is not detrimental to align datasets to make them register data along the same *m*/*z* or time axis, even if not every workflow will take advantage of this. In general, alignment should result in fewer missing values and greater ease in choosing parameters that are dependent on uncertainty in *m*/*z* or retention time. Bio.tools currently contains 17 different software tools for *chromatographic* alignment, including packages such as PyMS,[18] OpenMS, XCMS and MZmine 2.

### 2.1.3 Peak Detection

Many, though far from all, algorithms for interpreting mass spectra require one form of peak or feature detection. In jargon, practitioners of mass spectrometry contrast full profile with line, or centroid, spectra, where the former retain all *m*/*z* channels and display full peak profiles and the latter a list of *m*/*z* and intensity values corresponding to detected peaks. Fundamentally, peak detection is a statistical process. In principle, each hypothetical peak has, in addition to *m*/*z* and intensity, also the probability of the observation given that the null hypothesis – observation is noise – is true. These probabilities are not always reported by software as such, however. Practically, peak detection is a way to reduce tens of thousands to millions of data points to a few hundred or thousand meaningful features that can be identified, quantified and compared across spectra and samples. All vendor software, as well as many third-party tools (34 of which can be found on bio.tools, including the already mentioned msconvert[19]), perform the operation of peak detection in raw, full profile, mass spectra. Though conceptually fairly simple, peak detection is far from trivial. Determining the intensity, as signal-to-noise, requires accurate determination of the noise. A related problem is the accurate determination of background – meaning the background at the *m*/*z* of the peak detected that should be subtracted from the intensity for accurate comparisons and ratios of peak intensity. This is particularly important for peaks of low intensity, which are often overestimated, leading to compressed ratios in relative quantitation schemes. In peak detection, there is always a trade-off between sensitivity (loss of information) and specificity (only detecting true peaks, and their *m*/*z* span). Increasing sensitivity reduces information loss but can negatively affect quantitative accuracy, if false peaks are included in abundance or ratio calculations. A number of software packages take a comprehensive approach, using modelling of the noise distribution, integrating consecutively acquired spectra, such as in LC-MS, and isotopic distribution using average elemental compositions for the analytes expected in the sample, most famously the THRASH algorithm[20] by Horn *et al.* With the increasing popularity of ion mobility as an additional dimension of separation between chromatography and mass spectrometry, leading to raw data file sizes in the hundreds of gigabytes or even terabytes, efficient and robust multidimensional peak detection becomes critical.

### 2.1.4 Identification

Except for targeted methods, where the identity is presumed known and the intensity monitored in specific, pre-determined, *m*/*z* and retention time windows (often using tandem mass spectrometry), identification is a core component in most mass spectrometry data analysis workflows. Small molecules are often identified using spectral libraries from databases. Peptides can also be identified from spectral libraries, though more commonly by comparing experimental spectra against those predicted from a list of sequences, enabling identification of peptides never previously observed. For both metabolites and peptides, the elemental composition is relatively easily and unambiguously identified. For metabolites such as lipids, there can be remaining uncertainty in the exact location of double bonds. For peptides, the localization and structure of post-translational modifications can remain uncertain. However, even with some remaining ambiguity, the analytes are considered identified at this stage, and can be recorded with their elemental composition, retention time and tandem mass spectra, allowing calculation of correct masses for calibration, matching between analyses and quantification. Except for purely *de novo* methods, the availability of spectral and sequence databases is important for identification of analytes from mass spectrometry data. Secondary and specialized analysis tools can refine the identification, including localizing and characterizing modifications and deal with chimeric spectra – tandem mass spectra superimposing fragments from two or more distinct peptides. Currently, bio.tools contains more than 60 tools for peptide identification from mass spectrometry data.

### 2.1.5 Calibration

Mass filters and analyzers in a mass spectrometer require calibration. These calibrations are typically performed using standards, calibration mixtures or background ions before measuring unknown samples. Most QC pipelines and robust data analysis workflow contains at least an inspection of the mass measurement uncertainty, and whether there is any additional uncertainty caused by poor calibration, beyond the inherent mass measurement precision limit of the mass analyzer. Though it is difficult to correct for an offset precursor isolation window, mass spectra – in particular the precursor or MS1 spectra – can be calibrated after the measurement. The simplest approach is to globally correct for the average mass measurement error, typically as a function of $m/z$, for all spectra. This approach is sometimes referred to as a statistical recalibration, as it is based on this mass measurement statistic. On bio.tools there 21 tools for mass spectra calibration, many using analyzer-specific calibration functions and calibrating each spectrum locally rather than finding one global calibration. This is particularly relevant for ion trapping instruments where space charge has an effect on the mass measurement, including Orbitrap and FTICR mass analyzers. Many of the software tools listed on bio.tools also perform other functions. It is, therefore, worth looking into libraries and multifunctional tools to see if they can recalibrate the mass spectra of a particular type from a specific mass analyzer or instrument. Some tools, *e.g.* LCMSWARP[21] and software built on this tool calibrate mass spectra as a step toward aligning LC-MS datasets based on accurate masses, further emphasizing the similarity between retention time alignment and mass measurement calibration. For use in workflows, tools that write out the recalibrated data in open formats such as mzXML or mzML are ideal, as the tool can then be inserted as a recalibration module in any mass spectrometry data analysis workflow.[22] It should also be mentioned that calibration can also refer to the mapping of system response as a function of analyte abundance or concentration, establishing calibration curves for absolute quantification of known analytes. Calibration in this sense can be seen as a prerequisite for or integral part of the quantification operation.

### 2.1.6 Quantification

Quantitative information on analytes can be generated by mass spectrometry in many different ways. Very broadly, methods either use some form of label or not. The latter are often referred to as 'label-free' and are subdivided further into intensity based methods and spectral counting.[23] As the spectral count, or number of peptide-spectrum matches generated in a given time, increases with each new instrument generation, counting methods become increasingly attractive, even though the counts are still far from those generated by RNA-Seq on the same time scale. Labeling methods differ in the choice of label as well as when and where it is applied, from metabolic labeling using labeled growth media to chemical labeling incorporating one or more heavy isotopes distinguishable by mass spectrometry. Some software for quantitative mass spectrometry enable the user to generically specify the label, other tools are dedicated to a particular labeling scheme. Most of the time, the quantitative information is assigned to one or a few identified analytes, though it is possible to quantify unidentified features defined by their $m/z$ and possibly retention time.

Bio.tools currently contains 23 tools for labeled quantification and 18 for label-free quantification, with eight able to perform both. Some 46 tools are annotated as performing protein quantification. As new labeling schemes continue to be developed, new software tools for quantification appear regularly. In proteomics, the problem of protein inference and how to assign quantitative information on the peptide level, where many peptides are shared between one or more proteins or proteoforms, is still an active area of research and the protein inference problem itself was specifically investigated in a recent ABRF iPRG study.[24] As with peak integration, there is a trade-off between sensitivity (using all available quantitative information) and specificity (only using that quantitative information which can be uniquely attributed to a single analyte). To make things more complicated, this depends on assumptions made about the sample, manifested by the set of sequences or the spectral library the spectra are matched against. Quantification tools that output their results in an open format such as mzQuantML, or as tab-delimited text files, can be assembled in workflows.

### 2.1.7 Quality Control

In routine work and studies involving a large number of samples, automated system suitability tests and quality control (QC) methods are essential. There now exists a very large number of software tools specifically designed to help quality control mass spectrometry data. Some system suitability tools refer to specific standards and instruments, such as the recent triple knock-out or TKO Visualization Tool[25] and the original MSQC.[26] These have the advantage that few or no (as in the case of the TKO QC tool) parameters need to be specified by the user. Typical QC tools monitor global metrics such as the number of peptide-spectrum matches, total ion intensity, trap fill times and mass measurement errors, and the retention time (and possibly mass measurement error) of internal standards added to the samples. These quality metrics are themselves important metadata that should be taken into account when integrating data from different laboratories and instruments. Though bio.tools annotates 168 tools on the topic of data quality management, only 11 are explicitly within proteomics or metabolomics, and fewer still are modular or easily combined with other tools. More commonly, QC methods contain an entire data analysis workflow, including statistical analysis and visualization methods aimed at revealing appropriate QC metrics. As with quantification, important QC decisions are integral to the experimental design, such as the scheduled analyses of standards or additions of internal standards to the samples. The PSI Quality Control workgroup is chartered to develop the qcML format[27] for storing and sharing mass spectrometry QC metrics and define associated controlled vocabulary terms.

### 2.1.8 Statistical Analysis

Though statistical models are found in many software tools in the above categories (with the possible exception of formatting), most quantitative mass spectrometry workflows in proteomics and metabolomics contain an explicit statistics component. The purpose of all statistics is to extract meaningful patterns from noisy data. In mass spectrometry data analysis workflows, statistics components extract patterns from the results for the data analysis thus far, for example differentially abundant, modified or spatiotemporally distributed proteins and metabolites between different biological conditions, where each is sampled with several biological replicates. The exact configuration of the statistical analysis is dependent on the design and aims of the study. It is therefore not uncommon to use a scripting language, such as R, to define exactly what is to be compared and how, in data analysis workflows.[28]

The importance of attention to statistics can hardly be overstated. When combining multiple operations in sequence, as in an automated workflow, it is also important to look at the assumptions in and consequences of applying multiple statistical tests. Bayesian approaches have been used to connect the statistics used in several of the above categories and remove unnecessary assumptions or thresholds.[29] As for the software, statistical methods are often bundled with appropriate visualization tools, *e.g.* Perseus.[30] As in genomics studies, gene ontology and pathway enrichment analyses from lists of differential proteins are commonplace, though many are gene-centric and less suitable for direct analysis of post-translational modification data.

### 2.1.9 Visualization

The most common end-point of a data analysis workflow is some form of data presentation or visualization. This may be a static table or figure, or an interactive data browser or visualization. For QC purposes, simple visualizations can be produced in intermediate steps in the workflow, enabling the user to monitor the data processing and check for errors or unexpected behavior in real time as well as retrospectively. Nearly 900 software tools for visualization are annotated on bio.tools. These cover a wide range of generic and specific visualization software, as well as tools with visualization functionality. Chapter 15 in this book shows how to combine several of the operations described above, including formatting, identification, quantitation and statistical analyses, with visualization of the results, as R Markdowns. Such Markdowns integrate in a single document the executable code, parameters and documentation, well suited as supplemental information accompanying manuscripts including mass spectrometry data analyses.

### 2.1.10 Deposition

A parallel end-point of data analysis workflows is the deposition of raw data, results and study and sample metadata in public repositories such as PRIDE[31] or MassIVE (http://massive.ucsd.edu). Though this step typically require some manual interaction, especially in providing the metadata and organizing what is to be uploaded, some software now help the user in automating this process to some degree. This is particularly useful when dealing with large numbers of files, in studies involving large numbers of samples or fractions thereof. Many of the steps above are mandatory for full submissions to PRIDE, including formatting and identification. Proper calibration and thorough quality control should also be encouraged before submission, as more examples of poorly calibrated data of poor quality are already in the repositories than what is needed by developers of software tools for calibration and quality control.

## 2.2 Workflows

Data analyses in mass spectrometry-based proteomics and metabolomics usually multistep processes including many of the operations described above. Knitting together separate software modules into an executable pipeline is the task of workflow managers. These typically also provide a graphical user interface, record past executions in a database and facilitate documentation and sharing of workflows on various platforms. The graphical user interfaces enable biologists with limited programming expertise to repeat analyses and reorganize the workflows.

Taverna[32] was the first scientific workflow manager to be used for large-scale mass spectrometry data analysis, at least in proteomics[22] and on the cloud.[33] In metabolomics, KNIME[34] is probably currently the most popular workflow engine, and has been used for GC-MS and LC-MS data[35] as well as NMR-based metabolomics.[36] The Galaxy project is widely popular in genomics and has the largest and most active user and developer base.[37] Galaxy is also used for combining RNA-Seq and mass spectrometry data in proteomics,[38] proteogenomics,[39] metaproteomics,[40] affinity proteomics,[41] and multiomics analyses,[42] and is more than a workflow engine – it is a community and an infrastructure for collaboration and large-scale data analysis. Other workflow managers have also been used, and simpler workflow engines are now incorporated in many software projects, including vendor software for processing mass spectrometry data. In the end, the choice of workflow engine depends as much on personal preferences and experience as on the application. There are ongoing efforts in facilitating sharing workflows across platforms through the Common Workflow Language, or CWL.[43–45]

Data analysis pipelines can, and have, been implemented in almost any scripting language or environment. Jupyter notebooks[46–48] and R Markdown[49–51] have become increasingly popular for containing, documenting and disseminating data processing workflows, as well as teaching bioinformatics and computational metabolomics and proteomics.[52] The 2016 ABRF iPRG study on proteoform inference provided examples and actively encouraged submissions as either an R Markdown or Jupyter notebook.[24] Jupyter has recently also been integrated with Galaxy.[37] Uses of R Markdowns in mass spectrometry based proteomics data analysis are described in detail with specific examples in Chapter 15 in this book.

The challenge of automatic workflow composition has been revisited in the context of mass spectrometry-based proteomics,[53] leveraging recent developments in composition algorithms and semantic software annotation in software registries such as bio.tools. When combined with software containers, automatic workflow composition can generate large numbers of executable, semantically similar workflows, to test the robustness of data analyses and interpretation, repair broken workflows by identifying replacement modules or services, and optimize the workflows themselves.[54] The process of taking a semantic workflow definition and generating an executable workflow is illustrated in Figure 2.1 below. De Bruin *et al.*[22] started from basic peptide identification workflows in Taverna and built more advanced workflows including chromatographic alignment, merging datasets, mass calibration and peptide identification with iterative refinement. These workflows and their modules are still available on http://ms-utils.org (https://www.ms-utils.org/Taverna/).

Interactive software is great for interacting with mass spectrometry data, drilling down from proteins to peptides to spectra to individual peaks in order to better understand the nature and quality of the data, as well as troubleshooting the experimental protocol or mass spectrometer. However, assembling open-source tools, each designed to perform one operation and perform it well, into complex but modular workflows to accomplish larger data analysis missions, is an attractive alternative to monolithic, typically closed-source, software. Everything is recorded in a form that can be shared and reproduced, promoting openness and ultimately increasing confidence in our scientific output. Scripts and workflows are automated and easily scale to analyses of hundreds or thousands of datasets, with many solutions already deployed or deployable in cloud environments.[38,40,42,55] Workflows are adaptable – a new data format typically only necessitates a conversion or filter to be compatible with existing workflows. The active

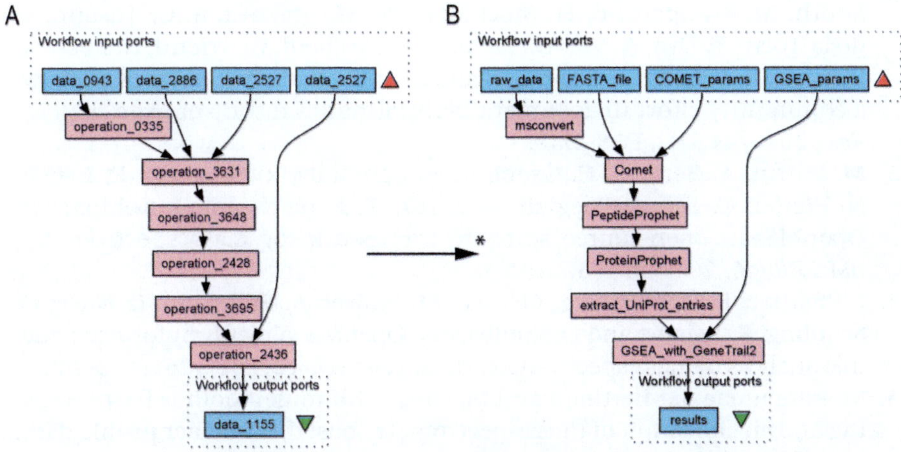


**Figure 2.1** Generic workflow (A) for processing LC-MS/MS data (EDAM data:0943) in Thermo.raw format (EDAM format:3712) all the way to enriched Reactome pathways (EDAM data:1155). The workflow, here visualized in the 'Taverna'[32] style, proceeds through format conversion (operation:0335), peptide-spectrum matching (operation:3631), protein identification (operation:2428), extraction of protein IDs (a type of filtering, operation:3695) and functional enrichment analysis (operation:2436). The composition of workflows combining actual software (B), indicated by the asterisk, is the task of workflow composers[53] and relies on rich semantic and functional annotation of software, as provided by bio.tools.

communities formed around the most popular workflow managers provide resources such as examples, tutorials and courses, paving a clear path to learning about and adopting scientific workflows in proteomics or metabolomics data analysis.

# References

1. J. Ison, M. Kalas, I. Jonassen, D. Bolser, M. Uludag, H. McWilliam, J. Malone, R. Lopez, S. Pettifer and P. Rice, EDAM: an ontology of bioinformatics operations, types of data and identifiers, topics and formats, *Bioinformatics*, 2013, **29**(10), 1325–1332.
2. J. Ison, K. Rapacki, H. Menager, M. Kalas, E. Rydza, P. Chmura, C. Anthon, N. Beard, K. Berka, D. Bolser, T. Booth, A. Bretaudeau, J. Brezovsky, R. Casadio, G. Cesareni, F. Coppens, M. Cornell, G. Cuccuru, K. Davidsen, G. D. Vedova, T. Dogan, O. Doppelt-Azeroual, L. Emery, E. Gasteiger, T. Gatter, T. Goldberg, M. Grosjean, B. Gruning, M. Helmer-Citterich, H. Ienasescu, V. Ioannidis, M. C. Jespersen, R. Jimenez, N. Juty, P. Juvan, M. Koch, C. Laibe, J. W. Li, L. Licata, F. Mareuil, I. Micetic, R. M. Friborg, S. Moretti, C. Morris, S. Moller, A. Nenadic, H. Peterson, G. Profiti, P. Rice, P. Romano, P. Roncaglia, R. Saidi, A. Schafferhans, V. Schwammle, C.

Smith, M. M. Sperotto, H. Stockinger, R. S. Varekova, S. C. Tosatto, V. de la Torre, P. Uva, A. Via, G. Yachdav, F. Zambelli, G. Vriend, B. Rost, H. Parkinson, P. Longreen and S. Brunak, Tools and data services registry: a community effort to document bioinformatics resources, *Nucleic Acids Res.*, 2016, **44**(D1), D38–D47.

3. M. Sturm, A. Bertsch, C. Gropl, A. Hildebrandt, R. Hussong, E. Lange, N. Pfeifer, O. Schulz-Trieglaff, A. Zerck, K. Reinert and O. Kohlbacher, OpenMS–an open-source software framework for mass spectrometry, *BMC Bioinf.*, 2008, **9**, 163.
4. J. Pfeuffer, T. Sachsenberg, O. Alka, M. Walzer, A. Fillbrunn, L. Nilse, O. Schilling, K. Reinert and O. Kohlbacher, OpenMS–A platform for reproducible analysis of mass spectrometry data, *J. Biotechnol.*, 2017, **261**, 142–148.
5. M. Katajamaa, J. Miettinen and M. Oresic, MZmine: toolbox for processing and visualization of mass spectrometry based molecular profile data, *Bioinformatics*, 2006, **22**(5), 634–636.
6. T. Pluskal, S. Castillo, A. Villar-Briones and M. Oresic, MZmine 2: modular framework for processing, visualizing, and analyzing mass spectrometry-based molecular profile data, *BMC Bioinf.*, 2010, **11**, 395.
7. C. A. Smith, E. J. Want, G. O'Maille, R. Abagyan and G. Siuzdak, XCMS: processing mass spectrometry data for metabolite profiling using nonlinear peak alignment, matching, and identification, *Anal. Chem.*, 2006, **78**(3), 779–787.
8. N. G. Mahieu, J. L. Genenbacher and G. J. Patti, A roadmap for the XCMS family of software solutions in metabolomics, *Curr. Opin. Chem. Biol.*, 2016, **30**, 87–93.
9. P. G. Pedrioli, Trans-proteomic pipeline: a pipeline for proteomic analysis, *Methods Mol. Biol.*, 2010, **604**, 213–238.
10. E. W. Deutsch, L. Mendoza, D. Shteynberg, J. Slagel, Z. Sun and R. L. Moritz, Trans-Proteomic Pipeline, a standardized data processing pipeline for large-scale reproducible proteomics informatics, *Proteomics: Clin. Appl.*, 2015, **9**(7–8), 745–754.
11. R. Adusumilli and P. Mallick, Data Conversion with ProteoWizard msConvert, *Methods Mol. Biol.*, 2017, **1550**, 339–368.
12. N. Hulstaert, J. Shofstahl, T. Sachsenberg, M. Walzer, H. Barsnes, L. Martens and Y. Perez-Riverol, ThermoRawFileParser: Modular, Scalable, and Cross-Platform RAW File Conversion, *J. Proteome Res.*, 2019, **19**(1), 537–542.
13. P. G. Pedrioli, J. K. Eng, R. Hubley, M. Vogelzang, E. W. Deutsch, B. Raught, B. Pratt, E. Nilsson, R. H. Angeletti, R. Apweiler, K. Cheung, C. E. Costello, H. Hermjakob, S. Huang, R. K. Julian, E. Kapp, M. E. McComb, S. G. Oliver, G. Omenn, N. W. Paton, R. Simpson, R. Smith, C. F. Taylor, W. Zhu and R. Aebersold, A common open representation of mass spectrometry data and its application to proteomics research, *Nat. Biotechnol.*, 2004, **22**(11), 1459–1466.
14. E. Deutsch, mzML: a single, unifying data format for mass spectrometer output, *Proteomics*, 2008, **8**(14), 2776–2777.

15. L. Pasa-Tolic, C. Masselon, R. C. Barry, Y. Shen and R. D. Smith, Proteomic analyses using an accurate mass and time tag strategy, *BioTechniques*, 2004, **37**(4), 621–624, 626–633, 636 passim.
16. E. Nevedomskaya, R. Derks, A. M. Deelder, O. A. Mayboroda and M. Palmblad, Alignment of capillary electrophoresis-mass spectrometry datasets using accurate mass information, *Anal. Bioanal. Chem.*, 2009, **395**(8), 2527–2533.
17. M. Palmblad, D. J. Mills, L. V. Bindschedler and R. Cramer, Chromatographic alignment of LC-MS and LC-MS/MS datasets by genetic algorithm feature extraction, *J. Am. Soc. Mass Spectrom.*, 2007, **18**(10), 1835–1843.
18. S. O'Callaghan, D. P. De Souza, A. Isaac, Q. Wang, L. Hodkinson, M. Olshansky, T. Erwin, B. Appelbe, D. L. Tull, U. Roessner, A. Bacic, M. J. McConville and V. A. Likic, PyMS: a Python toolkit for processing of gas chromatography-mass spectrometry (GC-MS) data. Application and comparative study of selected tools, *BMC Bioinf.*, 2012, **13**, 115.
19. W. R. French, L. J. Zimmerman, B. Schilling, B. W. Gibson, C. A. Miller, R. R. Townsend, S. D. Sherrod, C. R. Goodwin, J. A. McLean and D. L. Tabb, Wavelet-based peak detection and a new charge inference procedure for MS/MS implemented in ProteoWizard's msConvert, *J. Proteome Res.*, 2015, **14**(2), 1299–1307.
20. D. M. Horn, R. A. Zubarev and F. W. McLafferty, Automated reduction and interpretation of high resolution electrospray mass spectra of large molecules, *J. Am. Soc. Mass Spectrom.*, 2000, **11**(4), 320–332.
21. N. Jaitly, M. E. Monroe, V. A. Petyuk, T. R. Clauss, J. N. Adkins and R. D. Smith, Robust algorithm for alignment of liquid chromatography-mass spectrometry analyses in an accurate mass and time tag data analysis pipeline, *Anal. Chem.*, 2006, **78**(21), 7397–7409.
22. J. S. de Bruin, A. M. Deelder and M. Palmblad, Scientific workflow management in proteomics, *Mol. Cell. Proteomics*, 2012, **11**(7), M111 010595.
23. W. Zhu, J. W. Smith and C. M. Huang, Mass spectrometry-based label-free quantitative proteomics, *J. Biomed. Biotechnol.*, 2010, **2010**, 840518.
24. J. Y. Lee, H. Choi, C. M. Colangelo, D. Davis, M. R. Hoopmann, L. Kall, H. Lam, S. H. Payne, Y. Perez-Riverol, M. The, R. Wilson, S. T. Weintraub and M. Palmblad, ABRF Proteome Informatics Research Group (iPRG) 2016 Study: Inferring Proteoforms from Bottom-up Proteomics Data, *J. Biomol. Tech.*, 2018, **29**(2), 39–45.
25. J. P. Gygi, Q. Yu, J. Navarrete-Perea, R. Rad, S. P. Gygi and J. A. Paulo, Web-Based Search Tool for Visualizing Instrument Performance Using the Triple Knockout (TKO) Proteome Standard, *J. Proteome Res.*, 2019, **18**(2), 687–693.
26. P. A. Rudnick, K. R. Clauser, L. E. Kilpatrick, D. V. Tchekhovskoi, P. Neta, N. Blonder, D. D. Billheimer, R. K. Blackman, D. M. Bunk, H. L. Cardasis, A. J. Ham, J. D. Jaffe, C. R. Kinsinger, M. Mesri, T. A. Neubert, B. Schilling, D. L. Tabb, T. J. Tegeler, L. Vega-Montoto, A. M. Variyath, M. Wang, P. Wang, J. R. Whiteaker, L. J. Zimmerman, S. A. Carr, S. J. Fisher, B. W. Gibson, A. G. Paulovich, F. E. Regnier, H. Rodriguez, C. Spiegelman, P. Tempst, D. C.

Liebler and S. E. Stein, Performance metrics for liquid chromatography-tandem mass spectrometry systems in proteomics analyses, *Mol. Cell. Proteomics*, 2010, **9**(2), 225–241.

27. M. Walzer, L. E. Pernas, S. Nasso, W. Bittremieux, S. Nahnsen, P. Kelchtermans, P. Pichler, H. W. van den Toorn, A. Staes, J. Vandenbussche, M. Mazanek, T. Taus, R. A. Scheltema, C. D. Kelstrup, L. Gatto, B. van Breukelen, S. Aiche, D. Valkenborg, K. Laukens, K. S. Lilley, J. V. Olsen, A. J. Heck, K. Mechtler, R. Aebersold, K. Gevaert, J. A. Vizcaino, H. Hermjakob, O. Kohlbacher and L. Martens, qcML: an exchange format for quality control metrics from mass spectrometry experiments, *Mol. Cell. Proteomics*, 2014, **13**(8), 1905–1913.
28. M. Choi, C. Y. Chang, T. Clough, D. Broudy, T. Killeen, B. MacLean and O. Vitek, MSstats: an R package for statistical analysis of quantitative mass spectrometry-based proteomic experiments, *Bioinformatics*, 2014, **30**(17), 2524–2526.
29. M. The and L. Kall, Integrated Identification and Quantification Error Probabilities for Shotgun Proteomics, *Mol. Cell. Proteomics*, 2019, **18**(3), 561–570.
30. S. Tyanova, T. Temu, P. Sinitcyn, A. Carlson, M. Y. Hein, T. Geiger, M. Mann and J. Cox, The Perseus computational platform for comprehensive analysis of (prote)omics data, *Nat. Methods*, 2016, **13**(9), 731–740.
31. P. Jones, R. G. Cote, L. Martens, A. F. Quinn, C. F. Taylor, W. Derache, H. Hermjakob and R. Apweiler, PRIDE: a public repository of protein and peptide identifications for the proteomics community, *Nucleic Acids Res.*, 2006, **34**(Database issue), D659–D663.
32. D. Hull, K. Wolstencroft, R. Stevens, C. Goble, M. R. Pocock, P. Li and T. Oinn, Taverna: a tool for building and running workflows of services, *Nucleic Acids Res.*, 2006, **34**(Web Server issue), W729–W732.
33. Y. Mohammed, E. Mostovenko, A. A. Henneman, R. J. Marissen, A. M. Deelder and M. Palmblad, Cloud parallel processing of tandem mass spectrometry based proteomics data, *J. Proteome Res.*, 2012, **11**(10), 5101–5108.
34. M. R. Berthold, N. Cebron, F. Dill, T. R. Gabriel, T. Kötter, T. Meinl, P. Ohl, C. Sieb, K. Thiel and B. Wiswedel, KNIME: The Konstanz Information Miner, in *Data Analysis, Machine Learning and Applications. Studies in Classification, Data Analysis, and Knowledge Organization*, ed. C. Preisach, H. Burkhardt, L. Schmidt-Thieme and R. Decker, Springer, Berlin, Heidelberg, 2008.
35. S. Liggi, C. Hinz, Z. Hall, M. L. Santoru, S. Poddighe, J. Fjeldsted, L. Atzori and J. L. Griffin, KniMet: a pipeline for the processing of chromatography-mass spectrometry metabolomics data, *Metabolomics*, 2018, **14**(4), 52.
36. A. Verhoeven, M. Giera and O. A. Mayboroda, KIMBLE: A versatile visual NMR metabolomics workbench in KNIME, *Anal. Chim. Acta*, 2018, **1044**, 66–76.
37. E. Afgan, D. Baker, B. Batut, M. van den Beek, D. Bouvier, M. Cech, J. Chilton, D. Clements, N. Coraor, B. A. Gruning, A. Guerler, J. Hillman-Jackson,

S. Hiltemann, V. Jalili, H. Rasche, N. Soranzo, J. Goecks, J. Taylor, A. Nekrutenko and D. Blankenberg, The Galaxy platform for accessible, reproducible and collaborative biomedical analyses: 2018 update, *Nucleic Acids Res.*, 2018, **46**(W1), W537–W544.

38. G. M. Sheynkman, J. E. Johnson, P. D. Jagtap, M. R. Shortreed, G. Onsongo, B. L. Frey, T. J. Griffin and L. M. Smith, Using Galaxy-P to leverage RNA-Seq for the discovery of novel protein variations, *BMC Genomics*, 2014, **15**, 703.
39. P. D. Jagtap, J. E. Johnson, G. Onsongo, F. W. Sadler, K. Murray, Y. Wang, G. M. Shenykman, S. Bandhakavi, L. M. Smith and T. J. Griffin, Flexible and accessible workflows for improved proteogenomic analysis using the Galaxy framework, *J. Proteome Res.*, 2014, **13**(12), 5898–5908.
40. C. Blank, C. Easterly, B. Gruening, J. Johnson, C. A. Kolmeder, P. Kumar, D. May, S. Mehta, B. Mesuere, Z. Brown, J. E. Elias, W. J. Hervey, T. McGowan, T. Muth, B. Nunn, J. Rudney, A. Tanca, T. J. Griffin and P. D. Jagtap, Disseminating Metaproteomic Informatics Capabilities and Knowledge Using the Galaxy-P Framework, *Proteomes*, 2018, **6**(1), 7.
41. P. A. Stewart, B. M. Kuenzi, S. Mehta, P. Kumar, J. E. Johnson, P. Jagtap, T. J. Griffin and E. B. Haura, The Galaxy Platform for Reproducible Affinity Proteomic Mass Spectrometry Data Analysis, *Methods Mol. Biol.*, 2019, **1977**, 249–261.
42. J. Boekel, J. M. Chilton, I. R. Cooke, P. L. Horvatovich, P. D. Jagtap, L. Kall, J. Lehtio, P. Lukasse, P. D. Moerland and T. J. Griffin, Multi-omic data analysis using Galaxy, *Nat. Biotechnol.*, 2015, **33**(2), 137–139.
43. A. Peter, R. C. Michael, T. Nebojša, C. Brad, C. John, H. Michael, K. Andrey, L. Dan, M. Hervé, N. Maya, S. Matt, S.-R. Stian and S. Luka, *Common Workflow Language, v1.0*, 2016.
44. M. Kotliar, A. V. Kartashov and A. Barski, CWL-Airflow: a lightweight pipeline manager supporting Common Workflow Language, *GigaScience*, 2019, **8**(7), 1–8.
45. P. K. Korhonen, R. S. Hall, N. D. Young and R. B. Gasser, Common workflow language (CWL)-based software pipeline for *de novo* genome assembly from long- and short-read data, *GigaScience*, 2019, **8**(4), 1–16.
46. A. Ressa, M. Fitzpatrick, H. van den Toorn, A. J. R. Heck and M. Altelaar, PaDuA: A Python Library for High-Throughput (Phospho)proteomics Data Analysis, *J. Proteome Res.*, 2019, **18**(2), 576–584.
47. L. Malmstrom, Computational Proteomics with Jupyter and Python, *Methods Mol. Biol.*, 2019, **1977**, 237–248.
48. J. Willforss, A. Chawade and F. Levander, NormalyzerDE: Online Tool for Improved Normalization of Omics Expression Data and High-Sensitivity Differential Expression Analysis, *J. Proteome Res.*, 2019, **18**(2), 732–740.
49. P. Navarro, J. Kuharev, L. C. Gillet, O. M. Bernhardt, B. MacLean, H. L. Rost, S. A. Tate, C. C. Tsou, L. Reiter, U. Distler, G. Rosenberger, Y. Perez-Riverol, A. I. Nesvizhskii, R. Aebersold and S. Tenzer, A multicenter study benchmarks software tools for label-free proteome quantification, *Nat. Biotechnol.*, 2016, **34**(11), 1130–1136.

50. N. Aben, J. A. Westerhuis, Y. Song, H. A. L. Kiers, M. Michaut, A. K. Smilde and L. F. A. Wessels, iTOP: inferring the topology of omics data, *Bioinformatics*, 2018, **34**(17), i988–i996.
51. E. C. Considine and R. M. Salek, A Tool to Encourage Minimum Reporting Guideline Uptake for Data Analysis in Metabolomics, *Metabolites*, 2019, **9**(3), 43.
52. B. Batut, S. Hiltemann, A. Bagnacani, D. Baker, V. Bhardwaj, C. Blank, A. Bretaudeau, L. Brillet-Gueguen, M. Cech, J. Chilton, D. Clements, O. Doppelt-Azeroual, A. Erxleben, M. A. Freeberg, S. Gladman, Y. Hoogstrate, H. R. Hotz, T. Houwaart, P. Jagtap, D. Lariviere, G. Le Corguille, T. Manke, F. Mareuil, F. Ramirez, D. Ryan, F. C. Sigloch, N. Soranzo, J. Wolff, P. Videm, M. Wolfien, A. Wubuli, D. Yusuf, N. Galaxy Training, J. Taylor, R. Backofen, A. Nekrutenko and B. Gruning, Community-Driven Data Analysis Training for Biology, *Cell Syst.*, 2018, **6**(6), 752–758 e1.
53. M. Palmblad, A. L. Lamprecht, J. Ison and V. Schwammle, Automated workflow composition in mass spectrometry-based proteomics, *Bioinformatics*, 2019, **35**(4), 656–664.
54. S. Holl, Y. Mohammed, O. Zimmermann and M. Palmblad, Scientific workflow optimization for improved peptide and protein identification, *BMC Bioinf.*, 2015, **16**, 284.
55. J. Slagel, L. Mendoza, D. Shteynberg, E. W. Deutsch and R. L. Moritz, Processing shotgun proteomics data on the Amazon cloud with the trans-proteomic pipeline, *Mol. Cell. Proteomics*, 2015, **14**(2), 399–404.

CHAPTER 3

# Metabolomics

DAVID S. WISHART*[a,b], JOANNA GODZIEN[c,d], ALBERTO GIL-DE-LA-FUENTE[c], RUPASRI MANDAL[a], RAHMATOLLAH RAJABZADEH[a], HAMED PIRIMOGHADAM[a], CAROL LADNER-KEAY[a], ABRAHAM OTERO[c] AND CORAL BARBAS[c]

[a]Department of Biological Sciences, University of Alberta, Edmonton, AB, Canada T6G 2E8; [b]Department of Computing Science, University of Alberta, Edmonton, AB, Canada T6G 2E9; [c]Centro de Excelencia en Metabolomica y Bioanalysis, Univesidad San Pablo, Madrid, Spain; [d]Clinical Research Centre, Medical University of Bialystok, Poland
*E-mail: dwishart@ualberta.ca

## 3.1 Introduction to Metabolomics

Metabolomics is a field of 'omics' research concerned with comprehensive characterization of the small molecule metabolites in the metabolome.[1] Just as the genome represents the complete collection of genes, and the proteome represents the complete collection of proteins, the metabolome is formally defined as the complete collection of all small molecules (<1500 Da) found in a specific cell, organ or organism.[2] These small molecules include endogenous metabolites such as small peptides, amino acids, lipids, nucleic acids, fatty acids, carbohydrates, organic acids, vitamins and minerals. They also include exogenous chemicals or xenobiotics such as plant phytochemicals, food additives, drugs, pesticides, herbicides, cosmetic chemicals, dyes, pollutants and just about any other chemical that a plant or animal can produce, synthesize, ingest, absorb or to which it can be exposed.

New Developments in Mass Spectrometry No. 8
Processing Metabolomics and Proteomics Data with Open Software: A Practical Guide
Edited by Robert Winkler

Published by the Royal Society of Chemistry, www.rsc.org

Small molecules are vital to life. They act as the bricks and mortar for cells. They serve as the building blocks for all of the cell's larger molecules including proteins, RNA, DNA, oligosaccharides and all other biopolymers that give cells their ability to move, divide or interact with their environment. Small molecules also act as the fuel to power cellular processes, the buffers to protect against environmental insults, the walls to keep cells intact, and the messengers for most intracellular and intercellular events. Small molecules are clearly important as the cell's building materials. However, their small size means that they can diffuse or be transported quickly, and they can be produced or destroyed on very short notice. This molecular 'nimbleness' means that small molecules are perfect for sensing their environment, for signaling larger (slower) molecules or for amplifying cellular events and responses. As a result, small molecules are sometimes referred to as the 'canaries' of the genome. Just as tiny, caged canaries served as sensitive indicators of poisonous gases for coal miners working in coal mines, small molecule metabolites can serve as sensitive sentinels of problems in the genome. Indeed, a single base change in a gene can lead to a 10 000-fold change in the concentration of a metabolite in the metabolome. Metabolite levels are not only exquisitely sensitive to the state of the genome, they are also very sensitive to the environment, including the types of foods or beverages we eat, our age, our gender, our level physical activity or the time of day.[3]

Because metabolites are effectively the end-products of complex interactions occurring inside the cell (the genome) and events, exposures or phenomena occurring outside the cell (the environment), it means that the comprehensive measurement of metabolites (*i.e.* metabolomics) allows one to measure interactions between genes and the environment. This means that metabolomics offers an ideal route for measuring the phenotype – the end product of *gene* × *environment* interactions.[4] Metabolomics is also an ideal route for probing processes as they occur in real time or over the time scale of seconds or minutes. This ability to probe what is occurring inside a cell, a tissue or an organism is what makes metabolomics fundamentally different to genomics. In particular, metabolomics tells you *what is happening*, while genomics tells you *what might happen*. This ability to probe an organism's phenotype or to chemically 'read' what is going on inside an organism is the reason why metabolomics is increasingly being used in biomedical research, drug testing, food and nutritional analysis, animal health studies and many kinds of exploratory biological studies.[5–8] As a result, the field of metabolomics has experienced very rapid growth: in 1999 only two papers were published on the subject. In 2018, there were more than 5000.

A diagram depicting the standard workflow for a metabolomics experiment is shown in Figure 3.1. Typically, after a biological sample (tissue, organ, plant, or cell culture) is collected, it is metabolically quenched (with liquid nitrogen) and extracted or homogenized to produce a liquid mixture containing hundreds of metabolites. In many cases, it is easier to collect a biofluid, such as blood, urine, tree sap or cell growth media, as this avoids the tissue extraction process. Once the appropriate biofluid has been obtained, it can

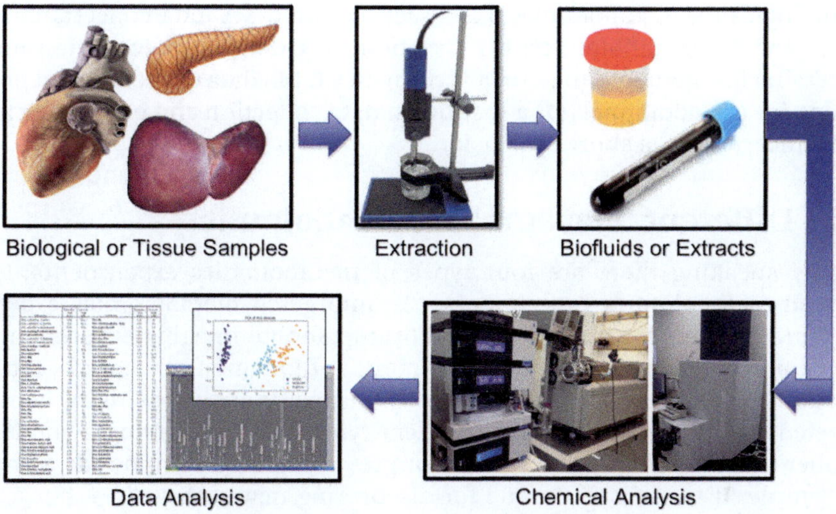


**Figure 3.1** A diagram illustrating the standard workflow for a metabolomics experiment. Samples (tissues or organs) are initially taken and then the metabolites are extracted using sonicators or solvent extraction methods. Alternately, biofluids that bathe the tissues (such as urine or blood) may be obtained. Once the metabolite mixtures (extracts or biofluids) are obtained, the samples are analyzed *via* GC-MS, CE-MS or LC-MS methods. The resulting spectra are then analyzed and the identified metabolites annotated.

be run through one or more analytical chemistry platforms. These platforms may be mass spectrometers (MS) equipped with liquid chromatography (LC), capillary electrophoresis (CE) or gas chromatography (GC) systems, or they may be nuclear magnetic resonance (NMR) instruments.

Our focus in this chapter will be on MS-based metabolomics and the software used to facilitate it. MS-based metabolomics may employ liquid chromatography mass spectrometry (LC-MS), gas chromatography mass spectrometry (GC-MS) or capillary electrophoresis mass spectrometry (CE-MS) systems. These analytical systems are capable of separating, detecting and characterizing hundreds to thousands of chemicals in complex chemical mixtures. In almost all cases, these instruments produce chromatograms and spectra consisting of thousands of peaks or features. The primary challenge in metabolomics, therefore, is having the appropriate software tools to determine which peaks in these spectra match with which chemical compounds. The secondary challenge is having the appropriate software to determine which compounds or spectral peaks have changed significantly and why.

This chapter will describe the software tools, databases and workflows needed to perform and analyze metabolomics (and lipidomics) experiments using LC-MS, GC-MS and CE-MS technologies. It is organized into eight sections: (1) a brief introduction to metabolomics; (2) an overview of the different 'flavours' of metabolomics; (3) a short summary of MS-based technologies for metabolomics; (4) a section on LC-MS data collection and processing for metabolomics; (5) a section on GC-MS data collection and processing for metabolomics; (6) a section on CE-MS data collection and processing for metabolomics; (7) a section on data collection and processing for lipidomics; and (8) a short conclusion.

## 3.2 Different 'Flavours' of Metabolomics

Broadly speaking there are four types of metabolomics experiments: (1) targeted metabolomics experiments; (2) untargeted metabolomics experiments; (3) fluxomics experiments and (4) metabolite imaging experiments. The choice of the metabolomic experiment is fundamentally based on the questions being asked and the availability of specific analytical instruments. Targeted metabolomics focuses on identifying and quantifying a modest number (50–500) of known, expected or pre-determined compounds within the sample. It is not a technique for discovering new metabolites, but it is frequently used for hypothesis testing and biomarker development. Targeted metabolomics is also very amenable to high-throughput, kit-based systems using LC-MS, GC-MS or CE-MS equipment and appropriate software. Targeted metabolomics is widely used in clinical and medical applications.[8,9] On the other hand, untargeted metabolomics uses LC-MS, GC-MS or CE-MS to attempt to characterize as many metabolites or putative metabolites as possible (often 10 000+ features). In MS-based metabolomics the term 'feature' usually means a unique mass-to-charge ($m/z$) ratio and a unique chromatographic retention/elution time. As a general rule, untargeted metabolomics is not particularly quantitative or reproducible. However, untargeted metabolomics is ideal for metabolite discovery and hypothesis generation.[10] While untargeted metabolomics is often very labour intensive, it has led to the discovery of a number of novel metabolites or metabolite classes including oncometabolites and atherotoxins.[8] Both targeted and untargeted metabolomics can be applied to characterize water soluble and water-insoluble metabolites – including lipids. The specific study of lipids and lipid mixtures represents a branch of metabolomics called lipidomics.[11] Because lipids represent such a large, interesting and chemically diverse class of metabolites (>80 000 lipid molecules exist in humans), the field of lipidomics is thriving and many elegant targeted and untargeted lipidomic studies are now appearing.[12,13]

Regardless of whether one uses targeted or untargeted metabolomics, both methods require the use of internal standards, solvent blanks and quality control (QC) samples. Internal standards are key to the accurate quantification and long-term reproducibility in metabolomics. Internal standards are known reference chemicals or metabolites at known concentrations added to the biological sample of interest. Internal standards may consist of single, pure compounds or mixtures of pure compounds and they may be isotopically labeled or non-isotopically labeled, depending on the instrument, the purpose or the method being used. Solvent blanks are simply samples containing a standard elution buffer. Columns often 'shed' chemical material, solvents contain chemical contaminants and mass spectrometers often generate ubiquitous ions that lead to contaminant signals. Solvent blanks allow researchers to easily detect and remove generic background or contaminant signals that are unrelated to the biological sample(s) being studied. Internal standards are different than QC samples. QC samples are artificial or specially prepared samples that are periodically added to a multi-sample or multi-day metabolomics experiments. They are used to assess the reproducibility and correct the performance of the instrument(s) or method(s). QC standards may be defined a chemical mixtures, well-characterized biological samples or pooled biological samples. Because of their importance in obtaining high quality metabolomic data, much has been written about internal standards, blank samples as well as QC standards and protocols.[14]

The third type of metabolomics experiment is called metabolic flux measurement or simply fluxomics. Fluxomics is a branch of targeted metabolomics that measures metabolite reaction rates and monitors the movement of isotopic labels through various metabolic intermediates using either LC-MS or NMR.[15] Fluxomics is used to understand the dynamics of metabolism and has revealed important insights into both cellular and physiological metabolism.

The fourth class of metabolomics experiments, metabolite imaging, is a newly emerging field of metabolomics that involves *in vivo* or *in vitro* detection and visualization of metabolites in tissues. Metabolite imaging may be done *via* NMR or by MS. The two most popular imaging MS systems are matrix-assisted laser-desorption (MALDI) and desorption electrospray ionization (DESI).[16] The field of imaging that uses different kinds of MS platforms to visualize the tissue location of small molecules, lipids and even proteins, is called imaging mass spectrometry or IMS. In metabolite imaging only a small (<20) number of metabolites are typically identified.[16] Metabolite imaging is primarily used to explore and assess cell-, tissue- and organ-specific metabolism and has even been used to guide surgical operations.

Metabolomics experiments are often done as 'case *versus* control' or 'pre-intervention *versus* post-intervention' studies that involve large groups of individuals.[9] As a result, metabolomic data is typically characterized by large tables containing dozens to hundreds of individuals (patients or samples) linked to dozens to hundreds of metabolite/feature identifiers and their absolute or relative concentrations. To make sense of this large volume of data, researchers must employ multivariate statistics and a variety of clustering and classification techniques.[17] These techniques help scientists select the most important metabolite changes (for targeted metabolomics) or the most important feature changes (for untargeted metabolomics) along with other key trends in the data. The most common clustering technique is called principle component analysis (PCA) while the most common classification technique is partial least squares discriminant analysis (PLS-DA).[17] These two methods, along with other multivariate statistical methods, can be used to simplify metabolomic MS data, extract or group metabolite biomarkers, distinguish among metabolically distinct groups or visualize important data trends.[17] When combined with additional biological knowledge found in pathway databases or when processed using tools such as metabolite set enrichment analysis,[18] these data can be used to gain key insights into how certain metabolites contribute to specific biological processes or important physiological phenomena.

## 3.3 Technologies for Metabolomics

Metabolomics only became possible in the late 1990's as a result of technological breakthroughs in small molecule separation and identification. These include the development of very high-resolution MS instruments for precise mass determination, the widespread deployment of high-resolution, high throughput NMR spectrometers, the invention of high-performance liquid chromatography or ultra-high performance liquid chromatography (HPLC or UPLC) and the development of other innovative systems for rapid compound separation.[19] In this section we will briefly review the technologies used in metabolomics, with a special focus on the three most common MS-based platforms: LC-MS, GC-MS and CE-MS. This discussion is needed so that readers can better appreciate the type of data that a typical MS-based metabolomics experiment generates and the type of software needed to interpret it.

### 3.3.1 LC-MS and LC-MS/MS for Metabolomics

LC-MS is a hybrid analytical technique that couples liquid-based chromatographic separation (using HPLC or UPLC) with compound detection *via* MS. Among all the different ionization methods in mass spectrometry, electrospray ionization (ESI) is the most widely used.[20] LC-MS is commonly employed for both proteomics and metabolomics and only minor modifications are needed to convert an LC-MS instrument to perform one or the other omics technique. Liquid chromatography separates individual analytes (peptides or metabolites) from a liquid mixture based on their differential adsorption or physicochemical interactions as the liquid mixture moves through an immobile or stationary matrix aided by a mobile phase solvent that is pushed through the column.

HPLC and UPLC are the two most commonly used liquid chromatography techniques in metabolomics. Both employ high or very high pressures and very small particle sizes to accelerate the separation process while at the same time improving the resolution and reproducibility of the separation. In HPLC and UPLC, the stationary matrix or stationary phase is usually composed of specially coated microscopic (1.7–50 microns) particles placed in a hollow column about 2–5 mm wide and 30–250 mm long. Most HPLC and UPLC separations use a gradient of two or more mobile phases to enhance the separation and improve peak shapes. The typical pressures used range from 5800 psi or 400 bar for HPLC to 12 000 psi or 800 bar for UPLC.

In proteomics the column matrix or the column particles are almost always non-polar, being coated with aliphatic chains about 18 carbons long (C18). On the other hand, because the chemical diversity is so great in metabolomics, different types of column matrices must be used to separate different classes of chemicals. If the column particles are coated with polar molecules such as diols, amides or sulfonic acids, they are best used for separating polar molecules such as nucleosides, nucleotides, amino acids and sugars. This form of chromatography is called hydrophilic interaction liquid chromatography or HILIC. If the column particles are coated with aliphatic chains (ranging in carbon length from C8–C18), they are best suited for separating non-polar molecules such as lipids, sterols and fatty acids (reversed phase chromatography or RP). Currently, the most commonly used form of LC for metabolomics studies is reversed phase LC (RPLC), which is a standard tool to separate semi-polar and non-polar analytes. RPLC uses a non-polar stationary phase inside the column and an aqueous, moderately polar mobile phase to separate and elute the loaded analytes. The mobile phase is typically composed of water with an organic solvent, typically acetonitrile (ACN) or methanol (MeOH), and a few ionic additives such as formic acid or ammonium acetate, which is pumped at high pressures through the column.

Depending on the chemical properties of the column matrix, the size of the particles in the matrix, the solvents used to elute the analytes in the starting mixture and the flow rates, different levels of compound separation can be achieved. Typical metabolomic HPLC runs last 20–30 minutes while UPLC runs can be as short as a few minutes. The time at which a given compound comes off or elutes from a HPLC/UPLC column is called the retention time (RT). In principle, each compound has a unique RT (for a given column and a given set of elution conditions) and if the HPLC/UPLC column and elution protocols are carefully standardized, the RT can be used to help identify compounds in LC-MS based metabolomics. However, UPLC/HPLC RTs are not particularly reproducible and vary considerably with column types, column age and solvent conditions. For this reason, RT standards (non-biological chemicals that elute over a range of times) are often used to help calibrate or recalibrate measured RTs.

LC-MS allows metabolomics researchers to characterize complex chemical mixtures by first separating the chemicals (*via* LC) and then detecting and identifying individual compounds according to their molecular weight (*via* MS). As readers should already know, MS does not technically measure the molecular weights of molecules, but rather it measures the mass-to-charge ratio ($m/z$) of molecular ions. To perform a MS experiment (whether for proteomics or metabolomics), a compound must first be ionized and then accelerated through either a magnetic field or an electric field (under a strong vacuum) where its velocity or rate of curvature can be measured. Depending on the chemical structure of the molecule and its characteristic chemical groups (amines, carboxylates, hydroxyl groups, *etc.*), different molecules can be given either positive or negative charges. Electrospray ionization (ESI) instruments ionize molecules by passing them through a spraying device under a strong electric field.

In proteomics, almost all ESI work is done in the positive ion mode, whereas in metabolomics, both positive and negative ionization modes must be used to accommodate the wide chemical diversity (cationic and anionic nature) of metabolites. Furthermore, just as with peptides in proteomics, metabolites can often form adducts with salt ions present in the biological matrix or the column buffers. Occasionally metabolites (just like peptides) can acquire multiple charges through the ionization of different groups on the molecule. As a result, a single metabolite is often represented by multiple ionic features (isotopes, salt adducts, multiply charged species). In fact, depending on its abundance, a single metabolite can generate up to 10 or more ionic features in a single mass spectrum.

Many different kinds of mass spectrometer designs exist,[19] with each having a specific name based on the ion separation technology and/or the mass analyzer being used. These include magnetic sector instruments, time-of-flight instruments (TOF), quadrupole time-of-flight (QTOF), linear ion trap (LIT), triple quadrupole (QqQ), linear quadrupole trap (LQT), orbitrap and Fourier transform ion cyclotron resonance (FT-ICR) instruments. TOF mass spectrometers measure the time it takes for ions to pass through a long drift tube of a defined distance. Orbitrap spectrometers use a specially designed cell to trap the ions and spin them around a central spindle-like electrode at a high frequency, with the frequency being used to determine the $m/z$ of the ions. QqQ mass spectrometers are tandem mass spectrometers (also called MS/MS) consisting of two quadrupole (Q) electrodes for accelerating ions and one central quadrupole (q) for colliding ions to produce smaller molecular fragments. Each mass spectrometer configuration has its specific strengths and weaknesses with regard to robustness, mass resolution, sensitivity and speed (see Table 3.2 for more details).

If a mass spectrometer is very precise or is designed for high-resolution work (such as a QTOF, FT-ICR or an Orbitrap instrument), it is possible to determine the mass of a molecule with an accuracy of four or five decimal places (~1 ppm). This level of accuracy/precision is often sufficient to determine the exact molecular formula (but not necessarily the structure) of a small molecule. If additional information can be obtained about how the molecular ion fragments (as a result of passing it through a quadrupole collision cell or another MS instrument), it is also possible to determine the molecule's structure. This technique of connecting two mass analyzers together is called tandem mass spectrometry or MS/MS. The triple quadrupole mass spectrometer (QqQ-MS) is an example of an MS/MS instrument that is commonly used in metabolomics, however, it is also possible to convert TOF, Orbitrap and linear ion trap mass spectrometers into tandem MS instruments.

LC-MS and LC-MS/MS are particularly well suited for analyzing liquid mixtures such as biofluids or tissue extracts. An example of an LC-MS spectrum of human urine is shown in Figure 3.2. This figure shows what is called a base peak chromatogram (BPC), a specialized chromatogram displays the most intense MS peak in each MS spectrum over the course of the liquid chromatographic elution period. BPC plots are useful but in many cases each

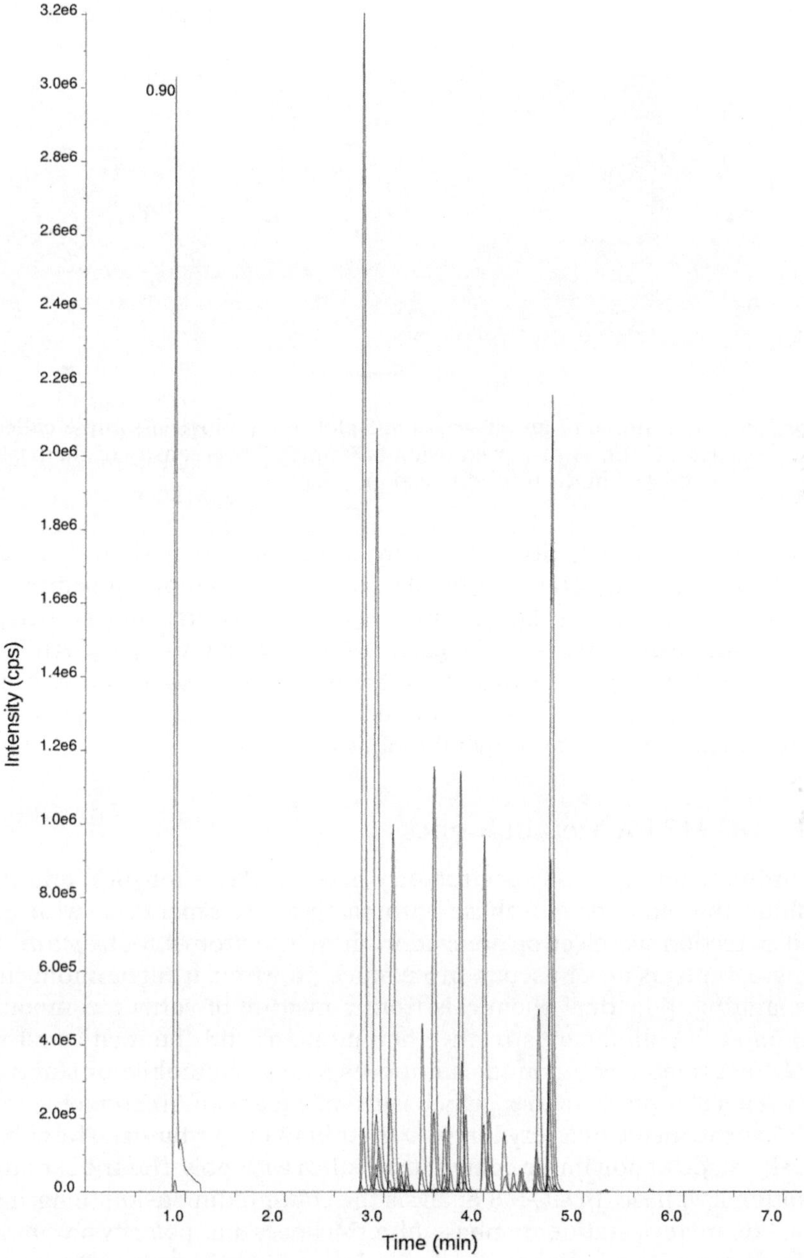


**Figure 3.2** An LC-MS spectrum (corresponding to a base peak chromatogram or BPC) collected from human urine.

peak in the BPC represents many other compounds that are 'hidden' behind the most intense peaks. As a result, two-dimensional or 2D RT *versus* *m*/*z* plots (also known as RT-*mz* or feature plots) that show both *m*/*z* and retention time (as well as intensity) are more frequently used. Figure 3.3 illustrates a

**Figure 3.3** An example of an RT *versus* *m*/*z* plot of a biological sample collected *via* LC-MS, as displayed using MZMine2. The intensity of the peaks is displayed in the third dimension.

2D plot (with peak intensity in the third dimension) that shows an example of an RT-*mz* plot, generated by MZmine2[39] or a biological sample run through a high-resolution LC-MS system. LC-MS was first used to perform metabolomics experiments in the early 2000's and due to its superb sensitivity and high-throughput capacity, LC-MS is now the most commonly used technique in metabolomics research. Each year more than 2500 papers are published using LC-MS based metabolomics.

### 3.3.2 GC-MS for Metabolomics

Gas chromatography mass spectrometry (like LC-MS) is a hybrid analytical technique that couples gas-phase chromatographic separation with compound detection *via* electron ionization mass spectrometry (EI-MS).[21] Gas chromatography is never used in proteomics. However, it is commonly used for separating individual chemicals from a mixture of volatile compounds based on each component's relative boiling point and the relative adsorption of each volatile compound as it moves past an immobile or stationary matrix. This stationary matrix, which is usually made of hydrophobic materials, lines the inside of a very long (10 m) thin (2 mm) column. The column is usually made of polyimide coated fused silica with polysiloxane serving as the stationary phase. In GC-MS analysis the column dimension (length and internal diameter), stationary phase film thickness and polarity are important parameters. For instance, thicker column films increase RTs, sample capacity and column bleed. A carrier gas (typically 99.999% pure helium) instead of a liquid solvent (as in LC) is used to push the chemical mixture through the column. The gas is normally supplied by a gas cylinder or an in-line gas generator and the pressure is regulated by an electronic pressure controller (EPC). To facilitate separation the GC column is normally placed in a temperature-programmable oven that allows the temperature to vary from 20 °C to 400 °C. On elution from the column, the carrier gas and analytes pass into the MS detector where ionization (either chemical ionization (CI) or electron ionization (EI)) occurs and the molecules fragment into ions. These ions are separated based on their mass to charge ratio by the mass analyzer (usually a single quadrupole or a time of flight instrument) and the signals are collected to form a mass trace at a given elution time.

Compared to conventional LC, GC produces better separation, and often requires smaller sample sizes. GC is also much more reproducible and far more standardized than LC.[19] In other words, nearly every compound that can be run through a GC system has a unique or characteristic elution time that is often the same (to within a few seconds) from run to run or instrument to instrument. This allows the normalized elution time (called the retention index – RI or Kovats RI) to be used in compound identification.

GC is ideal for separating mixtures of volatile or gaseous molecules. However, by chemically derivatizationing non-volatile chemicals with trimethylsilane (TMS) it is possible to convert non-volatile compounds into volatile compounds for GC-based separation. This makes GC quite useful for analyzing biofluids. Coupling a GC with a mass spectrometer (*i.e.* a GC-MS) allows analytical chemists to characterize complex chemical mixtures by first separating the chemicals (*via* GC) and then detecting and identifying individual compounds according to their mass spectral fragmentation patterns generated (*via* EI-MS).

Unlike ESI, which is a soft ionization technique, EI is a hard ionization technique that uses fast moving electrons (70 eV) to shatter the parent molecule into smaller charged fragments. Each parent molecule will have a unique fragmentation pattern. Therefore, by looking for characteristic fragments and matching EI-MS patterns of known molecules to the observed EI-MS patterns, it is often possible to accurately identify compounds. It is because of this that EI is the preferred method for GC-MS metabolomics with the vast majority of studies using EI. In contrast to EI, chemical ionization (CI) is a softer ionization technique. Analytes are ionized through a chemical reaction with a charged reagent gas plasma (usually methane). As a result, mass spectrum produced by CI shows much less fragmentation than an EI spectrum. Indeed, it is often possible to see the parent ion in a CI-MS spectrum whereas it is quite rare to see the parent ion in an EI-MS spectrum.

An example of a GC-MS spectrum of human urine is shown in Figure 3.4. To collect this spectrum, the urine sample had to be extracted and chemically derivatized with TMS before running it through the GC-MS system. An inherent limitation of GC-MS is that the maximum *m*/*z* that can typically be detected is about 500–600 Da. GC-MS was first used to perform metabolomics experiments in the early 2000's[22] and is often used to analyze plant samples, food products and urine. Each year more than 400 papers are published using GC-MS based metabolomics.

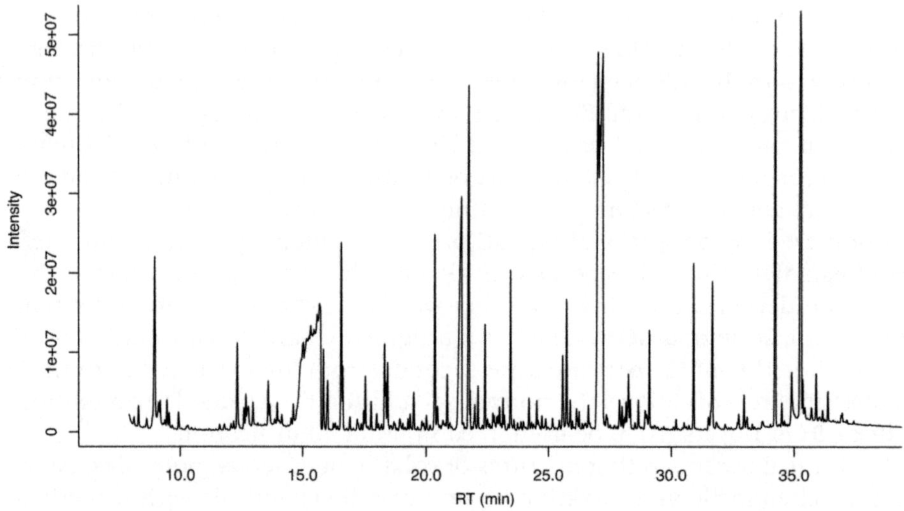


**Figure 3.4** A GC-MS spectrum collected from human urine. The sample was derivatized with TMS and the collection conditions optimized to detect organic acids.

### 3.3.3 CE-MS for Metabolomics

CE-MS was first developed in 1987 and has been used for both proteomics and metabolomics.[23] Unlike GC-MS, which uses gas chromatography for compound separation or LC-MS, which uses liquid chromatography for compound separation, CE-MS uses electrophoresis to separate molecules. More specifically, capillary electrophoresis is a separation technique that uses a strong electric field (5–20 kV) to produce electro-osmotic flow in a thin (<100 μm inside diameter), long (20–90 cm) glass capillary tube filled with a liquid buffer, to separate charged ions. The ions of interest migrate from one end of the capillary to other based on their charge and size as well as the buffer viscosity. Because CE separates on the basis of charge, CE-MS is limited to analyzing polar or ionic metabolites. In CE-MS the electrophoretic buffer (or electrolyte) that is used in the capillary must be compatible with both CE separation and ESI operation. However, the buffer can be modified with various additives or 'selectors' such as organic solvents, cyclodextrins, micelles or neutral polymers to change its viscosity or modify its chiral selectivity. Most CE-MS systems use a sheath flow system to interface between the CE and the MS instrument. A sheath flow system has a sheath liquid flowing coaxially in a metal capillary tube surrounding the glass capillary tube. The coaxial metal tube is use to apply the high voltage needed for ESI. Most often the sheath liquid is 1 : 1 mixture of water–methanol with 0.1% acetic acid or formic acid.

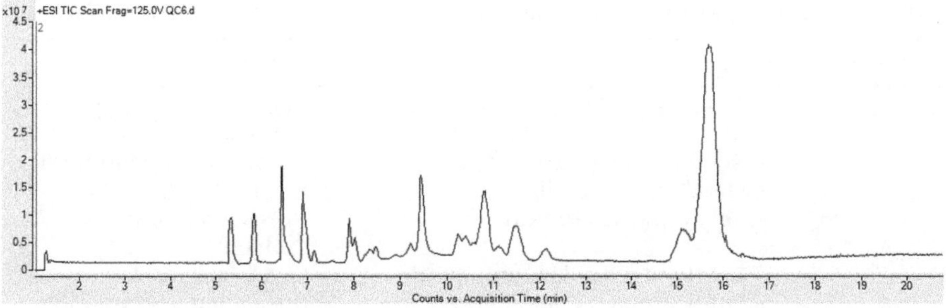


**Figure 3.5** A CE-MS spectrum (corresponding to the total ion chromatogram or TIC) collected from human plasma.

CE-MS is most similar to LC-MS in terms of MS instrumentation and general concepts. That is, CE-MS systems use ESI, just as LC-MS systems to ionize and introduce samples into the MS instrument. They also employ similar kinds of MS instruments (single quadrupole, TOF, QqQ, QTOF), they have similar kinds of separation times (5–40 minutes), they do not require any prior chemical derivatization and they produce similar kinds of data (extracted ion chromatograms (EICs), migration time (MT) *versus* $m/z$ plots, *etc.*). An example of electropherogram obtained for plasma is shown in Figure 3.5. However, CE-MS systems have several unique features that make them particularly appealing for certain kinds of metabolomic applications. For instance, CE-MS is excellent for separating highly polar and highly charged molecules that are not easily separated (or even seen) with LC-MS. Furthermore, CE-MS is ideal for working with tiny volumes (5–10 nL), which is very useful for sample-limited metabolomic studies, such as single cell or insect studies. CE-MS can also be used to detect and quantify metal ions such as $Na^+$, $K^+$ and $Mg^{++}$, which is not possible with LC-MS or GC-MS.[24] Because of the wide diversity of metabolites that can be readily separated *via* capillary electrophoresis, CE-MS methods have been used to measure up to 300 different metabolites in certain studies.[25]

As a result of the electrophoretic nature of CE-MS it is possible to do a few sample handling 'tricks' that are not possible with GC-MS or LC-MS. For instance, samples can be pre-concentrated directly within the capillary (in-line concentration) through electrophoretic techniques such as field amplified sample stacking.[26] This pre-concentration step can greatly improve sensitivity. Additionally, CE-MS can also be adopted to perform multi-segment injection (MSI-CE-MS), which allows up to seven different sample injections to be loaded into a single capillary.[27] MSI-CE-MS can significantly improve the throughput of CE-MS, almost matching the speed of direct-infusion MS methods.

Despite these many advantages, CE-MS is not widely used in metabolomics. This is because the technology is not as robust or as reliable

**Table 3.1** A comparison between MS-based metabolomic technology platforms.

| Technology | Advantages | Disadvantages |
|---|---|---|
| CE-MS | - Good sensitivity (10 nM) | - Destructive to sample |
| | - Minimal sample volume (10 nL) | - Expensive |
| | - Requires no derivatization | - Not routinely quantitative |
| | - Good reproducibility | - Requires separation |
| | - Very rapid separation | - Novel compound ID is difficult |
| | - Excellent for polar compounds | - Less robust instrumentation |
| | - Fairly broad metabolite coverage | - Does not detect volatiles |
| | - Excellent resolving power | - Limited to ionic/polar compounds |
| GC-MS | - Robust, mature technology | - Destructive to sample |
| | - Relatively inexpensive | - Requires sample derivatization |
| | - Small sample volumes (50 μL) | - Not routinely quantitative |
| | - Good sensitivity (100 nM) | - Requires separation |
| | - Excellent reproducibility | - Novel compound ID is difficult |
| | - Detects volatile compounds | - Limited mass range (<500 Da) |
| | - Fairly broad metabolite coverage | - Limited mass resolution |
| | - Excellent resolving power | |
| LC-MS | - Excellent sensitivity (1 nM) | - Destructive to sample |
| | - Minimal sample volume (5 μL) | - Expensive |
| | - Very flexible technology | - Not routinely quantitative |
| | - Can be done without separation | - Modest reproducibility |
| | - Excellent for lipid analysis | - Less robust instrumentation |
| | - Very broad metabolite coverage | - Does not detect volatiles |
| | - Supports large mass range | - Novel compound ID is difficult |
| | - Excellent mass resolution | |

as LC-MS or GC-MS. In particular, the CE interface is unusually delicate and frequently clogs or needs replacing. Furthermore, CE-MS does not have the same concentration sensitivity as LC-MS and the level of separation reproducibility is less than that of GC-MS. Likewise, there is only one major CE-MS vendor (Agilent), which obviously limits user choice and pricing options. Typically, there are only 15–20 papers published each year in the field of CE-MS-based metabolomics, although this number is steadily growing.

Table 3.1 provides a brief description of the advantages and disadvantages of the three major technologies (LC-MS, GC-MS and CE-MS) discussed here. As already noted, CE-MS is capable of detecting (and occasionally quantifying) up to 300 compounds and has a lower detection limit of about 20 nM.[25,28] CE-MS supports rather broad metabolite coverage with amino acids,

**Table 3.2** Comparison of different type of mass analyzers.

| | Mass analyzer | | | | | |
|---|---|---|---|---|---|---|
| Feature | Magnetic sector | TOF | QQQ | QIT and LIT | Orbitrap | FT-ICR |
| Resolution | Very high | High | Low | Low | Very high | Highest |
| Mass accuracy | Very high | High | Low | Medium | Very high | Very high |
| *m*/*z* range | Medium | Very high | Low | Medium | Low | Medium |
| Sensitivity | High | High | High | High | Medium | Medium |
| Dynamic range | Very high | Medium | High | Medium | Medium | Medium |
| Quantification | Very good | Medium | Very good | Poor | Medium | Medium |
| Speed | Slow | Fast | Fast | Fast | Slow | Slow |
| Application in metabolomics | N/A | Untargeted | Targeted | Targeted | Untargeted and targeted | Untargeted |

biogenic amines, nucleosides, nucleotides, metal cations, organic acids, and fatty acids being routinely detectable. GC-MS is capable of detecting between 50 and 150 compounds (depending on the sample), with a lower sensitivity limit of about 100 nM.[19,29] GC-MS provides relatively broad metabolite coverage with amino acids, sugars, organic acids, phosphorylated compounds, fatty acids and even cholesterol being routinely detected. Because of the exquisite sensitivity of today's MS instruments, LC-MS methods can routinely detect 1000's of 'features'.[21,30] However, the number of compounds that can be positively identified by LC-MS is typically much less (~400). LC-MS methods are particularly useful in targeted metabolomic studies of lipids, where up to 1000 different kinds of lipids and fatty acids can be detected and quantified.[31] Recent studies have shown that the combination of multiple detection technologies gives a far more complete picture of the metabolome than just a single detection technology.[29,32] Readers wishing to learn more about GC-MS, LC-MS or CE-MS or to acquire more details about how these technologies can be used in metabolomics are encouraged to read the following ref. 20, 21 and 28.

## 3.4 LC-MS Processes and Software for Metabolomics

As noted in Section 3.2, there are two general approaches for metabolomic analysis of biofluids or tissues: untargeted and targeted. Untargeted approaches require the use of high resolution (HRMS) instruments such as QTOF, OrbiTrap and FT-ICR mass spectrometers coupled to HPLC or UPLC systems while targeted approaches typically (but not always) use lower resolution QqQ or LIT instruments that may or may not be coupled to LC systems. Fundamentally different types of software and workflows are needed to perform untargeted *versus* targeted metabolomics. In the following sections we will discuss the general software and workflow requirements for both types of metabolomic assays separately.

### 3.4.1 Untargeted LC-MS Metabolomics Tools and Workflows

Untargeted LC-MS metabolomic data may be acquired in different ways depending on the instruments and intent of the experiments. The simplest untargeted LC-MS experiment is only aimed at identifying unique peaks or features and determining the retention time and the exact mass (or highest resolution mass) for that *m*/*z*/RT feature. Different software tools can be used to select, merge or eliminate peaks so that a single monoisotopic mass for the feature (or features) is determined. A relative intensity of the combined peaks is often used as a proxy for the concentration of the feature. After the feature has been identified and the monoisotopic *m*/*z* value determined a variety of software tools can then be used to determine the molecular formula of that peak or feature. Alternatively, database searches against large metabolomics/chemical databases can be used to determine the possible identity of the feature based on its accurate monoisotopic mass. This simple approach allows one to putatively identify and quantify many compounds relatively quickly. Occasionally it can lead to the identification of novel metabolites or the necessary clues to formally identify novel metabolites. However, the data generated is usually of low quality and often results in the assignment of single features with multiple compound matches, misidentifications or ambiguous results, unreliable quantification and a large fraction of unidentifiable peaks. In a more formal sense, this approach only achieves a metabolomics standards initiative (MSI)[33] level 3 or level 4 grade of metabolite identification (see Table 3.3). The MSI evaluation scheme for metabolite identification/annotation shown in Table 3.3 has been used by the metabolomics community for more than a decade and it gives a reasonably good indication of the quality and reliability of most published or deposited metabolomics data.

More sophisticated LC-MS experiments employ instruments and methods to collect the MS/MS spectra of selected features. These MS/MS spectra must typically be collected on an appropriately configured mass spectrometer such as QTOF, OrbiTrap or FT-ICR instruments. The resulting MS/MS or $MS^n$ product ion spectra may be collected in a data dependent manner (DDA – data dependent acquisition) or in a data independent manner (DIA). DDA methods take manual user input or automatically select certain mass windows to choose individual ions for further fragmentation and the collection of MS/MS spectra. DIA methods essentially allow the collection of MS/MS spectra for essentially every identifiable peak. The availability of RT data, accurate *m*/*z* data as well as MS/MS or $MS^n$ fragment spectra, greatly

**Table 3.3** MSI standards for metabolite identification.

| Rating | Label and data requirement |
|---|---|
| Level 0 | *Characterized compounds* – isolated, purified and stereochemically characterized compounds with determined unambiguous 3D structure. |
| Level 1 | *Identified compounds* – a minimum of two independent and orthogonal data relative to an authentic compound analyzed under identical experimental conditions for confident 2D structure determination (*e.g.* retention time/index and mass spectrum, retention time and NMR spectrum, accurate mass and tandem MS, accurate mass and isotope pattern). |
| Level 2 | *Putatively annotated compounds* – identified without chemical reference standards, based upon spectral similarity with public/commercial spectral libraries with determination of probable structure. |
| Level 3 | *Putatively characterized compounds or compound classes* –identified based upon characteristic physicochemical properties of a chemical class of compounds, such as by formula matching or by accurate mass matching for possible structure or class determination. |
| Level 4 | *Unknown compounds* – not identified or identifiable or classifiable *via* mass matching, formula matching or spectral matching. Although unidentified or unclassified these metabolites can still be differentiated and quantified based upon spectral data. |

improves the likelihood of identifying a metabolite. However, sophisticated untargeted metabolomics approaches are slower, more manually intensive, need sophisticated software and require appropriate MS/MS spectral databases. Nevertheless, untargeted metabolomics that employs MS/MS instruments and MS/MS fragment analysis enables compounds to be identified and quantified more confidently, often achieving a MSI level of 2 for metabolite identification (see Table 3.3). If users have authentic standards of the putatively identified compound, to confirm the identification, then MSI level 1 identification can be achieved.

### 3.4.1.2 *Data Preprocessing and Formula Generation Software*

Most high-resolution mass spectrometers (HRMS) come with their own software to generate *m*/*z versus* RT plots (Figure 3.3). A typical, unfiltered *m*/*z versus* RT plot in a HRMS metabolomics study can have >20 000 features. However, many of these features are noise artifacts or redundant peaks arising from multiple adducts or isotope peaks. The objective in data pre-processing is to reduce the number of peaks or features in these raw data plots to a reasonable number. In some (simple) situations these raw *m*/*z versus* RT plots may be manually inspected and 'true' peaks picked and merged through visual inspection. However, for most metabolomics studies this is usually far too onerous, tedious and generally too user-dependent to be done manually. As a result, a number of software packages have been developed to facilitate the process of peak picking, peak filtering and peak grouping (called data preprocessing). The entire workflow involves seven steps: (1) construction of the extracted ion chromatograms (EIC); (2) feature or peak detection; (3) feature filtering; (4) feature grouping (*via* de-isotoping, adduct merging, identification of dimers and trimers, removal of in-source fragments, *etc.*); (5) feature alignment; (6) statistical selection of significant features or feature groups and (7) feature identification *via* formula generation or mass matching. Both commercial tools and freeware packages are available to perform these steps. Most commercial packages are instrument-dependent and offer a well-integrated pipeline for performing all seven steps. These commercial packages, such as MassHunter Profinder (Agilent), ProfileAnalysis and BioScape (Bruker), Sieve (Thermo-Fisher), and Progenesis QI (Waters) must be purchased or leased (on an annual basis) separately from the instrument vendors and are not typically bundled with the instrument.

Because the algorithms and methods used in commercial programs are not generally known or made public, we will focus on discussing freely available packages where the algorithms have been described. Among the available software tools are both comprehensive freeware packages that will perform all seven steps as well as more specialized software packages that perform just a few of the steps. The comprehensive freeware packages include XCMS,[34,35] XAlign,[36] OpenMS/Knime,[37,38] MZmine and MZMine2[39,40] and MSDial.[41] The more specialized packages or webservers include programs for MS data preprocessing such as MetAlign;[42] programs for feature filtering and grouping such as AMDIS (Automated Mass Spectrometry Deconvolution and Identification System),[43,44] MSClust,[45] CAMERA,[46] QUICS,[47] Parafac2[48] and RAMClustR;[49] as well as web tools or programs for formula generation such as MS-Finder,[50] ChemCalc[51] and Sirius.[52]

**3.4.1.2.1 Construction of EICs.** The first step in LC-MS data analysis for metabolomics is the construction of the EICs. This is done to eliminate noise and reduce the size of the raw LC-MS data files. The EIC construction process involves analyzing the raw LC-MS files and creating chromatographic plots or strips for each identifiable *m*/*z* value. In an EIC, the total intensity or base peak intensity within a mass tolerance window around a particular analyte's mass-to-charge ratio is plotted at every point over the chromatogram. EIC construction can be done by binning the raw mass trace data along the *m*/*z* axis of the raw LC-MS data (which resembles a noisy *m*/*z versus* RT plot) and creating multiple mass traces or EICs for the entire mass range that was scanned.[53] For high-resolution instruments (such as OrbiTraps and QTOFs) the *m*/*z* bins must be relatively narrow (0.01–0.02 Da) while for lower resolution instruments (such as QqQs) the *m*/*z* bins should be much wider (0.4–0.5 Da). While binning is simple and fast, specifying the optimal bin size can be difficult and can lead to inappropriate merging or separation of peaks. An alternative approach to binning is to identify regions of interest (ROIs) on the mass trace data.[53] This is a much more flexible approach and avoids the problem of bin selection and instrument adjustment. It also allows *m*/*z* slices to be compared and merged as necessary.[34] Many programs use this ROI approach including XCMS, MetAlign and MZMine2.

**3.4.1.2.2 Peak Picking.** The second step in LC-MS analysis involves peak or feature detection from the reconstructed EICs. Peak picking helps to reduce the noise in raw LC-MS data sets by zeroing in on signals of relevance and identifying peaks that pass an appropriate signal-to-noise threshold. Peak picking can be done using matched filters or second-derivative Gaussian filters *via* the matchedFilter algorithm found in XCMS.[34] This approach convolutes the suspected peak with a second-derivative Gaussian function and effectively performs a background subtraction around the peak. Peak picking may also be done *via* centroid selection using the centroidPicker algorithm found in MZMine.[39] However, these days the preferred route to peak picking for most LC-MS metabolomics data is through the centWave algorithm.[53] The centWave algorithm uses a continuous wavelet transformation or CWT (similar to a Fourier transformation) to help with peak identification. CWT is widely used in signal processing and pattern recognition and can reliably detect chromatographic peaks with differing widths. The centWave algorithm is available in XCMS, MZMine2 and MZed. The result of feature or peak picking is a clean or relatively clean RT *versus* *m*/*z* plot (see Figure 3.3).

**3.4.1.2.3 Feature Filtering and Grouping.** The third and fourth steps in LC-MS data analysis involve feature filtering and feature grouping. In feature filtering, peaks that have been identified in step 2 are further analyzed to see if they pass certain criteria related to their signal-to-noise, their width (*i.e.* are they too broad or too narrow?) and their abundance. These steps are often user-defined and are highly dependent on the instrument and the sample. Feature grouping, on the other hand, is something that can be highly automated. The purpose of feature grouping is to identify *m*/*z* peaks with the same or similar RT value (as identified in steps 1 and 2) and to determine if they are multiply charged species, adducts, neutral loss fragments or cluster ions. This step is required since molecules eluting from the HPLC system at a specific RT can be modified as they are ionized and passed through the mass spectrometer. This can include spontaneous fragmentation (*i.e.* in-source fragments and neutral losses), pairing up with salt ions (*i.e.* adducts), aggregation (*i.e.* forming charge clusters or dimers and trimers) or the acquisition of extra charges (*i.e.* multiply charged species). By using known rules or tables regarding adduct mass additions (22.989 Da for $Na^+$, 38.963 Da for $K^+$) or neutral loses (loss of H, $CH_3$, $NH_2$, OH, CO, $CH_2O$ or $H_2O$) it is also possible to identify those peaks that correspond to adducts and neutral losses from the parent molecule.

Other pattern searching algorithms can assist with the identification of multiply charged species, dimers and trimers as well as in-source fragments. Low-intensity isotope peaks can also lead to multiple peaks appearing at the same RT. By looking for peaks that differ from each other by precisely 1.0033 Daltons and looking for the appropriate intensity ratios it is possible to identify peaks arising from the presence of $^{13}C$, $^{15}N$, $^{2}H$ or $^{18}O$ atoms in the parent molecule (*i.e.* isotope peaks). Co-eluting peaks may also be identified by comparing EICs, peak shapes and peak intensities and then using graph theory to distinguish them.[46] Given that many compounds have 5–10 'daughter' peaks, the removal, labeling or merging of these peaks greatly reduces the spectral complexity of an untargeted LC-MS experiment. Feature filtering and feature grouping can be done by a number of specialized programs such as AMDIS,[44] MSClust,[45] CAMERA,[46] QUICS,[47] Parafac2[48] and RAMClustR.[49] These programs, or variations on them, have also been implemented in most of the comprehensive LC-MS data processing programs including XCMS,[34,35] XAlign,[36] OpenMS/Knime,[37,38] MZmine and MZMine2[39,40] and MSDial.[41]

**3.4.1.2.4 Feature Alignment.** The fifth step involves feature alignment. Metabolomics experiments typically involve the analysis of multiple samples. The result is the generation of dozens to hundreds of *m*/*z versus* RT data sets. Each of them, including technical replicates, will be subtly different because HPLC or UPLC systems experience systematic drift over repeated runs. As a result, it is often necessary to align RT chromatograms to ensure that the same peaks are being identified from sample to sample. This alignment can only be done once the features in the *m*/*z versus* RT plots have been identified as features (steps 1–4). A variety of algorithms exist for feature or chromatogram alignment (along the RT axis) including correlation optimized warping (COW),[54,55] dynamic time warping (DTW)[55] and constrained correlation optimized warping (C–COW).[56] All three techniques use dynamic programming, a technique commonly used to align DNA and protein sequences. These algorithms have been implemented in a number of specialized programs and webservers including ChromA,[57] VIPER,[58] MetAlign[42] and ChromAligner.[56] Most of the comprehensive LC-MS programs such as OpenMS, XCMS and XAlign use these dynamic programming alignment approaches as well. OpenMS and XAlign use linear versions while the others use non-linear shifting to correct any distortion in the RT dimension. The exceptions are MS-DIAL and MZMine, which use a joint alignment algorithm called JointAligner. This algorithm has 4 steps: (1) making a reference table of peaks, (2) fitting each sample peak table to the reference peak table, (3) filtering aligned peaks, and (4) interpolating missing values.

All peak alignment algorithms take advantage of the more precisely measured *m*/*z* dimension to group corresponding features and to estimate the underlying warping function needed to align peaks in the RT dimension. Depending on the software package and user choices, feature alignment can be done in combination with, after or before feature filtering and feature grouping. It may also be done iteratively to further assist with peak filtering and grouping. The main objective for feature alignment is to help reduce the number of spurious peaks being identified or included in an untargeted LC-MS analysis. To help with this process, blank samples are often prepared to serve as quality control or reference samples in untargeted metabolomics studies. Peaks that only show up in blank samples or which do not show up in repeated technical replicates can be eliminated through this alignment process.

The combined process of peak filtering, peak grouping and peak alignment can often lead to a significant reduction in the number of peaks reported in an untargeted LC-MS experiment. Beginning with an initial set of 20 000 peaks in a raw positive ion mode LC-MS spectrum, alignment can typically reduce the number of peaks by 20–25%. The removal of adducts can further reduce the number of peaks by 15–20%. Removal of multiply charged species, dimers, trimers and other charge clusters can reduce the total number of peaks by another 20%, while the removal neutral loss fragments, noise and isotope peaks can eliminate another 40–50% of peaks. The net result is that steps 1–5 in a well-constructed data pre-processing pipeline can allow one to go from 20–000 peaks to as few as 2000 to 2500 'true' peaks in any given sample – an 85–90% reduction in the total number of peaks.

**3.4.1.2.5 Statistical Selection.** The sixth step in untargeted LC-MS data processing involves the statistical selection of significant features or feature groups from the 'true' peak list. This 'true' peak list will consist of peak intensities, *m*/*z* values and RT values that have survived the filtering and merging steps done in the previous five data processing steps. If these steps have been done correctly, the typical number of peaks that will need to be compared and statistically assessed will be of the order of a few hundred to a few thousand for each of the metabolomic samples being analyzed. In untargeted metabolomics, the peak intensities and peak areas are often (but not always) proportional to compound concentrations. Therefore, if one peak is consistently 2–3 times higher or has an integrated area that is 2–3 times greater in one set of samples (*i.e.* disease cases) compared to another set of samples (*i.e.* healthy controls), then this peak or feature can be assumed to be significantly more concentrated in the disease cases than in the controls. In most metabolomic studies, the distribution of peak intensities or peak areas for any given metabolite over a large number of biological samples will follow a Gaussian or normal distribution. By comparing the mean and the shape of the distribution of the intensities for a common peak shared between one group (*i.e.* cases) and another group (*i.e.* say controls) it is possible to determine if that peak is significantly different using a Student's *t*-test.

A *t*-test reports a *p*-value, which is the probability that the two (peak intensity) distributions in two groups have identical means. If a *p*-value is $<0.05$, then it is highly likely that the peak or feature is significantly different in its average concentration between the two groups. If the peak intensity distributions are not normally distributed or cannot be transformed or scaled to be normally distributed, then one must use a non-parametric test such as a Mann–Whitney U or Wilcoxon rank sum test (to perform pairwise comparisons). If one is comparing more than two groups, then analysis of variance (ANOVA) must be used. If one is comparing hundreds or thousands of peaks across two groups, then it is important to correct for the false discovery rate (FDR) and to report the q-value (the FDR-corrected *p*-value). Once again, if the *q*-value is $<0.05$ the peak or feature is significantly different in its average concentration.

In most metabolomic studies only a few to a few dozen metabolites or features typically pass this threshold of being significantly changed. The remaining peaks can be assumed to be biologically uninteresting and therefore do not need to be identified or further analyzed. The net effect of combining peak filtering, peak grouping, peak alignment and statistical significance testing (*i.e.* steps (1)–(6)) is that one can reduce a data set which may have had 20 000 noisy, unknown and un-annotated features to a tiny table that may only have a few dozen 'significant' features with well-defined *m*/*z* values and retention times. This represents an enormous reduction in complexity. Many different programs support these kinds of statistical analyses including XCMS,[34,35] OpenMS/Knime,[37,38] MZmine and MZMine2,[39,40] MSDial[41] and MetaboAnalyst.[59]

**3.4.1.2.6 Mass Matching and Formula Generation.** The final step in the untargeted LC-MS metabolomic workflow is feature identification *via* chemical formula generation or mass matching. Mass matching involves comparing the masses obtained from step (6) against a database of known compounds with known masses. Identifying compounds *via* mass matching using a list of a few dozen highly confident *m*/*z* values is far easier than attempting to identify compounds from an ambiguous and noisy list of 20 000 *m*/*z* values. This is why so much effort is typically put into the first six steps in untargeted metabolomics studies. It is also important to remember that mass matching is only useful if mass values can be confidently measured with a precision of a few ppm (typically to the third or fourth decimal place). A large number of databases support mass matching for metabolite identification, including HMDB,[60] METLIN,[61] KEGG,[62] ECMDB,[63] YMDB[64] and PubChem.[65] Mass matching will not always identify a compound. In some cases, multiple mass matches are possible, which makes compound identification ambiguous. It is also possible that even if a strong, single match is found, the compound being identified is something entirely different. In about 50–80% of cases no mass match will be found, which also makes it impossible to identify the compound.

When performing mass matching it is important to choose a search database wisely. If one is studying *E. coli* metabolism, then it is best to search against an *E. coli*-specific database (ECMDB). If one is studying human metabolism it is best to search against a human database (HMDB). It is generally not a good idea to search against a large chemical (non-metabolite) database such as PubChem[65] as 99.9% of the compounds in PubChem are not metabolites and have never left the synthetic lab they were made in. Likewise searching general (non-organism specific) metabolism databases (such as KEGG, METLIN and LIPID MAPS[66]) can also lead to errors wherein users inadvertently identify insect-specific metabolites in plants or mammalian lipids in fungi or human drugs in Antarctic fish. While these results may be theoretically possible, they are highly improbable and likely reflect a careless misidentification.

If mass matches cannot yield a useful hit or a putative match, it is always possible to generate a molecular formula. Molecular formulas are more informative than simple, unidentified masses. This is because molecular formulas can sometimes suggest possible structures or unexpected atomic compositions (*i.e.* the presence of a Cl, a Br or an S atom in the molecule). Molecular formulas can also be used as a constraint to identify a compound if one has the corresponding MS/MS spectra. As noted earlier, chemical formula generation is only possible with high resolution MS data, where mass values can be confidently measured with a precision of a few ppm or to the third or fourth decimal place. Formula generation can be aided by including information on isotope peak patterns and intensities in the LC-MS data. The process can be further aided by placing constraints on the types of atoms that might be expected in different molecules given the biological system(s) being analyzed. In most metabolomic studies the atoms of interest are H, C, N, O, P, S, I, and Br. This is because few other atom types are found in living systems. Other rules pertaining to the feasibility of structures given various atom-type ratios can also be employed to further limit the formula list. For instance, it is impossible to have more than four hydrogens per heavy atom. Many of these formula generation rules or constraints have been described in a paper known as the Seven Golden Rules[67] and are available in programs such as Seven Rules (https://fiehnlab.ucdavis.edu/projects/seven-golden-rules), MSFinder,[50] ChemCalc[51] and Sirius.[52] The final result of an untargeted metabolomic study using high resolution LC-MS data is typically a list of 20–30 putatively identified compounds that 'may' have been significantly altered due to some treatment or disease. The fact that one has gone from 20 000 features to less than a few dozen putatively identified compounds is remarkable. However, this exercise in data reduction also highlights how much is not known about the metabolome. About 98% of the initial set of peaks in an untargeted LC-MS study are not identifiable.[68] In addition, a large quantity of data and a considerable amount of work often generates comparatively few metabolite identifications in return. Indeed, untargeted metabolomics is very labor and computer intensive, with many dozens of hours needed to inspect, process and handle LC-MS spectra. Furthermore, untargeted LC-MS metabolomic studies do not yield confident annotations nor do they yield confident metabolite concentrations but instead only relative concentrations.

### 3.4.1.3 Data Processing for Untargeted LC-MS/MS Data

It has long been recognized that untargeted LC-MS metabolomics workflows where only *m/z* and RT data are generated are non-optimal for metabolite annotation. Metabolite identification is often incomplete, incorrect or ambiguous. However, if additional physico-chemical information about the features of interest could be acquired then this information could greatly reduce this ambiguity. In particular, the inclusion of ESI-MS/MS data in untargeted LC-MS metabolomic workflows can make a significant difference to metabolite identification. This is because MS/MS fragmentation patterns uniquely reflect the structure and molecular composition of small molecules.[69–71] As noted earlier, MS/MS data can be acquired on many types of MS instruments including QTOFs, OrbiTraps, FT-ICR instruments, LTQs, QqQs and LITs. They may also be acquired in a DIA or a DDA. Regardless of how the data is acquired, MS/MS data are almost always associated with a single parent ion with a known *m*/*z* value. So, LC-MS/MS data can look a lot like LC-MS data with *m*/*z* data being plotted against RT data. However, all the peaks (or at least many of the peaks) in the *m*/*z versus* RT plot are also associated with a well-defined MS/MS spectrum that is 'hidden' from view.

As a result, the workflow for analyzing and processing untargeted LC-MS/MS data is essentially identical to untargeted LC-MS data processing (see above). Indeed, the first six steps are largely the same including the: (1) construction of the extracted ion chromatograms (EIC); (2) feature or peak detection; (3) feature filtering; (4) feature grouping (*via* de-isotoping, adduct merging, identification of dimers and trimers, and removal of in-source fragments); (5) feature alignment; and (6) statistical selection of significant features or feature groups. However, instead of attempting to perform feature identification *via* formula generation or mass matching (step (7)), one uses the acquired MS/MS spectra to conduct a search against an MS/MS database to identify the features. Obviously, one does not need to throw out the *m*/*z* data of the parent ion or the calculated molecular formula or even the list of putative matches, which can often come for free with an LC-MS/MS experiment. These data provide useful constraints about what the structure should be or could be. However, having high quality MS/MS data (especially over multiple collision energies) can easily disambiguate ambiguous annotations.

MS/MS spectral searching can be performed against a variety of online databases that have either experimental or computationally generated MS/MS spectra. Typically, one only needs to provide the list of *m*/*z* values and their peak intensities to conduct these MS/MS searches. Among the more popular databases for MS/MS searching are MoNA or the MassBank of North America,[50] METLIN,[61] HMDB[60] and GNPS.[72] Another program/webserver called CFM-ID[73] also supports MS/MS searching against both the HMDB and KEGG data sets. However, if the query MS/MS spectrum does not match with any one of the MS/MS spectra in the above-mentioned databases then one will not be able to identify the compound through simple MS/MS searching.

However, thanks to the emergence of various MS/MS spectral prediction tools such as CFM-ID,[73] MagMA,[74] MetFrag,[75] LipidBLAST[76] and MS-Finder[50] it is now possible to investigate whether an observed MS/MS spectrum might match the predicted MS/MS spectrum for a suspected compound that is not in any spectral database. Using these tools, users typically generate a list of suspected metabolites and then have the MS/MS predictors generate a set of predicted MS/MS spectra. Visual inspection of these predicted MS/MS spectra against the observed MS/MS spectrum can greatly aid in the compound's identification. In addition to going from structures to spectra, it is also possible to go the opposite way. In particular spectral-to-compound 'translation' tools such as CSI-FingerID[77] and IOKR[78] are now available to the metabolomics community. These programs look for features or patterns in the MS/MS spectra that they have 'learned' to identify, which typically correspond to characteristic fragment ions (sugars, benzene rings, *etc.*). This allows them to assign observed MS/MS spectra to known molecules that do not have any available experimental or referential MS/MS spectra. These programs have proven to be quite accurate in the identification of known metabolites in untargeted LC-MS/MS studies.

### 3.4.2 Targeted LC-MS Metabolomics Tools and Workflows

Targeted LC-MS or LC-MS/MS metabolomics is fundamentally different than untargeted LC-MS metabolomics. Targeted LC-MS metabolomics generates confident metabolite identifications (MSI Level 1) and absolute quantifications while untargeted metabolomics generates putative metabolite identifications (MSI Level 2 or 3) and relative quantifications. Key to the success of targeted LC-MS metabolomics is the design of a specific workflow to identify a defined set of metabolites. That set can range from a few to a few hundred different compounds. Each metabolite to be identified *via* a targeted LC-MS assay must have an authentic standard and its corresponding isotope-labeled standard (often $^{13}C$-, $^{15}N$- or $^{2}H$-labeled) or an analogue that looks and behaves almost identically. Each standard must be run through the LC-MS/MS to look for its characteristic elution time and its characteristic mass fragments. This information must eventually be programmed into the MS instrument's software to conduct a targeted LC-MS/MS study. As a result, targeted metabolomic assays may require weeks of testing and validation before it is ready to be run. However, once it is ready, it can be performed quickly, automatically and repeatedly for many years on thousands of samples.

Targeted LC-MS experiments are typically performed on a triple-quadrupole (QqQ) and quadrupole-linear ion trap (QTrap) MS instruments using a technique known as selected reaction monitoring (SRM) or multiple reaction monitoring (MRM). More recently, targeted metabolomic experiments have also been conducted on OrbiTrap and QTOF mass spectrometers using a method known as parallel reaction monitoring (PRM). In SRM studies, which are most often conducted on QqQ MS instruments, a predefined precursor ion is selected in the first quadrupole (Q1), then fragmented in the second quadrupole (Q2) at a defined collision energy. This second quadrupole serves as a collision cell. After fragmentation, a predefined set of fragment ions (previously determined from studying the compound of interest) are filtered in the third quadrupole (Q3) and these are then sent to the detector. The peak area of one or two of the fragment ions (or transitions) of the precursor ion can then be integrated and used for quantification. If a known concentration of a $^{13}C$, $^{15}N$ or $^{2}H$-labeled internal standard is spiked into the sample, absolute quantification of the metabolite can be achieved. The choice of standards, fragment ions, collision energies and other conditions often requires some trial and optimization. For instance, the selection of the precursor ion and the optimization of the ion source parameters must be carefully done to maximize the transmission from the ion source to Q1. Once a precursor ion has been selected, a declustering potential (DP) must be optimized to maximize the precursor ion signal. DP is the potential applied to the orifice to prevent or eliminate ion clustering. DP is a term used in ABSciex instruments. The same term is called S-lens by Thermo Fisher instruments, fragmentor in Agilent instruments and cone voltage in Waters instruments. The fragment ions or transitions must be selected on the basis of their selectivity (ability to distinguish the compound of interest from all other similar compounds). Two fragments are typically selected, one as a quantifier (quantitation) and the other as a qualifier (confirmation). Often the collision energy has to be iteratively optimized for each transition to achieve the highest fragment selectivity and signal sensitivity.

In PRM studies, which are most often conducted on Orbitrap and QTOF instruments, a predefined precursor ion is selected in the quadrupole and transferred *via* the collision trap in either the high energy collision dissociation (HCD) cell (for Orbitraps) or the quadrupole cell (for QTOFs) for fragmentation. After fragmentation, the fragment ions are injected and analyzed in the Orbitrap or TOF mass analyzer. As with SRM, the peak area of two of the fragment ions (or transitions) of the precursor ion can be integrated and used for identification and quantification. If a known concentration of an isotopically labeled internal standard is spiked into the sample, absolute quantification of the metabolite can be achieved. As with SRM, some optimization of the experimental conditions needs to be done, including optimizing the collision energies to obtain maximum sensitivity and selectivity. A key advantage of PRM over SRM is the fact that the precursor masses and the MS/MS spectra are obtained at much high resolution (5–10 ppm or accuracies of 3–4 decimal places) than with QqQ instruments (500 ppm or single digit mass precision). This can make PRM studies more precise, more selective and more comprehensive than SRM studies.

Once a SRM/MRM or PRM assay has been developed, figures of merit for both the instrument and the method have to be evaluated. This includes obtaining limits of detection (LOD), limit of quantitation (LOQ), linear dynamic range (LDR), accuracy, repeatability (precision) and recovery of each metabolite at different concentrations. LC-MS/MS coupled to SRM (or MRM) and PRM is an excellent approach for analyzing hydrophobic or non-polar metabolites. For quantitative analysis of polar metabolites (*i.e.* amino acids, biogenic amines, carnitines) chemical derivatization can substantially improve chromatographic retention/separation and ionization efficiency. For example, phenylisothiocyanate (PITC) derivatization is often used for enhanced detection of primary and secondary amine-containing metabolites such as amino acids and biogenic amines. There are now several papers that describe methods for metabolite derivatization followed by SRM/MRM analysis for metabolite quantification.[79,80]

PRM and SRM assays and software are specific to each instrument and to the instrument's operating software. As far as we are aware, there is no open source software that can be used for targeted PRM or SRM studies in metabolomics. However, very good commercial, instrument-specific, SRM/MRM software packages are available, and these include programs such as Multiquant (ABSciex), Pinpoint (Thermo Fisher), MassLynx and Verify$^{E}$

(Waters), MassHunter (Agilent Technologies), and PACER (Bruker). In addition, Thermo Scientific offers the TraceFinder software for analyzing PRM data through extracted ion chromatograms (EICs) obtained from Orbitrap spectrometers. BioCrates, an Austrian company that sells MRM (p180) and PRM (p400) kits for LC-MS-based quantitative metabolomics, also has its own software. This software interfaces with vendor-specific software to perform automated or semi-automated metabolite quantification *via* targeted metabolomics.

The standard result from a targeted LC-MS/MS metabolomic run is a concentration table consisting of dozens to hundreds of identified metabolites for dozens to hundreds of samples. These concentration tables can then be analyzed using the same statistical software used in untargeted metabolomics. Rather than looking at peak intensities and their distributions, one looks at concentrations and their distributions. As before, *t*-tests, ANOVA, Mann–Whitney *U* tests and FDR corrections must be made. The same tests, the same statistical criteria ($p < 0.05$ for significance) and the same use of FDR corrections ($q < 0.05$ for significance) should be used as described previously in Section 4.1.2.6. The most popular tool for analyzing targeted metabolomic data is MetaboAnalyst[59] (Figure 3.6). This user-friendly web server offers a wide array of options for scaling and normalizing data and for identifying statistically significant compounds.

## 3.5 GC-MS Metabolomics Tools and Workflows

GC-MS metabolomic data closely resembles the data collected in untargeted LC-MS/MS metabolomic studies. In particular, metabolomic GC-MS data consists of hundreds to thousands of individual mass traces that are collected at different GC retention times. However, GC-MS data is also slightly different. Generally, the separation *via* GC is better than LC so the peaks are sharper and there are far fewer overlapping peaks. However, the mass resolution for GC-MS instruments is very modest (at best one decimal place but usually only 1 Da resolution) compared to Orbitrap or QTOF instruments. Likewise, GC-MS instruments automatically generate fragment ion spectra (using EI) and so a single parent ion is usually not detectable or seen in a GC-MS mass trace. As a result, two-dimensional RT *versus* $m/z$ plots are not typically generated in GC-MS studies. Instead, mass traces corresponding to each chromatographic peak are kept in separate files. However, some of the same data pre-processing ideas used in LC-MS/MS can be used in GC-MS. These include: (1) feature or peak detection; (2) feature separation or deconvolution; (3) feature identification *via* spectral matching and (4) statistical selection of significant features.

There are both commercial and freeware programs for analyzing GC-MS spectra. Commercial packages include AnalyzerPro (Spectral Works), ChromaTOF (LECO), MassFrontier/TraceFinder (Thermo Fisher) and ChemStation (Agilent). Freeware packages include AMDIS (distributed through NIST),[44] GC-AutoFit (http://gc-autofit.wishartlab.com/), MetaboliteDetector,[81] PyMS[82] and ADAP-GC.[83] Most of these packages support feature or peak

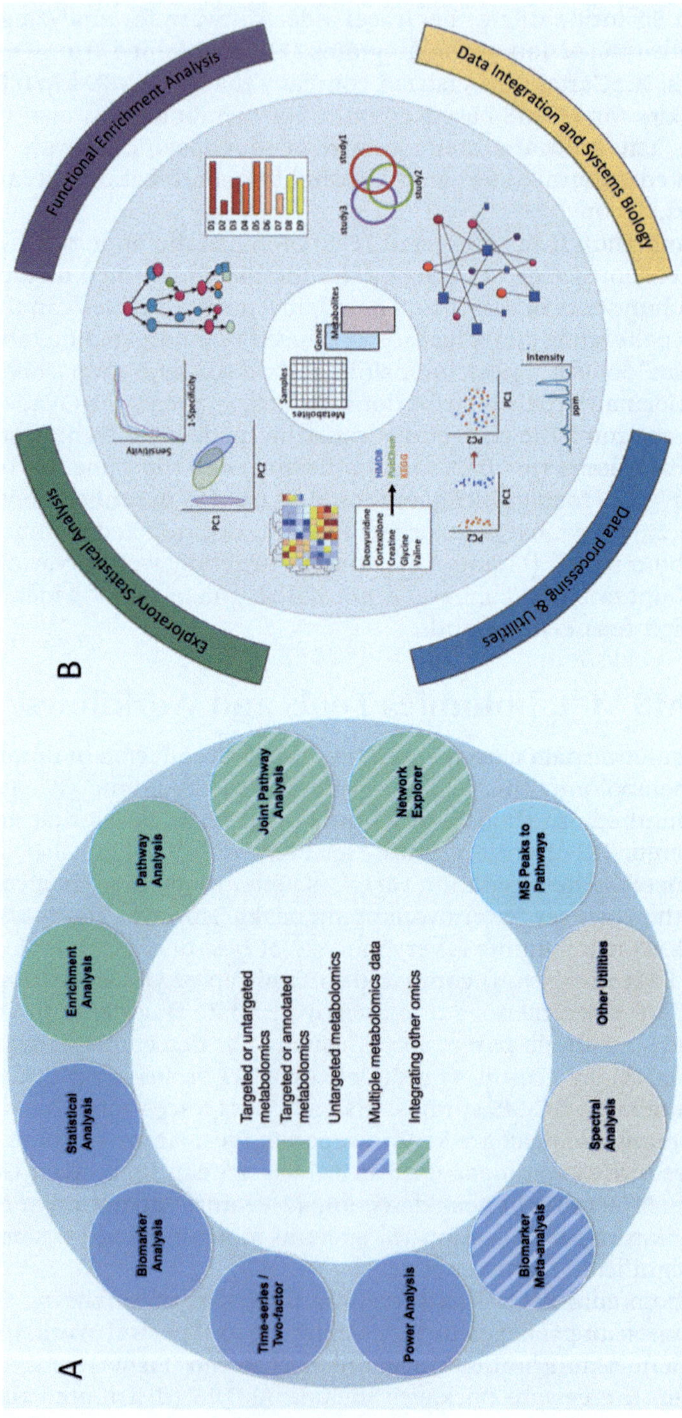


**Figure 3.6** A screenshot of MetaboAnalyst showing the multiple modules for statistical analysis of metabolomics data.

detection, feature separation or deconvolution and feature identification. The most widely used freeware package is AMDIS and this is what we will use to describe the typical data processing workflow.

As noted before, GC-MS chromatogram peaks are narrow and often well separated. However, some peaks consist of overlapping metabolites. The biggest challenge in GC-MS is computationally separating co-eluting components and creating a pure EI-MS spectrum for each component. This process is called 'deconvolution'. Having a 'pure' EI-MS spectrum without any other contaminant *m*/*z* peaks (arising from overlapping compounds) is key to the proper identification of compounds *via* EI-MS spectral matching. The deconvolution process in AMDIS consists of four sequential steps: (1) noise analysis, (2) peak picking, (3) peak shape determination and (4) spectrum deconvolution.

In the noise analysis step, AMDIS looks for sample regions across the EIC (extracted ion chromatogram) that do not have obvious peaks. The peak-free regions are then used to calculate the noise or noise factor. In the peak picking step, AMDIS uses the noise factor calculated in step (1) to look for segments in the EIC that have intensities that are five times greater than the noise factor and which also satisfy certain characteristics of a standard GC-MS chromatogram peak. The next step involves modeling peaks according to an idealized peak shape. The idealized peak shape is determined from a set of 'sharp' peaks found elsewhere in the chromatogram. In this way peak shoulders can be detected, and two nearly overlapping peaks can be separated. The last step extracts the 'purified' or clean MS spectra from individual ion chromatograms. This deconvolution process decomposes potentially shared *m*/*z* peaks from the two overlapping EI-MS spectra yielding an EI-MS spectrum that is identical or nearly identical to the EI-MS spectrum of the pure compound. This four-step peak picking and spectral purification process is also used by MetaboliteDetector as well as ADAP-GC, but with slight variations.[84] It is believed that most commercial programs use a similar process for peak selection and deconvolution.

Once the EI-MS spectra have been extracted (a typical GC-MS chromatogram may have 100+ extractable EI-MS spectra) it is possible to attempt to identify the compounds corresponding to these spectra through spectral matching. Many GC-MS spectral processing programs are linked to the NIST database, a commercial database containing many thousands of experimentally collected EI-MS spectra. The most recent version of the NIST database, NIST17 (Figure 3.7), has more than 306 000 EI-MS spectra for more than 267 000 compounds. The algorithm that is typically used in EI-MS matching is based on a match factor (MF) formula. This measures the similarity of the EI-MS spectrum of the query to the EI-MS spectrum in the reference database. Strictly speaking it is defined as the normalized dot product between the query and the reference spectra. The maximum MF score for a perfect match is 1000. Generally, an MF score of >600 is considered sufficient for a match, however, MF scores of >800 are preferred. The NIST database was not specifically developed for metabolomics and contains

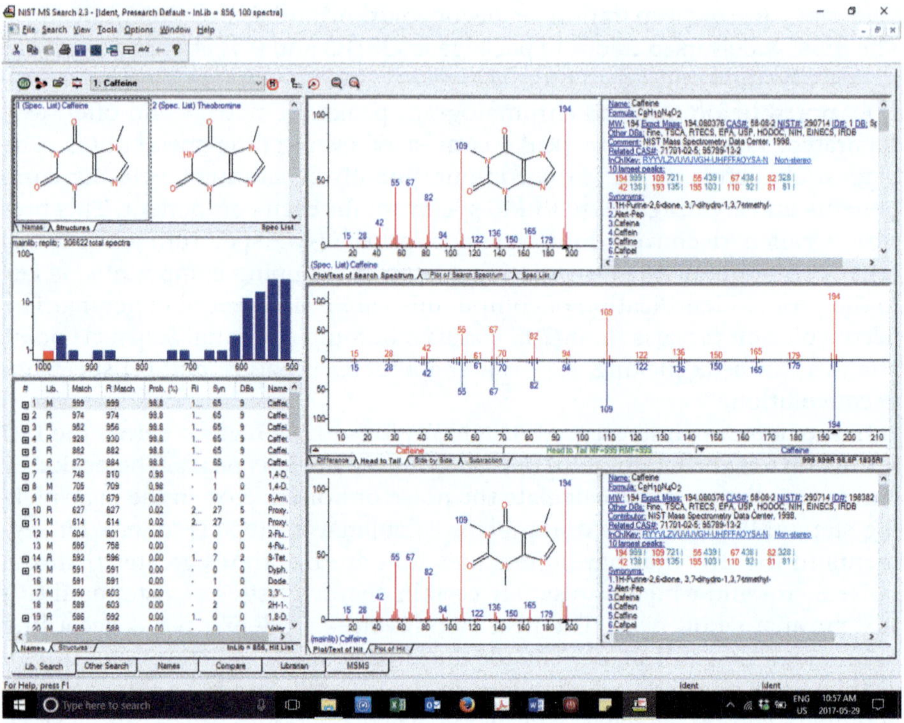


**Figure 3.7** A screenshot of a recent version of the NIST Database, NIST17.

many non-metabolites. There are other EI-MS spectral libraries which were developed for metabolomics and which are freely or commercially available. These include the Golm Metabolome Database (GMD),[85] and the FiehnLib database.[86] These and other commercial EI-MS libraries can be added to the NIST library through the NIST MS Search software. If the EI-MS spectra from the measured compound is not found to match any compound in the existing libraries, it is always possible to compare the observed EI-MS spectrum to predicted EI-MS spectra. A recent version of CFM-ID is able to predict EI-MS spectra with relatively high accuracy.[87] This tool has been used to predict EI-MS spectra for tens of thousands of metabolites (and their TMS derivatives) in the most recent release of the HMDB.[60] Therefore, EI-MS searches against the HMDB's set of predicted EI-MS spectra may offer a useful option when all else fails.

RI data in GC-MS is very reproducible and uniform across instruments and often Kovats RI values for each GC-chromatogram peak are determined. Retention indices are calculated by scaling or calibrating the retention time data to a set of pre-run alkane standards. Large libraries of RI values for many thousands of compounds have been archived in databases. In particular, the NIST17 database has retention indices for more than 72 000 compounds measured in non-polar, semi-polar or polar GC columns. These RI values can be used in compound identification with almost the same precision or utility as accurate mass determination and mass matching can be used to determine the identity of a compound *via* HRMS techniques. In other words, by combining RI values with MF scores it is possible to greatly increase the certainty by which a compound has been identified.

The general workflow in GC-MS metabolomics is to perform metabolite identification first and then use the measured GC-chromatogram intensities or actual concentrations to determine which metabolites have been significantly increased or decreased across samples. This workflow is similar to targeted LC-MS/MS metabolomics where metabolites are first identified and quantified in each sample and then the resulting concentration tables are used to determine which compounds have been significantly altered between groups of samples. Quantification of compound concentrations *via* GC-MS typically requires the running of authentic compound standards over a range of concentrations to generate a 5–7-point calibration curve. Typically, once a set of calibration curves has been created on a GC-MS instrument with a given column and a given set of running conditions, those calibration curves can be used for several weeks or even months without re-running the calibration set.

Using the concept of long-lasting calibration curves, relatively automated, fully quantitative GC-MS metabolomics has been made possible using a program called GC-AutoFit (http://gc-autofit.wishartlab.com). This open-access webserver is specifically designed to identify and quantify ~100 organic acids in urine *via* GC-MS analysis (Figure 3.8). Prior to using the software,

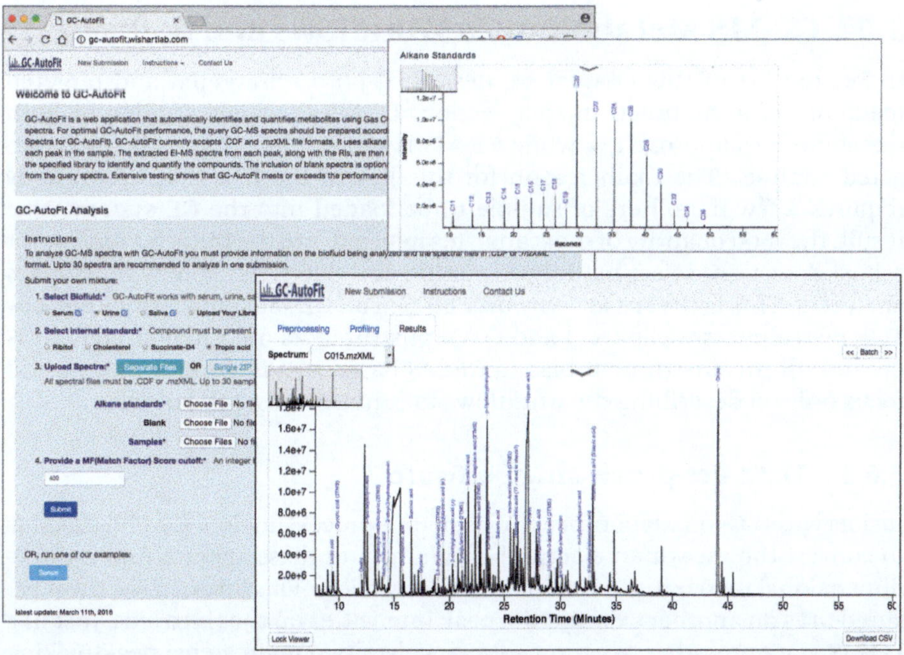


**Figure 3.8** A screenshot for GC-AutoFit, which allows for the near automated analysis of GC-MS spectra.

calibration files for each of the 100 organic acids standards must be generated. This can take several days of work. However, once this is done, the software can be run over and over again for hundreds to thousands of samples without modifying the calibration set. To run the software users must upload three types of input files (in mzXML or netCDF format), a set of standard alkane standards, a blank sample file and one or more urine sample files. Once the files have been uploaded, users must upload the calibration library files (for quantification) and then the program can be launched. GC-AutoFit will perform RI calibration, peak alignment (if multiple GC-MS spectra have been uploaded), peak identification, peak integration and concentration calculation. It takes approximately 60 seconds per spectrum and up to 110 compounds can be identified and quantified in a given urine sample with about 96% accuracy. The average urine sample typically yields 50–60 organic acids.

The standard result from a GC-MS based metabolomic study is a concentration (or relative concentration) table consisting of dozens of identified metabolites for dozens to hundreds of samples. These concentration tables can then be analyzed using the same statistical software used in both untargeted and targeted LC-MS metabolomics. As before, *t*-tests, ANOVA, Mann–Whitney *U* tests and multiple testing corrections must be made as described previously in Section 3.4.1.2.6. The most popular tool for analyzing GC-MS metabolomic data is MetaboAnalyst.[88] This user-friendly web server offers a wide array of options for scaling and normalizing data and for identifying statistically significant compounds.

## 3.6 CE-MS Metabolomics Workflows and Software

In Section 3.2 of this chapter we described both untargeted and targeted metabolomics. As noted in that section, LC-MS is widely applied to both metabolomics approaches, while CE-MS is more frequently used for untargeted analysis. The main reason for this lies in the fact that CE-MS only requires a few nanoliters of sample to be loaded into the CE system. As a result, the total quantity of most analytes injected into the MS is far too low for robust detection, selection, fragmentation and quantification by most mass analyzers. This is the reason why over 85% of all CE-MS applications employ TOF instruments while QqQ and QTOF instruments are rarely used89. It is because of this instrument bias in CE-MS-based metabolomics that we will focus only on describing the workflows for untargeted applications.

### 3.6.1 Data Pre-processing Software

Just as with LC-MS, data processing, the pre-processing of CE-MS data aims to convert the measured electropherograms and mass spectra into a three-dimensional matrix that contains data for the *m*/*z* values on one axis, the measured MTs on another axis and the peak intensities on the third axis. Just like LC-MS, data pre-processing in CE-MS involves two main steps: peak-picking and peak alignment. Because of the general similarity between CE-MS and LC-MS data, the same data pre-processing tools can be used to analyze and process both types of data. However, because of the differences in CE-MS peak shapes and peak widths, along with the high level of in-source fragmentation found with CE-MS combined with the relatively significant MT drift seen in CE-MS runs, peak-picking and alignment must be performed carefully and handled somewhat differently than with LC-MS.

### 3.6.1.1 Peak Picking for CE-MS

The peak peaking process for CE-MS (and LC-MS) is challenging because of the presence of multiple peaks that originate from the same molecule. In particular, each measured metabolite produces several ions called ionization products. These ionization products include ions formed by different adducts: $[M+H]^+$, $[M+Na]^+$, $[M+NH_4]^+$; neutral losses: $[M+H-H_2O]^+$, $[M+H-CO_2]^+$; multiple charges: $[M+H]^+$, $[M+2H]^{2+}$; multimers: $[2M+H]^+$, $[3M+H]^+$; as well as a large number of in-source fragments (which are a particularly problematic feature of CE-MS spectra).

Peak picking for CE-MS data can largely be performed using the same tools as for LC-MS data (see Section 3.4.1.2.2). However, as mentioned above, the peak shape and peak width of CE-MS peaks tends to differ somewhat from LC-MS peaks, therefore any standard LC-MS parameters related to these aspects must be adjusted specifically to suit the unique characteristics of CE-MS data so as not to miss any signals. It is also important to remember that the overall signal intensity obtained *via* CE-MS is often lower than what is seen for LC-MS data (due to the very small sample volumes used in CE-MS). This requires appropriate adjustment of thresholds and noise cut-offs, so as not to eliminate low abundance signals.

The two most frequently used software tools for processing CE-MS data are MasterHands[90] – an open source tool, and MassHunter Qualitative/MassHunter Profinder, a commercial product from Agilent Technologies. Approximately 50% of the published research papers involving CE-MS metabolomics use these two software packages, while the remaining CE-MS publications tend to employ in-house algorithms developed *via* Matlab[91] or R[92] environments. XCMS (which is written in R) is another software tool that is gaining traction within the CE-MS community as it reads both mzXML[93] and netCDF[94,95] file formats, which can be easily generated from CE-MS data. Because Agilent is the main commercial supplier of today's CE-MS systems, Agilent's MassHunter software is very popular and it offers several excellent solutions exclusively dedicated to the analysis of CE-MS data. However, because our focus in this chapter is on non-commercial software, we will limit our discussion to the MasterHands software package. MasterHands was initially developed at Keio University for internal use only, but over time it has been extended, refined and distributed to other CE-MS labs, making it a widely used, stand-alone CE-MS data analysis program.[96–98] Because MasterHands was specifically designed for processing of CE-MS data, all of its algorithms and parameters are optimized for CE-MS derived data. The next few sections will describe how MasterHands can be used in the processing of CE-MS data.

### 3.6.1.2 Feature Filtering and Grouping

As with LC-MS, identified peak features in CE-MS have to be filtered and grouped. However, this process is generally more challenging for CE-MS data since both multiple adducts form and in-source fragmentation occur on a much larger scale in CE-MS as compared to LC-MS. These ionization and matrix-dependent phenomena generate large numbers of extra MS peaks or redundant compound peaks in CE-MS spectra. The vast majority of these redundant peaks can be identified simply by looking for particular differences in the masses between co-eluting *m/z* features (see Section 3.4.1.2.3 on Feature Filtering and Grouping). Furthermore, the presence of adducts and neutral losses for the same parent ions allows the 'triangulation' of the parent-ion monoisotopic mass by comparing the experimental distance between two seemingly redundant features with a given theoretical *m/z* distance between two known adducts.[99] However, such an approach to *m/z* determination is not always sufficient, and to prevent the overestimation of adducts and isotopes it should be supported with other methods such as peak-shape and peak–abundance correlation. Both peak-shape and peak abundance correlation methods help to prove (or disprove) the relationship between co-eluting signals before annotating them to the same metabolite.

Once co-eluting and correlated peak intensities have been found they can be merged into a single feature and represented as a single value (as in case of MassHunter[100,101]) or these redundant signals can be removed, keeping only the most abundant ion (as in the case of MasterHands[102,103]). In most cases the most abundant ion signal is the $[M+H]^+$ peak in the positive ion mode or the $[M-H]^-$ peak in the negative ion mode.

Most peak picking/filtering/grouping tools can easily handle different ions, charge states, neutral losses or the presence of dimers, trimers or multimers. This is typically done using a list of expected ions, expected charge states and expected neutral losses, which allows co-eluting and co-related ions to be easily found. However, it is often quite challenging to identify signals arising from in-source fragmentation. This is not trivial for CE-MS, as this technique is particularly well suited for separating and detecting small ionic molecules, which tend to dissociate, in source, very easily. Moreover, only 10% of the metabolites in the METLIN database (with an experimental MS/MS spectrum) do not dissociate in-source.[99] The MasterHands software allows users to remove in-source fragment ions from the spectra but only those which are already known and provided as a library.[104,105] There are also a number of other software tools that can deal with in-source fragmentation, although they are not exclusively designed for CE-MS. For instance, CAMERA works in the same way as MasterHands, using a predefined list of common neutral losses, AStream finds fragments based on the pairwise/partial correlation,[46] xMSannotator employs weighted correlation network analysis[106] while MZmine2[40] and RAMClust[49] uses hierarchical clustering-based approaches and spectral matching with low-energy MS/MS spectra acquired in the same run (MZmine2) or with spectra from public libraries (RAMClust) to help identify in-source fragments.

### 3.6.1.3 Feature Alignment

Once CE-MS features have been found and curated, they have to be aligned between different samples. This is very challenging or even problematic for CE-MS data due to the relatively high MT shift variance that can be seen across samples. Similar to the alignment of LC-MS data, CE-MS strategies for alignment use spectral warping algorithms with various modifications including dynamic time warping (DTW) and correlation optimized warping (COW) to look for the corresponding (matching) features across the samples and to produce aligned spectra. Other warping methods may also be used, such as the OBI-Warp algorithm (ordered bijective interpolated warping) is used in XCMS[95,107] or reference peak warping (RPW), which is a part of MsXelerator.[108]

The main challenge in the alignment of CE-MS data is caused by the shift in the MT between different analysis. MT shifts can occur between runs but are also not constant for all molecules, since metabolites with greater electrophoretic mobility tend to have a smaller MT shift than compounds with a smaller electrophoretic mobility.[109] To minimize this problem the MT can be corrected prior to the actual alignment. This correction can be performed using a reference molecule(s), that can be either a spiked standard(s) or a set of known molecules detected across all samples. Then the difference between the MT of each reference compound across different runs can be used to correct and adjust the MTs for the entire electropherogram. Another strategy, implemented in MassHunter Profinder, uses a reference sample (for example a QC sample) to perform polynomial interpolation to correct the MTs across samples. Another very elegant solution is used by the CEqualizer software, which quantifies the MT shift from two orthogonal measures: the field strength and/or migration length differences and electro-osmotic flow (EOF) differences between runs.[110]

## 3.6.2 Statistical Analysis

Statistical analysis of pre-processed CE-MS data can be performed in the same way as for any other LC-MS platform (see Section 3.4.1.2.5 on Statistical Selection). Typically, the density (or total number of detected peaks per unit of time) of CE-MS data is a little different to LC-MS data due to the lower number of detected molecules and the lack of sample carry over between runs. A significant advantage of CE-MS over LC-MS is the constant ionization conditions that CE-MS affords. Constant ionization produces consistent ion intensities that are more proportional to their abundance or concentrations. In LC-MS, chromatographic gradients are commonly used to facilitate separation and these compositional changes of the mobile phase along the gradient significantly affect ionization efficiency. This makes signal comparison (and signal quantification) particularly difficult for LC-MS. For this reason, CE-MS data is well suited for performing ratio analysis or correlation analysis, especially with regard to an increasingly popular technique known as epi-metabolite analysis.[111]

### 3.6.3 Metabolite Annotation

Metabolite annotation and identification in CE-MS is not very different from LC-MS, as long as the CE system is coupled with an MS system that supports MS/MS analysis. Unfortunately, for most CE-MS based metabolomic applications this is not the case. As a result, most CE-MS researchers use other sources of information to confirm or reject the identity of detected metabolites. The vast majority (around 70%) of annotations performed in CE-MS studies make use of authentic standards and, therefore, can attain MSI Level 1 metabolite annotation. The remaining 30% typically achieve MSI Level 2 or Level 3 annotation.[89] Level 3 MSI annotation refers to the metabolites annotated based only on accurate mass and isotopomer intensity information which is then used for chemical formula determination and confirmation. Available software supporting chemical formula generation have already been discussed for LC-MS (see Section 3.4.1.2 on Data Preprocessing and Formula Generation Software) and they can be easily adapted for CE-MS data as well.

Level 2 MSI annotation refers to metabolites annotated based on a comparison of their measured physico-chemical properties (MS spectra, NMR spectra, RT, MT, or RI) to corresponding physico-chemical property libraries of authentic standards. Usually these refer to spectral comparisons of the experimental MS/MS spectrum with an MS/MS spectrum of an authentic standard from a (commercial or open-access) spectral library. However, in the majority of cases in CE-MS metabolomic studies, the CE system is coupled to TOF mass spectrometers, as opposed to QTOF instruments, which do not offer the possibility of performing tandem mass analysis. This limitation can be overcome by making use of the in-source fragmentation phenomenon characteristic of CE-MS to gain structural information about the metabolites of interest. The fragmentor voltage (also known as the cone voltage, declustering potential or S-lens) in CE-MS instruments can be enhanced leading to more significant in-source fragmentation at the ion source and, therefore, more detailed information about the structure of the molecule.[112] Since all the molecules are fragmented together (without prior precursor isolation) the right product ions have to be assigned to the right precursors. To achieve this assignment correlation analysis can be used. Correlation analysis assumes that a very strong positive correlation in peak intensity (with a correlation coefficient over 0.9) reflects an in-source fragmentation correlation that can be used to find ions originating from the same molecule. Software such as MasterHands, AStream, xMSannotator, MZmine2 and RamClust, support this kind of analysis. Therefore, instead of MS/MS data, in source fragmentation data obtained with an enhanced fragmentor setting can be used to confirm the identity of metabolites based on their in-source fragmentation patterns.

MSI Level 2 annotation also includes compounds annotated based (in part) on the information about their measured MT. MT is directly proportional to the charge and inversely proportional to the size of the molecule. For two molecules with the same mass, the one with greater charge will migrate faster along the capillary, and, therefore, will have a shorter MT. The MT of a given molecule can be affected by any changes in the length of the capillary, the composition of the buffer, the applied voltage or even changes in the composition of the sample. Relative to other separation techniques (such as GC or LC), the CE-measured MT is the least reproducible and tends to vary significantly between samples, between instruments and between runs. Although, the use of MT as an orthogonal parameter in the CE-MS annotation process could provide additional information, its low reproducibility has generally limited widespread use of this approach. However, new approaches are now appearing that could potentially overcome these limitations.

In CE the movement of ions through the capillary tube can be determined or predicted based on several analytical variables including the pH of the buffer and various molecular descriptors (MDs) concerning the analyte ions. These MDs can include structural and physico-chemical properties expressed as numeric values, such as the net charge or pKa. Migration calculations have been previously used to help evaluate and annotate metabolites manually,[100] but there are now elegant solutions that perform these calculations automatically.[113] For example, a software package called Simul uses absolute mobility and pKa to accurately predict the electro-migration of metabolites.[114] In the Simul software package, the Hubbard–Onsager model is used to estimate absolute mobility from the molecular volume and the intrinsic valence charge. Another type of molecular migration predictor that has been developed employs support vector regression to predict the MT based not only on the MDs but it also makes use of accurate mass and isotopic pattern distributions to help identify metabolites.[103] In this particular program, the charge is determined from the pKa and the buffer pH while the MDs are generated through the molecular operating environment (MOE).[115] MOE is also used to determine MDs in another software package for molecule mobility prediction, wherein artificial neural networks are employed to predict the MT from the structure of molecular cations.[113] This program uses information about the net charge, the pKa and MDs of molecules, whereupon the MT is first determined and then standardized to a relative MT (RMT) using methionine sulfone as an internal standard.[113]

Tools for the MT prediction tend to differ in the descriptors and algorithms they use, however, they all require a large set of an authentic standards as a training set, to allow machine learning methods to acquire the proper knowledge about the behavior of known molecules under defined conditions and to then robustly predict the MTs of novel molecules under similar conditions. An alternative approach to machine learning approaches employs RMT instead of an absolute MT calculation.[98,116,117] In this case the MT of each molecule is compared to the MT of a set of standards. This normalization method allows one to significantly reduce inter-batch MT variability and even eliminate intra-batch MT variation. However, the RMT approach works only in terms of MT matching for already known or previously measured molecules and cannot be used for predicting the MT of novel molecules. Another very interesting approach for molecular mobility comparison has been implemented in a program called ROMNCE (RObust Metabolomic analysis with Normalized CE).[118] This program aims to convert the original migration time scale into an effective mobility scale for CE-MS data. Because effective mobility (EM) only depends on the nature of the chosen background electrolyte and temperature, it is possible to overcome the intrinsic variability of MT with EM and, therefore, improve the quality of compound matches to reference EM values.

While most of our focus has been on tools and techniques to facilitate MSI Level 2 annotation with CE-MS, it is important to remember that MSI Level 1 annotation should be the primary goal of most metabolomics studies and that the majority of published CE-MS metabolomic studies actually achieve MSI Level 1 annotation. As noted earlier, MSI Level 1 annotation can only be achieved by analyzing an authentic standard under identical analytical conditions as the metabolite of interest. However, authentic standards can be used in different ways and, therefore, they tend to provide different levels of confidence. Often authentic standards are only analyzed once, and information obtained through such a one-off analysis is incorporated into an in-house library, which is then used for subsequent identification purposes. Other approaches to achieve MSI Level 1 annotation involve the analysis of standard mixtures at the beginning of each batch run.[119] These standardized mixtures have a precisely known chemical composition and the collection of this standard mixture data at the beginning of each CE-MS batch run ensures that the reference MS spectra and reference MTs will be essentially identical to those detected in the experimental run. This batch-based, defined mixture standardization provides a higher level of confidence than one-off approaches since the analytical variation related to the length of the capillary or the performance of the MS instrument is minimized. Despite the generally high level of confidence in metabolite identification that these methods should/could achieve, both methods still can lead to mis-identifications. For CE-MS this is often due to the large MT inter-batch shift, which can be affected by sample matrix effects, sample composition effects and differences in the complexity between the biological sample and the pure standard(s). Because of this, the highest level of confidence is generally achieved by directly spiking the biological sample with a pure form of the standard of interest and comparing the area of the peaks before and after spiking.[100,120]

## 3.7 Lipidomics Workflows and Software Tools

Lipids constitute a very large and diverse group of molecules with a broad spectrum of physical and chemical properties and a very wide range of concentrations in biological samples. Because of their chemical diversity, their biological importance and their crucial role in health and disease the field of lipidomics (which is formally defined as the high throughput, dedicated characterization of lipids) has emerged to become one of the fastest growing sub-disciplines in all of metabolomics. As might be expected, many of the metabolomic techniques described in Sections 3.2–3.4 can be applied to the analysis of lipids. Indeed, many standard metabolomics workflows routinely detect and/or quantify a number of lipids, fatty acids or lipid derivatives as part of their usual workflow. However, there are a number of unique features about lipids and lipid analysis that make a separate treatment of lipidomics worthwhile. Furthermore, with the growing interest in lipid characterization, specialized lipidomic assays that detect and/or quantify only lipids (or very large numbers of lipids) are becoming more and more popular.

Lipidomics methodologies can be classified into three main categories: (1) LC-MS approaches; (2) direct-infusion MS (DIMS) or shotgun approaches; and (3) lipid imaging MS. Among the three approaches LC-MS lipidomics methods cover a very broad range of applications since chromatographic separations can be performed based on several different types of chromatography (normal phase, reverse phase, and HILIC), different mobile phases and different mobile phase modifiers. Moreover, LC-MS lipidomics can be performed in a targeted, semi-targeted or an untargeted manner. Depending on the chromatography method being used, and the type of mass analyzer being employed, LC-MS lipidomic methods can either provide very wide metabolite coverage (but imprecise quantitation) or very precise quantitative information (with more limited metabolite coverage). While our focus here is on LC-MS metabolomics, (as LC separations dominate the field of lipid analysis) it does not mean that GC or CE cannot be successfully used in the analysis of lipids. Indeed, several very nice examples of GC-MS and CE-MS-based lipidomics or fatty-acid studies have been published. These include the high-throughput screening of fatty acids by multisegment injection-non-aqueous CE-MS[121] and application of high resolutions GC-MS for the analysis of fatty acid methyl esters (FAMEs), polyaromatic hydrocarbons (PAHs) and saturated hydrocarbons.[122] However, since the vast majority of lipidomics applications employ LC-MS, we will mostly limit our discussion here to this technique.

Shotgun or DIMS lipidomics is a rapid lipidomics or lipid characterization technique that avoids the use of LC separation entirely. This approach can also provide relatively broad coverage as well as quantitative or semi-quantitative information about lipids. However, the coverage of DIMS or shotgun lipidomics is significantly lower than with LC-MS lipidomics. Indeed, the lack of a separation step prior to detection requires careful sample preparation that extracts the lipids of interest and removes all other sample components, including lipids belonging to other lipid classes.

Imaging lipidomics or imaging mass spectrometry (IMS) is a technique of acquiring spatial information about lipid composition across thin tissue sections. It uses MALDI (matrix assisted laser desorption ionization), SIMS (secondary ion mass spectrometry) or DESI (desorption electrospray ionization) to generate MS spectra from the surface of carefully prepared tissue samples. Thousands of MS spectra are systematically collected over a grid defined by thousands of small (10–50 micron-sized) spots on the tissue sample. The collected MS spectra serve as pixels to define the image and are analyzed to identify the 'likely' lipids (*via* mass matching) and approximate abundance (by intensity). The resulting images can be artificially colored to display the lipids or their abundance, in much the same way that histological stains can display certain cellular or chemical features in tissue slices

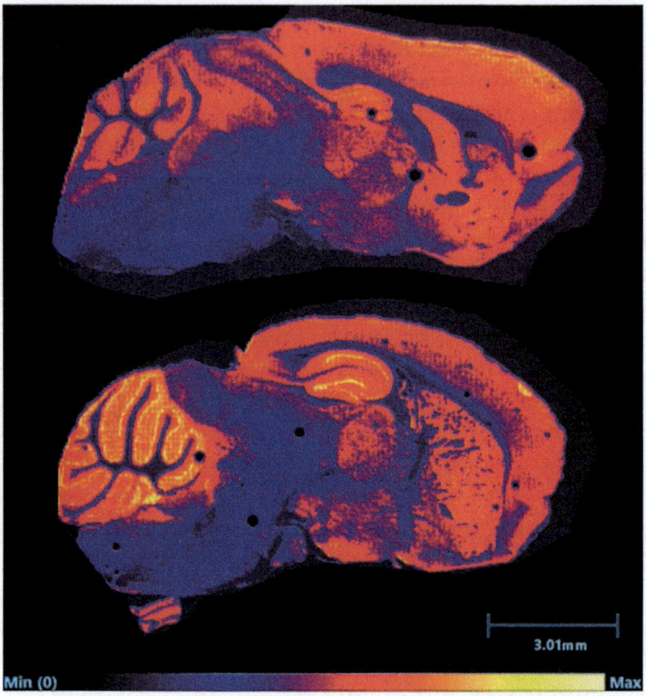


**Figure 3.9** An example of an IMS lipidomics image generated on mouse brain using a MALDI Synapt MS instrument. The ions imaged here correspond to two different glycerosphosphocholines.

(Figure 3.9). Imaging lipidomics does not provide (as much) quantitative information (as DIMS or LC-MS methods do) and it offers relatively low coverage, however, it does offer information and generates stunning images that show the distribution of lipids across tissue samples. Indeed, through imaging lipidomics the spatial distribution of lipids can be visualized without the need to extract, purify, separate or label the sample.

The selection of a particular lipidomics methodology depends on the level of quantification that one wishes to achieve, the level of desired metabolite coverage and the overall lipid concentrations that are known or expected in the measured samples. Some of these decisions are determined not only by the platform or method, but also on the decision to pursue targeted, semi-targeted and untargeted lipidomics methods. Targeted methods lead to the absolute quantification of already-known or targeted lipids, where each lipid is quantified using an individual $^{13}C$ or $^{2}H$ labelled standard. In lipidomics, this is often very hard to achieve as only a few dozen commercially available isotopically labeled lipid standards exist (*versus* tens of thousands of individual lipids). As a result, most lipid quantification is done *via* semi-targeted approaches. Semi-targeted methods allow lipids to be approximately or reliably quantified. The concentrations of individual lipids are determined using a small number of representative, isotopically labelled lipid standards for a group or class of lipids (acylcarnitines, phosphatidylcholines, cholesterol esters, *etc.*). Typically, one or two lipid standards are used to quantify a given group of lipids. Quantification by these methods take advantage of the fact that most members of a given lipid class behave almost identically in terms of ionization efficiency or matrix effects. Untargeted methods do not provide information about the absolute concentration of lipids as no isotopically labeled standards are used in the workflow. Rather, untargeted lipidomics methods permit relative quantification and are intended to identify (as opposed to quantify) as many lipids as possible *via* accurate mass matching and/or fragment mass matching. Untargeted methods may also be used in the identification of novel or unexpected lipids, which have no chemical standards.

Targeted lipidomic methods are typically performed using QqQ mass spectrometers and multiple reaction monitoring (MRM), semi-targeted approaches employ QqQ or QTOF mass spectrometers and employ single reaction monitoring (SRM), precursor ion scanning (PIS) or neutral loss scanning (NLS). Untargeted lipid profiling requires the use of very accurate, high resolution mass spectrometers such as QTOF, TOF, Orbitrap of FTICR mass spectrometers and typically use either DDA or DIA acquisition techniques. For truly comprehensive lipidomics analyses one should use several different MS platforms with several different analytical approaches and a variety of lipid extraction or purification protocols. The use of a wide number of internal reference lipid standards (regardless of whether one is using targeted, semi-targeted or untargeted methods) is also highly recommended.

### 3.7.1 LC-MS Lipidomics Software

The widest range of lipidomics applications have been developed using LC-MS techniques. This is partly because the same methodologies used in LC-MS-based metabolomics can be adapted to LC-MS-based lipidomics and so the same software tools can also be used. However, there are a number of software packages and databases that have been specifically designed to facilitate lipidomic analyses. Some of these are commercial and some are open source or open access tools. Some of the better-known commercial packages available for LC-MS lipidomics include LipidAnnotator from Agilent, ClinPro from Bruker, Progenesis from Waters, LipidSearch from ThermoFisher and LipidView from Sciex. As this book is primarily concerned with describing open source or open access software, we will focus on these tools here. In particular, we will describe seven popular software tools that are widely used for lipid identification *via* spectral (MS/MS) matching. These include, MS-DIAL,[41] LipoSTAR,[123] LipiDex,[124] LipidMatch,[125] LipidMatch Normalizer,[126] LIQUID[127] and LipidMiner.[128]

MS-DIAL (Mass Spectrometry–Data Independent AnaLysis), which has already been briefly described in Section 3.4, is a general metabolomics software package that supports the identification and quantification of metabolites *via* MS or MS/MS techniques.[41] MS-DIAL gets its name from the fact that it supports data analysis from a number of different DIA modalities (which are commonly used in lipidomics) including SWATH, All-Ions MS/MS, $MSc^2$ and all-ion fragmentation. MS-DIAL is uniquely configured for lipidomics work as it performs mass spectral deconvolution and spectral matching against LipidBlast.[76] LipidBlast[76] is a comprehensive database of experimental and predicted lipid MS/MS spectra, which is discussed in more detail in Section 3.7.1.1.

Lipostar is another useful software package that supports both targeted and untargeted LC-MS lipidomics.[123] Lipostar supports vendor-neutral MS data processing, multivariate data analysis and advanced lipid identification. Lipid identification in Lipostar can be performed by searching against databases of fragmented lipids or by searching for fragments that are lipid-class-specific in database-independent manner. Lipostar employs a full metabolite identification (MetID) procedure,[123] which is particularly crucial for drug safety applications and clinical translational studies. LipiDex is another very useful integrated software package for lipid identification and quantification.[124] It uses flexible *in silico* fragmentation templates and lipid-optimized MS/MS spectral matching routines to help robustly identify known lipid species and to track unknown compounds. In addition to its strong lipid annotation capabilities LipiDex also supports lipid quantification, including the quantification of co-isolated and co-eluting isobaric lipids. LipiDex also supports data filtration to remove ionization artefacts, which is particularly important in lipidomics studies to reduce the number of peak redundancies.

Another useful package for lipidomic analysis is LipidMatch, an R-based tool designed for lipid identification *via* LC-MS lipidomic studies. LipidMatch uses a rule-based identification scheme that exploits a large set of *in silico* fragmentation libraries covering a wide variety of lipid molecules such as oxidized lipids, bile acids and sphingolipids. LipidMatch provides a good deal of flexibility, allowing users to choose the rules for identifying each lipid class and to edit or correct any of LipidMatch's initial annotations. A recent extension to LipidMatch is LipidMatch Normalizer.[126] This tool helps to standardize and automate relative lipid quantitation by using a ranking system to evaluate the quality of lipid assignments based on the lipid standards that users employ in their lipidomic studies.

LIQUID (LIpid QUantification and IDentification) is yet another lipidomics software tool developed for targeted lipid identification *via* LC-MS-based techniques.[127] LIQUID is designed to support semi-automated processing and visualization of LC-MS lipidomics data. Lipid identification in LIQUID is performed using a customizable target library and a scoring model adjusted for each project. LIQUID allows users to visualize of multiple lines of spectral evidence for each lipid identification and to score lipid spectra based on expected fragment peak intensities learned from training data. In LIQUID the score of any given spectrum match is the sum of log likelihood scores for each MS fragment.

Another popular tool for targeted LC-MS lipidomics is called LipidMiner.[128] LipidMiner has a graphical user interface written in Python and most of its core functions are written in C#. It is composed of three functional modules:

(1) the ion detection module; (2) the feature alignment module and (3) the library match module. The ion fetection module is used to detect and quantify lipid features from raw LC-MS data files and to assign lipid classes to any detected lipid features. The feature alignment module performs chromatographic alignment of the detected lipid features across multiple data files while the library match module identifies lipid features through accurate mass matching against the LIPID MAPS library. MS/MS information is used to confirm the identity of tentatively assigned lipids based on a user-defined list of signature product ions (from product ion scans or PIS) or neutral loss scans (NLS).

A separate group of more specialized lipidomics software tools has emerged over the past few years, which is dedicated to the analysis of particular groups or classes of lipids. These include Greazy[129] and LipidHunter.[130] Greazy is a tool designed specifically for the automated annotation of phospholipids.[129] It employs user-provided parameters to build a phospholipid search space and associated theoretical phospholipid MS/MS spectra. The experimental MS/MS spectra are scored against these theoretical phospholipid MS/MS spectra and then another tool (called LipidLama) filters the results *via* mixture modelling and density estimation to determine the most probable hits.

Another tool designed specifically for the annotation of phospholipids is LipidHunter.[130] This is a stand-alone software package for phospholipids annotation, based on MS/MS data analysis using defined fragmentation rules for each phospholipid (PL) class. It is designed for DDA MS applications and matches fragments and neutral losses obtained by collision-induced dissociation to a user-defined list of expected fatty acid fragments and phospholipid class-specific fragments. Proposed structures of identified lipids are tested against elemental composition data and reference data provided *via* the LIPID MAPS database. LipidHunter can be used for both LC-MS lipidomics and shotgun lipidomics. A 'spin-off' of LipidHunter, called LPPtiger, can be used for the annotation of oxidized phospholipids.[131] LPPtiger combines three different algorithms to first predict the oxidized phospholipidome, then to generate oxidized phospholipid spectral libraries, and then to identify oxidized phospholipids based on the observed MS/MS data through a complex scoring and annotation process.

#### *3.7.1.1 Lipid Databases and Lipid MS/MS Databases*

Lipids are inherently modular molecules composed of 'head' groups consisting of glycerol, modified glycerols or inositol derivatives and fatty acid 'tails'. These fatty acid tails may be linked to the head groups *via* acyl or ether linkages. Because there are dozens of different fatty acids and because lipid head groups can be attached to between one and four different fatty acid groups, it is chemically (and biochemically) feasible to generate hundreds of thousands of different lipids. As a result, lipids are, without a doubt, the most numerous class of compounds in biology. However, the fact that lipids have a modular character to them also makes them among the easiest compounds to identify (*via* MS-based lipidomics) and it is also relatively easy to predict their MS/MS spectra. Several lipid-specific databases and computer programs have recently appeared that predict, or contain, predicted lipid spectra. LipidBlast[76] is both a program and a database containing computer-generated MS/MS for more than 100 000 lipids from 26 lipid compound classes, including phospholipids, glycerolipids, bacterial lipoglycans and plant glycolipids. LipidBlast uses a rule-based algorithm, learned by inspecting hundreds of experimentally collected lipid spectra, to accurately predict MS/MS spectra of lipids (merged over multiple collision energies). A number of the lipidomics programs mentioned in Section 3.7.1 make use of LipidBlast's predicted MS/MS spectra to help make spectral assignments and to perform lipid annotations.

Another software tool for predicting lipid MS/MS spectra is CFM-ID 3.0.[132] This is a widely used MS/MS spectral prediction program that uses machine learning and *in silico* fragmentation to predict both the intensities and masses/structures of compounds undergoing collision-induced dissociation in QTOF mass spectrometers. CFM-ID can also be used to annotate fragment ions from experimental MS/MS spectra and to annotate experimental MS/MS spectra *via* spectral matching against a large database of both experimentally collected (and CFM-ID predicted) MS/MS spectra. The latest version of CFM-ID uses rule-based techniques (similar to those used in LipidBlast) to predict both fragment ions and fragment ion intensities for >20 classes of lipids at three different collision energies (10, 20 and 40 eV). CFM-ID 3.0 has also been used to generate theoretical MS/MS spectra for all of the lipids (>90 000) contained in the Human Metabolome Database.[60] This large collection of MS/MS spectra makes the HMDB (and the CFM-ID database) particularly useful for lipid annotation and identification, especially for mammalian lipidomics studies.

Another popular MS/MS predictor and MS/MS annotation tool is MetFrag.[75] MetFrag is an *in silico* fragmenter that uses a form of 'rationalized' combinatorial fragmentation to generate predicted MS/MS data. Like CFM-ID, MetFrag not only predicts MS/MS spectra but it can also be used to annotate fragment ions (from MS/MS data) and to annotate experimental MS/MS spectra *via* spectral matching. MetFrag also makes use of experimentally collected MS/MS spectra to improve its performance in spectral matching and compound annotation. A recent extension to MetFrag is LipidFrag.[133] LipidFrag includes a number of improvements in the reliability of MetFrag's *in silico* fragmentation of lipids. Some of these improvements include the incorporation of the LIPID MAPS database into MetFrag's spectral and compound library along with lipid-class specific classifiers to calculate the probabilities for lipid class assignments.

The LIPID MAPS database has already been mentioned a number of times in this section. It is one of the largest and most widely used lipid databases in the lipidomics community. Originally started in 2007 by Edward Dennis as part of the LIPID MAPS consortium this database includes more than 39 000 lipid structures along with detailed information about lipid nomenclature and lipid classification schemes. It also contains a number of useful tools for lipid rendering, as well as lipid concentration data as measured in several biological systems and a number of reference MS/MS spectra for common lipids. While not exclusively dedicated to lipids, the HMDB[60] provides a great deal of detailed information about all known human metabolites (including lipids). Each compound in the HMDB contains detailed chemical descriptions, chemical taxonomy data, structural and functional ontology, physical and biological properties as well experimentally measured or computationally predicted NMR, GC-MS and MS/MS spectra. Moreover, with the latest update of the HMDB, the vast majority (>90%) of metabolites in the database are actually lipids (>90 000 lipid molecules), making the HMDB a larger resource for lipids than even LIPID MAPS.

Another useful lipid resource is the AOCS Lipid Library (http://lipidlibrary.co.uk), which contains an enormous amount of well-written and well-researched material on lipids and lipid derivatives, their structures, their biochemistry, their chemistry and their analytical measurement. The Lipid Library is continuously being updated and it is particularly useful for individuals who are new to lipid science or who want to learn more about the rapidly expanding field of lipidomics. Another lesser-known lipid resource is LipidBank.[134] While no longer regularly updated, LipidBank actually provided the original source material for the LIPID MAPS database and LipidBank still contains a tremendous amount of very useful information about a wide range of lipids found in microbes, plants, insects and animals.

### 3.7.2 Shotgun Lipidomics

Software tools for shotgun lipidomics can generally be divided into two groups: (1) those that can be used for lipid identification only and (2) those that can perform both lipid identification and quantification. There are two shotgun lipidomics software packages that fall into the first category: ALEX[135] and LipidInspector[136] and three software packages that fall into the second category: LipidXplorer,[137,138] AMDMS-SL[139] and LIMSA.[140]

We will briefly describe ALEX and LipidInspector and then go into more detail describing the other category programs (LipidXplorer, AMDMS-SL and LIMSA) as these are more widely used. ALEX (analysis of lipid experiments) was originally developed to assist in the processing, management and visualization of shotgun lipidomics datasets.[135] It supports the automated identification of lipids and the export of lipid intensity data directly from proprietary mass spectral data files. Unfortunately, ALEX does not perform lipid quantification (either relative or absolute). LipidInspector is another lipid-annotation-only software package. Written in Python, LipidInspector annotates lipids based on the information collected from shotgun DDA experiments.[136] Lipid identification is performed based on the list of fragments of head groups and fatty acid moieties. The list of $m/z$ values for these specific fragments and their neutral losses is used to identify the fragment ions whose $m/z$ values match to within a specified mass tolerance, to determine the lipid class of the fragmented precursor.

LipidXplorer[137,138] is one of the most widely used tools in shotgun lipidomics as it supports both lipid identification and lipid quantification. LipidXplorer is able to process data collected on any type of mass spectrometer and does not use any kind of database for lipid identification. Instead lipid identification routines are user-defined through a structured query language called the declarative molecular fragmentation query language (MFQL). LipidXplorer identifies and quantifies molecular species of any ionizable lipid class by considering any known or assumed molecular fragmentation pathway independently of any reference mass spectral database. This independence allows LipidXplorer to avoid any user biases. It also allows it to recognize uncommon lipid classes or species, such as those consisting of unconventional fatty acid moieties or lipid classes that may produce suboptimal spectra.

Two other lesser-known shotgun lipidomics programs are AMDMS-SL and LIMSA. AMDMS-SL (Automation of MultiDimensional Mass Spectrometry-based Shotgun Lipidomics)[139] is a shotgun lipidomics software package designed for both the identification and relative quantification of individual lipid species through array analysis of multi-dimensional mass spectrometry-based shotgun lipidomics (MDMS-SL). These are lipid data that are acquired directly from lipid extracts after direct infusion and intra-source separation. AMDMS-SL performs its lipid identification from a built-in database of 36 000 different lipid species constructed using known lipid building blocks. LIMSA (LIpid Mass Spectrum Analysis) was one of the first software tools developed specifically for the quantitative analysis of lipid MS spectra. It has a convenient user interface and can be applied to any type of data collected from either LC-MS/MS lipid experiments or shotgun lipidomics experiments.[140] LIMSA finds and integrates peaks in a mass spectrum, then matches the peaks with a user-supplied list of expected lipids. LIMSA also corrects for overlaps in their isotopic patterns and quantifies the identified lipid species according to internal standards.

### 3.7.3 Imaging Lipidomics of Mass Spectrometry Imaging

Mass spectrometry imaging (MSI) is growing rapidly in popularity. It offers a powerful way of performing precise, high-resolution chemical imaging of biological materials. In most cases the only molecules that can be routinely and robustly detected by MSI are lipids and so this is why it is sometimes called Imaging Lipidomics. MSI experiments generate very complex, multidimensional datasets that require robust and sophisticated data analysis tools. The majority of available software programs for MSI support data visualization and data reprocessing as well spectral annotation of the measured metabolites. Some of these software tools also provide utilities for statistical analysis and the statistical selection of discriminating molecules.

One of the more popular programs in the MSI field is Cardinal.[141] Cardinal is an R-based package designed to handle MALDI and DESI MSI data. It is designed to support both image segmentation and image classification. Cardinal works by partitioning a tissue section into regions of homogeneous chemical composition, selecting the number of segments and the most informative subset of (usually lipid) ions. Once this is done, the locations on the tissue associated with pre-defined chemical classes are assigned, the subset of informative ions are selected and classification error by (cross-) validation are estimated.

Another MSI package developed specifically for lipid imaging is massPix. It is an R-based program that supports the annotation and interpretation of data from MSI studies of lipids.[142] massPix supports many aspects of the standard MSI workflow including data processing (single ion images can be created), data analysis (multivariate statistics) and annotation (putative lipid IDs based on accurate mass matching against various lipid libraries). massPix also supports the classification or grouping of tissue regions with high spectral similarly. This clustering can be carried out by principal components analysis (PCA) or *k*-means clustering and is particularly useful for conducting visual comparisons between multiple MSI images of multiple tissue sections.

Yet another newly developed open source MSI software package is MSiReader.[143] This tool was written in MATLAB but the standalone version does not require a MATLAB license.[143] Originally introduced in 2013, MSiReader has recently undergone substantial updates. These updates have expanded the utility of MSiReader beyond simple visualization of molecular distributions to include the ability to analyze polarity switching files (without the need for data parsing), to perform image overlays, and to generate mass measurement accuracy (MMA) heatmaps across the entire image for quality assurance assessments. The latest version of MSiReader also includes a feature called MSiQuantification. With this tool users can calculate absolute concentrations from quantification MSI experiments.

SpectralAnalysis[144] is another popular MSI package that covers the entire MSI analysis workflow, from pre-processing of raw data to multivariate analysis for data sets acquired from single experiments to large multi-instrument, multi-center studies. It offers a wide range of methods to smooth, perform baseline correction, normalize, and generate images. The SpectralAnalysis software package is also capable of handling multiple spectral imaging modalities, each of which capture different information about the sample. Additionally, SpectralAnalysis supports a wide range of multivariate statistical analyses including principal component analysis (PCA), non-negative matrix factorization (NMF), maximum autocorrelation factor analysis (MAF), and probabilistic latent semantic analysis (PLSA). Another program worth mentioning is called Clustergrammer, which is not a stand-alone program but a web server.[145] Although, Clustergrammer is not a tool exclusively devoted to handling or manipulating MSI data, it is particularly useful for visualizing biological changes in a heatmap format as it is able to retain the high-dimensionality of the original data (including MSI data). In addition to these open access and open source packages, there are also a number of commercial IMS software packages including MALDIVision from Premier Biosoft, SCiLS from Bruker, ImageQuest from Thermo Fisher and High Definition Imaging from Waters. Many of these commercial packages are quite excellent and integrate readily with the instruments for which they were designed.

## 3.8 Conclusion

This chapter has described the software tools, data processing software, databases and workflows needed to perform and analyze both general metabolomics and more specialized lipidomics experiments using LC-MS, GC-MS and CE-MS technologies. In particular, it has provided a brief introduction to metabolomics/lipidomics along with a general overview of the different 'flavours' of metabolomics, including the critical distinctions between targeted and untargeted metabolomics. In addition to giving a broad overview of metabolomics and lipidomics, this chapter was also designed to provide more detailed descriptions and a short background of the different MS-based technologies and MS platforms that are commonly used for metabolomics. These preliminary sections were intended to provide the necessary baseline knowledge for readers to understand the general theory and common terminology associated with MS-based metabolomics. Individual sections were then provided that elaborated on the workflows, software, data processing and database needs for each of the main MS platforms commonly used for modern MS-based metabolomic analysis, including LC-MS, GC-MS, CE-MS and lipidomics. The focus in these sections was, in almost all cases, on free, open-source or open-access software or databases, although a number of commercial products were also mentioned. The intent of this chapter was to provide readers with a general understanding of how to conduct MS-based metabolomics experiments and how to handle, process or interpret the resulting data. Detailed procedures, standard operating protocols (SOPs) or 'recipes' for metabolomic data analysis are best obtained by reading the original papers or by looking at various protocol papers that have been referenced in this chapter.

## References


1. D. S. Wishart, *Am. J. Transplant.*, 2005, **5**, 2814–2820.
2. D. S. Wishart, D. Tzur, C. Knox, R. Eisner, A. C. Guo, N. Young, D. Cheng, K. Jewell, D. Arndt, S. Sawhney, C. Fung, L. Nikolai, M. Lewis, M. A. Coutouly, I. Forsythe, P. Tang, S. Shrivastava, K. Jeroncic, P. Stothard, G. Amegbey, D. Block, D. D. Hau, J. Wagner, J. Miniaci, M. Clements, M. Gebremedhin, N. Guo, Y. Zhang, G. E. Duggan, G. D. Macinnis, A. M. Weljie, R. Dowlatabadi, F. Bamforth, D. Clive, R. Greiner, L. Li, T. Marrie, B. D. Sykes, H. J. Vogel and L. Querengesser, *Nucleic Acids Res.*, 2007, **35**, D521–D526.
3. C. M. Slupsky, K. N. Rankin, J. Wagner, H. Fu, D. Chang, A. M. Weljie, E. J. Saude, B. Lix, D. J. Adamko, S. Shah, R. Greiner, B. D. Sykes and T. J. Marrie, *Anal. Chem.*, 2007, **79**, 6995–7004.
4. O. Fiehn, *Plant Mol. Biol.*, 2002, **48**, 155–171.
5. E. Holmes, I. D. Wilson and J. K. Nicholson, *Cell*, 2008, **134**, 714–717.
6. S. Kim, J. Kim, E. J. Yun and K. H. Kim, *Curr. Opin. Biotechnol.*, 2016, **37**, 16–23.
7. M. R. Viant, *Mol. BioSyst.*, 2008, **4**, 980–986.

8. D. S. Wishart, *Nat. Rev. Drug Discovery*, 2016, **15**, 473–484.
9. J. Xia, D. I. Broadhurst, M. Wilson and D. S. Wishart, *Metabolomics*, 2013, **9**, 280–299.
10. A. C. Schrimpe-Rutledge, S. G. Codreanu, S. D. Sherrod and J. A. McLean, *J. Am. Soc. Mass Spectrom.*, 2016, **27**, 1897–1905.
11. M. R. Wenk, *Nat. Rev. Drug Discovery*, 2005, **4**, 594–610.
12. J. Hartler, A. Triebl, A. Ziegl, M. Trötzmüller, G. N. Rechberger, O. A. Zeleznik, K. A. Zierler, F. Torta, A. Cazenave-Gassiot, M. R. Wenk, A. Fauland, C. E. Wheelock, A. M. Armando, O. Quehenberger, Q. Zhang, M. J. O. Wakelam, G. Haemmerle, F. Spener, H. C. Köfeler and G. G. Thallinger, *Nat. Methods*, 2017, **14**, 1171–1174.
13. O. Quehenberger, S. Dahlberg-Wright, J. Jiang, A. M. Armando and E. A. Dennis, *J. Lipid Res.*, 2018, **59**, 2436–2445.
14. W. B. Dunn, I. D. Wilson, A. W. Nicholls and D. Broadhurst, *Bioanalysis*, 2012, **4**, 2249–2264.
15. S. Klein and E. Heinzle, *Wiley Interdiscip. Rev.: Syst. Biol. Med.*, 2012, **4**, 261–272.
16. A. Buck, M. Aichler, K. Huber and A. Walch, *Adv. Cancer Res.*, 2017, **134**, 117–132.
17. D. S. Wishart, *Methods Mol. Biol.*, 2010, **593**, 283–313.
18. J. Xia and D. S. Wishart, *Nucleic Acids Res.*, 2010, **38**, W71–W77.
19. W. B. Dunn, N. J. Bailey and H. E. Johnson, *Analyst*, 2005, **130**, 606–625.
20. Z. Z. Fang and F. J. Gonzalez, *Arch. Toxicol.*, 2014, **88**, 1491–1502.
21. S. Naz, D. C. Moreira dos Santos, A. García and C. Barbas, *Bioanalysis*, 2014, **6**, 1657–1677.
22. R. Hall, M. Beale, O. Fiehn, N. Hardy, L. Sumner and R. Bino, *Plant Cell*, 2002, **14**, 1437–1440.
23. J. A. Olivares, N. T. Nguyen, C. R. Yonker and R. D. Smith, *Anal. Chem.*, 1987, **59**, 1230–1232.
24. J. J. Corr and J. F. Anacleto, *Anal. Chem.*, 1996, **68**, 2155–2163.
25. K. Sasaki, H. Sagawa, M. Suzuki, H. Yamamoto, M. Tomita, T. Soga and Y. Ohashi, *Anal. Chem.*, 2019, **91**, 1295–1301.
26. C. X. Zhang and W. Thormann, *Anal. Chem.*, 1996, **68**, 2523–2532.
27. N. L. Kuehnbaum, A. Kormendi and P. Britz-McKibbin, *Anal. Chem.*, 2013, **85**, 10664–10669.
28. R. Ramautar, G. W. Somsen and G. J. de Jong, *Electrophoresis*, 2019, **40**, 165–179.
29. S. Bouatra, F. Aziat, R. Mandal, A. C. Guo, M. R. Wilson, C. Knox, T. C. Bjorndahl, R. Krishnamurthy, F. Saleem, P. Liu, Z. T. Dame, J. Poelzer, J. Huynh, F. S. Yallou, N. Psychogios, E. Dong, R. Bogumil, C. Roehring and D. S. Wishart, *PLoS One*, 2013, **8**, e73076.
30. D. S. Wishart, *Bioanalysis*, 2011, **3**, 1769–1782.
31. T. Cajka and O. Fiehn, *TrAC, Trends Anal. Chem.*, 2014, **61**, 192–206.
32. N. J. Andreas, M. J. Hyde, M. Gomez-Romero, M. A. Lopez-Gonzalvez, A. Villaseñor, A. Wijeyesekera, C. Barbas, N. Modi, E. Holmes and I. Garcia-Perez, *Electrophoresis*, 2015, 2269–2285.

33. L. W. Sumner, A. Amberg, D. Barrett, M. H. Beale, R. Beger, C. A. Daykin, T. W. Fan, O. Fiehn, R. Goodacre, J. L. Griffin, T. Hankemeier, N. Hardy, J. Harnly, R. Higashi, J. Kopka, A. N. Lane, J. C. Lindon, P. Marriott, A. W. Nicholls, M. D. Reily, J. J. Thaden and M. R. Viant, *Metabolomics*, 2007, **3**, 211–221.
34. C. A. Smith, E. J. Want, G. O'Maille, R. Abagyan and G. Siuzdak, *Anal. Chem.*, 2006, **78**, 779–787.
35. R. Tautenhahn, G. J. Patti, D. Rinehart and G. Siuzdak, *Anal. Chem.*, 2012, **84**, 5035–5039.
36. D. S. Wishart, R. F. Boyko and B. D. Sykes, *Comput. Appl. Biosci.*, 1994, **10**, 687–688.
37. S. Aiche, T. Sachsenberg, E. Kenar, M. Walzer, B. Wiswedel, T. Kristl, M. Boyles, A. Duschl, C. G. Huber, M. R. Berthold, K. Reinert and O. Kohlbacher, *Proteomics*, 2015, **15**, 1443–1447.
38. H. L. Röst, T. Sachsenberg, S. Aiche, C. Bielow, H. Weisser, F. Aicheler, S. Andreotti, H. C. Ehrlich, P. Gutenbrunner, E. Kenar, X. Liang, S. Nahnsen, L. Nilse, J. Pfeuffer, G. Rosenberger, M. Rurik, U. Schmitt, J. Veit, M. Walzer, D. Wojnar, W. E. Wolski, O. Schilling, J. S. Choudhary, L. Malmström, R. Aebersold, K. Reinert and O. Kohlbacher, *Nat. Methods*, 2016, **13**, 741–748.
39. M. Katajamaa, J. Miettinen and M. Oresic, *Bioinformatics*, 2006, **22**, 634–636.
40. T. Pluskal, S. Castillo, A. Villar-Briones and M. Oresic, *BMC Bioinf.*, 2010, **11**, 395.
41. H. Tsugawa, T. Cajka, T. Kind, Y. Ma, B. Higgins, K. Ikeda, M. Kanazawa, J. VanderGheynst, O. Fiehn and M. Arita, *Nat. Methods*, 2015, **12**, 523–526.
42. A. Lommen, *Anal. Chem.*, 2009, **81**, 3079–3086.
43. S. E. Stein, *J. Am. Soc. Mass Spectrom.*, 1999, **10**, 770–781.
44. M. R. Meyer, F. T. Peters and H. H. Maurer, *Clin. Chem.*, 2010, **56**, 575–584.
45. Y. M. Tikunov, S. Laptenok, R. D. Hall, A. Bovy and R. C. de Vos, *Metabolomics*, 2012, **8**, 714–718.
46. C. Kuhl, R. Tautenhahn, C. Böttcher, T. R. Larson and S. Neumann, *Anal. Chem.*, 2012, **84**, 283–289.
47. C. D. Dehaven, A. M. Evans, H. Dai and K. A. Lawton, *J. Cheminf.*, 2010, **2**, 9.
48. H. A. L. Kiers, J. M. F. ten Berge and R. Bro, *J. Chemom.*, 1999, **13**, 275–294.
49. C. D. Broeckling, F. A. Afsar, S. Neumann, A. Ben-Hur and J. E. Prenni, *Anal. Chem.*, 2014, **86**, 6812–6817.
50. H. Tsugawa, T. Kind, R. Nakabayashi, D. Yukihira, W. Tanaka, T. Cajka, K. Saito, O. Fiehn and M. Arita, *Anal. Chem.*, 2016, **88**, 7946–7958.
51. L. Patiny and A. Borel, *J. Chem. Inf. Model.*, 2013, **53**, 1223–1228.
52. S. Böcker, M. C. Letzel, Z. Lipták and A. Pervukhin, *Bioinformatics*, 2009, **25**, 218–224.
53. R. Tautenhahn, C. Böttcher and S. Neumann, *BMC Bioinf.*, 2008, **9**, 504.
54. N.-P. Nielsen, J. M. Carstensen and J. Smedsgaard, *J. Chromatogr. A*, 1998, **805**, 17–35.

55. G. Tomasi, F. van den Berg and C. Andersson, *J. Chemom.*, 2004, **18**, 231–241.
56. S. Y. Wang, T. J. Ho, C. H. Kuo and Y. J. Tseng, *Bioinformatics*, 2010, **26**, 2338–2339.
57. N. Hoffmann and J. Stoye, *Bioinformatics*, 2009, **25**, 2080–2081.
58. M. E. Monroe, N. Tolić, N. Jaitly, J. L. Shaw, J. N. Adkins and R. D. Smith, *Bioinformatics*, 2007, **23**, 2021–2023.
59. J. Xia, N. Psychogios, N. Young and D. S. Wishart, *Nucleic Acids Res.*, 2009, **37**, W652–W660.
60. D. S. Wishart, Y. D. Feunang, A. Marcu, A. C. Guo, K. Liang, R. Vázquez-Fresno, T. Sajed, D. Johnson, C. Li, N. Karu, Z. Sayeeda, E. Lo, N. Assempour, M. Berjanskii, S. Singhal, D. Arndt, Y. Liang, H. Badran, J. Grant, A. Serra-Cayuela, Y. Liu, R. Mandal, V. Neveu, A. Pon, C. Knox, M. Wilson, C. Manach and A. Scalbert, *Nucleic Acids Res.*, 2018, **46**, D608–D617.
61. C. A. Smith, G. O'Maille, E. J. Want, C. Qin, S. A. Trauger, T. R. Brandon, D. E. Custodio, R. Abagyan and G. Siuzdak, *Ther. Drug Monit.*, 2005, **27**, 747–751.
62. M. Kanehisa, Y. Sato, M. Kawashima, M. Furumichi and M. Tanabe, *Nucleic Acids Res.*, 2016, **44**, D457–D462.
63. T. Sajed, A. Marcu, M. Ramirez, A. Pon, A. C. Guo, C. Knox, M. Wilson, J. R. Grant, Y. Djoumbou and D. S. Wishart, *Nucleic Acids Res.*, 2016, **44**, D495–D501.
64. M. Ramirez-Gaona, A. Marcu, A. Pon, A. C. Guo, T. Sajed, N. A. Wishart, N. Karu, Y. Djoumbou Feunang, D. Arndt and D. S. Wishart, *Nucleic Acids Res.*, 2017, **45**, D440–D445.
65. S. Kim, J. Chen, T. Cheng, A. Gindulyte, J. He, S. He, Q. Li, B. A. Shoemaker, P. A. Thiessen, B. Yu, L. Zaslavsky, J. Zhang and E. E. Bolton, *Nucleic Acids Res.*, 2019, **47**, D1102–D1109.
66. M. Sud, E. Fahy, D. Cotter, A. Brown, E. A. Dennis, C. K. Glass, A. H. Merrill, R. C. Murphy, C. R. Raetz, D. W. Russell and S. Subramaniam, *Nucleic Acids Res.*, 2007, **35**, D527–D532.
67. T. Kind and O. Fiehn, *BMC Bioinf.*, 2007, **8**, 105.
68. R. R. da Silva, P. C. Dorrestein and R. A. Quinn, *Proc. Natl. Acad. Sci. U. S. A.*, 2015, **112**, 12549–12550.
69. T. Kind, H. Tsugawa, T. Cajka, Y. Ma, Z. Lai, S. S. Mehta, G. Wohlgemuth, D. K. Barupal, M. R. Showalter, M. Arita and O. Fiehn, *Mass Spectrom. Rev.*, 2018, **37**, 513–532.
70. S. Neumann and S. Böcker, *Anal. Bioanal. Chem.*, 2010, **398**, 2779–2788.
71. A. Vaniya and O. Fiehn, *TrAC, Trends Anal. Chem.*, 2015, **69**, 52–61.
72. M. Wang, J. J. Carver, V. V. Phelan, L. M. Sanchez, N. Garg, Y. Peng, D. D. Nguyen, J. Watrous, C. A. Kapono, T. Luzzatto-Knaan, C. Porto, A. Bouslimani, A. V. Melnik, M. J. Meehan, W. T. Liu, M. Crüsemann, P. D. Boudreau, E. Esquenazi, M. Sandoval-Calderón, R. D. Kersten, L. A. Pace, R. A. Quinn, K. R. Duncan, C. C. Hsu, D. J. Floros, R. G. Gavilan, K. Kleigrewe, T. Northen, R. J. Dutton, D. Parrot, E. E. Carlson, B. Aigle, C. F. Michelsen, L. Jelsbak, C. Sohlenkamp, P. Pevzner, A. Edlund, J. McLean, J. Piel, B. T. Murphy, L. Gerwick, C. C. Liaw, Y. L. Yang, H. U. Humpf, M.

Maansson, R. A. Keyzers, A. C. Sims, A. R. Johnson, A. M. Sidebottom, B. E. Sedio, A. Klitgaard, C. B. Larson, C. A. B. P, D. Torres-Mendoza, D. J. Gonzalez, D. B. Silva, L. M. Marques, D. P. Demarque, E. Pociute, E. C. O'Neill, E. Briand, E. J. N. Helfrich, E. A. Granatosky, E. Glukhov, F. Ryffel, H. Houson, H. Mohimani, J. J. Kharbush, Y. Zeng, J. A. Vorholt, K. L. Kurita, P. Charusanti, K. L. McPhail, K. F. Nielsen, L. Vuong, M. Elfeki, M. F. Traxler, N. Engene, N. Koyama, O. B. Vining, R. Baric, R. R. Silva, S. J. Mascuch, S. Tomasi, S. Jenkins, V. Macherla, T. Hoffman, V. Agarwal, P. G. Williams, J. Dai, R. Neupane, J. Gurr, A. M. C. Rodríguez, A. Lamsa, C. Zhang, K. Dorrestein, B. M. Duggan, J. Almaliti, P. M. Allard, P. Phapale, L. F. Nothias, T. Alexandrov, M. Litaudon, J. L. Wolfender, J. E. Kyle, T. O. Metz, T. Peryea, D. T. Nguyen, D. VanLeer, P. Shinn, A. Jadhav, R. Müller, K. M. Waters, W. Shi, X. Liu, L. Zhang, R. Knight, P. R. Jensen, B. O. Palsson, K. Pogliano, R. G. Linington, M. Gutiérrez, N. P. Lopes, W. H. Gerwick, B. S. Moore, P. C. Dorrestein and N. Bandeira, *Nat. Biotechnol.*, 2016, **34**, 828–837.

73. F. Allen, A. Pon, M. Wilson, R. Greiner and D. Wishart, *Nucleic Acids Res.*, 2014, **42**, W94–W99.
74. L. Ridder, J. J. van der Hooft, S. Verhoeven, R. C. de Vos, R. van Schaik and J. Vervoort, *Rapid Commun. Mass Spectrom.*, 2012, **26**, 2461–2471.
75. C. Ruttkies, E. L. Schymanski, S. Wolf, J. Hollender and S. Neumann, *J. Cheminf.*, 2016, **8**, 3.
76. T. Kind, K. H. Liu, D. Y. Lee, B. DeFelice, J. K. Meissen and O. Fiehn, *Nat. Methods*, 2013, **10**, 755–758.
77. K. Dührkop, H. Shen, M. Meusel, J. Rousu and S. Böcker, *Proc. Natl. Acad. Sci. U. S. A.*, 2015, **112**, 12580–12585.
78. C. Brouard, H. Shen, K. Dührkop, F. d'Alché-Buc, S. Böcker and J. Rousu, *Bioinformatics*, 2016, **32**, i28–i36.
79. J. Han, R. Higgins, M. D. Lim, K. Atkinson, J. Yang, K. Lin and C. H. Borchers, *Anal. Chim. Acta*, 2018, **1037**, 177–187.
80. J. Zheng, R. Mandal and D. S. Wishart, *Anal. Chim. Acta*, 2018, **1037**, 159–167.
81. K. Hiller, J. Hangebrauk, C. Jäger, J. Spura, K. Schreiber and D. Schomburg, *Anal. Chem.*, 2009, **81**, 3429–3439.
82. S. O'Callaghan, D. P. De Souza, A. Isaac, Q. Wang, L. Hodkinson, M. Olshansky, T. Erwin, B. Appelbe, D. L. Tull, U. Roessner, A. Bacic, M. J. McConville and V. A. Likić, *BMC Bioinf.*, 2012, **13**, 115.
83. Y. Ni, M. Su, Y. Qiu, W. Jia and X. Du, *Anal. Chem.*, 2016, **88**, 8802–8811.
84. X. Du and S. H. Zeisel, *Comput. Struct. Biotechnol. J.*, 2013, **4**, e201301013.
85. J. Kopka, N. Schauer, S. Krueger, C. Birkemeyer, B. Usadel, E. Bergmüller, P. Dörmann, W. Weckwerth, Y. Gibon, M. Stitt, L. Willmitzer, A. R. Fernie and D. Steinhauser, *Bioinformatics*, 2005, **21**, 1635–1638.
86. T. Kind, G. Wohlgemuth, D. Y. Lee, Y. Lu, M. Palazoglu, S. Shahbaz and O. Fiehn, *Anal. Chem.*, 2009, **81**, 10038–10048.
87. F. Allen, A. Pon, R. Greiner and D. Wishart, *Anal. Chem.*, 2016, **88**, 7689–7697.

88. J. Xia, I. V. Sinelnikov, B. Han and D. S. Wishart, *Nucleic Acids Res.*, 2015, **43**, W251–W257.
89. A. García, J. Godzien, Á. López-Gonzálvez and C. Barbas, *Bioanalysis*, 2017, **9**, 99–130.
90. M. Sugimoto, M. Kawakami, M. Robert, T. Soga and M. Tomita, *Curr. Bioinf.*, 2012, **7**, 96–108.
91. E. Allard, D. Bäckström, R. Danielsson, P. J. Sjöberg and J. Bergquist, *Anal. Chem.*, 2008, **80**, 8946–8955.
92. C. Ibáñez, C. Simó, V. García-Cañas, A. Gómez-Martínez, J. A. Ferragut and A. Cifuentes, *Electrophoresis*, 2012, **33**, 2328–2336.
93. R. Ramautar, E. Nevedomskaya, O. A. Mayboroda, A. M. Deelder, I. D. Wilson, H. G. Gika, G. A. Theodoridis, G. W. Somsen and G. J. de Jong, *Mol. BioSyst.*, 2011, **7**, 194–199.
94. J. M. Busnel, B. Schoenmaker, R. Ramautar, A. Carrasco-Pancorbo, C. Ratnayake, J. S. Feitelson, J. D. Chapman, A. M. Deelder and O. A. Mayboroda, *Anal. Chem.*, 2010, **82**, 9476–9483.
95. C. Ibáñez, C. Simó, P. J. Martín-Álvarez, M. Kivipelto, B. Winblad, A. Cedazo-Mínguez and A. Cifuentes, *Anal. Chem.*, 2012, **84**, 8532–8540.
96. A. Hirayama, E. Nakashima, M. Sugimoto, S.-i. Akiyama, W. Sato, S. Maruyama, S. Matsuo, M. Tomita, Y. Yuzawa and T. Soga, *Anal. Bioanal. Chem.*, 2012, **404**, 3101–3109.
97. T. Kimura, K. Yasuda, R. Yamamoto, T. Soga, H. Rakugi, T. Hayashi and Y. Isaka, *Sci. Rep.*, 2016, **6**, 26138.
98. K. Maekawa, A. Hirayama, Y. Iwata, Y. Tajima, T. Nishimaki-Mogami, S. Sugawara, N. Ueno, H. Abe, M. Ishikawa, M. Murayama, Y. Matsuzawa, H. Nakanishi, K. Ikeda, M. Arita, R. Taguchi, N. Minamino, S. Wakabayashi, T. Soga and Y. Saito, *J. Mol. Cell. Cardiol.*, 2013, **59**, 76–85.
99. X. Domingo-Almenara, J. R. Montenegro-Burke, H. P. Benton and G. Siuzdak, *Anal. Chem.*, 2018, **90**, 480–489.
100. M. Ciborowski, E. Adamska, M. Rusak, J. Godzien, J. Wilk, A. Citko, W. Bauer, M. Gorska and A. Kretowski, *Electrophoresis*, 2015, **36**, 2286–2293.
101. N. L. Kuehnbaum, J. B. Gillen, A. Kormendi, K. P. Lam, A. DiBattista, M. J. Gibala and P. Britz-McKibbin, *Electrophoresis*, 2015, **36**, 2226–2236.
102. M. Sugimoto, D. T. Wong, A. Hirayama, T. Soga and M. Tomita, *Metabolomics*, 2010, **6**, 78–95.
103. M. Sugimoto, A. Hirayama, M. Robert, S. Abe, T. Soga and M. Tomita, *Electrophoresis*, 2010, **31**, 2311–2318.
104. M. Tsuruoka, J. Hara, A. Hirayama, M. Sugimoto, T. Soga, W. R. Shankle and M. Tomita, *Electrophoresis*, 2013, **34**, 2865–2872.
105. Y. Torii, Y. Kawano, H. Sato, K. Sasaki, T. Fujimori, J.-i. Kawada, O. Takikawa, C. K. Lim, G. J. Guillemin, Y. Ohashi and Y. Ito, *Metabolomics*, 2016, **12**, 84.
106. K. Uppal, D. I. Walker and D. P. Jones, *Anal. Chem.*, 2017, **89**, 1063–1067.
107. A. Hirayama, K. Kami, M. Sugimoto, M. Sugawara, N. Toki, H. Onozuka, T. Kinoshita, N. Saito, A. Ochiai, M. Tomita, H. Esumi and T. Soga, *Cancer Res.*, 2009, **69**, 4918–4925.

108. M. G. Kok, M. M. Ruijken, J. R. Swann, I. D. Wilson, G. W. Somsen and G. J. de Jong, *Anal. Bioanal. Chem.*, 2013, **405**, 2585–2594.
109. M. Sugimoto, A. Hirayama, T. Ishikawa, M. Robert, R. Baran, K. Uehara, K. Kawai, T. Soga and M. Tomita, *Metabolomics*, 2010, **6**, 27–41.
110. J. C. Reijeng, J. H. Martens, A. Giuliani and M. Chiari, *J. Chromatogr. B: Anal. Technol. Biomed. Life Sci.*, 2002, **770**, 45–51.
111. J. Godzien, Á. López-Gonzálvez,A. García and C. Barbas, in *The Handbook of Metabolic Phenotyping*, ed. J. C. Lindon, J. K. Nicholson and E. Holmes, Elsevier, 2019, pp. 171–204.
112. J. Godzien, E. G. Armitage, S. Angulo, M. P. Martinez-Alcazar, V. Alonso-Herranz, A. Otero, A. Lopez-Gonzalvez and C. Barbas, *Electrophoresis*, 2015, **36**, 2188–2195.
113. M. Sugimoto, S. Kikuchi, M. Arita, T. Soga, T. Nishioka and M. Tomita, *Anal. Chem.*, 2005, **77**, 78–84.
114. R. Lee, A. S. Ptolemy, L. Niewczas and P. Britz-McKibbin, *Anal. Chem.*, 2007, **79**, 403–415.
115. S. Vilar, G. Cozza and S. Moro, *Curr. Top. Med. Chem.*, 2008, **8**, 1555–1572.
116. X. Y. Qin, F. Wei, M. Tanokura, N. Ishibashi, M. Shimizu, H. Moriwaki and S. Kojima, *PLoS One*, 2013, **8**, e82860.
117. Y. Y. Zhao, H. L. Wang, X. L. Cheng, F. Wei, X. Bai, R. C. Lin and N. D. Vaziri, *Sci. Rep.*, 2015, **5**, 12936.
118. V. González-Ruiz, Y. Gagnebin, N. Drouin, S. Codesido, S. Rudaz and J. Schappler, *Electrophoresis*, 2018, 1222–1232.
119. T. Saito, M. Sugimoto, K. Igarashi, K. Saito, L. Shao, T. Katsumi, K. Tomita, C. Sato, K. Okumoto, Y. Nishise, H. Watanabe, M. Tomita, Y. Ueno and T. Soga, *Metabolism*, 2013, **62**, 1577–1586.
120. E. V. Alves-Ferreira, J. S. Toledo, A. H. De Oliveira, T. R. Ferreira, P. C. Ruy, C. F. Pinzan, R. F. Santos, V. Boaventura, D. Rojo, Á. López-Gonzálvez, J. C. Rosa, C. Barbas, M. Barral-Netto, A. Barral and A. K. Cruz, *PLoS Neglected Trop. Dis.*, 2015, **9**, e0004018.
121. S. Azab, R. Ly and P. Britz-McKibbin, *Anal. Chem.*, 2019, **91**, 2329–2336.
122. J. B. Powers and S. R. Campagna, *J. Am. Soc. Mass Spectrom.*, 2019, 2369–2379.
123. L. Goracci, S. Tortorella, P. Tiberi, R. M. Pellegrino, A. Di Veroli, A. Valeri and G. Cruciani, *Anal. Chem.*, 2017, **89**, 6257–6264.
124. P. D. Hutchins, J. D. Russell and J. J. Coon, *Cell Syst.*, 2018, **6**, 621–625. e625.
125. J. P. Koelmel, N. M. Kroeger, C. Z. Ulmer, J. A. Bowden, R. E. Patterson, J. A. Cochran, C. W. W. Beecher, T. J. Garrett and R. A. Yost, *BMC Bioinf.*, 2017, **18**, 331.
126. J. P. Koelmel, J. A. Cochran, C. Z. Ulmer, A. J. Levy, R. E. Patterson, B. C. Olsen, R. A. Yost, J. A. Bowden and T. J. Garrett, *BMC Bioinf.*, 2019, **20**, 217.
127. J. E. Kyle, K. L. Crowell, C. P. Casey, G. M. Fujimoto, S. Kim, S. E. Dautel, R. D. Smith, S. H. Payne and T. O. Metz, *Bioinformatics*, 2017, **33**, 1744–1746.

128. D. Meng, Q. Zhang, X. Gao, S. Wu and G. Lin, *Rapid Commun. Mass Spectrom.*, 2014, **28**, 981–985.
129. M. A. Kochen, M. C. Chambers, J. D. Holman, A. I. Nesvizhskii, S. T. Weintraub, J. T. Belisle, M. N. Islam, J. Griss and D. L. Tabb, *Anal. Chem.*, 2016, **88**, 5733–5741.
130. Z. Ni, G. Angelidou, M. Lange, R. Hoffmann and M. Fedorova, *Anal. Chem.*, 2017, **89**, 8800–8807.
131. Z. Ni, G. Angelidou, R. Hoffmann and M. Fedorova, *Sci. Rep.*, 2017, **7**, 15138.
132. Y. Djoumbou-Feunang, A. Pon, N. Karu, J. Zheng, C. Li, D. Arndt, M. Gautam, F. Allen and D. S. Wishart, *Metabolites*, 2019, **9**, 72.
133. M. Witting, C. Ruttkies, S. Neumann and P. Schmitt-Kopplin, *PLoS One*, 2017, **12**, e0172311.
134. K. Watanabe, E. Yasugi and M. Oshima, *Trends Glycosci. Glycotechnol.*, 2000, **12**, 175–184.
135. P. Husen, K. Tarasov, M. Katafiasz, E. Sokol, J. Vogt, J. Baumgart, R. Nitsch, K. Ekroos and C. S. Ejsing, *PLoS One*, 2013, **8**, e79736.
136. D. Schwudke, J. Oegema, L. Burton, E. Entchev, J. T. Hannich, C. S. Ejsing, T. Kurzchalia and A. Shevchenko, *Anal. Chem.*, 2006, **78**, 585–595.
137. R. Herzog, D. Schwudke and A. Shevchenko, *Curr. Protoc. Bioinf.*, 2013, **43**, 14.12.11–14.12.30.
138. R. Herzog, K. Schuhmann, D. Schwudke, J. L. Sampaio, S. R. Bornstein, M. Schroeder and A. Shevchenko, *PLoS One*, 2012, **7**, e29851.
139. K. Yang, H. Cheng, R. W. Gross and X. Han, *Anal. Chem.*, 2009, **81**, 4356–4368.
140. P. Haimi, A. Uphoff, M. Hermansson and P. Somerharju, *Anal. Chem.*, 2006, **78**, 8324–8331.
141. K. D. Bemis, A. Harry, L. S. Eberlin, C. Ferreira, S. M. van de Ven, P. Mallick, M. Stolowitz and O. Vitek, *Bioinformatics*, 2015, **31**, 2418–2420.
142. N. J. Bond, A. Koulman, J. L. Griffin and Z. Hall, *Metabolomics*, 2017, **13**, 128.
143. M. T. Bokhart, M. Nazari, K. P. Garrard and D. C. Muddiman, *J. Am. Soc. Mass Spectrom.*, 2018, **29**, 8–16.
144. A. M. Race, A. D. Palmer, A. Dexter, R. T. Steven, I. B. Styles and J. Bunch, *Anal. Chem.*, 2016, **88**, 9451–9458.
145. N. F. Fernandez, G. W. Gundersen, A. Rahman, M. L. Grimes, K. Rikova, P. Hornbeck and A. Ma'ayan, *Sci. Data*, 2017, **4**, 170151.

CHAPTER 4

# *Proteomics*

MARC VAUDEL*[a,b]

[a]Center for Diabetes Research, Department of Clinical Science, University of Bergen, Norway; [b]Center for Medical Genetics and Molecular Medicine, Haukeland University Hospital, Bergen, Norway
*E-mail: marc.vaudel@uib.no

## 4.1 The Proteome: Dimensions, Scales, and Complexity

Proteins are major components of biological systems, acting collectively to achieve complex molecular tasks. As illustrated in Figure 4.1, the amino acid sequence of a protein is derived from the exons of a gene. The different sequences that can be obtained from the combination of the transcribed exons through alternative splicing are termed *isoforms*. Subsequently, the sequence of a given isoform is processed, yielding different forms of the original protein, called *proteoforms*.[1] The entire collection of all proteoforms in a biological system, *e.g.* an individual, an organ, a cell, or a subcellular compartment, is called *proteome*, and the scientific discipline studying the proteome is called *proteomics*. The study of the proteome of biological systems that do not originate from a single organism but rather a population of multiple species, *e.g.* when sampling an ecosystem or a microbiome, is referred to as *metaproteomics*.

The joint study of the genome and the proteome is called *proteogenomics*. As can be anticipated from Figure 4.1, genomic sequence variation directly impacts the protein, resulting in variation in expression, splicing, and amino

New Developments in Mass Spectrometry No. 8
Processing Metabolomics and Proteomics Data with Open Software: A Practical Guide
Edited by Robert Winkler

Published by the Royal Society of Chemistry, www.rsc.org

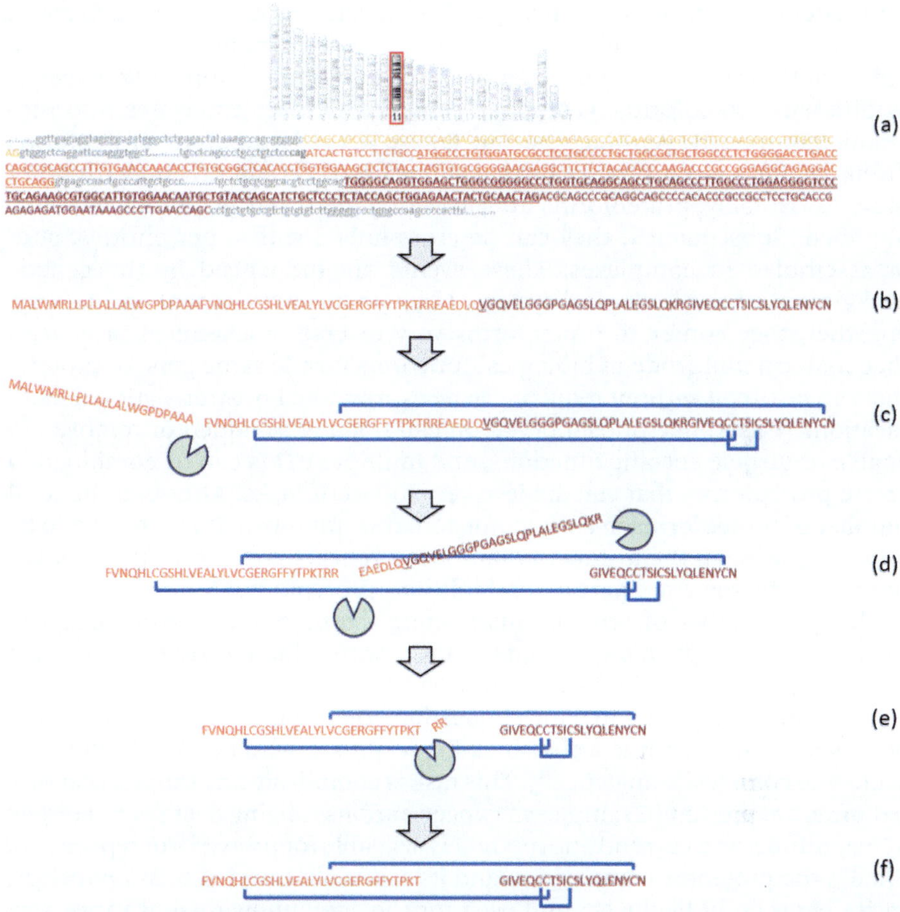


**Figure 4.1** The production of human insulin from the transcript INS-201 according to Ensembl release 97.[2] (a) The transcript is coded by three exons on chromosome eleven, coloured in yellow, orange, and dark red. The translated sequence is underlined. Sequence variation in the translated sequence can result in sequence variations in the amino acid sequence, and hence different proteoforms. (b) The sequence obtained after translation represents the raw proteoform of insulin, called preproinsulin, which requires post-translational maturation to obtain the mature form of insulin. Amino acids are coloured according to the coding exons and the residue overlapping splice site is underlined. (c) The signal peptide is cleaved, yielding proinsulin, and cysteines cross-linked by disulphide bonds, making a new proteoform of insulin. (d) Proteases cleave a large fragment of the sequence, the C-peptide. The C-peptide is often used as a proxy to measure insulin production. (e) Proteases cleave pairs of amino-acids, yielding the mature form of insulin. (f) The mature form of insulin consists of two cross-linked peptides. It can be further modified, yielding even more proteoforms. Adapted from https://en.wikipedia.org/wiki/Insulin#/media/File:Ideogram_human_chromosome_11.svg, Public Domain image.

acid sequence. If proteins coded by different versions of a gene are present, *e.g.* for genes coded on chromosome pairs, samples pooled from multiple individuals, or metaproteomic samples, the different versions might coexist as different proteoforms. Genomics complexity hence translates into proteomic complexity. As illustrated in Figure 4.1, the different proteoforms produced for a given protein further differ based on interaction with other proteins. Notably, proteoforms are cleaved, folded, and post-translationally modified. Subsequently, they can be cross-linked with other entities, and/or assembled in complexes. These events are influenced by the genetic background of the biological system, but also its environment and history. Together, they confer to proteoforms very diverse biochemical properties that enable a multitude of biological functions for the same gene or protein, that can be tuned without requiring gene expression. Post-translational modifications (PTMs) like phosphorylation can notably be added or removed to enable or disable specific functions, and multiple PTMs can be combined to create proteoforms that can achieve very different tasks. Although the total number of proteoforms in a proteome remains unknown, it is expected to be much higher than the number of its coding genes: current estimates for the human proteome anticipate several millions proteoforms.[3]

The vast number of proteoforms making the proteome allows biological systems to accomplish a multitude of tasks with a limited set of genes, it is therefore an advantage biologically, and characterizing the proteome holds the key to many biological and clinical questions that would remain unanswered otherwise.[4] However, it is a challenge for scientists, who need to capture the proteome complexity analytically. This task is complicated by the fact that proteoforms are present in samples in concentrations ranging over several orders of magnitude, and currently no method is available for proteoform replication. Finally, the proteome composition and its assembly, referred to as *proteotype*, varies between individuals,[5] and even for the same individual, it varies spatially,[6] between cells,[7] cell compartments,[8] and over time.[9]

The development of ionization techniques for large biomolecules[10–12] started the era of peptide and protein analysis by mass spectrometry. Since then, mass spectrometry has become the method of choice to characterize proteomes, and multiple protocols and methods have been established to analyse the very different proteomes found in nature. The mass spectrometers used in proteomics are characterized by (1) high acquisition rates, allowing the detection of many biomolecules for higher sample coverage, (2) high resolution, allowing the discrimination of species with near-equal masses, and (3) multi-stage acquisition, allowing the fragmentation of biomolecules and the analysis of their fragments to confirm their identity. Such instruments produce large volumes of complex data, that need to be carefully processed to provide a system-wide view on the proteome, and integrated with other types of data, *e.g.* genomic, transcriptomic, or metabolomic data. To solve this problem, bioinformatic research is constantly providing new solutions to help better understand the proteome, in all its dimensions, scales, and complexity. This chapter proposes an introduction to the most common bioinformatic challenges in mass spectrometry-based proteomics and associated solutions. It is by no means exhaustive, and more details on the methods and their implementations are provided in other chapters.

## 4.2 Proteomic Experiments and Data Life Cycle

In order to maximize the analytical performance in proteomics, multiple experimental designs, methods, techniques, and instrumentation have been established, some aiming at a general characterization of proteomes from various types of samples, others focusing on *sub-proteomes*, like the membrane proteome, terminal proteome, phosphoproteome, glycoproteome, to name a few.[13] Similarly, bioinformatics methods and tools were developed to adapt to the various samples, protocols, and data acquisition strategies. Due to the intrinsic complexity of the proteome, in the lab as well as in data interpretation, there is currently no *one size fits all* solution.

While detailing the different strategies available to characterize the proteome is beyond the scope of this chapter, three approaches will be distinguished in the following:

1. *Discovery versus targeted* Depending on the biological question, proteomics experiments can be discovery or targeted. Discovery studies attempt to characterize as many proteoforms as possible in an unbiased fashion – a *cast net* approach. Targeted studies attempt to characterize a set of predefined proteoforms – a *harpoon* approach. Note that the higher sample coverage of discovery studies comes at the price of lower sensitivity, stochasticity in the identified proteoforms, and increased complexity in data processing, as detailed in the following.
2. *Bottom-Up versus Top-Down* Due to the high diversity in proteoform physicochemical properties, it is challenging to conduct experiments with proteome-wide coverage. To overcome this problem, in bottom-up proteomics, proteoforms are broken down into peptides, usually *via* proteolysis, and the peptides are used as proxies to infer the presence and abundance of proteoforms. Smaller molecules fall within a narrower range of physicochemical properties, hence allowing higher proteome coverage for a given experimental setup. Moreover, for proteoforms yielding multiple peptides, even if some remain beyond the detection capabilities of the experiment, *e.g.* transmembrane domains, others might become detectable, hence enabling the presence of the proteoform to be inferred even with limited sequence coverage. The increased proteome coverage comes at the price of ambiguity in the results, because peptides might map to different proteoforms, as detailed in the following. Conversely, top-down approaches analysing entire proteins do not suffer from this, but require the characterization of much more complex molecules.

3. *DIA versus DDA* During multi-stage analysis, when selecting biomolecules for fragmentation and downstream mass analysis, the selection can be guided, *e.g.* a fragmentation window is set around peaks of highest intensity in a prior scan called *survey scan*, or unsupervised, *e.g.* by fragmenting everything or using sliding mass windows, often referred to as *swath*, hence not necessarily requiring survey scans. The guided and unsupervised selection modes are termed *data dependent acquisition* and *data independent acquisition*, abbreviated to DDA and DIA, respectively. Conditioning the fragmentation on a survey scan greatly simplifies the downstream analysis, at the cost of overlooking molecules and introducing stochasticity.

In addition to their analytical performance, it is important to note that these proteomic strategies have different levels of adoption and maturity, and consequently the degree of development of the associated bioinformatics environments differ. For example, most bioinformatic methods and tools established to date are dedicated to discovery bottom-up DDA experiments. This means that some analyses can be conducted routinely using professional-grade user-friendly bioinformatic tools, while others will require more bioinformatic effort, expertise, and maybe even some coding. Different experiments will, therefore, require different bioinformatic tools and skills, and it is not rare to see projects stall once the data have been acquired, or proteomic data being inadequately or incorrectly interpreted.[14] It is thus extremely important to make sure that the bioinformatic resources needed to interpret the results of an experiment are in place *before* the data are generated. In the experience of the author, bioinformatic analyses fail more often and with much more adverse consequences due to inadequate experimental design than due to problems in the bioinformatic analysis. When envisioning an experiment, even before it is budgeted, consider the following questions: *Are the bioinformatic tools available for our use case? Can our hardware handle the amount of data in a reasonable time? Do we have the expertise in bioinformatics, data science, and biostatistics to process and interpret the data?* When a piece of the puzzle is missing, do not hesitate to reach out to the community for help – writing software to process the results of an experiment is very different from designing a prototype or proof of concept,[15] and the costs of long-term maintenance rapidly become overwhelming.[16]

It can be very challenging to estimate the feasibility of a bioinformatic analysis. For this, preliminary data are extremely valuable. Do not hesitate to run small pilot or benchmark experiments. Also, look at whether public data can be reused for this purpose.[17] One of the main parameters that will influence the feasibility and success of the experiment is the number of samples considered. Here again, preliminary and public data can help gathering preliminary estimates *e.g.* of means and variances, and numerous statistical tools are available to conduct power analyses. Other essential aspects of experimental design like finding the right controls, sample randomization, and blocking will also affect the data processing and interpretation, as detailed by Burzykowski, *et al.*[18] It is therefore important to verify upfront that the downstream data processing and interpretation pipeline can handle the complexity of the experimental design. For example, conducting a time series analysis of multiple conditions in replicates with fractionation is a daunting task, and processing steps performed otherwise routinely like the control of error rates become extremely challenging. When planning experiments, it is important to find a good balance between generating enough data to answer the biological question confidently, and maintaining technical feasibility. Then, the importance of restraining oneself to a single constrained question is paramount, convoluted experiments attempting to answer multiple complex biological questions in the same time often lead to unpleasant surprises during bioinformatic and biostatistics analyses.

In addition to tools helping design the experiments, bioinformatics also provides solutions to optimize their setup. For example, tools are available to optimize fractionation[19] and separation.[20] Finally, in order to maximize the outcome of a mass spectrometry run, bioinformatic methods and tools were established to guide acquisition and dynamically tune the mass spectrometer during acquisition.[21–24] Importantly, most modern mass spectrometers include embedded signal and data processing units that pre-process the data, *e.g.* conducting baseline reduction, noise filtering, isotope or charge inference, and the algorithms and their implementation are proprietary and rarely disclosed. After acquisition, the data produced by the instrument, called *raw data,* are available for processing and interpretation.

These data are the result of a long process: experimental design, sample collection, experimental workflow, and data acquisition. As detailed in the following, after careful quality control (QC), they are converted and processed with specialist tools to qualitatively and quantitatively characterize the proteome; the qualitative analysis aims to identify the species present in the sample, while the quantitative analysis aims to infer their abundance; the qualitative and quantitative results are then interpreted in light of a scientific question, and generally in combination with the results of other experiments, yield new knowledge. The new data and knowledge are finally disseminated to the community, *e.g.* through scientific publication, where they can be used to build new hypotheses, new questions, new experiments, and generate new data and knowledge. This data life cycle illustrated in Figure 4.2 can be virtuous, yielding more and better knowledge at each iteration, but scientific progress depends on the performance of each step, and it is dramatically impaired if data and knowledge do not flow seamlessly. This is why tremendous efforts were invested by the scientific community in creating the vast and complex bioinformatic environment available today in life sciences.

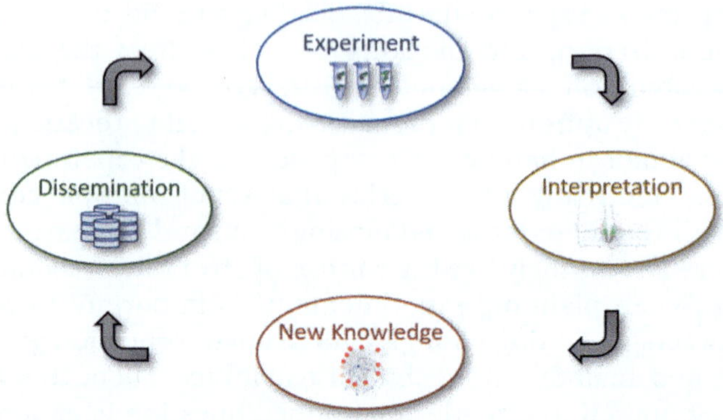


**Figure 4.2** Proteomics data life cycle. Data generated in an experiment are interpreted and new knowledge is generated. The new knowledge is disseminated allowing the design of new experiments and the generation of new knowledge. Reproduced from ref. 17, https://doi.org/10.1002/pmic.201500295, under the terms of a CC BY 4.0 license, https://creativecommons.org/licenses/by/4.0/.

## 4.3 Signal Processing

The raw data produced by instruments come in different formats, usually proprietary, that require specialized libraries to be processed. Unfortunately for high performance computing environments running on the Linux operating system, most of these libraries are only available for Microsoft Windows. However, these formats can be converted to open standard formats using Proteowizard.[25] In addition, Linux-compatible libraries are currently being developed paving the way for much simplified cross-platform operation.[26] Importantly, converters might need to process the data, *e.g.* conduct peak picking or infer the charge of some peaks.[27] For the sake of transparency and reproducibility, it is therefore important to keep all intermediate files and make them available upon publication,[28] and differences should be expected when using different converters.

The standard format for mass spectrometry files is mzml.[29] This format captures all the information of a mass spectrometry run, and while this is essential for transparency and archiving, it can be cumbersome in practice. Consequently, many pipelines use other intermediate mass spectrometry formats, which can be tailored to a specific application like the Mascot Generic File (mgf) format (http://www.matrixscience.com/help/data_file_help.html), or optimized for performance.[30]

The first task to conduct on the data is QC. This is achieved by visualizing diagnostic metrics and summary statistics. This can be achieved using software provided by the instrument vendor, and additional dedicated bioinformatic tools exist that can help identifying and diagnosing problems with the data.[31–33] The importance of running QC on all samples cannot be underestimated, as different levels of analytical performance will greatly impact the downstream analysis. In fact, even if not explicitly mentioned in the following, the data obtained at each step of the processing and interpretation process need to be thoroughly QCed.

Subsequently, the processing steps required for the raw data processing depend mainly on the instrument, acquisition mode, and downstream analysis. These can include baseline reduction, noise filtering, deisotoping, charge deconvolution, and peak picking, which can be achieved using vendor software or third-party tools.[34] Modern instruments however contain embedded signal processing units and produce ready-to-use raw files.

## 4.4 Qualitative Analysis

In discovery analyses, the composition of a proteome needs to be inferred from the data, this is the aim of qualitative proteomic analyses. Multiple bioinformatic approaches were established for this, depending on how the data were generated, as detailed in reviews.[35,36] On the other hand, targeted analyses, which measure the abundance of preselected analytes, are by design quantitative, the presence of a target is assessed by comparing its measured abundance to the level of detection of the experiment.

When a chain of amino acids is fragmented, the mass spectrum of its fragment ions creates series of ions differing in mass by one amino acid, as illustrated in Figure 4.3. Theoretically, it is therefore possible to sequence a series of amino acids by extracting these so-called *mass ladders* until reaching the total mass of the biomolecule that was fragmented. However, this approach requires the acquisition of spectra from a single amino acid sequence, and is very sensitive to noise and incomplete fragmentation. In the example shown in Figure 4.3, the three tailing amino acids at the C-terminus are not annotated with any fragment ion, any combination of amino acids making up the same mass would identically explain the spectrum. In addition, multiple charges, isotopes, or alternative products of fragmentation like neutral losses yield peaks that dramatically complexify the mass spectrum, and PTMs and their combinations combinatorially increase the number of possible mass differences between peaks. In practice, sequencing by mass spectrometry is, therefore, not used for whole proteome analysis but rather for protein characterization, *e.g.* for antibody sequencing.[37]

For the analysis of complex samples, a reference is used to match the experimental data against, hence reducing the number of molecules that can match a particular spectrum. As illustrated in Figure 4.4A, it is possible to use a library of spectra from known molecules, *e.g.* peptide spectra from the PeptideAtlas.[39] While immensely valuable, these libraries do not cover the full extent of the proteome. To broaden the space of discoverable molecules, sequence libraries are used instead, and the reference spectra used are generated *in silico*, as illustrated in Figure 4.4B. These libraries aim to cover the entire proteome, and are derived from genomic data. Such sequences

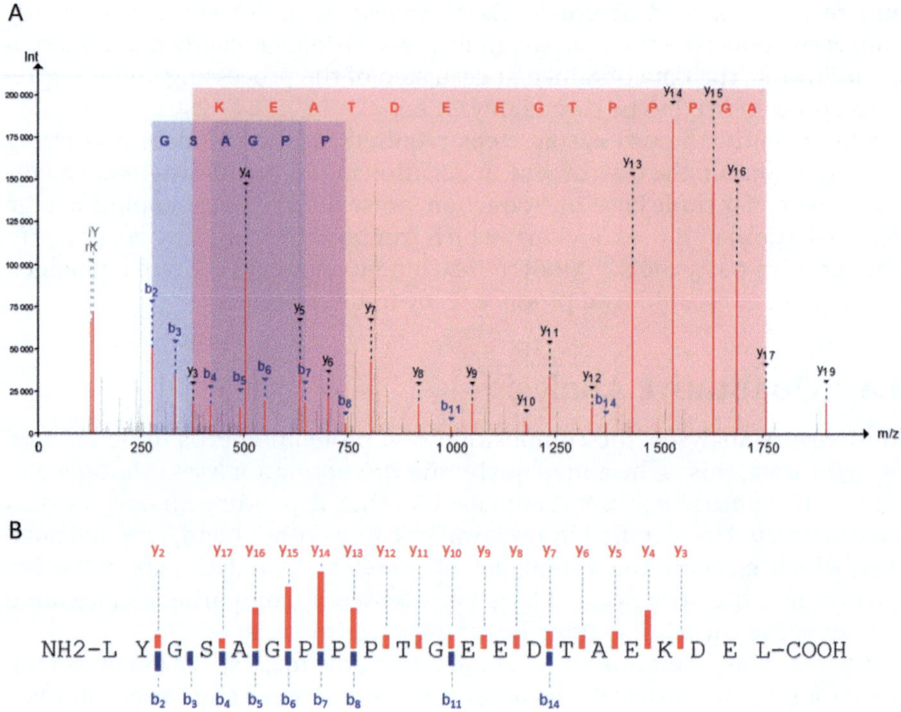


**Figure 4.3** Peptide sequencing. (A) The spectrum of a fragmented peptide with fragment ions annotated. Mass ladders corresponding to series of peaks separated by one amino acid are annotated at the top. (B) The sequence of the peptide identified in the spectrum with the fragment ions and the intensity of the peaks annotated. Illustrations were exported from PeptideShaker.[38]

are consolidated in knowledge bases like UniprotKB[40] or can be assembled from sequencing data.[41–43] Specialized bioinformatic tools called *database search engines* process these sequences to best model the way the data were generated: accounting for proteolytic cleavage of bottom-up data, possible modifications, and expected charges and fragmentation patterns. Multiple methods, algorithms, and tools have been made available throughout the years, offering a large palette of solutions for different use cases.[44] Finally, as illustrated in Figure 4.4C, using so-called *de novo* sequencing or *de novo* tagging, it is possible to generate all sequences or fragment of sequences that can be derived from a spectrum, and match these to a reference sequence database.[45,46]

Most methods, algorithms, and tools were developed to process bottom-up proteomics data. Due to the difference in complexity of spectra obtained from peptides and from entire proteoforms, specialist approaches are better suited to interpret spectra obtained *via* top-down analysis.[47] In addition, the strategies introduced above and illustrated in Figure 4.4 assume the presence

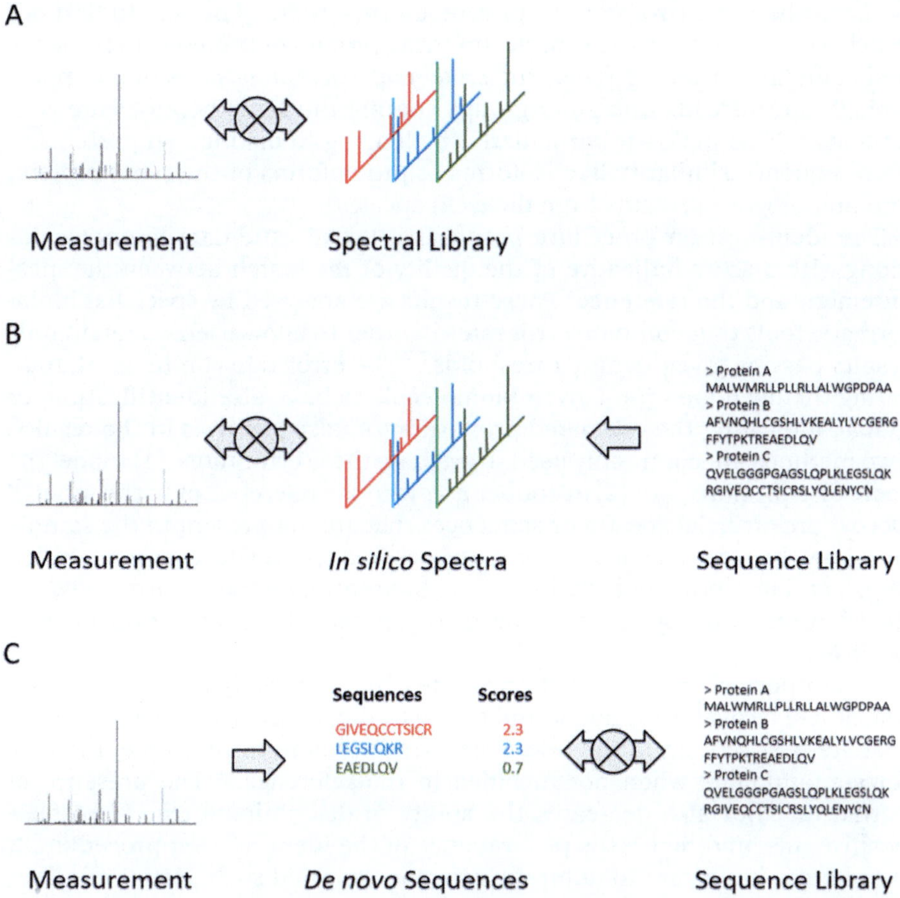


**Figure 4.4** Spectrum identification. (A) In spectral library searching, the measured spectrum is compared to a library of reference spectra. (B) In database searching, reference spectra are inferred *in silico* from a sequence library called database, and compared to the measured spectrum. (C) In *de novo* sequencing and tagging, possible sequences or fragments of sequence, also called tags, are inferred from the spectra, and compared to a sequence library.

of a single biomolecule per spectrum, which is practically challenging when analysing entire proteomes. To alleviate this, some tools designed for DDA analyses implement the search of co-isolated peptides, called *chimeric* spectral searches.[48] For DIA analyses, where multiple molecules are analysed simultaneously, spectra are either deconvoluted prior to identification,[49–51] or the identification procedure is conducted on the convoluted spectra using spectral libraries[52] or database searches.[53]

While bottom-up proteomic approaches present the advantage of reducing the size and complexity of the biomolecules to analyse, when a peptide is shared between two proteins or proteoforms, there is no way to find out which was originally in the sample. Inferring protein or proteoform presence and abundance from peptides, the *protein inference problem*, is often impossible,[54] and instead, ambiguity groups sharing the same peptides are constructed.[55] This makes it particularly challenging to distinguish proteins of high sequence similarity like isoforms or proteoforms of the same protein, and homologous proteins from different species.

The identification procedure produces a list of candidate biomolecules along with a score indicative of the quality of the match between the measurement and the reference. These results are analysed by specialist bioinformatic tools that compute error rates in order to allow users to retain only results passing given quality thresholds.[56] The error rate can be local, indicating the likelihood for a given biomolecule to be a false identification, or global, indicating the estimated proportion of false positives in the results. Two methods are commonly used to evaluate these error rates: (1) model the score distributions,[57] or (2) introduce *decoys* in the reference or in the input.[58] Decoys are artificial spectra or sequences that are not present in the sample but show the same characteristics as the original spectra or sequences, called *target*, and are indistinguishable by the bioinformatic tools. Ultimately, the distribution of decoy scores is used to estimate the distribution of false positives.

It is important to stress that error rates, and especially decoy-based estimates, can be overly optimistic and fooled by the presence of partial matches that are best scoring in the reference used.[59] This is, for example, the case for contaminants when not included in the reference.[14] The presence of partial matches also decreases the ability to discriminate true from false positive hits, and hence the performance of the identification procedure. It is therefore important to tailor the reference to avoid such partial matches, the prevalence of which grows with the size of the search space.[44] Finally, small differences in the sequence that can only be discriminated by specific fragment ions, like the localization of close modifications or amino acid permutations, are poorly accounted for by identification scores, and require specific scoring.[60]

## 4.5 Quantitative Analysis

When probing biological systems, evaluating the abundance of biomolecules can be more informative than asserting their presence. Therefore, quantitative proteomic studies aim to infer protein and proteoform abundances from mass spectrometry data. This is made challenging by the fact that biomolecule ionization is a competitive process: depending on the molecules that co-elute and are simultaneously ionized, even on the same instrumentation with exactly the same conditions, the same biomolecule at the same abundance in liquid phase can yield different number of ions in different charge states in gas phase. In practice, this means that it is not possible to directly infer or compare biomolecule abundances from their intensities, and that the same molecule can be present at different charge states. In addition, sample preparation can introduce inter-sample variation, and for bottom-up proteomics, variation between protein and peptide abundance. To circumvent these problems, various quantification strategies were developed.[61,62]

A first approach consists in spiking proteins or peptides containing heavy isotopes in known amounts in the samples. These spike-ins will remain indistinguishable until data interpretation, where a mass shift corresponding to the heavy isotopes will separate the endogenous version from the reference. Since this strategy requires a set of prior targets with heavy standards available, it is therefore used in targeted strategies. A library is built from a list of targets, containing the retention time (RT) and mass over charge ($m/z$) coordinates of the expected ions to monitor. This library can be built from previously acquired data, or predicted computationally.[63,64] The library is then used to tune the instrument and extract the intensities of the molecules of interest and their spiked counterpart. The quality of the quantification greatly depends on the analytical performance of the experiment, which impacts the quality of the matching between the experimental run and the library, it is therefore important to limit RT and $m/z$ deviation. The bioinformatic software can to some extent correct for drifts during the acquisition, and provide RT and $m/z$ tolerances when matching data to the library. These tolerances are set relative to the performance of the experimental setup. Increasing the tolerances improves the sensitivity but decreases the specificity of the matching between the data and the library.[65] For molecules where both endogenous and heavy versions are found, the software extracts the respective intensities, usually by fitting an elution profile to the detected intensities, and scales the spiked abundance using the ratio of intensities to obtain an estimate of the endogenous abundance, hence achieving so-called absolute quantification. In order for the scaling of the intensities to be accurate, it is therefore important that both heavy and endogenous molecules are above noise level, and in the linear detection range of the instrument.

In discovery studies, the complexity of the proteome precludes the creation of a list of targets, and instead the analytical setup attempts to measure as many molecules as possible. It is then possible to spike an entire proteome labelled with heavy isotopes. For cultured cells and organisms where it is possible to control the environment like bacteria, plants or small animals like mice, it is possible to label entire proteomes by providing heavy isotopes in amino acids used for feeding. A labelled proteome is then built that is as close as possible to the sample proteome and spiked-in to use as a reference. Since the abundance of the spiked molecules is unknown, it is not possible to estimate the abundance of their endogenous counterparts, but it is still possible to compare the intensity of *heavy* and *light* versions. This approach, named *Stable Isotope Labeling by Amino acids in Cell culture* (SILAC),[66] hence does not allow absolute quantification, but relative quantification of the detected molecules between heavy and light proteomes. When it is not possible to control the amino acid intake, *e.g.* with human or environmental samples, it is still possible to use other samples as a common reference, an approach called *super-SILAC*,[67] making it possible to compare many different samples using the proteins common to the samples and the super-SILAC mix. It is also possible to label proteins or peptides chemically *in vitro*, during sample preparation, allowing the multiplexing of two or more proteomes before ionization, hence allowing to control for variation in ionization efficiency, but not for variation introduced prior to labelling.[61] Here again, multiple samples can be compared to the same reference.

These techniques and the associated bioinformatic resources were designed for, and are primarily used in, DDA bottom-up experiments. In order to infer relative abundances of the measured peptides, the bioinformatic software needs to identify peptide intensity fingerprints in the *m*/*z versus* RT space, so-called features as illustrated in Figure 4.5, integrate their intensity profile, link heavy to light peptides, and infer their identity. The peptide features are detected by identifying elution and isotopic profiles in the RT and *m*/*z* dimensions,[68] respectively, the charge of the peptide is inferred by comparing the *m*/*z* distance between isotopes,[69] and the intensity integrated in both directions. Note that multiple methods exist to achieve these tasks. The features are matched to the identified peptides by linking the feature *m*/*z* and RT to the peptide mass and RT at which the peptide fragmentation spectrum was measured, respectively. Note that features detection can be conducted first, or after identification.[70] In the former case, the *m*/*z* of peptide features can be used to improve identification results.[71] In the latter case, the identification results can be used to guide feature detection. When multiple mass spectrometry measurements are matched, *e.g.* in a super-SILAC setup, if a

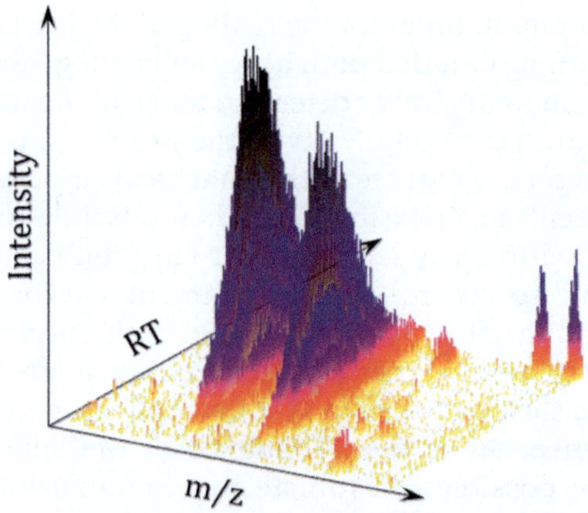


**Figure 4.5** Peptide features. Peptide features are characterized by a bell-shaped elution profile detected for multiple isotopes at different *m*/*z*. Here, two features are prominently visible in this zoom in a mass spectrometry acquisition. The two features could correspond to the heavy and light version of the same profile. Obtained with OpenMS,[34] adapted from Vaudel *et al.*[65] with permission from John Wiley & Sons, Ltd. Copyright 2010.

peptide was identified in one or more measurements but not all, it is possible to look for peptide features in measurements where the peptide was not found after retention time alignment of the different measurements. This procedure, called *match between runs* is extensively used to increase the proteome coverage in quantitative experiments.[72] Finally, features from heavy and light peptides are matched using the mass difference expected from the labelling strategy, and their intensities are compared to estimate their differential relative abundance.[73]

Processing peptide *m*/*z* *versus* RT maps is computationally intensive, moreover feature identification, integration, and matching with identified peptides is error prone. As illustrated in Figure 4.6, to alleviate this, peptides can instead be labelled using *isobaric mass tags*, allowing the labelling and multiplexing of multiple samples that remain undistinguishable until fragmentation, where the fragmentation of the tag will release sample-specific *reporter ions*.[21,74,75] These reporter ions are found directly in the spectrum used for identification, or in a fragmentation spectrum obtained from the spectrum used for identification, they are thus intrinsically linked to

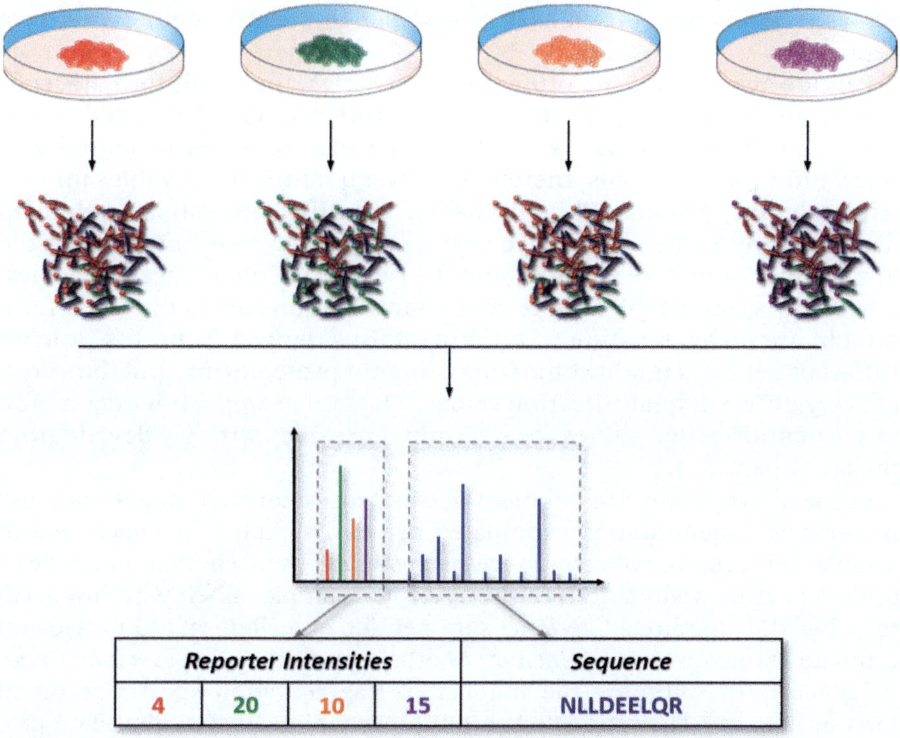


**Figure 4.6** Reporter ion quantification. After protein digestion, samples are labelled with isobaric mass tags and multiplexed. They are processed together until data acquisition, where sample-specific reporter ions (left) are used to estimate the relative abundance of the peptide in the samples, while other fragment ions are used to identify the peptide. Reproduced from ref. 79 with permission from the Royal Society of Chemistry.

the peptide, and the extraction of the intensity is computationally much simpler. As a consequence, quantification is, however, only available for identified spectra, introducing missing values when matching multiple measurements and impairing the ability to conduct match between runs. Also, it is important to note that when reporter ions are obtained directly in the peptide spectrum, all molecules that carry a tag and were co-isolated along with the peptide will contribute to the reporter ions intensities. This results in a relative intensity convergence towards 1 : 1, called *ratio compression*,[76,77] and the magnitude of this compression increases with the complexity of the sample.[78] Once the intensities specific to the different samples are extracted, their comparison allows the relative quantification of a peptide between all samples.[79]

With the progresses of instrumentation, the technical variation between runs has been dramatically reduced. Thus, it has become common for large scale quantitative studies to compare peptide features directly.[80] In these *label free* quantification studies, peptide intensity features are extracted and linked to peptides as described above, and match between runs is conducted to reduce the prevalence of missing values. The relative intensities across experiments is used as a measure of the relative abundance for a peptide.

Peptide abundances are often not informative, thus bioinformatic software attempt to aggregate them to the protein level. As detailed in the previous section, this task is complexified by the presence of shared peptides. Bioinformatic tools, therefore, deliver quantification values for protein ambiguity groups. While in theory, quantification values could help disentangling protein inference cases, it is in practice impaired by the low number of unique peptides and the variance of quantification values. Finally, it is important to note that quantification values obtained for a protein are an aggregation of all proteoforms derived from this protein. Different peptides might come from different proteoforms, and thus present very different quantification values.[54] It is for example not uncommon to see quantification values from terminal peptides strongly deviate from the other peptides.

In the above, due to the problem of competitive ionization, peptides and proteins were quantitatively compared between samples, but never within sample. Yet it can be valuable to have an estimate as to whether a peptide or protein is more abundant than the other in a sample, albeit with low accuracy. For this, methods like *iBAQ* sum peptide contributions to provide an estimate of total protein intensity.[81] Another approach called *spectrum counting* consists in summing the number of fragmentation spectra recorded for a protein.[82,83] The rationale behind this technique is that abundant proteins are more likely to generate high intensity peptides, that in turn have a higher probability to trigger a fragmentation event that leads to an identified spectrum. The abundance estimate is then typically normalized for protein length and uniqueness of peptides.[84]

The result of quantification studies are large tables containing the estimated relative abundances of the quantified biomolecules among the different samples. Extracting valuable information from these can rapidly become overwhelming. To handle these large matrices, data science and statistical programming software exist, with different levels of user friendliness and support for different methods. The value of learning how to process such tables programmatically and transparently using open source bioinformatic environments cannot be understated. The various tasks to accomplish include data visualization, quality control, handling of missing values, data normalization, evaluation of the significance of changes between samples, and evaluation of error rates.[85] Here again, there is no one-size-fits-all analysis strategy, and the different biostatistical analysis methods and their implementations are beyond the scope of this chapter.

## 4.6 Getting the Bigger Picture

Protein abundances alone are not always sufficient to answer a specific question. Proteomics is thus often completed with different readouts. Supplementing proteomics with genomics data enables a better characterization of sequence and splice variants, as well as the comparison of structural variants, gene expression, and protein abundance.[5,86–88] Conversely, metabolomics and lipidomics provide additional information for the study of metabolic networks.[89] Such multi-omics studies require advanced bioinformatic and systems biology methods and tools,[90,91] here again it is advised to carefully plan before conducting multi-omics experiments. Moreover, it is important to carefully assess the results in light of the limitations of the proteomic methods used. Notably, protein inference ambiguity and lack of proteoform-level annotation are difficult to take into account in multi-omics analyses.[92] Furthermore, while it is possible to characterize the 3D structural state of proteins,[93] their assembly in complexes,[94] and linking with other molecules like membranes, ions, or metabolites, using mass spectrometry, these approaches are still challenging to conduct routinely on a system-wide scale. Similarly, the proteome is dynamically affected by the history and the environment of biological systems, yet these are very difficult to monitor and account for in analyses.[95]

Due to their complexity, the computational representation of biological systems is challenging.[97] Currently, the state of the art is to represent them as biological networks, that are mathematically represented as graphs.[98] Then, entities are represented as a node in the network, or vertex in the graph, and an interaction between two entities is represented as a connection or edge in the network or graph, respectively. Such networks can be derived from protein–protein interaction (PPI) data, that can be obtained experimentally or from the literature, and notably from PPI databases and pathway knowledge bases.[99] Then, nodes represent proteins or genes connected when involved together in a biochemical reaction, and paths

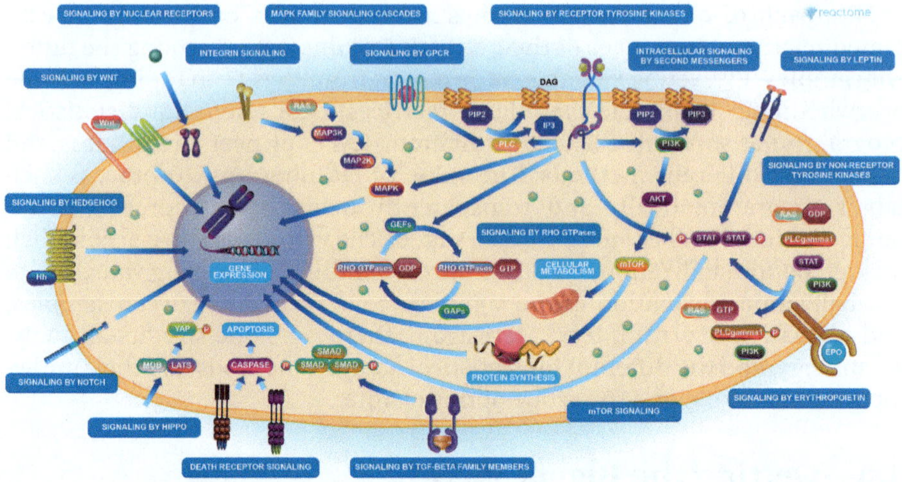


**Figure 4.7** Pathway representation. Representation of a biological pathway using Reactome.[96]

through the network represent biological pathways as illustrated in Figure 4.7. These approaches aim at identifying signals that are collectively meaningful, and would be lost when taken individually, and improving the modelling of biological systems in all their dimensions and scales is a very dynamic and promising scientific field.[100,101]

The biological and analytical challenges posed by the analysis of the proteome make it difficult to design strongly powered studies. Like in other scientific fields, meta-analyses are gaining momentum, that build on large data sets to extract new knowledge.[102–104] These strategies are notably enabled by the availability of the vast amounts of public data made available by the community.[105] Meta-analyses using large numbers of experiments increase the probability of detecting rare events, and are therefore a promising area of development for proteomics. The scientific and technical challenges posed by the analysis of these large data sets moreover present an exciting field of research for the interpretation of proteomics data.[106,107]

In conclusion, the complexity of the proteome, its ubiquity and versatility yielded a plethora of analysis strategies, and in turn large numbers of methods and tools to interpret the data were established. The rich bioinformatics environment available to the proteomics community is constantly improving, increasing our ability to characterize the proteome and generate new knowledge on biological systems. This complex ecosystem might seem overwhelming to wet-lab scientists and newcomers, but they should be assured that the bioinformatics community will be here to help when they reach out, and that investing in one's bioinformatic skills is of value throughout your working life.

## References


1. L. M. Smith, N. L. Kelleher and Consortium for Top Down Proteomics, *Nat. Methods*, 2013, **10**, 186–187.
2. D. R. Zerbino, P. Achuthan, W. Akanni, M. R. Amode, D. Barrell, J. Bhai, K. Billis, C. Cummins, A. Gall, C. G. Girón, L. Gil, L. Gordon, L. Haggerty, E. Haskell, T. Hourlier, O. G. Izuogu, S. H. Janacek, T. Juettemann, J. K. To, M. R. Laird, I. Lavidas, Z. Liu, J. E. Loveland, T. Maurel, W. McLaren, B. Moore, J. Mudge, D. N. Murphy, V. Newman, M. Nuhn, D. Ogeh, C. K. Ong, A. Parker, M. Patricio, H. S. Riat, H. Schuilenburg, D. Sheppard, H. Sparrow, K. Taylor, A. Thormann, A. Vullo, B. Walts, A. Zadissa, A. Frankish, S. E. Hunt, M. Kostadima, N. Langridge, F. J. Martin, M. Muffato, E. Perry, M. Ruffier, D. M. Staines, S. J. Trevanion, B. L. Aken, F. Cunningham, A. Yates and P. Flicek, *Nucleic Acids Res.*, 2018, **46**, D754–D761.
3. R. Aebersold, J. N. Agar, I. J. Amster, M. S. Baker, C. R. Bertozzi, E. S. Boja, C. E. Costello, B. F. Cravatt, C. Fenselau, B. A. Garcia, Y. Ge, J. Gunawardena, R. C. Hendrickson, P. J. Hergenrother, C. G. Huber, A. R. Ivanov, O. N. Jensen, M. C. Jewett, N. L. Kelleher, L. L. Kiessling, N. J. Krogan, M. R. Larsen, J. A. Loo, R. R. Ogorzalek Loo, E. Lundberg, M. J. MacCoss, P. Mallick, V. K. Mootha, M. Mrksich, T. W. Muir, S. M. Patrie, J. J. Pesavento, S. J. Pitteri, H. Rodriguez, A. Saghatelian, W. Sandoval, H. Schlüter, S. Sechi, S. A. Slavoff, L. M. Smith, M. P. Snyder, P. M. Thomas, M. Uhlén, J. E. Van Eyk, M. Vidal, D. R. Walt, F. M. White, E. R. Williams, T. Wohlschlager, V. H. Wysocki, N. A. Yates, N. L. Young and B. Zhang, *Nat. Chem. Biol.*, 2018, **14**, 206–214.
4. R. Aebersold and M. Mann, *Nature*, 2016, **537**, 347–355.
5. C. J. Ricketts, J. R. Forman, E. Rattenberry, N. Bradshaw, F. Lalloo, L. Izatt, T. R. Cole, R. Armstrong, V. K. A. Kumar, P. J. Morrison, A. B. Atkinson, F. Douglas, S. G. Ball, J. Cook, U. Srirangalingam, P. Killick, G. Kirby, S. Aylwin, E. R. Woodward, D. G. R. Evans, S. V. Hodgson, V. Murday, S. L. Chew, J. M. Connell, T. L. Blundell, F. Macdonald and E. R. Maher, *Hum. Mutat.*, 2010, **31**, 41–51.
6. P. J. Thul and C. Lindskog, *Protein Sci.*, 2018, **27**, 233–244.
7. M. J. Roth, B. A. Parks, J. T. Ferguson, M. T. Boyne and N. L. Kelleher, *Anal. Chem.*, 2008, **80**, 2857–2866.
8. P. J. Thul, L. Åkesson, M. Wiking, D. Mahdessian, A. Geladaki, H. Ait Blal, T. Alm, A. Asplund, L. Björk, L. M. Breckels, A. Bäckström, F. Danielsson, L. Fagerberg, J. Fall, L. Gatto, C. Gnann, S. Hober, M. Hjelmare, F. Johansson, S. Lee, C. Lindskog, J. Mulder, C. M. Mulvey, P. Nilsson, P. Oksvold, J. Rockberg, R. Schutten, J. M. Schwenk, Å. Sivertsson, E. Sjöstedt, M. Skogs, C. Stadler, D. P. Sullivan, H. Tegel, C. Winsnes, C. Zhang, M. Zwahlen, A. Mardinoglu, F. Pontén, K. von Feilitzen, K. S. Lilley, M. Uhlén and E. Lundberg, *Science*, **356**, eaal3321.
9. M. S. Robles, J. Cox and M. Mann, *PLoS Genet.*, 2014, **10**, e1004047.
10. M. Karas, D. Bachmann and F. Hillenkamp, *Anal. Chem.*, 1985, **57**, 2935–2939.

11. M. Karas, D. Bachmann, U. Bahr and F. Hillenkamp, *Int. J. Mass Spectrom. Ion Processes*, 1987, **78**, 53–68.
12. J. B. Fenn, M. Mann, C. K. Meng, S. F. Wong and C. M. Whitehouse, *Science*, 1989, **246**, 64–71.
13. K. Gevaert, P. Van Damme, B. Ghesquière, F. Impens, L. Martens, K. Helsens and J. Vandekerckhove, *Proteomics*, 2007, **7**, 2698–2718.
14. G. M. Knudsen and R. J. Chalkley, *PLoS One*, 2011, **6**, e20873.
15. B. A. Grüning, S. Lampa, M. Vaudel and D. Blankenberg, *GigaScience*, DOI: 10.1093/gigascience/giz054.
16. A. Nowogrodzki, *Nature*, 2019, **571**, 133–134.
17. M. Vaudel, K. Verheggen, A. Csordas, H. Raeder, F. S. Berven, L. Martens, J. A. Vizcaíno and H. Barsnes, *Proteomics*, 2016, **16**, 214–225.
18. T. Burzykowski, J. Claesen and D. Valkenborg, in *Mass Spectrometry of Proteins*, ed. C. A. Evans, P. C. Wright and J. Noirel, Springer New York, New York, NY, 2019, vol. 1977, pp. 181–197.
19. E. M. Solovyeva, A. A. Lobas, A. T. Kopylov, I. Y. Ilina, L. I. Levitsky, S. A. Moshkovskii and M. V. Gorshkov, *Anal. Bioanal. Chem.*, 2018, **410**, 3827–3833.
20. L. Moruz, P. Pichler, T. Stranzl, K. Mechtler and L. Käll, *Anal. Chem.*, 2013, **85**, 7777–7785.
21. D. K. Schweppe, J. K. Eng, D. Bailey, R. Rad, Q. Yu, J. Navarrete-Perea, E. L. Huttlin, B. K. Erickson, J. A. Paulo and S. P. Gygi, *bioRxiv*, DOI: 10.1101/668533.
22. F. Meier, P. E. Geyer, S. Virreira Winter, J. Cox and M. Mann, *Nat. Methods*, 2018, **15**, 440–448.
23. C. Wichmann, F. Meier, S. Virreira Winter, A.-D. Brunner, J. Cox and M. Mann, *Mol. Cell. Proteomics*, 2019, **18**, 982–994.
24. J. V. Olsen, L. M. F. de Godoy, G. Li, B. Macek, P. Mortensen, R. Pesch, A. Makarov, O. Lange, S. Horning and M. Mann, *Mol. Cell. Proteomics*, 2005, **4**, 2010–2021.
25. D. Kessner, M. Chambers, R. Burke, D. Agus and P. Mallick, *Bioinformatics*, 2008, **24**, 2534–2536.
26. N. Hulstaert, J. Shofstahl, T. Sachsenberg, M. Walzer, H. Barsnes, L. Martens and Yasset Perez-Riverol, *J. Proteome Res.*, 2020, **19**, 537–542.
27. W. R. French, L. J. Zimmerman, B. Schilling, B. W. Gibson, C. A. Miller, R. R. Townsend, S. D. Sherrod, C. R. Goodwin, J. A. McLean and D. L. Tabb, *J. Proteome Res.*, 2015, **14**, 1299–1307.
28. L. Martens, A. I. Nesvizhskii, H. Hermjakob, M. Adamski, G. S. Omenn, J. Vandekerckhove and K. Gevaert, *Proteomics*, 2005, **5**, 3501–3505.
29. L. Martens, M. Chambers, M. Sturm, D. Kessner, F. Levander, J. Shofstahl, W. H. Tang, A. Römpp, S. Neumann, A. D. Pizarro, L. Montecchi-Palazzi, N. Tasman, M. Coleman, F. Reisinger, P. Souda, H. Hermjakob, P.-A. Binz and E. W. Deutsch, *Mol. Cell. Proteomics*, 2011, **10**, R110.000133.
30. D. Bouyssié, M. Dubois, S. Nasso, A. Gonzalez de Peredo, O. Burlet-Schiltz, R. Aebersold and B. Monsarrat, *Mol. Cell. Proteomics*, 2015, **14**, 771–781.

31. P. Pichler, M. Mazanek, F. Dusberger, L. Weilnböck, C. G. Huber, C. Stingl, T. M. Luider, W. L. Straube, T. Köcher and K. Mechtler, *J. Proteome Res.*, 2012, **11**, 5540–5547.
32. C. Chiva, R. Olivella, E. Borràs, G. Espadas, O. Pastor, A. Solé and E. Sabidó, *PLoS One*, 2018, **13**, e0189209.
33. C. Bielow, G. Mastrobuoni and S. Kempa, *J. Proteome Res.*, 2016, **15**, 777–787.
34. H. L. Röst, T. Sachsenberg, S. Aiche, C. Bielow, H. Weisser, F. Aicheler, S. Andreotti, H.-C. Ehrlich, P. Gutenbrunner, E. Kenar, X. Liang, S. Nahnsen, L. Nilse, J. Pfeuffer, G. Rosenberger, M. Rurik, U. Schmitt, J. Veit, M. Walzer, D. Wojnar, W. E. Wolski, O. Schilling, J. S. Choudhary, L. Malmström, R. Aebersold, K. Reinert and O. Kohlbacher, *Nat. Methods*, 2016, **13**, 741–748.
35. A. I. Nesvizhskii, O. Vitek and R. Aebersold, *Nat. Methods*, 2007, **4**, 787–797.
36. M. Vaudel, A. Sickmann and L. Martens, *Expert Rev. Proteomics*, 2012, **9**, 519–532.
37. N. H. Tran, M. Z. Rahman, L. He, L. Xin, B. Shan and M. Li, *Sci. Rep.*, 2016, **6**, 31730.
38. M. Vaudel, J. M. Burkhart, R. P. Zahedi, E. Oveland, F. S. Berven, A. Sickmann, L. Martens and H. Barsnes, *Nat. Biotechnol.*, 2015, **33**, 22–24.
39. F. Desiere, *Nucleic Acids Res.*, 2006, **34**, D655–D658.
40. The UniProt Consortium, *Nucleic Acids Res.*, 2019, **47**, D506–D515.
41. M. C. Chambers, P. D. Jagtap, J. E. Johnson, T. McGowan, P. Kumar, G. Onsongo, C. R. Guerrero, H. Barsnes, M. Vaudel, L. Martens, B. Grüning, I. R. Cooke, M. Heydarian, K. L. Reddy and T. J. Griffin, *Cancer Res.*, 2017, **77**, e43–e46.
42. J. Crappé, E. Ndah, A. Koch, S. Steyaert, D. Gawron, S. De Keulenaer, E. De Meester, T. De Meyer, W. Van Criekinge, P. Van Damme and G. Menschaert, *Nucleic Acids Res.*, 2015, **43**, e29.
43. P. Cifani, A. Dhabaria, Z. Chen, A. Yoshimi, E. Kawaler, O. Abdel-Wahab, J. T. Poirier and A. Kentsis, *J. Proteome Res.*, 2018, **17**, 3681–3692.
44. K. Verheggen, H. Raeder, F. S. Berven, L. Martens, H. Barsnes and M. Vaudel, *Mass Spectrom. Rev.*, DOI: 10.1002/mas.21543.
45. T. Muth, F. Hartkopf, M. Vaudel and B. Y. Renard, *Proteomics*, 2018, **18**, e1700150.
46. E. Mørtz, P. B. O'Connor, P. Roepstorff, N. L. Kelleher, T. D. Wood, F. W. McLafferty and M. Mann, *Proc. Natl. Acad. Sci. U. S. A.*, 1996, **93**, 8264–8267.
47. T. K. Toby, L. Fornelli and N. L. Kelleher, *Annu. Rev. Anal. Chem.*, 2016, **9**, 499–519.
48. V. Dorfer, S. Maltsev, S. Winkler and K. Mechtler, *J. Proteome Res.*, 2018, **17**, 2581–2589.
49. C.-C. Tsou, D. Avtonomov, B. Larsen, M. Tucholska, H. Choi, A.-C. Gingras and A. I. Nesvizhskii, *Nat. Methods*, 2015, **12**, 258–264, 7 p. following 264.

50. M. Bern, G. Finney, M. R. Hoopmann, G. Merrihew, M. J. Toth and M. J. MacCoss, *Anal. Chem.*, 2010, **82**, 833–841.
51. S. Purvine, J.-T. Eppel, E. C. Yi and D. R. Goodlett, *Proteomics*, 2003, **3**, 847–850.
52. J. D. Egertson, B. MacLean, R. Johnson, Y. Xuan and M. J. MacCoss, *Nat. Protoc.*, 2015, **10**, 887–903.
53. B. Van Puyvelde, S. Willems, R. Gabriels, S. Daled, L. De Clerck, A. Staes, F. Impens, D. Deforce, L. Martens, S. Degroeve and M. Dhaenens, *bioRxiv*, DOI: 10.1101/681429.
54. L. V. Schaffer, R. J. Millikin, R. M. Miller, L. C. Anderson, R. T. Fellers, Y. Ge, N. L. Kelleher, R. D. LeDuc, X. Liu, S. H. Payne, L. Sun, P. M. Thomas, T. Tucholski, Z. Wang, S. Wu, Z. Wu, D. Yu, M. R. Shortreed and L. M. Smith, *Proteomics*, 2019, **19**, e1800361.
55. A. I. Nesvizhskii and R. Aebersold, *Mol. Cell. Proteomics*, 2005, **4**, 1419–1440.
56. A. I. Nesvizhskii, *J. Proteomics*, 2010, **73**, 2092–2123.
57. A. Keller, A. I. Nesvizhskii, E. Kolker and R. Aebersold, *Anal. Chem.*, 2002, **74**, 5383–5392.
58. J. E. Elias and S. P. Gygi, *Nat. Methods*, 2007, **4**, 207–214.
59. N. Colaert, S. Degroeve, K. Helsens and L. Martens, *J. Proteome Res.*, 2011, **10**, 5555–5561.
60. R. J. Chalkley and K. R. Clauser, *Mol. Cell. Proteomics*, 2012, **11**, 3–14.
61. M. Bantscheff, M. Schirle, G. Sweetman, J. Rick and B. Kuster, *Anal. Bioanal. Chem.*, 2007, **389**, 1017–1031.
62. M. Bantscheff, S. Lemeer, M. M. Savitski and B. Kuster, *Anal. Bioanal. Chem.*, 2012, **404**, 939–965.
63. L. K. Pino, B. C. Searle, J. G. Bollinger, B. Nunn, B. MacLean and M. J. MacCoss, *Mass Spectrom. Rev.*, DOI: 10.1002/mas.21540.
64. S. Gessulat, T. Schmidt, D. P. Zolg, P. Samaras, K. Schnatbaum, J. Zerweck, T. Knaute, J. Rechenberger, B. Delanghe, A. Huhmer, U. Reimer, H.-C. Ehrlich, S. Aiche, B. Kuster and M. Wilhelm, *Nat. Methods*, 2019, **16**, 509–518.
65. M. Vaudel, A. Sickmann and L. Martens, *Proteomics*, 2010, **10**, 650–670.
66. S.-E. Ong, B. Blagoev, I. Kratchmarova, D. B. Kristensen, H. Steen, A. Pandey and M. Mann, *Mol. Cell. Proteomics*, 2002, **1**, 376–386.
67. T. Geiger, J. Cox, P. Ostasiewicz, J. R. Wisniewski and M. Mann, *Nat. Methods*, 2010, **7**, 383–385.
68. L. Nilse, F. C. Sigloch, M. L. Biniossek and O. Schilling, *Proteomics: Clin. Appl.*, 2015, **9**, 706–714.
69. M. Mann, C. K. Meng and J. B. Fenn, *Anal. Chem.*, 1989, **61**, 1702–1708.
70. M. The and L. Käll, *bioRxiv*, DOI: 10.1101/488015.
71. J. Cox and M. Mann, *Nat. Biotechnol.*, 2008, **26**, 1367–1372.
72. J. Cox, M. Y. Hein, C. A. Luber, I. Paron, N. Nagaraj and M. Mann, *Mol. Cell. Proteomics*, 2014, **13**, 2513–2526.

73. J. Cox, I. Matic, M. Hilger, N. Nagaraj, M. Selbach, J. V. Olsen and M. Mann, *Nat. Protoc.*, 2009, **4**, 698–705.
74. P. L. Ross, Y. N. Huang, J. N. Marchese, B. Williamson, K. Parker, S. Hattan, N. Khainovski, S. Pillai, S. Dey, S. Daniels, S. Purkayastha, P. Juhasz, S. Martin, M. Bartlet-Jones, F. He, A. Jacobson and D. J. Pappin, *Mol. Cell. Proteomics*, 2004, **3**, 1154–1169.
75. A. Thompson, J. Schäfer, K. Kuhn, S. Kienle, J. Schwarz, G. Schmidt, T. Neumann, R. Johnstone, A. K. A. Mohammed and C. Hamon, *Anal. Chem.*, 2003, **75**, 1895–1904.
76. C. D. Wenger, M. V. Lee, A. S. Hebert, G. C. McAlister, D. H. Phanstiel, M. S. Westphall and J. J. Coon, *Nat. Methods*, 2011, **8**, 933–935.
77. L. Ting, R. Rad, S. P. Gygi and W. Haas, *Nat. Methods*, 2011, **8**, 937–940.
78. M. Vaudel, J. M. Burkhart, S. Radau, R. P. Zahedi, L. Martens and A. Sickmann, *J. Proteome Res.*, 2012, **11**, 5072–5080.
79. M. Vaudel, in *Proteome Informatics*, ed. C. Bessant, Royal Society of Chemistry, Cambridge, 2016, pp. 155–177.
80. J. M. Asara, H. R. Christofk, L. M. Freimark and L. C. Cantley, *Proteomics*, 2008, **8**, 994–999.
81. B. Schwanhäusser, D. Busse, N. Li, G. Dittmar, J. Schuchhardt, J. Wolf, W. Chen and M. Selbach, *Nature*, 2011, **473**, 337–342.
82. Y. Ishihama, Y. Oda, T. Tabata, T. Sato, T. Nagasu, J. Rappsilber and M. Mann, *Mol. Cell. Proteomics*, 2005, **4**, 1265–1272.
83. L. Florens, M. J. Carozza, S. K. Swanson, M. Fournier, M. K. Coleman, J. L. Workman and M. P. Washburn, *Methods*, 2006, **40**, 303–311.
84. A.-M. Hesse, V. Dupierris, C. Adam, M. Court, D. Barthe, A. Emadali, C. Masselon, M. Ferro and C. Bruley, *J. Proteome Res.*, 2016, **15**, 3896–3903.
85. S. Tyanova, T. Temu, P. Sinitcyn, A. Carlson, M. Y. Hein, T. Geiger, M. Mann and J. Cox, *Nat. Methods*, 2016, **13**, 731–740.
86. D. Wang, B. Eraslan, T. Wieland, B. Hallström, T. Hopf, D. P. Zolg, J. Zecha, A. Asplund, L.-H. Li, C. Meng, M. Frejno, T. Schmidt, K. Schnatbaum, M. Wilhelm, F. Ponten, M. Uhlen, J. Gagneur, H. Hahne and B. Kuster, *Mol. Syst. Biol.*, 2019, **15**, e8503.
87. G. Menschaert and D. Fenyö, *Mass Spectrom. Rev.*, 2017, **36**, 584–599.
88. P. Mertins, D. R. Mani, K. V. Ruggles, M. A. Gillette, K. R. Clauser, P. Wang, X. Wang, J. W. Qiao, S. Cao, F. Petralia, E. Kawaler, F. Mundt, K. Krug, Z. Tu, J. T. Lei, M. L. Gatza, M. Wilkerson, C. M. Perou, V. Yellapantula, K. Huang, C. Lin, M. D. McLellan, P. Yan, S. R. Davies, R. R. Townsend, S. J. Skates, J. Wang, B. Zhang, C. R. Kinsinger, M. Mesri, H. Rodriguez, L. Ding, A. G. Paulovich, D. Fenyö, M. J. Ellis, S. A. Carr and NCI CPTAC, *Nature*, 2016, **534**, 55–62.
89. J. A. Stefely, N. W. Kwiecien, E. C. Freiberger, A. L. Richards, A. Jochem, M. J. P. Rush, A. Ulbrich, K. P. Robinson, P. D. Hutchins, M. T. Veling, X. Guo, Z. A. Kemmerer, K. J. Connors, E. A. Trujillo, J. Sokol, H. Marx, M. S. Westphall, A. S. Hebert, D. J. Pagliarini and J. J. Coon, *Nat. Biotechnol.*, 2016, **34**, 1191–1197.

90. J. Boekel, J. M. Chilton, I. R. Cooke, P. L. Horvatovich, P. D. Jagtap, L. Käll, J. Lehtiö, P. Lukasse, P. D. Moerland and T. J. Griffin, *Nat. Biotechnol.*, 2015, **33**, 137–139.
91. C. Meng, B. Kuster, A. C. Culhane and A. M. Gholami, *BMC Bioinf.*, 2014, **15**, 162.
92. L. F. Hernandez Sanchez, B. Burger, C. Horro, A. Fabregat, S. Johansson, P. R. Njolstad, H. Barsnes, H. Hermjakob and M. Vaudel, *bioRxiv*, 2018, 375097.
93. J. D. Chavez, J. P. Mohr, M. Mathay, X. Zhong, A. Keller and J. E. Bruce, *Nat. Protoc.*, , DOI: 10.1038/s41596-019-0181-3.
94. A. J. R. Heck, T. P. Woerner, J. Snijder, A. D. Bennett, M. Agbandje-McKenna and A. Makarov, *bioRxiv*, DOI: 10.1101/717413.
95. N. D. Price, A. T. Magis, J. C. Earls, G. Glusman, R. Levy, C. Lausted, D. T. McDonald, U. Kusebauch, C. L. Moss, Y. Zhou, S. Qin, R. L. Moritz, K. Brogaard, G. S. Omenn, J. C. Lovejoy and L. Hood, *Nat. Biotechnol.*, 2017, **35**, 747–756.
96. K. Sidiropoulos, G. Viteri, C. Sevilla, S. Jupe, M. Webber, M. Orlic-Milacic, B. Jassal, B. May, V. Shamovsky, C. Duenas, K. Rothfels, L. Matthews, H. Song, L. Stein, R. Haw, P. D'Eustachio, P. Ping, H. Hermjakob and A. Fabregat, *Bioinformatics*, 2017, **33**, 3461–3467.
97. B. Burger, L. F. Hernández Sánchez, R. R. Lereim, H. Barsnes and M. Vaudel, *J. Proteome Res.*, 2018, **17**, 3801–3809.
98. A. R. Sonawane, S. T. Weiss, K. Glass and A. Sharma, *Front. Genet.*, 2019, **10**, 294.
99. D. Türei, T. Korcsmáros and J. Saez-Rodriguez, *Nat. Methods*, 2016, **13**, 966–967.
100. T. Rolland, M. Taşan, B. Charloteaux, S. J. Pevzner, Q. Zhong, N. Sahni, S. Yi, I. Lemmens, C. Fontanillo, R. Mosca, A. Kamburov, S. D. Ghiassian, X. Yang, L. Ghamsari, D. Balcha, B. E. Begg, P. Braun, M. Brehme, M. P. Broly, A.-R. Carvunis, D. Convery-Zupan, R. Corominas, J. Coulombe-Huntington, E. Dann, M. Dreze, A. Dricot, C. Fan, E. Franzosa, F. Gebreab, B. J. Gutierrez, M. F. Hardy, M. Jin, S. Kang, R. Kiros, G. N. Lin, K. Luck, A. MacWilliams, J. Menche, R. R. Murray, A. Palagi, M. M. Poulin, X. Rambout, J. Rasla, P. Reichert, V. Romero, E. Ruyssinck, J. M. Sahalie, A. Scholz, A. A. Shah, A. Sharma, Y. Shen, K. Spirohn, S. Tam, A. O. Tejeda, S. A. Trigg, J.-C. Twizere, K. Vega, J. Walsh, M. E. Cusick, Y. Xia, A.-L. Barabási, L. M. Iakoucheva, P. Aloy, J. De Las Rivas, J. Tavernier, M. A. Calderwood, D. E. Hill, T. Hao, F. P. Roth and M. Vidal, *Cell*, 2014, **159**, 1212–1226.
101. J.-F. Rual, K. Venkatesan, T. Hao, T. Hirozane-Kishikawa, A. Dricot, N. Li, G. F. Berriz, F. D. Gibbons, M. Dreze, N. Ayivi-Guedehoussou, N. Klitgord, C. Simon, M. Boxem, S. Milstein, J. Rosenberg, D. S. Goldberg, L. V. Zhang, S. L. Wong, G. Franklin, S. Li, J. S. Albala, J. Lim, C. Fraughton, E. Llamosas, S. Cevik, C. Bex, P. Lamesch, R. S. Sikorski, J. Vandenhaute, H. Y. Zoghbi, A. Smolyar, S. Bosak, R. Sequerra, L. Doucette-Stamm, M. E. Cusick, D. E. Hill, F. P. Roth and M. Vidal, *Nature*, 2005, **437**, 1173–1178.

102. M. Wilhelm, J. Schlegl, H. Hahne, A. M. Gholami, M. Lieberenz, M. M. Savitski, E. Ziegler, L. Butzmann, S. Gessulat, H. Marx, T. Mathieson, S. Lemeer, K. Schnatbaum, U. Reimer, H. Wenschuh, M. Mollenhauer, J. Slotta-Huspenina, J.-H. Boese, M. Bantscheff, A. Gerstmair, F. Faerber and B. Kuster, *Nature*, 2014, **509**, 582–587.
103. S. Gupta, K. Verheggen, J. Tavernier and L. Martens, *J. Proteome Res.*, 2017, **16**, 2204–2212.
104. P.-J. Volders, J. Anckaert, K. Verheggen, J. Nuytens, L. Martens, P. Mestdagh and J. Vandesompele, *Nucleic Acids Res.*, 2019, **47**, D135–D139.
105. J. A. Vizcaíno, E. W. Deutsch, R. Wang, A. Csordas, F. Reisinger, D. Ríos, J. A. Dianes, Z. Sun, T. Farrah, N. Bandeira, P.-A. Binz, I. Xenarios, M. Eisenacher, G. Mayer, L. Gatto, A. Campos, R. J. Chalkley, H.-J. Kraus, J. P. Albar, S. Martinez-Bartolomé, R. Apweiler, G. S. Omenn, L. Martens, A. R. Jones and H. Hermjakob, *Nat. Biotechnol.*, 2014, **32**, 223–226.
106. M. M. Savitski, M. Wilhelm, H. Hahne, B. Kuster and M. Bantscheff, *Mol. Cell. Proteomics*, 2015, **14**, 2394–2404.
107. K. Verheggen and L. Martens, *EuPa Open Proteomics*, 2015, **8**, 28–35.

CHAPTER 5

# *Statistics, Data Mining and Modeling*†

MIGUEL REBOIRO-JATO[a, b, c], DANIEL GLEZ-PEÑA[a, b, c] AND HUGO LÓPEZ-FERNÁNDEZ*[a, b, c]

[a]Department of Computer Science, University of Vigo, ESEI, Campus As Lagoas, 32004 Ourense, Spain; [b]The Biomedical Research Centre (CINBIO), Campus Universitario Lagoas-Marcosende, 36310 Vigo, Spain; [c]SING Research Group, Galicia Sur Health Research Institute (ISS Galicia Sur), SERGAS-UVIGO, Spain
*E-mail: hlfernandez@uvigo.es

## 5.1 Sample Comparison

This section aims to provide several strategies to perform different types of sample comparison. As in the remaining sections in this chapter, the input data used for these analyses are peak lists from MALDI-TOF MS experiments. This means that the raw spectra of each sample must have been pre-processed previously (as described in Chapter 2) in order to obtain peak lists that typically contain tens or hundreds of peaks. Once the peak lists have been obtained, a final pre-processing step called peak binning (or *m*/*z* matching) is required in order to allow direct comparison of the different

†The source code used in the examples in this chapter is publicly available at https://doi.org/10.5281/zenodo.2654335.

New Developments in Mass Spectrometry No. 8
Processing Metabolomics and Proteomics Data with Open Software: A Practical Guide
Edited by Robert Winkler

Published by the Royal Society of Chemistry, www.rsc.org

spectra. Before the peak binning, the same peak (representing a biological entity) can have different *m*/*z* values across the samples. The peak binning procedure finds a set of common *m*/*z* locations in several spectra and reassigns them so that all spectra will have the same *m*/*z* values for the same biological entities. Section 3.4.1.2.1 of Chapter 3 and Section 8.4.14 of Chapter 8 give brief explanations of this procedure, which in MALDIquant is implemented by the binPeaks function.

The examples in this section are illustrated using a dataset composed of sera from five patients with lymphoma, five patients with myeloma, and two healthy donors.[1] This dataset is further referenced as the Lymphoma/Myeloma dataset. Each sample has five technical replicates and the spectra have been pre-processed previously (*i.e.* each spectra is a peak list). This dataset is appropriate for this section because its reduced size shows easily how the different sample comparison techniques described work. As the practical examples are done using R, the code in the `load-cancer.R` file shows how to: (i) download this dataset from the internet, (ii) load the spectra into `MassPeak` objects using the `createMassPeaks` function from MALDIquant library, and (iii) perform the peak binning using the `binPeaks` function from MALDIquant. The data matrix is finally available at the `binnedPeaksMatrix` variable, which contains spectra in rows and peaks in columns. In addition, the `data` variable contains the source information: `data$datasetConditions`, which contains the labels of the conditions in the dataset; `data$datasetConditionsColors`, which contains the condition colors (healthy in red, lymphoma in blue and myeloma in green); `data$sampleNames`, which contains the names of the samples in the dataset; `data$sampleConditions`, which contains the condition labels associated to the samples in the dataset; `data$samplesColors`, which contains the condition color of each sample; `data$spectraNames`, which contains the spectra names; `data$spectraSampleNames`, which contains the sample names that correspond to each spectra; `data$spectra`, which contains the `MassPeak` objects; `data$spectraConditions`, which contains the class or condition of the corresponding sample; and `data$spectraColors`, which contains the condition color of each spectra.

The `binnedPeaksMatrix` object contains 60 rows, representing each technical replicate in the dataset. Sometimes it is useful or interesting to analyze the data at sample level. To do this, a consensus spectrum is created for each sample using its technical replicates spectra. In this process, the percentage of presence (POP) parameter is used to define the minimum number of replicates where a peak must be present (*i.e.* have an intensity greater than 0) in order to be considered a consensus peak.[2] Consensus peaks are created with the mean intensity of the peaks having an intensity greater than 0. The file `data-functions.R` defines a function named `toConsensusSpectraData` to convert source data containing replicates (available in the `data` object) into a new object containing a consensus spectra for each sample. This function also receives a `tolerance` parameter to perform the peak binning of the replicates, necessary to identify the common peaks of the replicates. The function is used to convert the data and then this data is binned to obtain a binned peaks matrix as in the previous case. This `consensusBinnedPeaksMatrix` variable of the `load-cancer.R` file contains the consensus spectra and is, in general, interchangeable with the `binnedPeaksMatrix` variable in the examples.

## 5.1.1 Distance Measures

This section explains the most common distance measures that can be used to compare spectra in pairs of two.

These measures are also useful in other tasks explained in different sections of this chapter, such as the hierarchical clustering analysis. The source code to reproduce all the examples in this section can be found in the `distance-measures.R` file.

### *5.1.1.1 Euclidean Distance*

The Euclidean distance comes from Euclidean geometry and is the ordinary straight-line distance between two points in the Euclidean space. In a 2D space, the Euclidean distance between two points *a* and *b* is the length of the straight line between *a* and *b*. This concept is extensible to *n*-dimensional spaces. Given two points (*a*, *b*) in an *n*-dimensional space, the Euclidean distance is given by:

$$\text{dist}(a,b) = \sqrt{\sum_{i}^{n}(a_i - b_i)^2} \tag{5.1}$$

Intuitively, the Euclidean distance compares all coordinates of a given point *a* with the values at the respective coordinates in the point *b*. The Euclidean distance is always a non-negative real number, with a 0 value for points in the same location. For example, given the two points in a 3D space:

$$a = (4, -7, 5);\ b = (3, 5, -1) \tag{5.2}$$

The distance is computed as follows:

$$\text{dist}(a,b) = \sqrt{(4-3)^2 + (-7-5)^2 + (5-(-1))^2} = 13.45362 \tag{5.3}$$

The Euclidean distance can be used to measure the difference between two spectra. Each spectrum is taken as a point, where the dimensions are the distinct masses and the intensities are the coordinates of the spectrum in that space. Thus, when comparing two spectra, the Euclidean distance indicates how different they are in terms of their intensities on each mass. In the following example, the Euclidean distance between the first two spectra of the Lymphoma/Myeloma dataset is computed.

```
spectrum_1 <- binnedPeaksMatrix[1,]
spectrum_2 <- binnedPeaksMatrix[2,]
dist(rbind(spectrum_1, spectrum_2), method = "euclidean")
#            spectrum_1
# spectrum_2  0.5017978
```

### 5.1.1.2 Manhattan Distance

As the Euclidean distance, the Manhattan distance is also a geometric distance, with the difference that the resulting distance is not the shortest path between two points, but the sum of the differences of all coordinates. The name of this distance comes from the idea that the path from one place to another in a city, which can be seen as a grid of house blocks, is not a straight line, but a "snake" path between the blocks. The length of this path is equal to the sum of the differences on *x*-axis and the *y*-axis. Given two points $(a, b)$ in an $n$-dimensional space, the Manhattan distance is given by:

$$\text{dist}(a,b) = \sum_{i}^{n} |a_i - b_i| \tag{5.4}$$

The Manhattan distance is always a non-negative real number, with a 0 value for points in the same location. If two points are the same, the distance is 0. For example, given the two points in a 2D space:

$$a = (4, -7, 5);\ b = (3, 5, -1) \tag{5.5}$$

The distance is computed as follows:

$$\text{dist}(a,b) = |4-3| + |-7-5| + |5-(-1)| = 1 + 12 + 6 = 19 \tag{5.6}$$

The Manhattan distance may be more intuitive than Euclidean in some cases. For example, to compare two spectra in terms of the presence/absence of peaks, their peaks can be encoded as 1 and 0 (1 = the peak is present, 0 = the peak is absent). This way, the number given by the Manhattan distance will be a natural number counting the peak discrepancies between the two spectra. Moreover, the Manhattan distance has been proven to give better results in data mining than Euclidean for high-dimensional data.[3] In the following example, the Manhattan distance between the first two spectra of the Lymphoma/Myeloma dataset is computed (note that, before distance calculation, the intensity matrix is converted into a presence/absence matrix using an auxiliary function called `asPresenceMatrix` defined in the `data-functions.R` file).

```
presenceBinnedPeaksMatrix <- asPresenceMatrix(binnedPeaksMatrix)
spectrum_1 <- presenceBinnedPeaksMatrix[1,]
spectrum_2 <- presenceBinnedPeaksMatrix[2,]
dist(rbind(spectrum_1, spectrum_2), method = "manhattan")
#              spectrum_1
# spectrum_2         41
```

### *5.1.1.3 Jaccard Index*

The Jaccard index computes the similarity between two non-empty sets. It is a very simple and intuitive measure for comparing sets. Given two sets of elements $A$ and $B$, the Jaccard index is the ratio of the size of the intersection of $A$ and $B$ (*i.e.* the number of common elements) to the size of the union of $A$ and $B$ (*i.e.* the total number of different elements found in $A$ and $B$):

$$\text{Jaccard}(A,B) = \frac{|A \cap B|}{|A \cup B|} \tag{5.7}$$

The Jaccard index takes values between 0 and 1. If $A$ and $B$ do not share any element, the Jaccard index is 0, whereas if $A$ and $B$ are exactly the same set of elements, the Jaccard index is 1. For example, given the following two sets:

$$A = \{\text{“apple”}, \text{“banana”}, \text{“tomato”}\}, B = \{\text{“apple”}, \text{“orange”}, \text{“tomato”}\} \tag{5.8}$$

$A$ and $B$ share “apple” and “tomato”, whereas “banana” and “orange” are exclusive elements of $A$ and $B$, respectively. The Jaccard index is calculated as follows:

$$\text{Jaccard}(A,B) = \frac{|A \cap B|}{|A \cup B|} = \frac{|\{\text{"apple","tomato"}\}|}{|\{\text{"apple","orange","tomato","banana"}\}|} = \frac{2}{4} = 0.5 \tag{5.9}$$

Thus, the Jaccard index can be used to measure how different two spectra are, considering each spectrum just as a “set of masses” and ignoring their intensities. In the following example, the Jaccard index between the first two spectra in the Lymphoma/Myeloma dataset is computed.

```
jaccard <- function(a, b) {
    length(intersect(a, b)) / length(union(a, b))
}

all_masses <- colnames(binnedPeaksMatrix)
masses_1 <- all_masses[binnedPeaksMatrix[1,] != 0]
masses_2 <- all_masses[binnedPeaksMatrix[2,] != 0]
jaccard(masses_1, masses_2)
# [1] 0.7588235
```

In order to make the Jaccard index give a sense of distance, `1 - jaccard` can be taken. This way, 0 means same sets (minimum distance) and 1 means totally different sets (maximum distance).

### 5.1.1.4 Pearson Correlation

The Pearson correlation coefficient is a statistical measure of the linear relation between two variables $X$ and $Y$. Given a set of data points $(x_i, y_i)$, this coefficient gives a measure of the linear correlation between values of $X$ and values of $Y$. This value and is computed as the mean of the product of the differences of each coordinate to its mean divided by the product of the standard deviations of $X$ and $Y$:

$$\text{Pearson}(X,Y) = \frac{\sum_{i}^{n}(x_i - \bar{x})(y_i - \bar{y})}{\sqrt{\sum_{i}^{n}(x_i - \bar{x})^2}\sqrt{\sum_{i}^{n}(y_i - \bar{y})^2}} \tag{5.10}$$

Where $\bar{x}$ and $\bar{y}$ are the mean values of the $X$ and $Y$ variables, respectively.

The Pearson correlation takes values between −1 and 1, where −1 means total negative correlation, 0 no correlation, and 1 total positive correlation. For example, given the following two vectors of the variables $X$ and $Y$:

$$x = (3, 5, 1), y = (4, 7, -5) \tag{5.11}$$

The Pearson correlation is:

$$\text{Pearson}(x,y) = \frac{(3-3)(4-2)+(5-3)(7-2)+(1-3)(-5-2)}{\sqrt{(3-3)^2+(5-3)^2+(1-3)^2}\sqrt{(4-2)^2+(7-2)^2+(-5-2)^2}} = 0.9607 \tag{5.12}$$

The Pearson correlation coefficient can also be used as a distance measure. When comparing two mass spectrometry spectra, $X$ and $Y$ are the sets of intensities of the two spectra. In order to use it as a distance metric, the distance between two spectra $X$ and $Y$ is defined as:

$$\text{Pearson.distance}\,(X, Y) = 1 - \text{Pearson}\,(X, Y) \tag{5.13}$$

This way, the possible values range from 0 (perfectly correlated, *i.e.* minimum distance) over 1 (uncorrelated) to 2 (perfectly anti-correlated, *i.e.* maximum distance). In the following example, the Pearson correlation distance between the first two spectra of the Lymphoma/Myeloma dataset is computed.

```
spectrum_1 <- binnedPeaksMatrix[1,]
spectrum_2 <- binnedPeaksMatrix[2,]
1 - cor(spectrum_1, spectrum_2, method = "pearson")
# [1] 0.02520005
```

### 5.1.1.5 *Visualizing Distances in Multiple Samples*

Given a set of spectra, the pairwise distances between all of them can be computed using any of the above distance metrics. This results is a dissimilarity matrix where each cell contains the distance between a pair of spectra. Note that this results in a symmetric matrix because the distance metrics are commutative operations (*i.e.* the distance between spectra $A$ and $B$ is equal to the distance between $B$ and $A$).

Computing the distance matrix using Euclidean and Manhattan distances is straightforward in R (note that in the Manhattan distance the presence/absence peaks matrix should be used):

```
distances <- dist(binnedPeaksMatrix, method="euclidean")

presenceBinnedPeaksMatrix <- asPresenceMatrix(binnedPeaksMatrix)
distances <- dist(presenceBinnedPeaksMatrix, method = "manhattan")
```

For the Jaccard distance, which is not a distance present in the R base packages, an auxiliary function for computing this matrix given a binned peaks matrix must be created. As explained previously, in order to make Jaccard give a sense of distance, `1 - jaccard` is taken, so that 0 means the same sets and 1 means totally different sets.

```
toSpectraList <- function(binnedPeaksMatrix) {
    spectraData <- list()

    for (i in 1:nrow(binnedPeaksMatrix)) {
        intensities <- as.numeric(binnedPeaksMatrix[i,])
        masses <- as.numeric(as.character(colnames(binnedPeaksMatrix)))

        masses <- masses[intensities > 0]
        intensities <- intensities[intensities > 0]

        spectrum <- createMassPeaks(mass = masses, intensity = intensities, metadata
= list(name=rownames(binnedPeaksMatrix)[i]))
        spectraData <- c(spectraData, spectrum)
    }
    spectraData
}

jaccard.dist  <- function(dataNames, spectraData) {
    result <- matrix(nrow = length(dataNames), ncol = length(dataNames))
    colnames(result) <- dataNames
    rownames(result) <- dataNames

    for(i in 1:length(spectraData)) {
        currentValues <- rep(NA, length(dataNames))
        for(j in 1:length(spectraData)) {
            massesA <- mass(spectraData[[i]])
            massesB <- mass(spectraData[[j]])
            currentValues[j] <- 1 - jaccard(massesA, massesB)
        }
        result[i,] <- currentValues
    }

    as.dist(result)
}

distances <- jaccard.dist(rownames(binnedPeaksMatrix),
toSpectraList(binnedPeaksMatrix))
```

Finally, to obtain a distance matrix with Pearson distances, a `pearson.dist` function is defined to create a distance object with the pairwise distances between all the samples in a peaks matrix object (`binnedPeaksMatrix` or `consensusBinnedPeaksMatrix`). As explained previously, the possible values range from 0 (perfectly correlated, *i.e.* minimum distance) over 1 (uncorrelated) to 2 (perfectly anti-correlated, *i.e.* maximum distance).

```
pearson.dist <- function(binnedPeaksMatrix) {
    result <- 1 - cor(t(binnedPeaksMatrix), t(binnedPeaksMatrix), method = "pearson")

    as.dist(result)
}
distances <- pearson.dist(binnedPeaksMatrix)
```

Once a distances object is calculated and assigned to the `distances` variable, the ggplot2 library can be used to create a false color image visualization. For instance, Figure 5.1 shows the visualization of the Pearson

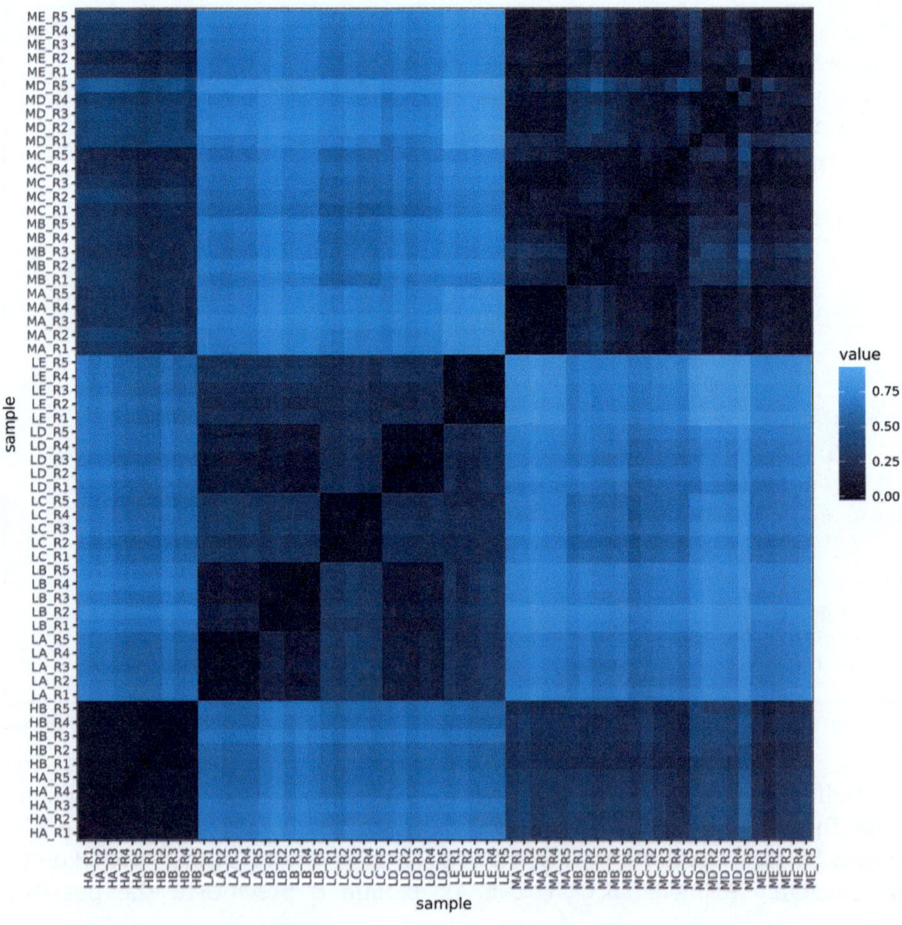


**Figure 5.1** Visualization of the Pearson correlation distance matrix as a false color image.

correlation distances matrix. Here, the darkest colors indicate smaller distances and lighter ones indicate greater distances. As can be seen from this figure, there are three main dark squares in the diagonal that indicate that the similarity between replicates from the same class is higher than between replicates from different classes. Interestingly, it seems also that there is some similarity between healthy and myeloma samples.

```
library("ggplot2")

distanceMatrix <- as.matrix(distances)
sampleNames <- rownames(binnedPeaksMatrix)
dataPlot <- data.frame(row.names = seq(length(sampleNames) * length(sampleNames)))
dataPlot$sampleA <- rep(sampleNames, each = length(sampleNames))
dataPlot$sampleB <- rep(sampleNames, length(sampleNames))
dataPlot$value <- as.vector(t(distanceMatrix))
library(ggplot2)
ggplot(data = dataPlot,
       aes(x = sampleA, y = sampleB, fill = value)) + geom_tile() +
       theme_bw() + xlab("sample") + ylab("sample") + theme(axis.text.x =
element_text(angle = 90, hjust = 1))
```

## 5.1.2 Multiple Sample Visualization

Apart from doing statistical and mathematical sample comparisons, it is also useful to graphically represent two or more samples in order to visually inspect and assess their differences. Note that, within the scope of this chapter, as explained in the introduction, samples consist of binned peak lists, typically containing dozens or hundreds of peaks. To this end, different types of visualizations are particularly useful. The source code to reproduce all the examples in this section can be found in the `multiple-sample-visualization.R` and `multiple-sample-visualization-functions.R` files.

The first one consists of plotting the intensities of the peaks using two or more samples (usually the entire dataset) in rows and the peaks in columns (Figure 5.2A). For this visualization, which can be called the intensity matrix, the `heatmap.2` function (from the gplots R package) can be used with the following settings: disable rows and column clusterings, disable dendrograms visualization, turn off trace lines inside the heat map, turn off the density plot inside the color legend and using a color palette starting with white (for the low peak intensity) and ending in black (for the highest peak intensity). The following code snippet defines a function that creates this visualization using a peaks matrix (*e.g.* `binnedPeaksMatrix`).

```
library("gplots")
plotIntensityMatrix <- function(peaksMatrix) {
    colorPalette <- colorpanel(n = 10, low = "white", high = "black");

    heatmap.2(
        peaksMatrix,    # peaks matrix
        main = "",         # heat map title
        notecol = "black",    # change font color of cell labels to black
        density.info = "none",    # turns off density plot inside color legend
        key = FALSE,      # turns off the color key
        trace = "none",        # turns off trace lines inside the heat map
        margins = c(12,9),    # widens margins around plot
        col = colorPalette,       # use on color palette defined earlier
        dendrogram = "none",         # disable dendogram
        Colv = "NA",         # disable columns clustering
        Rowv = "NA",        # disable rows clustering,
        lwid = c(0.5, 5), # adjust plot margins
        lhei = c(0.5, 5) # adjust plot margins
    )
}
plotIntensityMatrix(binnedPeaksMatrix)
```

As Figure 5.2A shows, this visualization allows the distribution of the most intense peaks across all the samples to be identified in a straightforward manner. In addition, more detailed plots (*e.g.* including specific samples or showing only specific peaks) can be created by simply subsetting the peaks matrix passed to the `heatmap.2` function.

The second visualization aims to compare two binned spectra in order to discover what the shared and unshared peaks are. For instance, this visualization can be useful to compare two reference or representative spectra with the goal of discovering whether the shared peaks are located at specific mass ranges and whether there are mass ranges that are unique to one of the spectra. Figure 5.2B shows an example of this visualization, where shared peaks are represented in red and unshared peaks in gray and green, depending on the spectra they belong to. As the following code snippet shows, this type of plot can be created easily in R using the `abline` function to plot one vertical line of the corresponding color at each peak location. The code automatically breaks the mass range into specified subranges (controlled by the `comparison.parts` variable), plotting one subplot for each one. The parameter `comparison.lwd` may need to be adjusted to reduce the width of the lines if peaks are too close to each other.

```
a <- binnedPeaksMatrix[1,]
b <- binnedPeaksMatrix[2,]

col.onlyA <- "#7285a5"
col.onlyB <- "#4b560e"
col.both <- "#ff7369"

comparison.parts <- 6
comparison.lwd <- 2
masses <- as.numeric(colnames(binnedPeaksMatrix))
min.mass <- min(as.numeric(colnames(binnedPeaksMatrix)))
max.mass <- max(as.numeric(colnames(binnedPeaksMatrix)))
comparison.range <- (max.mass - min.mass) / comparison.parts

par(mfrow = c(comparison.parts, 1), mar = c(2, 2, 2, 2), oma = c(4, 4, 4, 4))

for (part in 1:comparison.parts) {
    peak.start <- min.mass + (part - 1) * comparison.range
    peak.end <- min(max.mass, min.mass + part * comparison.range)

    indexes <- intersect(which(masses > peak.start), which(masses < peak.end))

    plot(NULL, xlim = c(peak.start, peak.end), ylim = c(0,1), xlab = "m/z", ylab =
NA, yaxt = 'n')

    for (index in 1:length(indexes)) {
        i <- indexes[index]
        if (a[i] > 0 && b[i] == 0) {
            abline(v = as.numeric(colnames(binnedPeaksMatrix)[i]), col = col.onlyA,
lwd = comparison.lwd)
        } else if (a[i] == 0 && b[i] > 0) {
            abline(v = as.numeric(colnames(binnedPeaksMatrix)[i]), col = col.onlyB,
lwd = comparison.lwd)
        } else if (a[i] > 0 && b[i] > 0) {
            abline(v = as.numeric(colnames(binnedPeaksMatrix)[i]), col = col.both,
lwd = comparison.lwd)
        }
    }
}
```

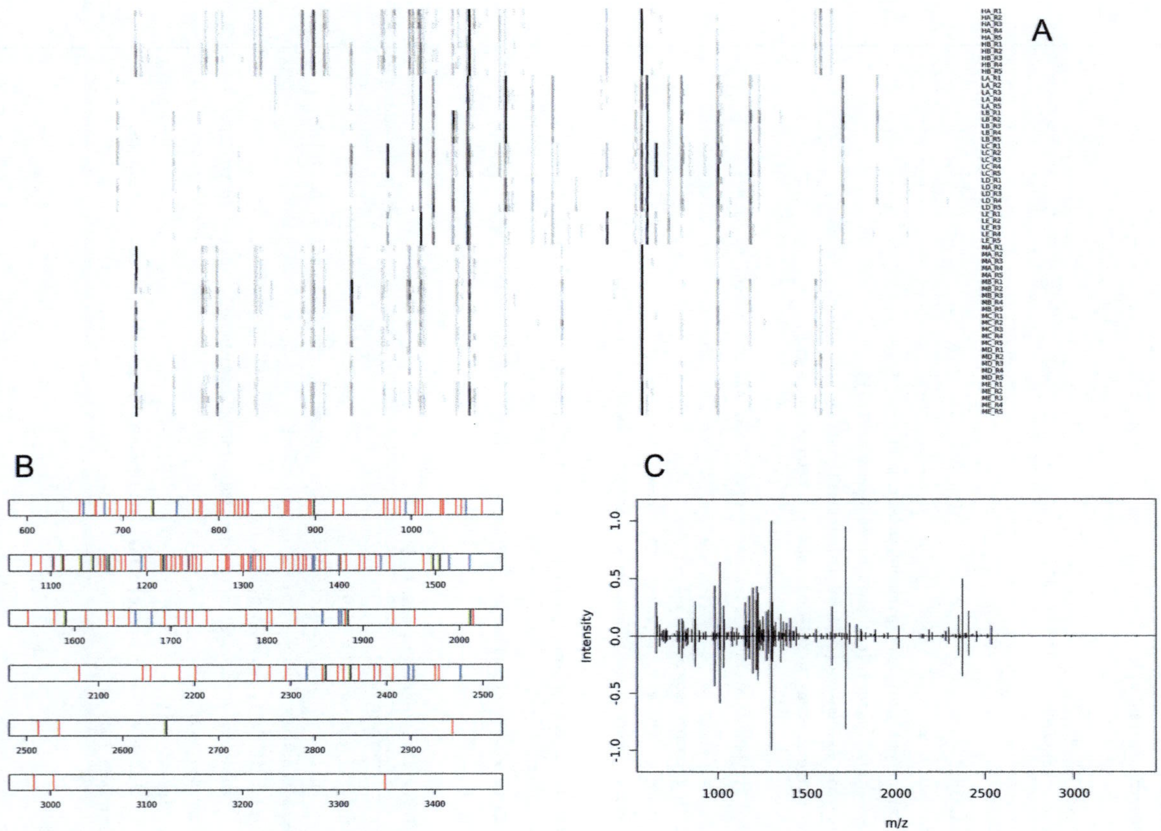


**Figure 5.2** (A) Intensity matrix visualization of the Lymphoma/Myeloma dataset. (B) Comparison of two binned spectra of the Lymphoma/Myeloma dataset in order to discover what the shared and unshared peaks are. (C) Inverse spectrum comparison of two spectra of the Lymphoma/Myeloma dataset.

```
positive <- data$spectra[[1]]
negative <- data$spectra[[2]]
plot(NULL, xlim = c(min(mass(positive), mass(negative)), max(mass(positive),
mass(negative))), ylim = c(-1.1,1.1), xlab = "m/z", ylab = "Intensity")
for (i in 1:length(positive)) {
    segments(mass(positive)[i], 0, y1 = intensity(positive)[i])
}
For (i in 1:length(negative)) {
    segments(mass(negative)[i], 0, y1 = -intensity(negative)[i])
}
abline(h = 0)
```

Finally, another useful type of visualization is the inverse spectrum comparison, where two spectra are plotted against each other in a mirror so that their intensities can be directly compared (Figure 5.2C). As in the previous visualization, this one can also be useful to compare two reference or representative spectra with the goal of discovering differences in their peak intensities. In addition, this type of visualization can be also used to compare two peak lists that have not been binned yet. These type of plots can be created easily in R using the `segments` function to plot vertical lines proportional to the peak intensities.

### 5.1.3 Outlier Detection

Sample comparison may be a useful tool to identify noisy samples in a dataset. By analyzing and comparing some characteristics (*e.g.* number of peaks, mean intensity of the peaks, *etc.*) of the samples it is possible to detect irregularities that may reveal that an error happened during sample analysis. In that case, samples should be discarded, as they may negatively affect the results.

The origin of noisy samples may be diverse, as sample analysis usually includes many steps where errors can occur, including sample acquisition, sample preparation, sample analysis and data analysis. The noise sources are diverse, but, in general, they can be categorized into (i) chemical noise, such as contaminants introduced during the sample handling or undesired fragmentation of peptides along their backbone caused by instability after receiving energy from the laser pulse in MALDI-TOF analyses, and (ii) electronic noise, that is implicit to electronic devices.[4] In general, noisy samples have their origin in chemical noise.

One simple but effective way to identify noisy samples is to analyze whether some of their characteristics can be considered as outliers when compared with other samples. An outlier observation is one that appears to deviate markedly from other related observations.[5] In some cases, this deviation is the result of an experimental error that leads to an abnormal observation value. When this happens, the observation should be removed from the data, as it will certainly affect the results of the experiment. However, in other cases, an outlier may be just an extreme manifestation of the random variability inherent in the data and, hence, it should be retained and processed in the same manner as the other observations. Therefore, outlier observations should not be automatically discarded and, instead, they should be carefully studied to determine if they were caused by an experimental error or if, conversely, they are legitimate values.

In this context, the characteristics to be compared must be selected previously depending on the type of analyses to be done. For example, when dealing with mass spectrometry data, one may want to analyze the number of peaks identified in each sample, the average intensity of the peaks or the lowest and highest peak value, as extreme values in these characteristics may be caused by experimental errors. In the ideal case, where the experimental design of the analysis includes technical or biological replicates, each sample can be compared with its corresponding replicates and the same values for the selected characteristics are expected. However, even if there are no replicates, the comparison of each sample with the remaining samples that comprise of the dataset can still be useful to detect outliers, although one should be more careful when interpreting the results.

Although there are several methods to detect outliers, one of the most commonly used is the Tukey's fences method,[6] based on the interquartile range (IQR). When using this method, the first quartile ($Q_1$) and third quartile ($Q_3$) are calculated as the values that hold 25% of the samples below or above them, respectively. Then, the IQR is calculated as follows:

$$\mathrm{IQR} = Q_1 - Q_3 \tag{5.14}$$

Thus, the IQR is the distance between both quartiles. On this basis, Tukey defines an outlier as any value that is 1.5 × IQR below $Q_1$ or above $Q_3$, and as "far out" when the distance is higher than 3 × IQR. Interestingly, when using this method the dataset is usually represented with a box and whiskers plot, as it facilitates the result interpretation. In this type of plot, a box is drawn using $Q_1$ and $Q_3$ as the bottom and top limits, respectively, and crossing it with the median value (*i.e.* the second quartile, $Q_2$). The "whiskers" are two lines extending from the box to the values that are nearest to the outlier limit without exceeding it. Sometimes, data appears grouped and the same plot includes several box-and-whiskers.

Fortunately, generating a box-and-whiskers plot and getting the corresponding outliers is very easy in R. As the Lymphoma/Myeloma dataset includes five technical replicates for each sample, box-and-whiskers plot grouping spectra by sample can be generated. The following example aims to look for outlier values in two characteristics: number of peaks and mean intensity.

This example generates the box-and-whiskers plot presented in Figure 5.3. As can be seen, there are some outlier values in both plots. Regarding the

```
# The spectra length is the number of peaks that it holds.
spectraLengths <- sapply(data$spectra, length);

# Calculate the mean intensity of the peaks in the spectra.
intensityMeans <- sapply(data$spectra, function(spectra) mean(intensity(spectra)));

# Data is grouped in a data frame, so that it can be used in the boxplot function.
dataFrame <- data.frame(
  "spectra" = data$spectraNames,
  "samples" = data$sampleNames,
  "spectraLengths" = spectraLengths,
  "intensityMeans" = intensityMeans,
  stringsAsFactors = FALSE
)

# This graphical configuration allows including two plots in the same window.
par(mfrow = c(1,2))

# Generation of the boxplot for the spectra lengths.
bpLength = boxplot(
  spectraLengths ~ samples, data = dataFrame,
  main = "Spectra Size", ylab = "Number of peaks", xlab = "Sample",
  col = data$samplesColors
)


# Generation of the boxplot for the mean peak intensity.
bpIntensity = boxplot(
  intensityMeans ~ samples, data = dataFrame,
  main = "Intensities", ylab = "Mean Intensity", xlab = "Sample",
  col = data$samplesColors
)
```

number of peaks, the outliers in samples HB and LA are notably far from the rest of the values, while the outliers in samples LB and MC are not too far. Apart from the outliers, it can also be observed that values in MB and MD are very disperse, which may also be indicative of some inconsistency in the data. On the other hand, the outliers in the mean intensity plot are all notably far from the rest of the values and a remarkable value dispersion in some samples can also be seen (*e.g.* LD or ME).

As noted before, detecting an outlier is not enough to discard an observation. A detailed analysis of the values should be done to determine if there was an error or if the values are valid. To do so, the outlier samples can be firstly identified by executing the following code right after generating the box-and-whiskers plot.

```
# The boxplot result includes the "out" attribute, a vector with the outlier values.
# The spectra with outlier values in the characteristics analyzed are extracted here.
dataFrame[dataFrame$spectraLengths %in% bpLength$out,]
dataFrame[dataFrame$intensityMeans %in% bpIntensity$out,]
```

Which gives the following results:

```
>dataFrame[dataFrame$spectraLengths %in% bpLength$out,]
spectra samples spectraLengths intensityMeans
HB_R2 HB_R2 HB 110 0.12663636
LA_R5 LA_R5 LA 89 0.11761798
LB_R1 LB_R1 LB 110 0.12050000
LB_R5 LB_R5 LB 103 0.11399029
MC_R2 MC_R2 MC 152 0.06509868
MC_R3 MC_R3 MC 162 0.06461728
MD_R2 MD_R2 MD 103 0.07598058
ME_R4 ME_R4 ME 103 0.11736893

> dataFrame[dataFrame$intensityMeans %in% bpIntensity$out,]
spectra samples spectraLengths intensityMeans
HA_R3 HA_R3 HA 116 0.11297414
HB_R2 HB_R2 HB 110 0.12663636
LA_R5 LA_R5 LA 89 0.11761798
LB_R4 LB_R4 LB 107 0.08464486
MB_R2 MB_R2 MB 121 0.11819008
MB_R3 MB_R3 MB 191 0.06543979
```

From these results, it can be concluded that replicates HB_R2 and LA_R5 are especially problematic, as they were marked as outliers for both characteristics. To further analyze this, the inverse spectrum comparison shown in Section 5.1.2 (implemented in the `compareSpectra` function at the `multiple-sample-visualization-functions.R` source file) can be used to compare each problematic spectra with the rest of the replicates. The following code compares the HB_R2 replicate against the other HB replicates.

The result can be seen in Figure 5.4, where no remarkable differences can be observed between spectrum HB_R2 (upper spectrum) and the other replicates. Taking into account that spectrum HB_R2 has a low number of peaks but with higher intensity when compared with other replicates, a possible explanation for this may be that, although sample acquisition and preparation were done correctly, there were some problems during sample analysis

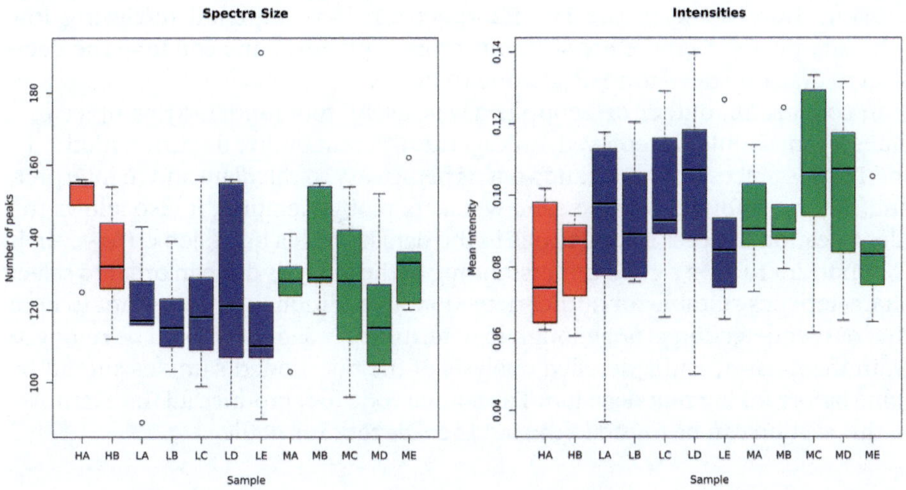


**Figure 5.3** Boxplots for each sample in the Lymphoma/Myeloma dataset showing the distributions of the number of peaks (left) and intensities (right).

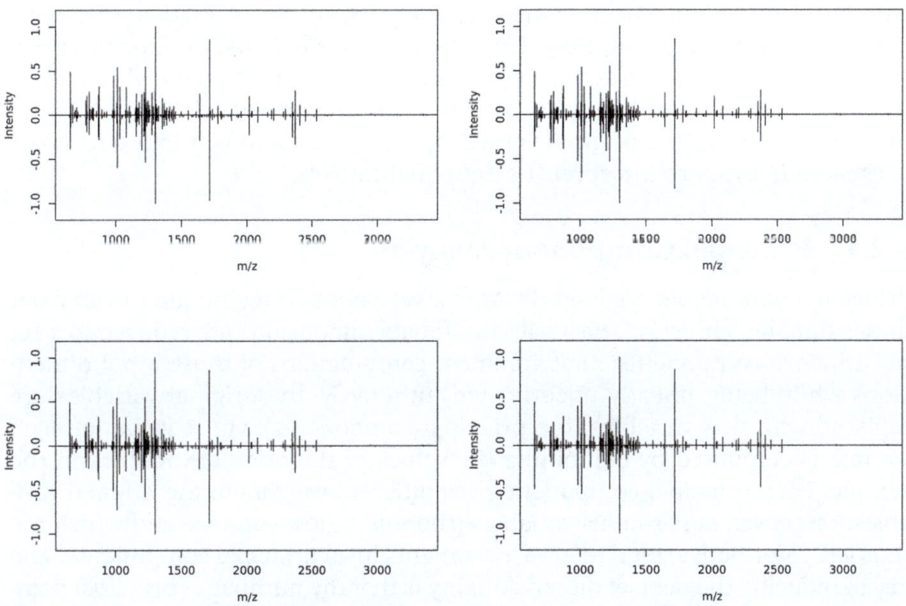


**Figure 5.4** Inverse spectrum comparisons between the replicates of the HB sample. The HB_R2 replicate is always the upper spectrum.

```
par(mfrow = c(2,2))
compareSpectra(data$spectra[[7]], data$spectra[[6]])
compareSpectra(data$spectra[[7]], data$spectra[[8]])
compareSpectra(data$spectra[[7]], data$spectra[[9]])
compareSpectra(data$spectra[[7]], data$spectra[[10]])
```

causing more noise in the HB_R2 spectrum that impeded retrieving low intensity peaks. As there are still four more replicates, one can take the decision to discard spectrum HB_R2 due to noise.

In conclusion, outlier detection is a very useful tool to identify samples in a dataset that should be removed. Among the different outlier detection methods, the Tukey's fences method stands out as being easy to calculate and to interpret, and, when combined with box-and-whiskers plot generation, it also allows the identification of other irregularities in the dataset with a quick look. However, it is important to have a good understanding of the analysis done in order to select characteristics suitable for outlier detection. In addition, it is important to note that outlier detection is not enough to determine if a sample should be removed from the dataset, and a detailed analysis of the questioned samples should be done before taking that decision. The source code to reproduce all the examples in this section can be found in the `outlier-detection.R` file.

## 5.2 Dimensionality Reduction

Mass spectrometry data is high dimensional, with each sample (*i.e.* spectrum) containing hundreds of peaks. It has been proved that very high dimensional data raises undesired effects in further analyses, such as statistical methods, machine learning models, clustering algorithms, *etc.* Thus, this section presents some techniques to reduce the dimensionality of the data which allow (i) simpler datasets to be obtained with a lower number of variables while preserving most of the original variability and (ii) data to be plotted in 2D or 3D spaces in order to interpret the data qualitatively.

### 5.2.1 Principal Component Analysis

Principal components analysis (PCA) is a very popular technique to carry out dimensionality reduction. Basically, input data dimensions are reduced to a set of "principal components" that are linear combinations of those input dimensions while being linearly uncorrelated. Intuitively, the original variables are replaced with new variables (the principal components) whose value for each sample is computed by combining the values of the original variables in the sample. Each principal component gives different weights to the original variables. Moreover, superfluous variables (those with low variance and which are thus less informative) tend to have low weights in all principal components. The key to reducing the dataset dimensionality is that the number of principal components is less than the number of input variables, while preserving most of the shape of the original dataset. This section aims to show how to do a PCA analysis with R in order to get a reduced representation of a mass spectrometry dataset, analyze which input variables (*i.e.* peaks) are more important and create 2D and 3D plots of high dimensional datasets. The source code to reproduce all the examples in this section can be found in the `pca.R` file.

In R, PCA is computed using the `prcomp` function, which returns an object with the computed principal components. For instance, the following example computes the PCA for the binned peaks matrix of the Lymphoma/Myeloma dataset:

The `scale = TRUE` normalizes scales of each input variable to have unit variance, which is recommended.

```
pca <- prcomp(binnedPeaksMatrix, scale = TRUE)
```

The PCA summary shown by invoking `summary(pca)` shows how much variance each principal component captures. For example, for the first seven principal components, it can be observed that a 55% of the variance is captured. This means that only seven variables capture more than a half of the variance of the original dataset which is composed of 464 variables (peaks). It is important to note that the principal components are sorted by how much variance they capture, so the first principal component can be seen as more important than the second, and so on.

```
                          PC1     PC2     PC3     PC4     PC5     PC6     PC7
Standard deviation     9.9538 6.74923 5.66932 4.94936 4.68678 4.18291 3.86606
Proportion of Variance 0.2135 0.09817 0.06927 0.05279 0.04734 0.03771 0.03221
Cumulative Proportion  0.2135 0.31170 0.38097 0.43376 0.48110 0.51881 0.55103
(more rows not shown)
```

The `pca` object returned by `prcomp` has three useful values. `pca$rotation` contains the principal component coefficients for each input variable, that is, how each principal component weights each input variable to compute its value for a given sample. `pca$sdev` contains the standard deviation of each principal component (the same as has been shown with `summary`). The cumulative proportion of variance of each principal component can be obtained with `cumsum(pca$sdev^2) / sum(pca$sdev^2)`. Finally `pca$x` contains the projected dataset into the new space of principal components.

Then, a dimension-reduced dataset containing only a subset of principal components can be obtained. For example, the reduced dataset capturing the 95% of variance can be obtained with the following instructions:

```
# Which is the first principal component accumulating the 95% of variance?
pc95 <- min(which(cumsum(pca$sdev^2) / sum(pca$sdev^2) >= 0.95))
# pca95 is the 43th component
binnedPeaksMatrix.pca.95 <- pca$x[, 1:pc95]
```

An interesting utility of reducing the dimensionality of the dataset is the possibility of creating 2D or 3D plots by using, for example, the first two or three principal components, respectively. The library pca3d contains the `pca2d` and `pca3d` functions which allow these sorts of plots to be created easily. For example, to create a 2D plot with the Lymphoma/Myeloma replicates dataset by using only the first two principal components and by coloring each point according to its corresponding condition (H = Healthy, L = Lymphoma, M = Myeloma), the `pca2d` function can be executed as follows:

```
pca2d(pca, group = data$spectraConditions, components = c(1, 2), legend =
"bottomright")
```

Figure 5.5 shows the resulting 2D plots for the replicates matrix (*A*) and for the samples matrix (*B*). As can be seen, samples of the same condition seem to be grouped in the same area of the plot, suggesting that samples of the same condition tend to have a strongly similar peak profile that is maintained with only the two first principal components.

In addition to the `pca2d` function, the `pca3d` function (from the pca3d package) allows an interactive 3D plot to be drawn with very similar parameters. Figure 5.6A shows a screenshot of the interactive 3D plot created with the following code:

```
pca3d(pca, group = data$spectraConditions, components = c(1, 2), legend =
"bottomright")
```

An interesting feature of `pca3d` and `pca2d` plots is the possibility of adding biplots by only setting `biplot = TRUE`, which are very useful to identify the most important variables for each component plotted. For example, adding a biplot to the previous 2D plot (Figure 5.5B) will produce the result shown in Figure 5.6B. As can be observed, the peak 1284.732 428 571 43 (bottom left in the figure) strongly drives PC1 (it has a clear direction in the *x*-axis). In fact, this peak is the variable with the higher coefficient in PC1, which means that samples that appear on the left side of the 2D plot (samples of condition H or M) tend to have higher values than those on the right part (samples of condition L). This situation can be checked by showing the value for this peak in all samples:

```
consensusBinnedPeaksMatrix[, "1284.73242857143"]
       HA        HB        LA        LB        LC        LD        LE        MA
0.1582000 0.1226000 0.0000000 0.0000000 0.0000000 0.0000000 0.0000000 0.1150000
       MB        MC        MD        ME
0.1473333 0.1135000 0.1406000 0.1244000
```

## 5.2.2 Self-organizing Maps

A self-organizing map (SOM) is a computational method for the visualization and analysis of high-dimensional data.[7] Specifically, a SOM is a type of artificial neural network (ANN) that maps the input samples onto a regular and, usually, two-dimensional output grid, preserving the topological properties of the input space. However, there are several characteristics that differentiate SOMs from other ANNs. Particularly SOMs are composed of only two layers: an input layer, with one node for each feature of the input samples, and an output layer, with a number of nodes configured by the user organized in a two-dimensional space. Each output node is fully connected to the nodes in the input layer, and each connection has a weight that is adjusted when the SOM is trained. The vector of weights for each output node is known as the codebook vector.

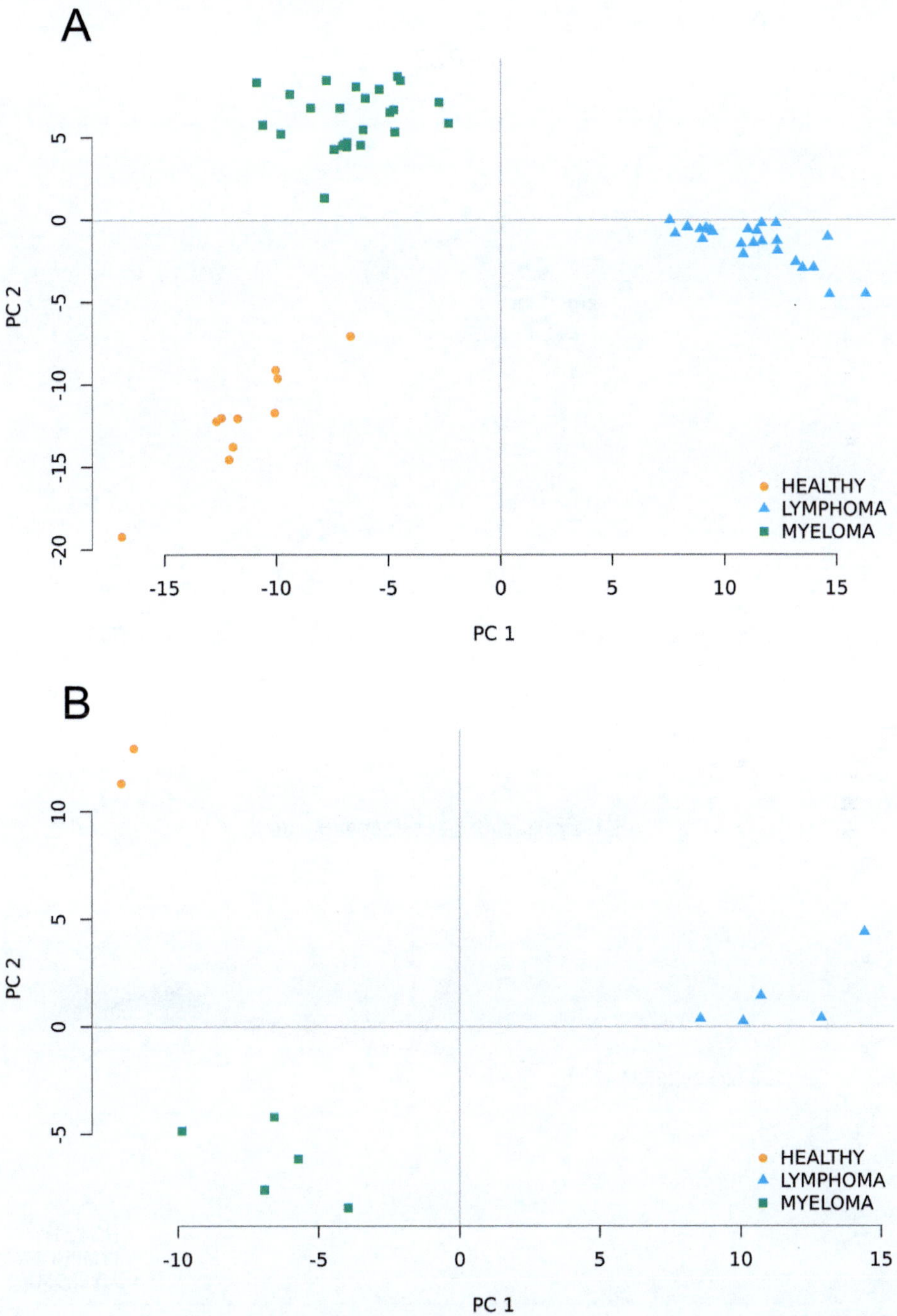


**Figure 5.5** Principal component analysis 2D plots for the replicates matrix (A) and for the samples matrix (B) of the Lymphoma/Myeloma dataset.

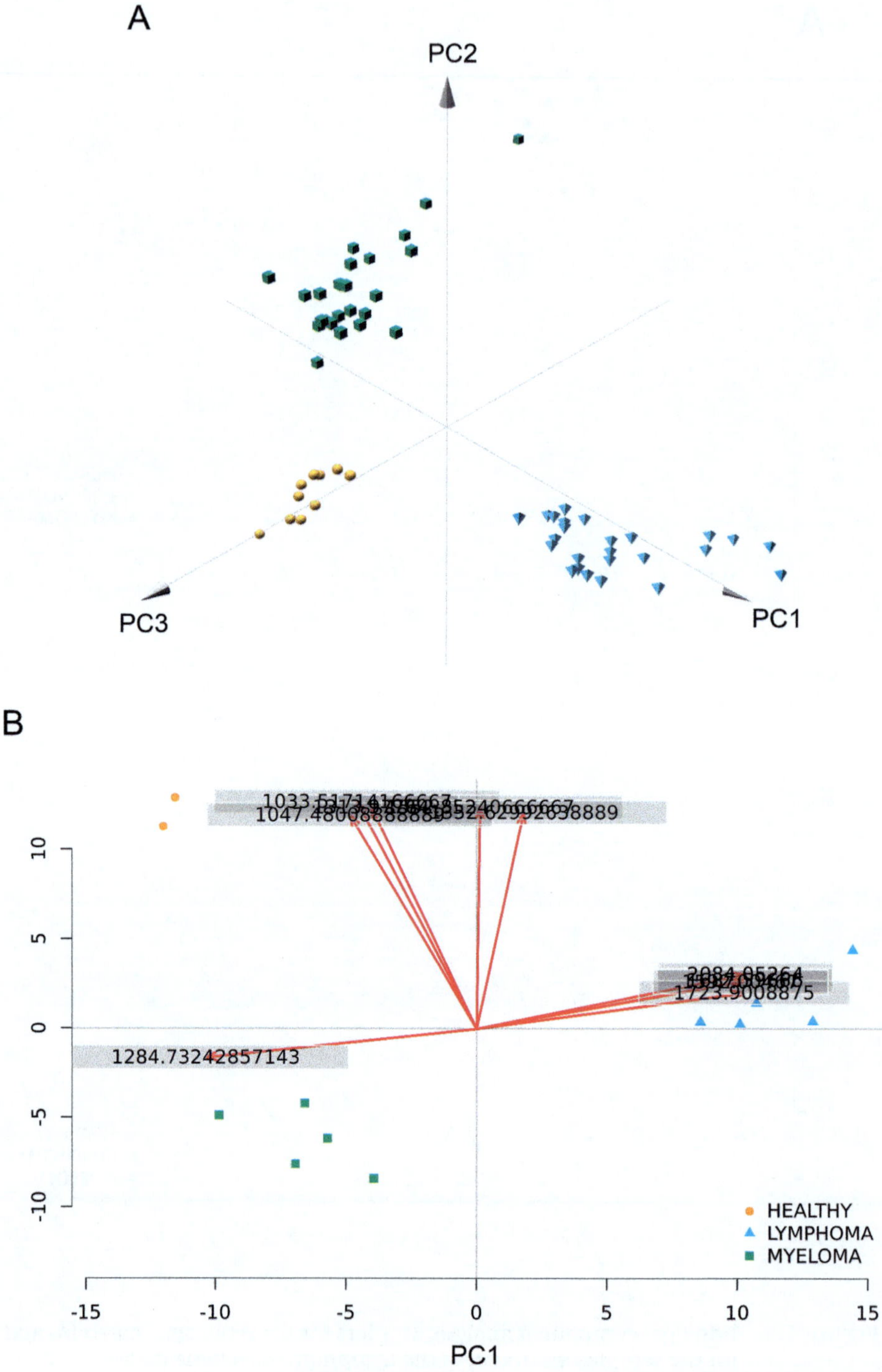


**Figure 5.6** (A) Principal component analysis 3D plot for the replicates matrix of the Lymphoma/Myeloma dataset. (B) Principal component analysis 2D plots for the samples matrix (B) of the Lymphoma/Myeloma dataset showing the biplots.

As for other ANNs, a SOM needs to be trained. Although there are several variants, the main training algorithm consists of the following steps:

1. Randomly initialize the weights of the codebook vectors of each output node.
2. For each sample of the training dataset:
   a. Calculate the Euclidean distance between the sample and each codebook vector (*i.e.* each output node).
   b. Select the output node with the smallest distance to the sample. This is the best matching unit (BMU).
   c. Update the codebook vectors of the neighbor nodes of the BMU to reduce their distance to the input sample.
3. Repeat step 2 until a predetermined number of iterations is reached.

The adjustment of the codebook vectors done in step 2(c) is controlled by an alpha ($\alpha$) parameter, which is usually reduced linearly during the training process. This way, the adjustments done in the first iterations are rougher than in the last iterations. On the other hand, the determination of the neighbor nodes depends on the output grid configuration, which can arrange the nodes in a rectangular or hexagonal space. In the first case, each node has four or eight neighbor nodes (according to whether the diagonal neighbors are excluded or not), while in the second case each node has six neighbor nodes. In the particular case of the nodes at the edges of the grids, they will have fewer neighbors, unless the output grid is configured as a toroidal space, where opposite edges are connected.

In the resulting SOM, the codebook vector's similarity is expected to decrease with the distance between the corresponding output nodes (*i.e.* each output node should be similar to near output nodes and different to far output nodes). In addition, the BMU for each training sample should be in different zones of the SOM, depending on its related class.

The kohonen library allows SOMs in R to be worked with. This library is named after Teuvo Kohonen, who originally proposed the SOM models, and includes functions to train SOMs, graphically represent them, and to perform prediction over new samples. The source code to reproduce all the examples in this section can be found in the `som.R` file.

When using this library, SOMs are created with the `supersom` function, which is an extension of SOMs to support multiple data layers, possibly with different numbers and types of variables.[8] When multiple layers are used, the distance to determine the BMU is calculated as a weighted sum of the distances calculated separately for each layer. This function admits several parameters to customize the training process, such as the number of iterations (`rlen`), the distance function to use on each layer (`dist.ftcs`), the alpha parameter (`alpha`), the configuration of the output grid (`grid`) or the weight of each layer (`user.weigths`). Nevertheless, only the training dataset (`data`) is mandatory.

In addition to the supersom function, the som and xyf functions are included. These are wrapper functions to create SOMs with one and two input layers, respectively. The following example shows how to create a SOM with a single layer using the Lymphoma/Myeloma dataset. Note that each spectrum is used as a sample in this example (*i.e.* the binnedPeaksMatrix is used).

```
library(kohonen)

set.seed(2019)
somresult <- som(
    scale(binnedPeaksMatrix),
    grid = somgrid(xdim = 13, ydim = 3, topo = "hexagonal"),
    alpha = c(0.5, 0.01),
    rlen = 4000
)
```

Before calling the som function, the kohonen library is imported and the random seed is first set to an arbitrary value, so that the results obtained are reproducible. When calling this function, the first parameter provided is the input dataset, which should be a matrix. As in PCA, it is recommended that the data matrix is scaled, but, since in this case there is not a parameter to tell the function to do it, the scale function should be applied to the data matrix. The grid parameter receives a configuration for the output layer created by the somgrid function. The xdim and ydim parameters of the somgrid function determine the number of columns and rows in the output grid, respectively, while the topo parameters determines the topology, which can be rectangular or hexagonal. The alpha parameter sets the initial and final value for the alpha parameter, which will decrease linearly with the iterations. Finally, the rlen parameter sets the number of iterations to perform. These are the main parameters of this function but, as noted before, there are more parameters that allow other aspects of the training process to be modified.

Regarding the size of the output grid, Vesanto and Alhoniemi[9] recommend a number of nodes approximately equal to $5\sqrt{N}$, where $N$ is the number of samples. Additionally, Tian *et al.*[10] recommend a ratio between the two dimensions of the output grid approximately equal to the ratio of the two largest eigenvalues of the training data's covariance matrix. This ratio can be calculated in R with the following code.

```
binnedPeaksMatrix.cov <- cov(binnedPeaksMatrix)
binnedPeaksMatrix.eigen <- eigen(binnedPeaksMatrix.cov)
binnedPeakMatrix.eigen.sorted <- sort(binnedPeaksMatrix.eigen$values, decreasing =
TRUE)
ratio <- binnedPeakMatrix.eigen.sorted[1] / binnedPeakMatrix.eigen.sorted[2]
```

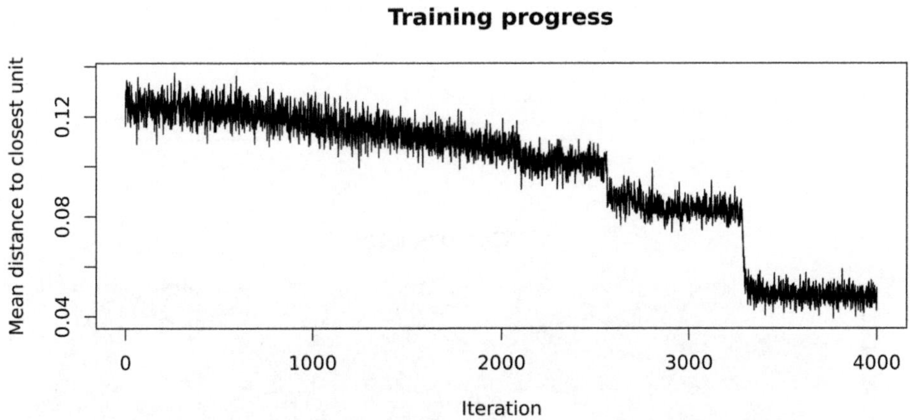


**Figure 5.7** Changes plot showing how the mean distance of each sample to the BMUs evolves during the training process.

The `somresult` object contains the codebook vectors for each output node, along with other relevant information of the SOM itself and the training process. The kohonen library includes several predefined visualizations for the information contained in this object. For example, it is possible to view how the mean distance of each sample to the BMUs evolves during the training process using a "changes" plot. Figure 5.7 presents the plot generated with the following command.

```
plot(somresult, type = "changes")
```

As can be seen from Figure 5.7, during the last part of the training process the variation in the mean distance is minimal, meaning that it is, essentially, fine-tuning.

Figure 5.8 present other plots that can be generated with the kohonen library. These plots are very useful to interpret the results of the SOM training. For example, the "counts" (top) and the "quality" (middle) plots represent the number of samples for which each node is the BMU and the distance between those samples and the codebook vectors of the BMUs, respectively. This information is very useful to evaluate the distribution of the samples in the map and the adjustment of the codebook vectors to the samples. On the other hand, the "dist.neighbours" plot (bottom), represents the sum of the distances between each node and their neighbors, and it is also known as the *U*-matrix plot. This last plot may help to identify the clusters that may be created. These plots can be generated with the following code.

```
plot(somresult, type = "counts")
plot(somresult, type = "quality")
plot(somresult, type = "dist.neighbours")
```

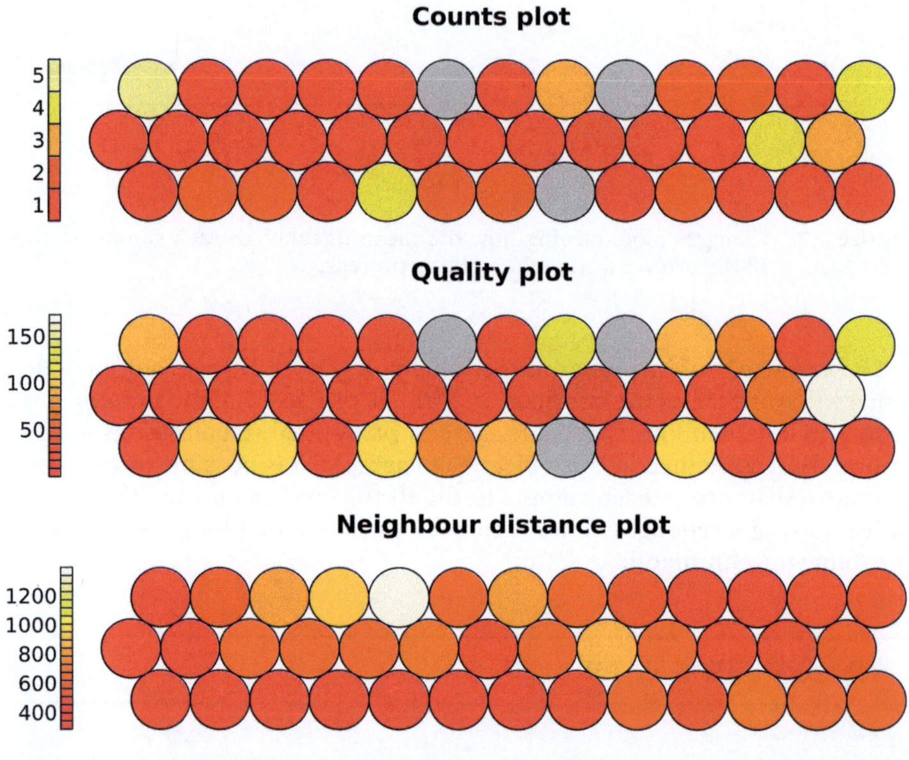


**Figure 5.8** Unsupervised SOM plots. The "counts" (top) and the "quality" (middle) plots represent the number of samples for which each node is the BMU and the distance between those samples and the codebook vectors of the BMUs, respectively. The "dist.neighbours" plot (bottom), represents the sum of the distances between each node and their neighbors, and it is also known as the U-matrix plot.

Within the information contained in the `somresult` object, the BMU of each input sample is, probably, the most important information, as it identifies the location of each sample in the output map. This information can be obtained with the following code.

```
classification <- somresult$unit.classif
names(classification) <- data$spectraNames
classification
HA_R1 HA_R2 HA_R3 HA_R4 HA_R5 HB_R1 HB_R2 HB_R3 HB_R4 HB_R5 LA_R1 LA_R2 LA_R3
33 34 34 34 6 7 6 21 7 20 25 25 11
LA_R4 LA_R5 LB_R1 LB_R2 LB_R3 LB_R4 LB_R5 LC_R1 LC_R2 LC_R3 LC_R4 LC_R5 LD_R1
25 25 36 36 24 37 37 26 26 26 12 13 10
LD_R2 LD_R3 LD_R4 LD_R5 LE_R1 LE_R2 LE_R3 LE_R4 LE_R5 MA_R1 MA_R2 MA_R3 MA_R4
23 10 22 9 39 38 39 39 39 5 5 19 5
MA_R5 MB_R1 MB_R2 MB_R3 MB_R4 MB_R5 MC_R1 MC_R2 MC_R3 MC_R4 MC_R5 MD_R1 MD_R2
5 30 17 31 29 28 3 3 16 4 18 2 15
MD_R3 MD_R4 MD_R5 ME_R1 ME_R2 ME_R3 ME_R4 ME_R5
1 2 14 27 27 27 27 27
```

The number associated with each spectra is the index of the output node, which are numerated left to right and bottom to top in the plots. Although relevant, this data can be difficult to interpret, as the neighborhood of each output node is not evident. However, it is possible to use a "mapping" plot to graphically view the location of each sample in the output grid. This plot type requires a `labels` parameter, that receives a vector with the label for each sample, or a `pchs` parameter, that receives a number that indicates the figure to use to represent the samples. In addition, the `col` parameter can be used to provide a color for each sample. In the following example, a circle is used to represent each sample and a different color for each condition. The `legend` function allows a legend to be added to the plot with the color assigned to each condition.

```
plot(somresult, type = "mapping", pchs = 1, col = data$spectraColors, main =
"Unsupervised SOM")
legend("bottom", legend = data$datasetConditions, col = data$datasetConditionsColors,
pch = 1, horiz = TRUE)
```

As can be seen from Figure 5.9, with this representation it is possible to view the spatial separation of the samples on the output grid. In this case, it can be observed that the samples from each condition are located in different spaces of the plot. Taking into account the neighbor distance plot of the Figure 5.9, it is possible to see some important details, such as that some healthy samples are near to some lymphoma samples.

Although SOMs are usually constructed without taking into account the condition of the samples, the kohonen library allows that information to be used to guide the training process. A convenient way to do this is by using the `xyf` function, which, as noted before, creates a SOM with two input layers. The following example demonstrates how to use this function with the Lymphoma/Myeloma dataset including the condition information.

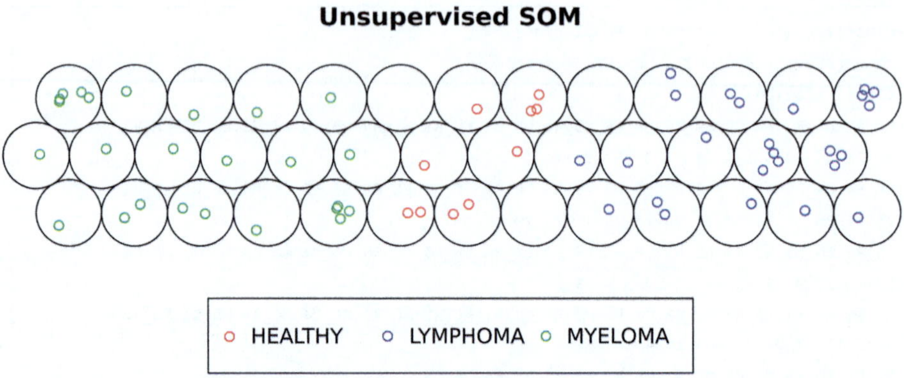


**Figure 5.9** Unsupervised SOM plot showing the location of each sample in the output grid. Samples are colored according to their conditions in order to facilitate the visual interpretation of the plot.

```
set.seed(2019)
somresult <- xyf(
    scale(binnedPeaksMatrix),
    factor(data$spectraConditions),
    grid = somgrid(xdim = 13, ydim = 3, topo = "hexagonal"),
    alpha = c(0.5, 0.01),
    rlen = 4000
)
```

As can be seen from this example, the only difference with the `som` example is that, in this case, a factorized condition for each spectrum is provided as the second parameter (*i.e.* the second input layer). The `xyf` function automatically converts the second parameter into a matrix with one column for each factor that takes value 1 for the sample's condition and 0 for other conditions. The result is a SOM with two input layers, where the distance of each sample to each output node is the weighted sum of distances from the two input layers. The inclusion of the conditions usually results in a clearer separation between samples from different conditions.

In this case, it is possible to use the "codes" plot to view the codebook vector of the conditions layer for each output node. The resulting plot can be seen in Figure 5.10, together with the neighbor distance plot. Both plots show that the samples of each condition are clearly separated and that the distances between samples of the same condition are lower that the distances between samples of different conditions.

## 5.3 Cluster Analyses

This section aims to show two of the most widely used clustering analyses to group samples: k-means clustering and hierarchical clustering analysis.

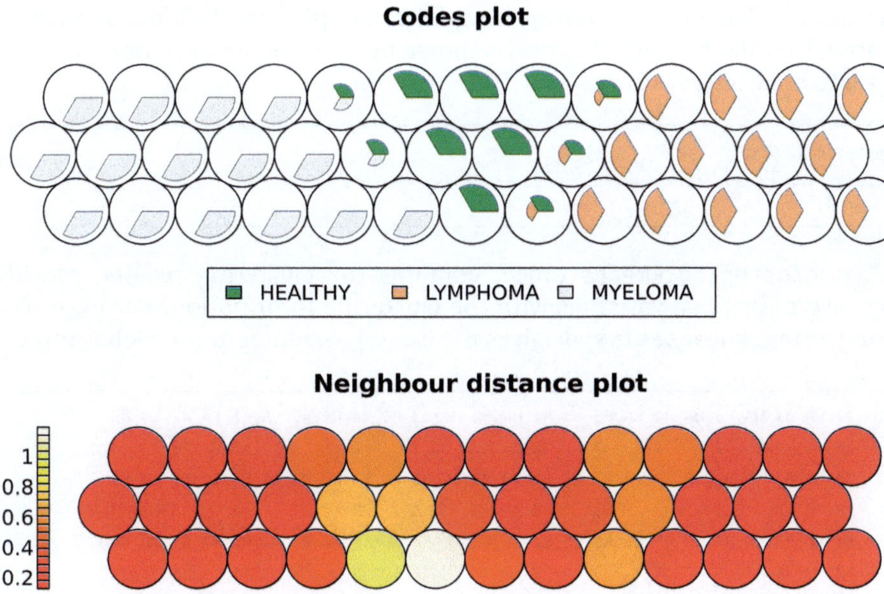


**Figure 5.10** Supervised SOM plots. The "codes" plot (top) shows the codebook vector of the conditions layer for each output node. The "dist.neighbours" plot (bottom), represents the sum of the distances between each node and their neighbors.

### 5.3.1 *K*-Means

*K*-means is a cluster analysis method in the category of partitioning methods, which divides the input dataset into a set of clusters where each data sample belongs to one, and exactly one, cluster.

The *k*-means method is based in the squared Euclidean distance and follows an iterative algorithm to partition samples into a predefined number of clusters ($k$), whose main steps are:

1. Initialize $k$ cluster centroids to random samples.
2. Compute the distance of each sample to the cluster centroids and assign the nearest centroid as it cluster.
3. Compute centroids as the mean of the samples belonging to the centroid.
4. Go back to step 2 until centroids do not change with respect to the previous iteration.

In R, the `kmeans` function allows a clustering with this algorithm to be performed. The source code to reproduce all the examples in this section can be found in the `kmeans-clustering.R` file. For instance, the following code performs a *k*-means clustering using the Lymphoma/Myeloma replicates matrix. In order to make the results shown here reproducible, a random seed is set to an arbitrary value.

```
set.seed(2019)
clustering.kmeans <- kmeans(binnedPeaksMatrix, centers = 3)
```

The `clustering.kmeans` object contains the clustering results, providing several internal variables with the clustering information. For example, `clustering.kmeans$cluster` gives the cluster assignment for each sample:

```
HA_R1 HA_R2 HA_R3 HA_R4 HA_R5 HB_R1 HB_R2 HB_R3 HB_R4 HB_R5 LA_R1 LA_R2 LA_R3
    2     2     2     2     2     2     2     2     2     2     3     3     3
LA_R4 LA_R5 LB_R1 LB_R2 LB_R3 LB_R4 LB_R5 LC_R1 LC_R2 LC_R3 LC_R4 LC_R5 LD_R1
    3     3     3     3     3     3     3     3     3     3     3     3     3
LD_R2 LD_R3 LD_R4 LD_R5 LE_R1 LE_R2 LE_R3 LE_R4 LE_R5 MA_R1 MA_R2 MA_R3 MA_R4
    3     3     3     3     3     3     3     3     3     1     1     1     1
MA_R5 MB_R1 MB_R2 MB_R3 MB_R4 MB_R5 MC_R1 MC_R2 MC_R3 MC_R4 MC_R5 MD_R1 MD_R2
    1     1     1     1     1     1     1     1     1     1     1     1     1
MD_R3 MD_R4 MD_R5 ME_R1 ME_R2 ME_R3 ME_R4 ME_R5
    1     1     1     1     1     1     1     1
```

As can be seen from this result, all healthy (H), lymphoma (L) and myeloma (M) replicates were assigned to cluster 2, 3 and 1, respectively. In this sense, this execution of *k*-means with 3 clusters fits exactly with the conditions. The same result, can be obtained at sample level by running the following statements:

```
clustering.kmeans.consensus <- kmeans(consensusBinnedPeaksMatrix, centers = 3)
clustering.kmeans.consensus$cluster
```

Which returns the following cluster assignments, showing that a perfect match between clusters and conditions is obtained again.

```
HA HB LA LB LC LD LE MA MB MC MD ME
 2  2  3  3  3  3  3  1  1  1  1  1
```

The *k*-means algorithm does not ensure that the cluster assignments is the optimum in terms of intra-cluster distance *versus* inter-cluster distances. Depending on the random initialization, the algorithm could converge to different clustering results. Moreover, the number of clusters ($k$) should be provided by the user, which is not easy to guess, since the underlying structure of the data is unknown (it is, in fact, what a clustering analysis like this one aims to uncover). In order to get some hints, the *k*-means clustering results have some variables which may be useful to assess if the clustering was good or not and if a different number of clusters should be tested. The `$betweenss` value gives the sum of squares between all pair of cluster centroids (between SS), that is, a value of the inter-cluster distance, which should be as high as possible. The `$withinss` gives the mean sum of squares of each cluster (within SS), which is the mean of the distances of the data points of each cluster to their corresponding centroid, which should be as low as possible. The specific weight of the inter-cluster among all the distances can be computed as `$betweenss / ($betweenss + sum($withinss))`. The higher this weight is, the better.

A possible way to calculate a good number of clusters is to try with several values of `k` and look at the value of `sum($withinss)`. By applying the Elbow method, a good value of `k` is one where adding another cluster ($k + 1$) does not produce a significant decrease within SS. With the following R code it is possible to get a plot which help to discover this good value for `k`.

```
clustering.kmeans.k_withinss <- vector()
for (c in 1:10) {
  clust <- kmeans(binnedPeaksMatrix, centers = c)
  clustering.kmeans.k_withinss[c] <- sum(clust$withinss)
}
plot(clustering.kmeans.k_withinss, type="b", xlab = "Number of clusters", ylab =
"total within_SS")
```

Figure 5.11A shows the total within SS at different numbers of clusters when clustering the replicates matrix. Figure 5.11B shows the same when clustering at samples level. As can be seen, when the number of clusters is four the total within SS does not decreases significantly, so, following the Elbow method, four should be a good number of clusters.

As was previously noted, the random initialization of the centroids could lead to different clustering results in different algorithm runs. In order to alleviate this undesired behaviour, it is possible to perform several runs of the algorithm and see which samples tend to be clustered together. Fortunately, there is an R package called ConsensusClusterPlus that allows more robust results to be obtained in several clustering algorithms with a random component. Although the package has many useful features, here we only show how to get a *k*-means "consensus" clustering using it.

```
library(ConsensusClusterPlus)
ccp <- ConsensusClusterPlus(t(binnedPeaksMatrix), maxK = 3, distance = "euclidean",
clusterAlg = "km", reps = 50, seed = 2019)
# show consensus clustering results for k = 3 clusters
ccp[[3]][["consensusClass"]]
```

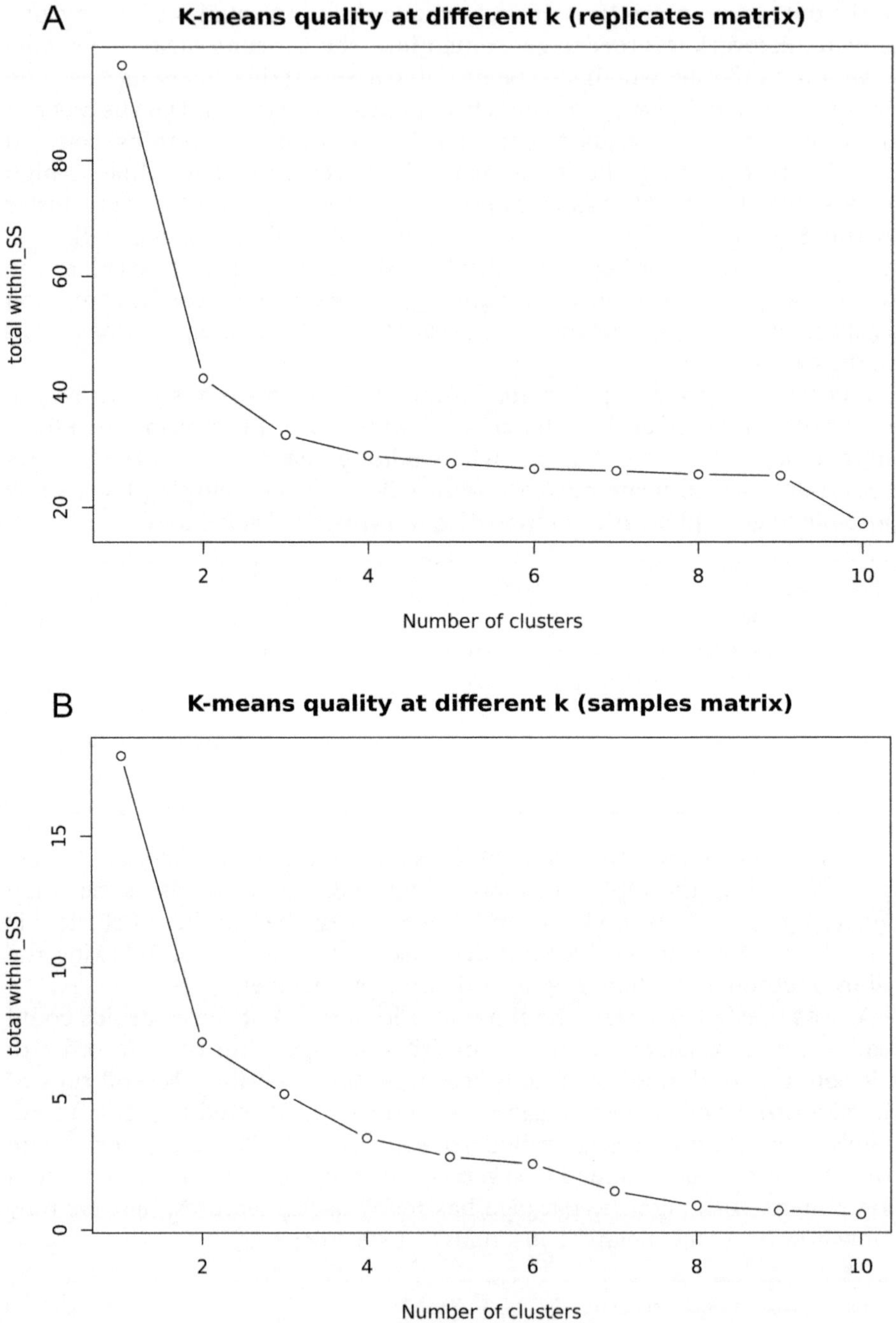


**Figure 5.11** Total within SS in k-means at different number of clusters when clustering the replicates matrix (A) and the samples matrix (B).

This will return the following cluster assignments:

```
HA_R1 HA_R2 HA_R3 HA_R4 HA_R5 HB_R1 HB_R2 HB_R3 HB_R4 HB_R5 LA_R1 LA_R2 LA_R3
    1     1     1     1     1     1     1     1     1     1     2     2     2
LA_R4 LA_R5 LB_R1 LB_R2 LB_R3 LB_R4 LB_R5 LC_R1 LC_R2 LC_R3 LC_R4 LC_R5 LD_R1
    2     2     2     2     2     2     2     2     2     2     2     2     2
LD_R2 LD_R3 LD_R4 LD_R5 LE_R1 LE_R2 LE_R3 LE_R4 LE_R5 MA_R1 MA_R2 MA_R3 MA_R4
    2     2     2     2     2     2     2     2     2     3     3     3     3
MA_R5 MB_R1 MB_R2 MB_R3 MB_R4 MB_R5 MC_R1 MC_R2 MC_R3 MC_R4 MC_R5 MD_R1 MD_R2
    3     3     3     3     3     3     3     3     3     3     3     3     3
MD_R3 MD_R4 MD_R5 ME_R1 ME_R2 ME_R3 ME_R4 ME_R5
    3     3     3     3     3     3     3     3
```

## 5.3.2 Hierarchical Clustering

Hierarchical clustering is a cluster analysis method to create hierarchies of clusters that groups similar observations together. In this section, we focus on clustering samples (*i.e.* spectra), although the same procedures can be applied to group peaks (*i.e.* each observation is a peak and its values across all samples).

There are two main strategies for performing hierarchical clustering: (i) agglomerative, where a bottom-up approach is followed: each sample starts in its own cluster and pairs of clusters are merged as one moves up the hierarchy until all clusters form a single cluster; (ii) divisive, where a top-down approach is followed: all samples start in one cluster and splits are performed recursively as one moves down the hierarchy until all samples have their own cluster. In R, the `hclust` function performs hierarchical cluster analysis following a bottom-up approach.

Hierarchical clustering relies on two computations to determine which clusters should be combined or split: (i) a measure of distance between pairs of samples (see Section 5.1.1) and (ii) a linkage criterion, which specifies the dissimilarity of clusters as a function of the pairwise distances between samples. The most common linkage criteria are: (i) minimum or single-linkage clustering (`method = "single"` in `hclust`), (ii) maximum or complete-linkage clustering (`method = "complete"` in `hclust`), (iii) mean or average clustering (`method = "average"` in `hclust`), (iv) median linkage-clustering (`method = "median"` in `hclust`), (v) centroid linkage clustering (`method = "centroid"` in `hclust`) and (vi) Ward's criterion (`method = "ward.D2"` in `hclust`). For instance, the default linkage criterion of the `hclust` function is `"complete"`, where the distance for two clusters $A$ and $B$ is calculated as follows:

$$D_{\text{complete}}(A, B) = \max\left(d(a_i, b_i)\right) \tag{5.15}$$

That is, the distance between two clusters is the distance between the two most distant members in each cluster.

Samples can be clustered with the `hclust` function using any of the distance objects computed in the previous section. For instance, the following code snippet uses the Jaccard distances (computed using the `binnedPeaksMatrix` variable of the Lymphoma/Myeloma dataset) to compute the clustering:

```
distances.jaccard <- jaccard.dist(rownames(binnedPeaksMatrix),
toSpectraList(binnedPeaksMatrix))
clustering.jaccard <- hclust(distances.jaccard)
```

Now, the `clustering.jaccard` object contains the hierarchical clustering results, providing several internal variables with the clustering information (`clustering.jaccard$merge`, `clustering.jaccard$height`, `clustering.jaccard$order`, `clustering.jaccard$labels`, `clustering.jaccard$method`, `clustering.jaccard$call` and `clustering.jaccard$dist.method`). A useful method for exploring the clustering results is the `cutree` function, which cuts the tree resulting from `hclust` into several groups, either by specifying the desired number(s) of groups or the cut height(s). As there are three conditions in this dataset (Lymphoma, Myeloma and Healthy), it makes sense to `call cutree(clustering.jaccard, k = 3)` to divide the tree into three clusters and know which samples are in each cluster. This produces the following output:

```
HA_R1 HA_R2 HA_R3 HA_R4 HA_R5 HB_R1 HB_R2 HB_R3 HB_R4 HB_R5 LA_R1 LA_R2 LA_R3
1     1     1     1     1     1     1     1     1     1     2     2     2
LA_R4 LA_R5 LB_R1 LB_R2 LB_R3 LB_R4 LB_R5 LC_R1 LC_R2 LC_R3 LC_R4 LC_R5 LD_R1
2     2     2     2     2     2     2     2     2     2     2     2     2
LD_R2 LD_R3 LD_R4 LD_R5 LE_R1 LE_R2 LE_R3 LE_R4 LE_R5 MA_R1 MA_R2 MA_R3 MA_R4
2     2     2     2     2     2     2     2     2     3     3     3     3
MA_R5 MB_R1 MB_R2 MB_R3 MB_R4 MB_R5 MC_R1 MC_R2 MC_R3 MC_R4 MC_R5 MD_R1 MD_R2
3     3     3     3     3     3     3     3     3     3     3     3     3
MD_R3 MD_R4 MD_R5 ME_R1 ME_R2 ME_R3 ME_R4 ME_R5
3     3     3     3     3     3     3     3
```

By examining the output carefully (or helped by using `which(cutree(-clustering.jaccard, k = 3) == 1`, for instance), it is possible to see that each cluster groups all the samples in each condition.

Although these objects provide all the information, the easiest way to look through the results is by plotting a dendrogram, a graphical representation that illustrates the arrangement of the clusters produced in the form of a tree. This dendrogram visualization can also be enhanced by adding rectangles around the branches of a dendrogram highlighting the clusters found with the `cutree` function, using the `rect.hclust` function. Figure 5.12A illustrates this dendrogram, which can be created easily using the `plot` and the `rect.hclust` methods as follows:

```
plot(clustering.jaccard, main = "Samples HCA (complete linkage)")
rect.hclust(clustering.jaccard, k = 3, border = c("blue", "red", "green"))
```

Figure 5.12B shows the equivalent clustering done at the samples level, which can be obtained using the following code:

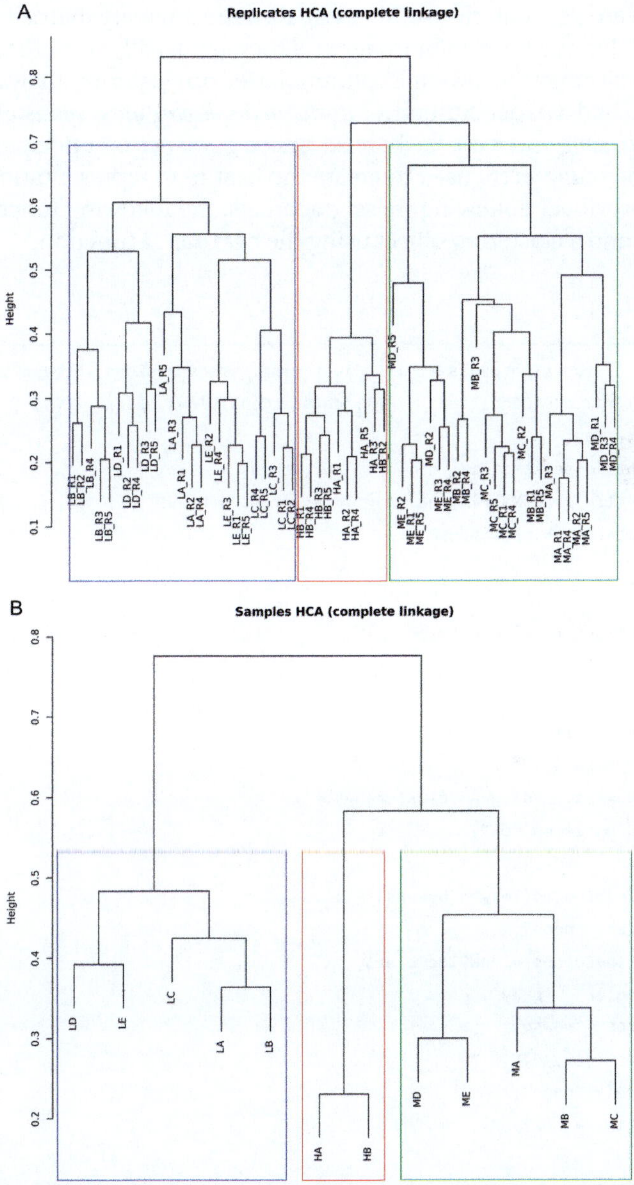


**Figure 5.12** Dendrograms corresponding to the hierarchical clustering for the replicates matrix (A) and for the samples matrix (B) of the Lymphoma/Myeloma dataset computed using Jaccard distances.

```
distances.jaccard.consensus <- jaccard.dist(rownames(consensusBinnedPeaksMatrix),
toSpectraList(consensusBinnedPeaksMatrix))
clustering.jaccard.consensus <- hclust(distances.jaccard.consensus)
plot(clustering.jaccard.consensus, main = "Samples HCA (complete linkage)")
rect.hclust(clustering.jaccard.consensus, k = 3, border = c("blue", "red", "green"))
```

Heat maps are particularly useful to visualize the intensity matrix formed by a group of samples as a false color image and they are usually combined with dendrograms to rearrange the rows and columns based on clustering analyses. Melissa Key has published a paper entitled *A tutorial in displaying mass spectrometry-based proteomic data using heat maps*[11] where advanced guidance on understanding and optimizing the parameters used to create the heat map representations using R packages is provided. Following these guidelines, the following function plots a peaks matrix and a clustering object using the `heatmap.2` function.

```
plotHeatmap <- function(peaksMatrix, clustering, spectraColors, breaksCount = 24,
rowsep = c(), datasetConditions = c(), datasetConditionsColors = c()) {
    breaks <- seq(
        quantile(peaksMatrix, na.rm = TRUE, probs = 0.001),
        quantile(peaksMatrix, na.rm = TRUE, probs = 0.999),
        length.out = breaksCount
    )

    rowSideColors <- spectraColors

    heatmap.2(
        peaksMatrix,
        Colv = FALSE,
        Rowv = as.dendrogram(clustering),
        Dendrogram = "row",
        Breaks = breaks,
        Col = bluered(length(breaks) - 1),
        Scale = "none",
        RowSideColors = rowSideColors,
        na.color = 'gray',
        rowsep = rowsep,
        lwid = c(3, 6),
        trace = "none",
        margins = c(12, 9),
    )
    If (length(datasetConditions) > 0 && length(datasetConditionsColors) > 0) {
        par(lend = 1)           # square line ends for the color legend
        legend("topright",      # location of the legend on the heatmap plot
            legend = datasetConditions,     # category labels
            col = datasetConditionsColors,  # color key
            lty = 1,            # line style
            lwd = 10            # line width
        )
    }
}
```

This function can be used to plot cluster analyses done using both replicates and samples matrices by using it as follows:

```
plotHeatmap(binnedPeaksMatrix, clustering.jaccard, data$spectraColors, rowsep = c(25,
35), datasetConditions = data$datasetConditions, datasetConditionsColors =
data$datasetConditionsColors)

plotHeatmap(consensusBinnedPeaksMatrix, clustering.jaccard.consensus,
consensusData$spectraColors, rowsep = c(5, 7), datasetConditions =
consensusData$datasetConditions, datasetConditionsColors =
consensusData$datasetConditionsColors)
```

Figure 5.13 shows the heat map representation of the samples matrix.

Finally, it is worth noting the pvclust package (https://cran.r-project.org/web/packages/pvclust/index.html) here. This R package is useful to assess the uncertainty in hierarchical cluster analysis,[12] as it allows *p*-values to be

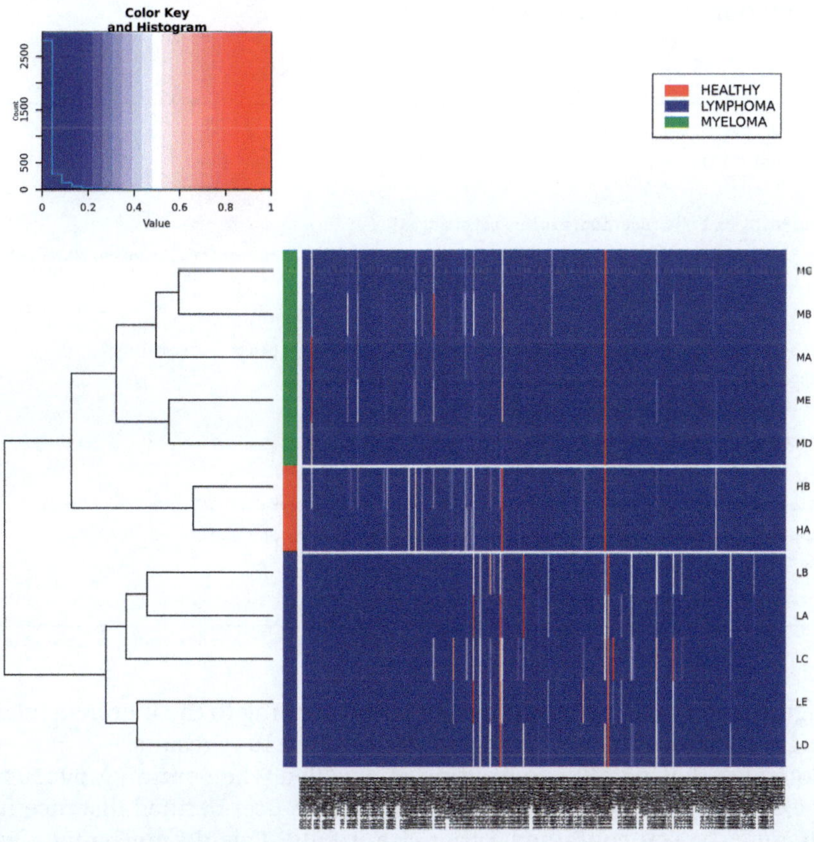


**Figure 5.13** Heat map representation with the hierarchical clustering dendrogram for the samples matrix of the Lymphoma/Myeloma dataset computed using Jaccard distances.

calculated for each cluster in a hierarchical clustering analysis *via* multiscale bootstrap resampling. The *p*-value of a cluster is a value between 0 and 1 indicating how strongly the cluster is supported by data. The `pvclust` function included in this package calculates two types of *p*-values: AU (approximately unbiased) *p*-value and BP (bootstrap probability) value. The AU *p*-value, which is computed by multiscale bootstrap resampling, is a better approximation to unbiased *p*-value than the BP value computed by normal bootstrap resampling. The `pvclust` function performs hierarchical cluster analysis using the `hclust` function and automatically computes *p*-values for all clusters contained in the clustering of the original data. In addition to the built-in distances supported by the `dist` function, it also implements the Pearson correlation distance as it was defined in the previous section. In addition, it can be also used with custom distance functions.

Following the previous examples in this subsection, the following code illustrates how to run `pvclust` for the cluster analyses of both matrices using Jaccard distances. Jaccard distance is implemented as a user-defined function, although it cannot be used directly by the `pvclust` function because it needs the matrix to be transposed. For this reason, a wrapper function is created.

```
library("pvclust")
set.seed(2019)

jaccard.dist.2 <- function(binnedPeaksMatrix) {
    jaccard.dist(rownames(t(binnedPeaksMatrix)), toSpectraList(t(binnedPeaksMatrix)))
}

res.pv.jaccard <- pvclust(t(binnedPeaksMatrix), method.hclust = "complete",
method.dist = jaccard.dist.2, nboot = 1000, parallel=F)
plot(res.pv.jaccard, hang = -1, cex = 0.5)
pvrect(res.pv.jaccard)

res.pv.jaccard.consensus <- pvclust(t(consensusBinnedPeaksMatrix), method.hclust =
"complete", method.dist = jaccard.dist.2, nboot = 1000, parallel = FALSE)
plot(res.pv.jaccard.consensus, hang = -1, cex = 0.5)
pvrect(res.pv.jaccard.consensus)
```

Figure 5.14 shows the dendrogram corresponding to the hierarchical clustering results for the samples matrix with the `pvclust` results.

Note also that parallel computing is disabled when running `pvclust` in this example. This is because to run in parallel, user-defined distance functions must be self-contained, otherwise it fails. This do not happen when

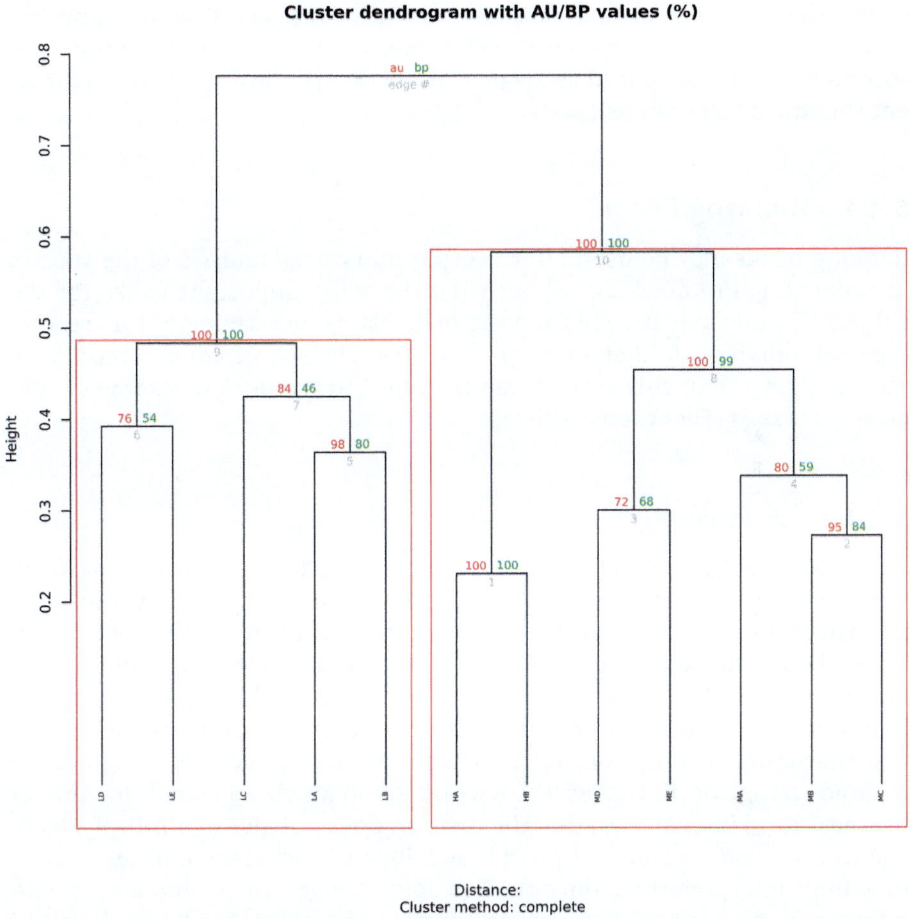


**Figure 5.14** Dendrogram corresponding to the hierarchical clustering results for the samples matrix of the Lymphoma/Myeloma dataset with the `pvclust` results.

using the build-in distances (*e.g.* Pearson correlation, Manhattan or Euclidean) and parallel computing can be enabled safely.

The file `hierarchical-clustering.R` provides the source to reproduce these analyses as well as to repeat them using the other three distance measures explained previously.

## 5.4 Important Variables

As seen previously, a pre-processed spectrum can have tens or even hundreds of peaks. However, it is likely that only a few of them are associated with the conditions being compared or worthy of further analyses. This section aims to provide strategies to find those important peaks. The first and simplest strategy is producing rankings of peaks based on some computations. The second and more advanced strategy is biomarker discovery, for which different statistical tests can be used.

### 5.4.1 Ranking Peaks

Ranking peaks may be useful in the exploratory data analysis of the spectra in order to gain knowledge about what the most important peaks for the subject of study may be. This subsection explains two strategies for creating the peak rankings: fold-changes and *t*-scores. The source code to reproduce all the examples in this section can be found in the `peak-rankings.R` and `peak-rankings-functions.R` files.

#### 5.4.1.1 *Fold-changes*

A fold-change represents the ratio between two values (*e.g.* two conditions). When it comes to a peak, a fold-change can be obtained by dividing the mean intensities in one condition by the mean intensities in another condition. Then, these peak fold-changes can be used to rank them. It is important to note that if a peak is missing in all samples of one condition, then there will be a zero in the denominator or the numerator of the division. In these cases, two alternatives can be considered: (i) ignore these peaks and assign NA as the fold-change or (ii) replace the 0 with a small amount (a small fraction of the maximum intensity) so that the division does not give 0 or infinity. Fold-changes are usually converted to a base-2 logarithmic scale in order to facilitate their interpretation, since the $\log_2$ fold-changes are centered on 0 with reductions represented by negative values and increments by positive values. In the $\log_2$ scale, each increment by one represents a real increment of two-fold with respect to the previous value. For instance: a two-fold increase (fold-change = 2, $\log_2$ fold-change = 1) between $A$ and $B$ means that $A$ is twice as big as $B$ (or $A$ is 200% of $B$) and a two-fold decrease (fold-change = 0.5, log2 fold-change = −1) between $A$ and $B$ means that $B$ is twice as big as $A$ (or $B$ is 200% of $A$).

To compute the fold-changes from a peaks intensity matrix easily, two auxiliary functions are defined (they can be found in the `peak-ranking-fold-change-functions.R` file): (i) `meanConditionSpectra`, to obtain the mean intensities of each peak in each condition, and (ii) `compareConditions`, to make the pairwise comparisons between the conditions. The latter function has a parameter named `zeroIncrease` which indicates the value to replace the 0's with when peaks are not present.

```
meanConditionSpectra <- function(conditions, peaksMatrix, rowsConditions) {
    avgPeaks <- c()

    for (condition in conditions) {
        spectraIndexes <- which(rowsConditions == condition)
        avgPeaks <- rbind(avgPeaks, apply(peaksMatrix[spectraIndexes,], 2, mean))
    }

    rownames(avgPeaks) <- conditions

    avgPeaks
}
compareConditions <- function(averagePeaksMatrix, zeroIncrease = 0) {
    avgPeaks <- averagePeaksMatrix
    comparisons <- combn(rownames(avgPeaks), 2)
    result <- matrix(nrow = ncol(avgPeaks), ncol = ncol(comparisons))
    rownames(result) <- colnames(avgPeaks)
    colnames(result) <- apply(comparisons, 2, function(x) paste0(x[1], " / ", x[2]))

    for (peak in colnames(avgPeaks)) {
        for (cmpIndex in 1:ncol(comparisons)) {
            a <- comparisons[1, cmpIndex]
            b <- comparisons[2, cmpIndex]

            avg.a <- avgPeaks[a, peak]
            if (avg.a == 0) {
                avg.a <- zeroIncrease
            }

            avg.b <- avgPeaks[b, peak]
            if (avg.b == 0) {
                avg.b <- zeroIncrease
            }

            avg <- avg.a / avg.b
            if (is.infinite(avg) | is.nan(avg)) {
                avg <- NA
            }
            result[peak, cmpIndex] <- avg
        }
    }

    result
}
```

For instance, these two functions can be used to obtain the fold-changes for the consensus data of the Lymphoma/Myeloma dataset and convert them to a base-2 logarithmic scale:

```
avgPeaks <- meanConditionSpectra(consensusData$datasetConditions,
consensusBinnedPeaksMatrix, consensusData$spectraConditions)
comparison <- compareConditions(avgPeaks)
comparison.log2 <- log2(comparison)
```

The `comparison.log2` object contains as many rows as peaks in the dataset and three columns, one with each possible comparison:

```
> head(comparison.log2)
                    HEALTHY / LYMPHOMA HEALTHY / MYELOMA LYMPHOMA / MYELOMA
600.430166666667                  -Inf                NA                 NA
605.26122                         -Inf                NA                 NA
618.430666666667                  -Inf                NA                 NA
635.3707                          -Inf                NA                 NA
640.40655                         -Inf                NA                 NA
656.173628571429                    NA         -1.018241               -Inf
```

If one is interested in the first comparison (healthy *versus* lymphoma), peaks can be sorted according to this column and then filtered out to remove peaks with infinite values or with a $\log_2$ fold-change lower than 1.5. Using the `barplot` function, a plot like the one in Figure 5.15 can be obtained, showing the peaks ranked according to their $\log_2$ fold-changes.

```
comparison.index <- 1
data <- sort(comparison.log2[, comparison.index])
data.plot <- data[!is.infinite(data) & abs(data) > 1.5]
barplot(data.plot, horiz = TRUE, names = substr(names(data.plot), 1, 8), las = 1,
space = 1.5, cex.names = 0.8, xlim = c(min(data.plot) - 1, max(data.plot) + 1), main
= colnames(comparison)[comparison.index])
```

### *5.4.1.2 t-scores*

Sebastian Gibb and Korbinian Strimmer proposed the use of standard *t*-scores for the identification of the most important class discriminating peaks and proposed different ways of using them to rank peaks.[13] For this, they propose the use of the SDA package (http://strimmerlab.org/software/sda/), which computes the *t*-score between each group (*i.e.* analysis condition) centroid and the pooled mean for each feature (*i.e.* peak) (refer to the documentation of the `catscore` function for more details: https://rdrr.io/cran/sda/man/catscore.html). Once *t*-scores have been computed, the overall ranking of a peak is determined by computing a summary score: (i) if there are two conditions, peaks are ranked according to the squared *t*-scores, and (ii) if there are more than two conditions, the way the overall score is computed is controlled by an option named `ranking.score`. The default setting of this option is "`entropy`", which means that mutual information between the response and the respective predictors (`ranking.score`) is used for ranking. This is equivalent to a weighted sum of squared *t*-scores across the classes. Another possibility is to employ the average of the squared *t*-scores for ranking (as suggested by Miika Ahdesmäki and Korbinian Strimmer[14])

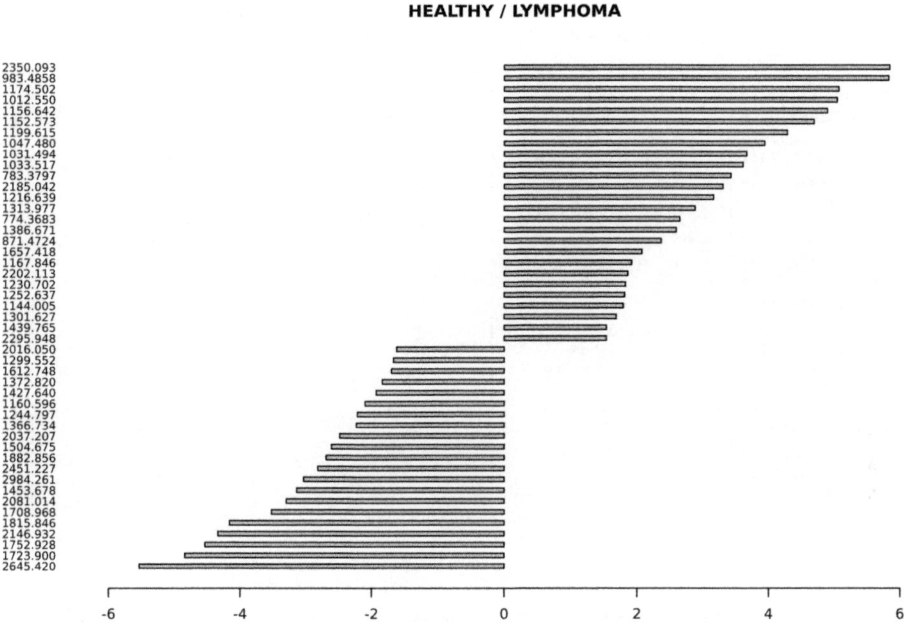


**Figure 5.15** Peaks of the Lymphoma/Myeloma dataset ranked according to their log2 fold-changes between Healthy and Lymphoma. Only peaks with a log2 fold-change greater or equal to 1.5 are shown.

by setting `ranking.score = "avg"`. A third option is to use the maximum of the squared *t*-scores across groups by setting `ranking.score = "max"`. This peak ranking is produced using `sda.ranking` function (https://rdrr.io/cran/sda/man/sda.ranking.html) from the SDA package.

The following example shows how this peak ranking can be obtained easily for the Lymphoma/Myeloma dataset used in previous examples.

```
ddar <- sda.ranking(Xtrain = consensusBinnedPeaksMatrix, L =
consensusData$spectraConditions, fdr = FALSE, diagonal = TRUE)
```

The `ddar` object contains a table with the peaks ordered by their ranking score and three additional columns, one for each condition containing the *t*-score associated with each one. Since the SDA package overrides the `plot` function for objects returned by the `sda.ranking` function, the `ddar` object can be plotted into a nice representation without effort. For example, the following command plots the top 20 features and produces the result shown in Figure 5.16.

```
plot(ddar, top = 20, arrow.col = "blue", zeroaxis.col = "red", ylab = "Peaks")
```

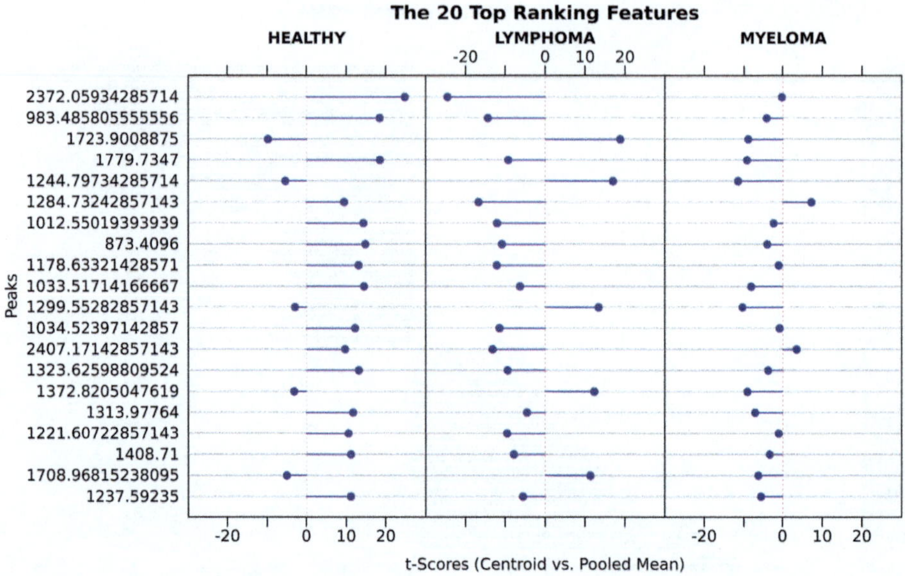


**Figure 5.16** Plot of the top 20 peaks in the Lymphoma/Myeloma dataset produced with the `sda.ranking` function of the SDA package.

Finally, it is important to note that the `ddar` object contains a column named `idx` which indicates the original position of the corresponding peak. This information facilitates the filtering of the peaks matrix to account only for the top ranked peaks. For instance, the following code shows how to perform a hierarchical clustering analysis using only these top 20 peaks:

```
consensusBinnedPeaksMatrix.topPeaks <- consensusBinnedPeaksMatrix[, sort(ddar[1:20,
"idx"])]
distances.jaccard.consensus <-
jaccard.dist(rownames(consensusBinnedPeaksMatrix.topPeaks),
toSpectraList(consensusBinnedPeaksMatrix.topPeaks))
clustering.jaccard.consensus <- hclust(distances.jaccard.consensus)
```

## 5.4.2 Biomarker Discovery

Biomarker discovery is one of the most valuable results of high-throughput data analysis. Basically, a biomarker is a measurable element in a biological sample that allows one to test if the sample has a condition of interest. For example, the blood LDL cholesterol levels is a commonly used biomarker to assess the cardiovascular risk.

One way to discover new biomarkers is to test one or several candidates over samples of different conditions, for example, case and control samples. For each candidate biomarker, one may try to see if it shows significant differences among samples of different conditions. Statistical tests are the most widely used tool to robustly guess if a given candidate biomarker shows statistically significant differences, instead of spurious differences that appear due to chance.

Statistical tests work by comparing two hypotheses: the null hypothesis and the alternative hypothesis. The null hypothesis is the one assuming that there are "no differences", whereas the alternative hypothesis is the opposite. When conducting a statistical test, one obtains a *p*-value that is the probability of obtaining the observed, or more extreme, data if the null hypothesis is true. In this sense, a small *p*-value is evidence against the null hypothesis, that is, it is evidence against assuming that there are "no differences". Normally, a *p*-value < 0.05 is enough to discard the null hypothesis saying that there are significant differences; with *p*-values < 0.01 it is said that there are very significant differences.

Some statistical tests make assumptions about the distribution of the data. Such tests are called parametric tests. An example of such tests is the Student's *t*-test, which assumes that the data is normally distributed. Conversely, non-parametric tests do not assume anything about the distribution of the samples. The Wilcoxon rank sum test is a non-parametric test that can be used with the same purpose as the Student's *t*-test, but in scenarios where the data is not normally distributed.

The procedure to discover biomarkers consists of applying a statistical test to each peak in order to find which ones show statistically significant differences. In this sense, one hypothesis for each peak is being tested. However, in this multiple-test scenario, it should be taken into account that as the number of tests increases, there is also a growing probability of getting small *p*-values simply by chance. Later on, a strategy to deal with this problem will be presented.

This section aims to show how to find biomarkers among spectra peaks with R. Different statistical tests (summarized in Table 5.1) will be used depending on the peak values (intensities *versus* presence/absence) and the number of conditions (two conditions *versus* more than two conditions).

The source code to reproduce all the examples in this section can be found in the `biomarker-discovery.R` file. In this section, two new datasets are used to illustrate the examples: (i) the Cancer Fiedler *et al.* 2009 dataset,[15] which contains 480 MALDI-TOF mass spectra from blood sera of 60 patients and 60 healthy controls (each sample has four technical replicates), and (ii)

**Table 5.1** Statistical tests used for biomarker discovery according to the number of conditions (2 or >2) and the way that peaks are considered (intensities or presence/absence).

| | | Peak values | |
|---|---|---|---|
| | | Intensities | Presence/absence |
| Conditions | 2 | Wilcoxon rank sum test | Fisher's exact test |
| | >2 | Kruskal–Wallis test | Chi-square independence |

the Species dataset, which contains spectra of four different bacteria species, each one represented by eight individual samples with three technical replicates by sample. Both datasets are available through the MALDIquantExamples package (http://www.github.com/sgibb/MALDIquantExamples/) and two scripts to load them (`load-maldiquant-cancer-fiedler. R` and `load-maldiquant-species.R`) are provided.

### 5.4.2.1 *Working With Peak Intensities*

Two different statistical tests can be used depending on whether two or more than two conditions are being compared. For two conditions, the Wilcoxon rank-sum test (also known as Mann–Whitney *U* test) will be used, whereas for more than two conditions, the Kruskal–Wallis test will be used.

**5.4.2.1.1 Two Conditions.** The Wilcoxon rank-sum test is the non-parametric alternative to the two sample Student's *t*-test. This test is suitable for looking for differences between two conditions as in the Cancer Fiedler *et al.* 2009 dataset where samples are grouped in two conditions: cancer and control. In R, the `wilcox.test` function performs the test by simply passing a formula. For example, the following code loads the dataset and performs this test in the first peak:

```
source("load-maldiquant-cancer-fiedler.R")

wilcox.test(binnedPeaksMatrix[,1] ~ binnedPeaksMatrix.conditions)
        Wilcoxon rank sum test

data:  binnedPeaksMatrix[, 1] by binnedPeaksMatrix.conditions
W = 809, p-value = 0.7811
alternative hypothesis: true location shift is not equal to 0
```

This function prints the result of the test, where the *p*-value (0.7811), among other information, can be observed. In addition, the function returns a list that contains this *p*-value. In order to perform a test over all peaks and obtain the *p*-values, the following code can be executed:

```
pvals <- c()
for (i in 1:ncol(binnedPeaksMatrix)) {
  pvals[colnames(binnedPeaksMatrix)[i]] <-
    wilcox.test(binnedPeaksMatrix[,i] ~ binnedPeaksMatrix.conditions)$p.value
}
```

Finally, since multiple tests have been conducted, it is necessary to adjust the *p*-values. R provides `p.adjust`, a very handy function to adjust *p*-values in multiple test scenarios with different methods. Some methods are conservative, such as "`bonferroni`", which is the less false–positive prone (at the cost of more false negatives). Other methods such as "`fdr`", are less conservative.

And now, for instance, it is possible to list those peaks with an adjusted *p*-value below a given threshold (*e.g.* < 0.01):

```
pvals <- p.adjust(pvals, method = "fdr")
```

For example, the peak 1450.336 830 952 67 has an adjusted *p*-value of 3.209 388 × $10^{-4}$. It may be interesting to plot the distribution of the intensities of this peak for each group.

```
pvals[pvals < 0.01]
1450.33683095267 1546.52890935556 2105.92981042575 2660.95514138932
    3.209388e-04     3.239048e-06     4.709899e-03     5.790161e-04
2769.96463053479 2933.16047176545 3192.67089815328 4091.97599528659
    1.244799e-03     1.515895e-03     1.189156e-03     3.775005e-03
4467.84851263507  6090.4588217156 8934.77896012354 9180.09938809373
    4.439140e-03     3.328785e-04     4.439140e-03     1.189156e-03
```

```
boxplot(binnedPeaksMatrix[, "1450.33683095267"] ~ binnedPeaksMatrix.conditions, ylab
= "intensity", xlab = "condition")
```

As can be seen from Figure 5.17, this peak in cancer patients has larger values than in control ones. The statistical test tells that these differences are significant.

**5.4.2.1.2 More than Two Conditions.** In a scenario where there are multiple conditions, the Kruskal–Wallis test can be used. This test is the non-parametric alternative to the one-way ANOVA test. Both tests give small *p*-values when at least one group differs from the others. As in the previous case, the `kruskal.test` function does the test by simply passing a formula. For example, the following code loads the Species dataset, performs this test for all the peaks and computes the adjusted *p*-values:

```
source("load-maldiquant-species.R")

pvals <- c()
for (i in 1:ncol(binnedPeaksMatrix)) {
  pvals[colnames(binnedPeaksMatrix)[i]] <-
    kruskal.test(binnedPeaksMatrix[, i] ~ binnedPeaksMatrix.conditions)$p.value
}
pvals <- p.adjust(pvals, method = "fdr")
```

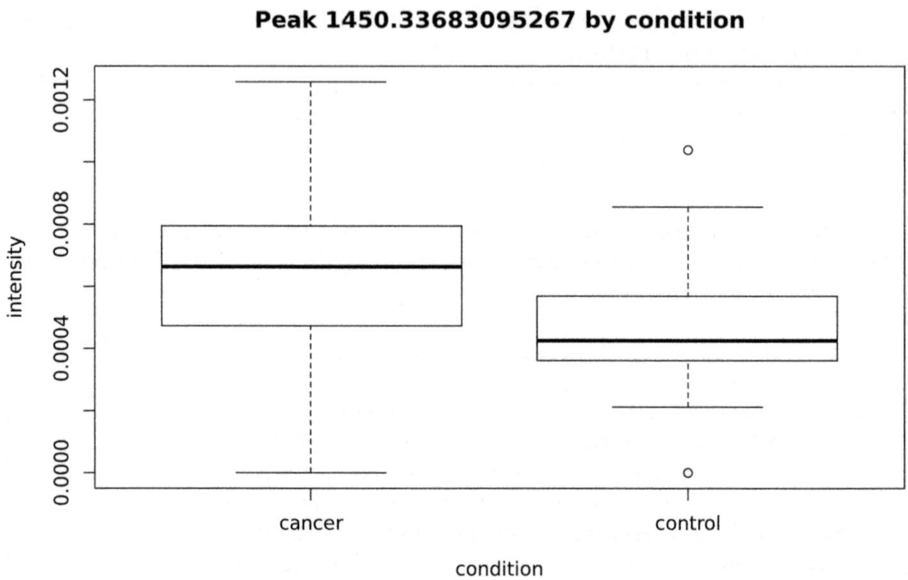


**Figure 5.17** Boxplot of the peak with *m*/*z* value 1450.336 830 952 67 in the two groups of the Cancer Fiedler *et al.* 2009 dataset. This peak has an adjusted *p*-value of $4.488\ 655 \times 10^{-6}$ in the Wilcoxon rank-sum test.

For example, the peak 1974.317 629 097 5 has an adjusted *p*-value of $9.043\ 241 \times 10^{-5}$. It may be interesting to plot the distribution of the intensities of this peak for each group.

```
boxplot(binnedPeaksMatrix[, "1974.3176290975"] ~ binnedPeaksMatrix.conditions, ylab =
"intensity", xlab = "species", main = "Peak 1974.3176290975 by species")
```

As can be seen from Figure 5.18, this peak shows clear differences among different species.

### *5.4.2.2 Working With Peak Presence/Absence*

It may be of interest to ignore the specific intensity of the peaks and consider only the presence of the peak in the sample. In this case, instead of working with decimal intensity values, nominal values (*e.g.*: "present", "absent") are used. To do this, an intensity matrix can converted into a presence/absence matrix using the auxiliary function `asPresenceMatrix` defined in the `data-functions.R` file (used also when applying the Manhattan distance).

Two different statistical tests can be used depending on whether two or more than two conditions are being compared. For two conditions, the Fisher's Exact Test will be used, whereas for more than two conditions, the Chi-squared test will be used.

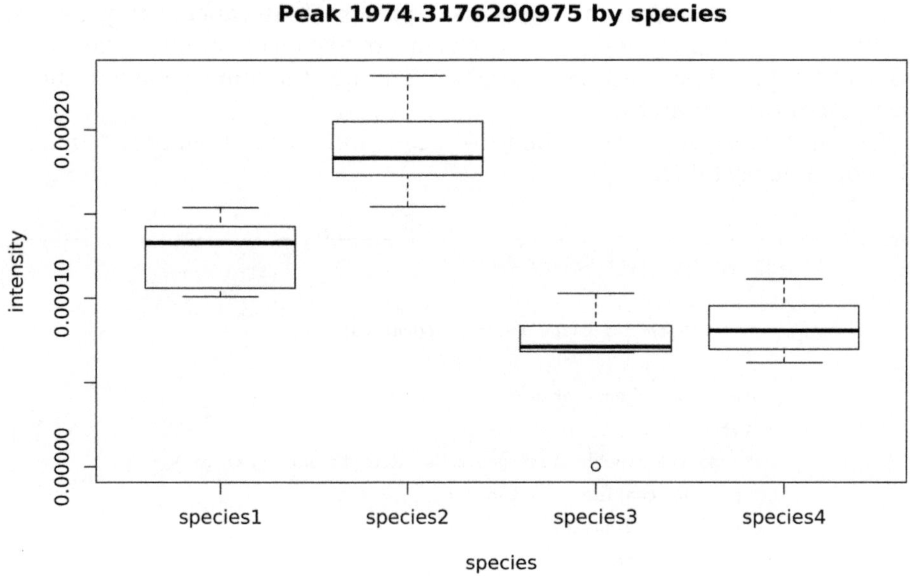


**Figure 5.18** Boxplot of the peak with *m*/*z* value 1974.317 629 097 5 in the four groups of the Species dataset. This peak has an adjusted *p*-value of $9.043\,241 \times 10^{-5}$ in the Kruskal–Wallis test.

**5.4.2.2.1 Two Conditions.** To illustrate the Fisher's Exact test, the Cancer Fiedler *et al.* 2009 dataset will be used again. Firstly, the dataset must be loaded and transformed into presence/absence data.

```
source("load-maldiquant-cancer-fiedler.R")
binnedPeaksMatrix <- asPresenceMatrix(binnedPeaksMatrix)
```

Secondly, for this statistical test, a contingency table showing the count of presences/absences on each condition must be created. The `table` function in R does this very easily. The following code creates the contingency table for the first peak:

```
contingencyTable <- table(factor(binnedPeaksMatrix[,1], c("present","absent")),
binnedPeaksMatrix.conditions)
contingencyTable
         binnedPeaksMatrix.conditions
          cancer control
  present     40      39
  absent      20      21
```

It can be observed that the first peak is present in 40 cancer samples and in 39 control samples, whereas it is absent in 20 cancer samples and in 21 control samples. It seems that this peak has almost the same behaviour independent of the condition.

To run the test, R provides the `fisher.test` function, which only requires the contingency table.

```
fisher.test(contingencyTable)

        Fisher's Exact Test for Count Data

data:  contingencyTable
p-value = 1
alternative hypothesis: true odds ratio is not equal to 1
95 percent confidence interval:
 0.4073241 2.1144415
sample estimates:
odds ratio
 0.9291499
```

As expected, looking at the contingency table the *p*-value is 1, indicating that there are no significant differences in the peak presence among conditions.

The test can be run for each peak in order to get the *p*-value and adjust it for multiple test as in the previous examples.

```
pvals <- c()
for (i in 1:ncol(binnedPeaksMatrix)) {
  pvals[colnames(binnedPeaksMatrix)[i]] <-
    fisher.test(table(factor(binnedPeaksMatrix[, i], c("present", "absent")),
binnedPeaksMatrix.conditions))$p.value
}
pvals <- p.adjust(pvals, method = "fdr")
```

And now, for instance, it is possible to list those peaks with an adjusted *p*-value below a given threshold (*e.g.* < 0.01):

```
pvals[pvals < 0.01]
2093.05826713668 2162.55556750713 2714.44332190603
    3.931561e-08     1.660302e-03     7.473357e-07
```

Finally, it is possible to show the contingency table of the first significant peak in order to observe its distribution:

```
table(factor(binnedPeaksMatrix[, "2093.05826713668"], c("present", "absent")),
binnedPeaksMatrix.conditions)
         binnedPeaksMatrix.conditions
          cancer control
  present     52      18
  absent       8      42
```

It can be seen that the presence of this peak seems to be associated with the cancer status (52 of the 60 cancer samples have the peak, whereas only 18 of the 60 have it). The statistical tests confirms that this association is statistically significant.

**5.4.2.2.2 More than Two Conditions.** To illustrate the Chi-Squared test, the Species dataset will be used again. Firstly, the dataset must be loaded and transformed into presence/absence data.

```
source("load-maldiquant-species.R")
binnedPeaksMatrix <- asPresenceMatrix(binnedPeaksMatrix)
```

This test works in the same way as the Fisher's Exact Test, thus a contingency table must first be created. As in the previous example, the following code creates the contingency table for the first peak:

```
contingencyTable <- table(factor(binnedPeaksMatrix[, 1], c("present", "absent")),
binnedPeaksMatrix.conditions)
contingencyTable
         binnedPeaksMatrix.conditions
          species1 species2 species3 species4
  present        2        1        3        2
  absent         6        7        5        6
```

This table allows us to see how many times the peak is present/absent among different species. To run the test, R provides the `chisq.test` function, which only requires the contingency table to work.

```
chisq.test(contingencyTable)
Pearson's Chi-squared test


data:  contingencyTable
X-squared = 1.3333, df = 3, p-value = 0.7212

Warning message:
In chisq.test(contingencyTable) :
  Chi-squared approximation may be incorrect
```

The test can be run for each peak in order to get the *p*-value and adjust it for multiple tests as in the previous examples.

```
pvals <- c()
for (i in 1:ncol(binnedPeaksMatrix)) {
  pvals[colnames(binnedPeaksMatrix)[i]] <-
    chisq.test(table(factor(binnedPeaksMatrix[, i], c("present", "absent")),
binnedPeaksMatrix.conditions))$p.value
}
pvals <- p.adjust(pvals, method = "fdr")
```

And now, for instance, it is possible to list those peaks with an adjusted *p*-value below a given threshold (*e.g.* < 0.01):

```
pvals[pvals < 0.01]
2008.24525043867 2013.87530331115 2038.62305818529 2047.36043593175
               4                5                7                8
2112.05086630122 2130.80965677743 2196.41368819683 2202.18490199803
              14               15               21               22
```

Finally, it is possible to show the contingency table of the first significant peak in order to observe its distribution:

```
table(factor(binnedPeaksMatrix[, "2008.24525043867"], c("present", "absent")),
binnedPeaksMatrix.conditions)
         binnedPeaksMatrix.conditions
          species1 species2 species3 species4
  present        8        4        8        0
  absent         0        4        0        8
```

As can be seen, this peak has a behaviour that changes among species. For example, all samples of `species1` and `species4` contain this peak, whereas none of the samples of `species4` do. This test tells us that this association is statistically significant.

### *5.4.2.3 Visualization*

**5.4.2.3.1 Filtered Heat Maps.** An interesting way of visualizing multiple biomarkers is the use of heat maps, where only a selection of significant peaks is taken. For example, the following code creates a heat map for the presence/absence version of the Cancer Fiedler *et al.* 2009 dataset, using only those peaks that are significant (adjusted *p*-value < 0.05):

```
mapConditionToColor <- function(conditions) {
    colorsVector = ifelse(conditions == "cancer", "#D95DA5", "#82D3A5")
    return(colorsVector)
}

topPeaksMatrixWithPvals <- t(binnedPeaksMatrix[, names(pvals[pvals < 0.05])])
rownames(topPeaksMatrixWithPvals) <- paste(rownames(topPeaksMatrixWithPvals)," (pval:
", pvals[pvals < 0.05], ")")

heatmap(
    topPeaksMatrixWithPvals,
    scale = "none",
    ColSideColors = mapConditionToColor(binnedPeaksMatrix.conditions),
    col = rev(terrain.colors(2))
)

legend(
    "left",
    Inset = .02,
    Title = "Peak presence",
    c("present", "absent"),
    fill = terrain.colors(2),
    horiz = TRUE,
    cex = 0.8
)
legend(
    "bottomleft",
    Inset = .02,
    Title = "Group",
    c("cancer", "control"),
    fill = c("#D95DA5", "#82D3A5"),
    horiz = TRUE,
    cex = 0.8
)
```

As can be seen from this figure, the six significant peaks are associated with the condition. By looking at the dendrogram generated taking into account only these six peaks, it can be observed that samples of different conditions tend to cluster together. For example, the first peak (adjusted *p*-value $< 0.04$) tends to be present in control cases (however, some cancer cases do have the peak too). Peaks with very significant *p*-values (such as 2714 and 2093) are more associated with the cancer condition (Figure 5.19).

**5.4.2.3.2 Volcano Plots.** Volcano plots are widely used to display the results of various omics data analyses when two conditions are being compared.[16,17] A volcano plot is a type of scatter plot that plots statistical significance *versus* magnitude of change (fold-change). More specifically, it is constructed by plotting the negative log of the *p*-value on the *y*-axis (usually base 10) and the log of the fold-change (between the two conditions) on the *x*-axis (usually base 2). Volcano plots are interesting because they allow one to easily identify points (*e.g.* *m*/*z* peaks) with large fold-changes that are also statistically significant: the most up-regulated points are located towards the

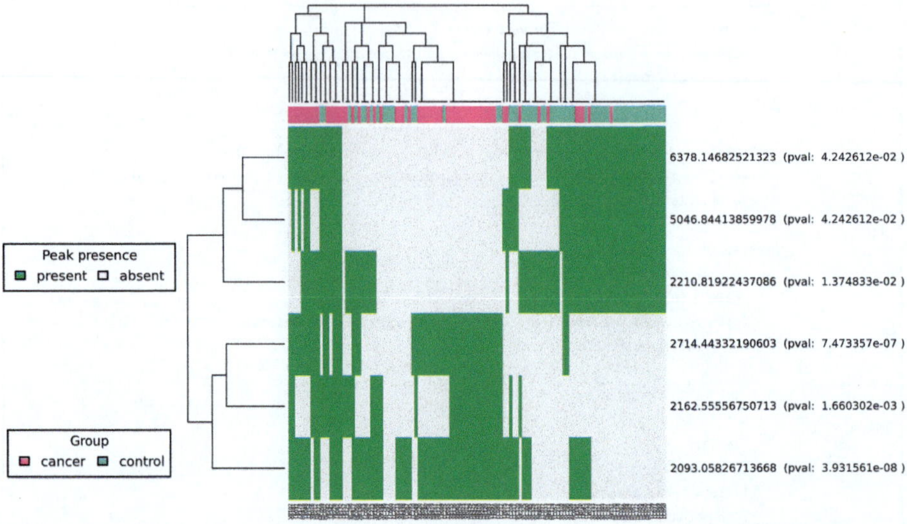


**Figure 5.19** Heat map representation with the hierarchical clustering dendrogram for the replicates matrix of the Fiedler *et al.* 2019 dataset computed using only those peaks that are significant (adjusted *p*-value < 0.05).

right, the most down-regulated points are located towards the left, and the most statistically significant genes are located towards the top.

Continuing with the two-condition presence/absence Cancer Fiedler *et al.* 2009 dataset, fold-changes between cancer and control conditions must be computed first using the auxiliary functions defined in the *5.4.1 Ranking Peaks/Fold-changes*.

```
source("peak-ranking-fold-change-functions.R")
avgPeaks <- meanConditionSpectra(unique(binnedPeaksMatrix.conditions),
binnedPeaksMatrix, binnedPeaksMatrix.conditions)
comparison <- compareConditions(avgPeaks)
comparison.log2 <- log2(comparison)
logFC <- comparison.log2[, 1]
```

Now, the `logFC` and `pvals` variables contain the data needed to create the volcano plot. Peaks are represented with black dots and some peaks are highlighted according to three possibilities: (i) red, if they have an absolute $\log_2$ fold-change greater than 1, (ii) orange, if they have an adjusted *p*-value lower than 0.05, and (iii) green, if they have both an absolute $\log_2$ fold-change greater than 1 and an adjusted *p*-value lower than 0.05. The following code produces the volcano plot shown in Figure 5.20:

```
library("calibrate")

abs_fcs <- cbind(as.integer(abs(logFC)))
o <- order(abs_fcs, decreasing = TRUE)
abs_fcs_order <- abs_fcs[o,]
limit <- abs_fcs_order[1] + 1

names.logFC.filter <- names(which(abs(logFC) > 1))
names.pValue.filter <- names(which(pvals < 0.05))
names.both.filter <- intersect(names.pValue.filter, names.logFC.filter)
plot(
    logFC, -log10(pvals),
    col = colors[1],
    pch = 20,
    xlim = c(-limit, limit),
    xlab = "log2(fold change)",
    ylab = "-log10(adjusted p-value)"
)

colors <- c("black", "red", "orange", "green")

points(logFC[names.logFC.filter], -log10(pvals[names.logFC.filter]), col = colors[2],
pch = 20)
points(logFC[names.pValue.filter], -log10(pvals[names.pValue.filter]), col =
colors[3], pch = 20)
points(logFC[names.both.filter], -log10(pvals[names.both.filter]), col = colors[4],
pch = 20)
legend(
    "bottomright",
    xjust = 1,
    yjust = 1,
    legend = c("|log2(fold change)|>1", "adjusted p-value < 0.05", "both"),
    pch = 20,
    cex = 1,
    col = c(colors[2], colors[3], colors[4])
)
```

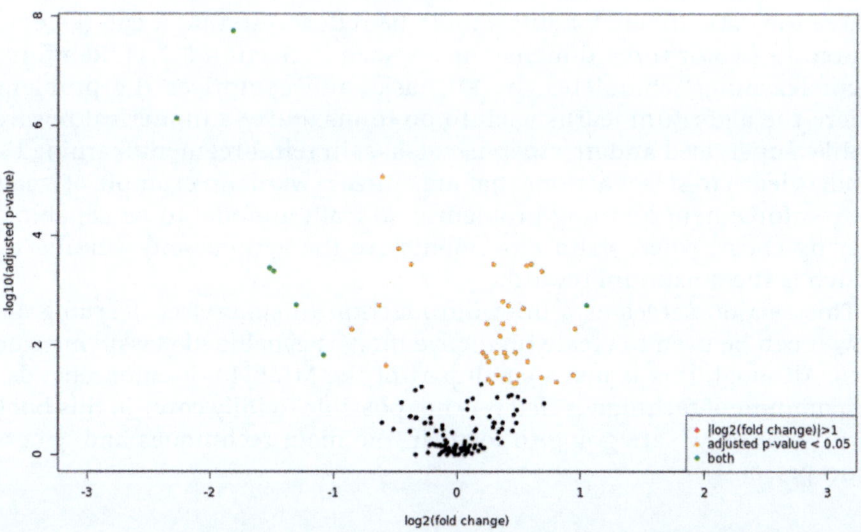


**Figure 5.20** Volcano plot of the Cancer Fiedler *et al.* 2009 dataset: *x*-axis shows the base-2 logarithm of the fold-changes and *y*-axis the base-10 logarithm of the adjusted *p*-values from the Wilcoxon rank-sum test.

# 5.5 Predictive Models

## 5.5.1 Machine Learning Introduction

Machine learning (ML) is a discipline that belongs to the artificial intelligence (AI) field, which focuses on creating algorithms or models that learn to solve problems without being explicitly programmed. ML models require a set of samples to learn from, called training sets, in a process where the internal parameters of the model are adjusted to represent the different patterns present in the training samples, so that they can recognize those patterns in future samples. This process is usually known as the training or learning phase. After this phase the trained model, or classifier, is expected to be able to classify or identify the category of new, previously unseen, samples, which comprise the test set. This ability to assign new samples to a category is known as generalization.

The process described above is, in fact, one of the three main tasks on which the ML discipline can be divided, depending on the characteristics of the training set and the output expected from the algorithm. This task is known as supervised learning, which is characterized by the need to know the expected output values for the samples in the training set. Supervised learning can be applied to solve (i) classification problems, where outputs are values from a finite set of possible values, and (ii) regression problems, where outputs are continuous values. In contrast to supervised learning, unsupervised learning does not require one to know the output expected for the samples in the training set. It is focused on solving problems such as (i) clustering, where similar samples should be grouped (as seen in Section 5.3), (ii) density estimation, where the distribution of samples in a dataset should be determined, or (iii) visualization, where the data dimensionality should be reduced so that it can be represented in two or three dimensions (as seen in Section 5.2.1). Reinforcement learning[18] completes the ML tasks and comprises the problems where the algorithm learns what to do to maximize a numerical reward. Unlike supervised and unsupervised tasks, in reinforcement learning the models learn to select actions that maximize reward. An example of a classic reinforcement learning problem is to train a model to be capable of playing chess, where valid movements are the actions and winning the match is the maximum reward.

This section serves as a brief introduction to supervised learning and how it can be used to create predictive models capable of classifying samples. Although this is just a small part of the ML field, it comprehends a vast amount of techniques that it is not possible to fully cover in this book. In this sense, we are going to focus in the main techniques and general concepts.

#### 5.5.1.1 *Classification Experimental Workflow*

Generating a classifier able to generalize is usually a complex task that requires a dataset with enough samples for a model to recognize patterns that characterize the different conditions into which the samples should be classified. In classification, a sample or instance represents an individual (*e.g.* patient, bacteria, tissue, *etc.*) that should be independent of other samples. This implies that, if several technical or biological replicates are used in an experiment for each sample, then they should be reduced to one single sample, for example, by combining them or by discarding all but one, when adding them to the dataset. Each sample is comprised of several variables, features or attributes that contain information about the sample, and one class or condition, which represents the expected output value. A variable is a single element of information about the sample (*e.g.* age, gender, peak intensity, *etc.*), that can be numeric (*e.g.* age), textual (*e.g.* nationality) or boolean (*e.g.* peak presence). On the other hand, the class of a sample is a categorical value (*e.g.* case or control, bacterial strain, cancer stage, *etc.*) that should be shared between several samples in the dataset.

The process to generate a classifier is usually divided into two main stages: (i) model selection, where several models and configurations are systematically evaluated with a training set to determine their performance and the best one is selected, and (ii) classifier testing, where a trained classifier is evaluated on the testing set to get a hint of its performance when used to predict new samples. Figure 5.21 represents this process, including the stages described and the steps to be followed. As can be seen from this figure, before the two main stages the base dataset is initially split into (i) the training set, used in model selection to evaluate the candidate models and in classifier testing to train the best model, and (ii) the testing set, used in classifier testing to evaluate the best classifier and get a hint of its performance when used to predict new samples.

When training a classifier, it is typical to try several models with different hyperparameter (*i.e.* a parameter set before training) configurations, as it may be difficult to know beforehand which one will perform best for the current problem. This can be done by simply configuring manually some models or, in more complex setups, by using algorithms that automatically explore the space of all possible configurations for a model, for example, using genetic algorithms guided by the model's performance. In any case, each selected model will be evaluated using the same procedure, so that the performance results obtained are comparable. Some of the most important models will be described in Section 5.5.2.

Model evaluation requires partitioning the training set to generate training and validation subsets, which will be used to train the model and to evaluate its performance on predicting the expected output for samples, respectively.

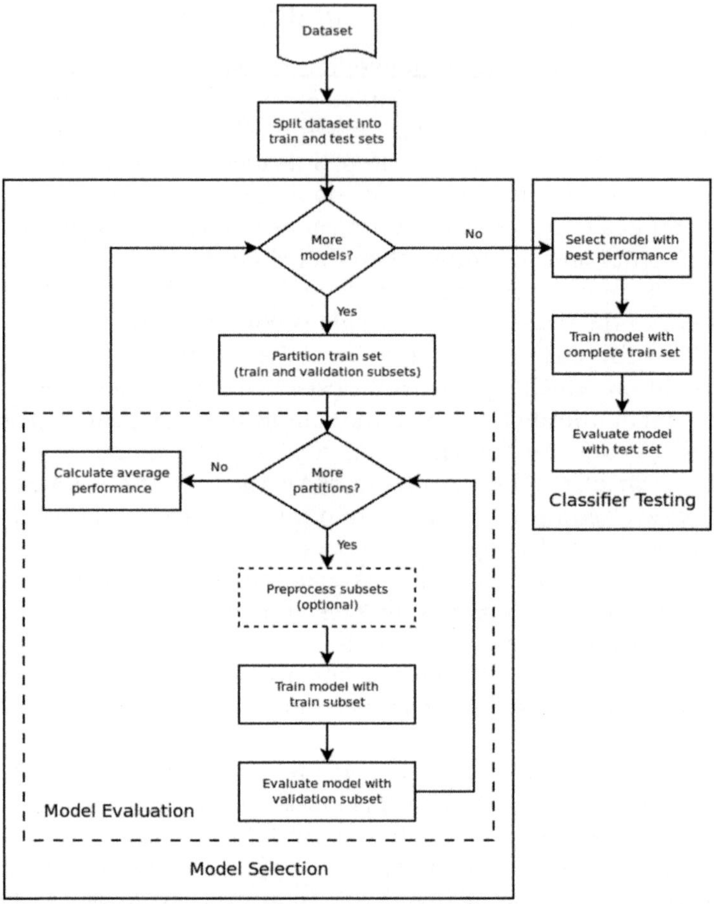


**Figure 5.21** General workflow to build a classifier, divided into two main stages: (i) model selection, where several models and configurations are systematically evaluated with a training set to determine their performance and select the best one, and (ii) classifier testing, where a trained classifier is evaluated on the testing set to get a hint of its performance when used to predict new samples.

There are several partitioning methods, which usually imply creating several partitions of the original training set to repeatedly exercise and evaluate the model, so that the final performance is calculated as an average of the performance achieved for each partition. Partitioning methods will be explained in Section 5.5.3.

In some cases, using the sample's data in its original form may cause some undesired effect. For example, if two features are using different scales (*e.g.* age ranging from 0 to 100 and salary ranging from 10 000 to 100 000), then they may implicitly have different weights for certain models, even if they really have the same importance for classifying a sample. To avoid these problems, data can be pre-processed before being fed to the classifier, in order to adjust feature values, remove features, generate new features, remove samples or add new samples. For example, one of the most common pre-processing operations is to scale and center the values of a feature, where the mean and standard deviation of that feature in the dataset are calculated to subtract the mean from the value of that feature on each sample and to divide the result by the standard deviation. It is important to note that, when using operations like these, where all the dataset information is needed, they must be applied independently for each training subset of each partition, without including the validation subset. Otherwise, the performance results may be overestimated, as some information from the validation subset is included in the training process, something which is not possible in a real case. In addition, the same operations applied to the training set should be applied to test samples before classification.

The training process continues until each candidate model has been evaluated on each training set partition and a performance result has been obtained. This performance result is decisive for selecting the final model, in that it is, usually, the model with the best performance. In the case when the best performance is below the expected, it is possible to repeat the process with more models and/or configurations or with more samples added to the training set. The main performance measurements will be described in detail in Section 5.5.4.

Once the best model reaches the expected performance, it is then trained with the complete training set to generate the final classifier. In addition, to get a measure of the expected performance of this classifier in a real environment, it should be evaluated in the test set reserved before starting the training process. The performance results obtained in the test set are the best indicator of the generalization capabilities of that classifier.

## 5.5.2 Supervised Learning Models

This subsection aims to give a brief overview of five widely-used supervised learning models and how to fit them in R. All models are available through different R packages and they will be used *via* the caret R package[19] (http://topepo.github.io/caret/index.html). This package is a very useful resource to build predictive models, as it contains numerous tools for developing them using the rich set of models available in R through a consistent interface. Caret is particularly focused on simplifying model training and tuning across a wide variety of techniques, also including different methods for pre-processing training data, calculating variable importance, and model visualizations.

The source code to reproduce all the examples in this section can be found in the `machine-learning.R` file. These examples use the Cancer Fiedler *et al.* 2009 dataset, which is prepared at the beginning of the script in order to: (i) replace missing values (`NA`) with 0 s, (ii) split the 120 samples in two datasets using the `createDataPartition` from the caret package: `train`, which contains the 70% of the original data (84 samples) and is used to fit the models, and `test`, which contains the remaining 30% of the data (36 samples) and is used to test the fitted models (see Section 5.5.3 for more details about dataset partitioning methods), and (iii) replace column names (initially containing the peaks) with $V_1$, $V_2$, ..., $V_n$, because some methods fail when dealing with feature names being numeric.

### 5.5.2.1 Logistic Regression

While (multiple) linear regression serves to predict continuous response variables, logistic regression is used for binary classification, that is, to predict binary outcomes. Logistic regression is a special case of a Generalized Linear Model (GLM), adapted to extend linear regression to other settings. In short, a logistic regression (or simply logit) model, models the log odds of the outcome as a linear combination of the predictor variables. Since for binary classification the interest is in the probability of the event instead of in the log odds of the event, the predicted values from the logit model (*i.e.* the log odds of the event) must be converted to probabilities.

In R, the `glm` function is used with the `family` parameter set to `binomial` to fit a logistic regression model. When using the caret package, this model is fitted using the `train` function with `method = "glm"`:

```
lr <- train(condition ~ ., data = train.fixedColnames, method = "glm", trControl =
trainControl(method = "none"))
```

Where the "`condition ~ .`" parameter is the formula that the model should fit: the condition of the samples should be explained using all the other variables (*i.e.* peaks) in the data.

Then, this model can be used to predict the classes of both train and test data using the `predict` function:

```
lr.train.result <- predict(lr, type = "raw")
lr.test.result <- predict(lr, type = "raw", newdata = test.fixedColnames)
```

And finally, the `confusionMatrix` function can be used to compare the predictions with the actual classifications:

```
> caret::confusionMatrix(data = lr.train.result, train.fixedColnames$condition)
Confusion Matrix and Statistics

          Reference
Prediction cancer control
   cancer      42       0
   control      0      42
               Accuracy : 1
                 95% CI : (0.957, 1)
    No Information Rate : 0.5
    P-Value [Acc > NIR] : < 2.2e-16

                  Kappa : 1
 Mcnemar's Test P-Value : NA

            Sensitivity : 1.0
            Specificity : 1.0
         Pos Pred Value : 1.0
         Neg Pred Value : 1.0
             Prevalence : 0.5
         Detection Rate : 0.5
   Detection Prevalence : 0.5
      Balanced Accuracy : 1.0

       'Positive' Class : cancer

> caret::confusionMatrix(data = lr.test.result, test.fixedColnames$condition)
Confusion Matrix and Statistics

          Reference
Prediction cancer control
   cancer      11       9
   control      7       9

               Accuracy : 0.5556
                 95% CI : (0.381, 0.7206)
    No Information Rate : 0.5
    P-Value [Acc > NIR] : 0.3089

                  Kappa : 0.1111
 Mcnemar's Test P-Value : 0.8026
            Sensitivity : 0.6111
            Specificity : 0.5000
         Pos Pred Value : 0.5500
         Neg Pred Value : 0.5625
             Prevalence : 0.5000
         Detection Rate : 0.3056
   Detection Prevalence : 0.5556
      Balanced Accuracy : 0.5556

       'Positive' Class : cancer
```

### 5.5.2.2 Decision Trees

Decision trees or tree models create models that can predict the value of a target variable based on several input variables by creating a set of if-then-else rules on them. This method has the advantage that it produces models that can be interpreted easily because these rules can be used as a visual tool for exploring the data, allowing which variables are important in the classification process to be discovered.

In R, the rpart package allows one to fit decision trees. The caret package allows one to fit decision trees using the rpart package *via* several training methods: `rpart`, `rpart1SE`, `rpart2`, and `rpartScore`. For instance:

```
tree <- train(condition ~ ., data = train.fixedColnames, method = "rpart1SE",
trControl = trainControl(method = "none"))
```

By simply typing the name of the variable (`tree`), basic information about the model is displayed:

```
> tree
CART


 84 samples
143 predictors
  2 classes: 'cancer', 'control'


No pre-processing
Resampling: None
```

Also, information about the underlying `rpart` model created, can be displayed by typing `tree$finalModel`:

```
> tree$finalModel
n= 84


node), split, n, loss, yval, (yprob)
      * denotes terminal node


1) root 84 42 cancer (0.50000000 0.50000000)
  2) V47>=8.75882e-05 48 12 cancer (0.75000000 0.25000000)
    4) V20< 0.0001381916 35  4 cancer (0.88571429 0.11428571)
      8) V93< 0.0005894743 28  0 cancer (1.00000000 0.00000000) *
      9) V93>=0.0005894743 7  3 control (0.42857143 0.57142857) *
    5) V20>=0.0001381916 13  5 control (0.38461538 0.61538462) *
  3) V47< 8.75882e-05 36  6 control (0.16666667 0.83333333)
    6) V49>=9.335664e-05 7  2 cancer (0.71428571 0.28571429) *
    7) V49< 9.335664e-05 29  1 control (0.03448276 0.96551724) *
```

As in the logistic regression example, this model can be used to predict the classes of both training and test data using the `predict` function and then use the `confusionMatrix` to compare the predictions with the actual classifications (this code is omitted to avoid redundancy, but it can be found in the `machine-learning-models.R` source file).

### 5.5.2.3 Random Forest

When it comes to prediction, it is known that aggregating the results from multiple decision trees is usually more powerful than just using a single decision tree. Random forest models were created with this idea in mind and it has been demonstrated that they always provide superior performance than single decision trees.[20] A random forest model is just a collection of decision trees where each decision tree is trained using a random subset of the training data and a random subset of the available features. This way, each single decision tree provides a partial vision of the available data and the overall model result is a combination of all these partial visions.

In R, the randomForest package allows to train random forest models. When using the caret package, this model is fitted using the train function with `method = "rf"`:

```
randomforest <- train(condition ~ ., data = train.fixedColnames, method = "rf",
trControl = trainControl(method = "none"))
```

Caret allows fine tuning of the `mtry` parameter, which specifies the number of variables tried at each split. For instance, to set this parameter to 15, the following parameter should be passed to the `train` function: `tuneGrid = expand.grid(mtry = 15)`.

As in the logistic regression example, this model can be used to predict the classes of both training and test data using the `predict` function and then using the `confusionMatrix` to compare the predictions with the actual classifications (this code is omitted to avoid redundancy, but it can be found in the `machine-learning-models.R` source file).

The interpretability that decision trees give is lost in random forest models. Nevertheless, random forest by default computes and gives the mean decrease in the Gini impurity score for all the nodes that were split on a variable. This is the score that the algorithm uses to measure how much improvement to the purity of the nodes that variable contributes. Thus, this information may be useful to understand what the more important variables (*i.e.* peaks) used in the classification are. These values are available at the `randomforest$finalModel$importance` variable (note that variables were renamed so that the model could be fit, using the following instruction to put back the original names: `rownames(randomforest$finalModel$importance) <- colnames(train)[-ncol(train)]`). Also, this information can be plotted easily using the `varImpPlot` function (Figure 5.22). By default, this function plots the top 30 variables sorted in descending order.

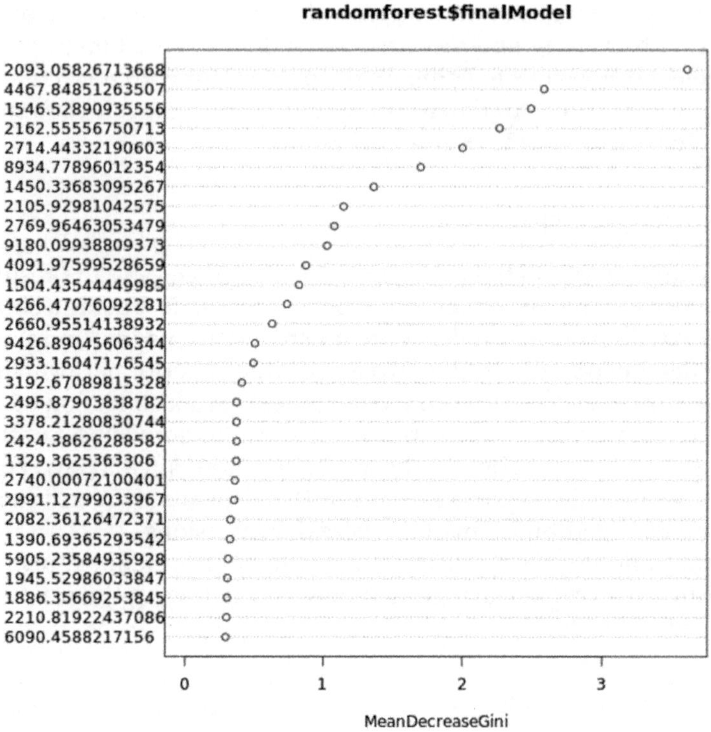


**Figure 5.22** Plot showing the importance of the variables in a Random Forest classifier generated using the `varImpPlot` function. By default, this function plots the top 30 variables sorted in descending order.

```
varImpPlot(randomforest$finalModel, type = 2)
```

### *5.5.2.4 Artificial Neural Networks*

In short, an artificial neural network (ANN) model is a network made up of individual elements called artificial neurons. Each artificial neuron receives one or more inputs and produces an output (neuron activation) computed by applying an activation function (*e.g.* Sigmoid, Rectified Linear Unit, *etc.*) to the received inputs. Each processing unit has a set of weights and a bias, used in the activation function to produce the output. Commonly, artificial neurons are distributed in layers, each one performing a different type of transformation on its inputs. The input data goes from the first layer (*i.e.* the input layer), to the last layer (*i.e.* output layer), passing through one or more hidden layers. Neurons in the output layer are responsible for making the predictions of the input data. In the training phase, the network learns from the train data by adjusting the weights to predict the correct class label of the given inputs using a backpropagation algorithm.

One of the most popular and simplest types of ANN is the multilayer perceptron (MLP). A MLP is formed by three layers of nodes: an input layer, a hidden layer and an output layer. In R, the RSNNS package allows one to fit MLP models using the `mlp` function. When using the caret package, this model is fitted using the train function with `method = "mlp"`:

```
nn <- train(condition ~ ., data = train.fixedColnames, method = "mlp", trControl =
trainControl(method = "none"), preProcess = c("center", "scale"))
```

Caret allows fine tuning of the `size` parameter, which specifies the number of hidden units (*i.e.* the number of neurons in the hidden layer). For instance, to set this parameter to 10, the following parameter should be passed to the `train` function: `tuneGrid = expand.grid(size = 10)`.

As in the logistic regression example, this model can be used to predict the classes of both train and test data using the `predict` function and then use the `confusionMatrix` to compare the predictions with the actual classifications (this code is omitted to avoid redundancy, but it can be found in the `machine-learning-models.R` source file).

### *5.5.2.5 Support Vector Machines*

In short, the objective of a Support Vector Machine (SVM) model is to find a hyperplane in an *N*-dimensional space (*N* is the number of features) that classifies the data points. Since there are many hyperplanes that might classify the data, the goal of the SVM model is to pick the best hyperplane: the one that has the largest separation (or margin) between the two classes and that is the hyperplane where the distance from it to the nearest data point on each side is maximized. Some advantages of SVMs are: (i) high dimensionality: SVM models are suitable for fitting models on high dimensional datasets, (ii) efficiency: SVM models use only a subset of the training points (the support vectors) when making predictions on new data, and (iii) versatility: apart from linear classification, SVMs can efficiently do on-linear classification using the so called kernel trick, by implicitly mapping the inputs into high-dimensional feature spaces. The main disadvantage is that the performance of SVM models require an appropriate kernel choice and fine tuning of the model hyperparameters (*e.g.* cost, epsilon, gamma, *etc.*).

In R, the e1071 package allows to fit SVM models by using the `svm` function. When using the caret package, a SVM with linear kernel is fitted using the `train` function with `method = "svmLinear"`:

```
svm.model <- train(condition ~ ., data = train, method = "svmLinear2", trControl =
trainControl(method = "none"))
```

Caret allows fine tuning of the cost parameter. Also, caret has other methods to fit other types of SVM, for instance: `method = "svmLinear"` fits a SVM with a linear kernel using the kernlab package; `method = "svmPoly"` fits a SVM with a polynomial kernel using the kernlab package; and `method = "svmRadial"` fits a SVM with radial basis function kernel using the kernlab package; among others.

As in the logistic regression example, this model can be used to predict the classes of both training and test data using the `predict` function and then use the `confusionMatrix` to compare the predictions with the actual classifications (this code is omitted to avoid redundancy, but it can be found in the `machine-learning-models.R` source file).

### 5.5.3 Dataset Partitioning Methods

When a model is evaluated in the model selection stage it is necessary to train a classifier with a training set and then exercise it by predicting the class of a new set of samples with a known class, so that the prediction can be compared to the expected output and performance can be measured. This is the basis of the model evaluation process, where several models are systematically evaluated on the training set in a way that their performance measurement is comparable. In order to be effective, it is important that the performance measurement obtained for each model is similar to the real performance of that model. A key point to achieve this is the method employed to split the training set into the training and validation subsets.

In this subsection the main partition methods used in classification problems are explained. Although each method has its own peculiarities, it is, in general, recommended to keep the proportion of samples of each class in the training and the validation subsets equal to the proportion in the training set, which is known as stratification. In addition, there should be no overlapping between samples in both subsets to avoid a bias in the performance measurements.

Although each method presented here has its advantages and disadvantages, they are not covered in great detail in this section as they have been thoroughly discussed in the literature.[21–24]

#### *5.5.3.1 Holdout*

Holdout is the most basic partition method, where a percentage of the samples in the training set are used as the training subset and the rest are used as the validation subset. Therefore, holdout guarantees that there is no overlapping between training and validation subsets.

The holdout method is a pessimistic estimator, as only a proportion of the training set is used for training. In addition, the more samples are left for the validation subset, the higher the bias will be. However, reducing the number of samples in the validation subset will increase the confidence interval of the estimation.[21]

#### 5.5.3.2 *Cross-validation*

Cross-validation or $k$-fold cross-validation is a method where the training dataset is randomly split into $k$ training and validation subsets (the folds), with the restriction that the validation subsets must be mutually exclusive and their union must include all the samples in the set. On each fold, the samples not included in the validation subset are used as the training subset. Interestingly, after the cross-validation procedure the model under evaluation will have predicted all the samples in the training set.

A special case of cross-validation is the leave-one-out cross-validation (LOOCV), where the validation subsets contain only one single sample. Therefore, as the training subset includes all the samples of the training set but one, the classifiers evaluated on each fold will be very similar and very close to a classifier trained with the whole training set.

#### 5.5.3.3 *Bootstrap*

In bootstrap, the training subset is created by sampling uniformly and with replacement $n$ instances from the training set, where $n$ is the size of the original training set. The trained classifier is evaluate against the training set or, alternatively, a validation subset may be created with the samples that were not included in the training subset. Based on probabilities, the number of different samples in the training subset should be, approximately, $0{,}632 \times n$, that is, roughly two thirds of the training set. Bootstrap is usually repeated several times, averaging the performance results.

### 5.5.4 Performance Measures

When evaluating the performance of different classification models, there are several measures that could be taken into account. Normally, the starting point is confusion matrix. For two conditions, a $2 \times 2$ confusion matrix (Table 5.2) where one condition is usually called the "positive" case (*e.g.*: "cancer") and the other the "negative" case (*e.g.*: "control").

Each cell contains how many samples having the condition of the corresponding column were predicted as having the condition in the corresponding row. In this sense, diagonal cells are correct predictions, whereas off-diagonal cells are mis-predictions.

**Table 5.2** Confusion matrix for two conditions.

| | | Real | |
|---|---|---|---|
| | | Positive | Negative |
| Predicted | Positive | True positive (TP) | False positive (FP) |
| | Negative | False negative (FN) | True negative (TN) |

With this matrix, the following measures can be computed:

- Accuracy: the proportion of correct predictions. The accuracy is given by:

$$\text{Accuracy} = \frac{\text{TP}+\text{TN}}{\text{TP}+\text{TN}+\text{FP}+\text{FN}} \tag{5.16}$$

- Sensitivity (or recall): the proportion of true positives that were predicted as such. For example: "What percentage of cancer cases are detected?". The sensitivity is given by:

$$\text{Sensitivity} = \text{recall} = \text{TP}/(\text{TP} + \text{FN}) \tag{5.17}$$

- Specificity: the proportion of true negatives that were predicted as such. For example: "What percentage of control cases are reported as negatives?". The specificity is given by:

$$\text{Specificity} = \text{Tn}/(\text{TN} + \text{FP}) \tag{5.18}$$

- Positive predictive value (or precision): the proportion of predicted positives that are real positives. For example: "What percentage of detected cancer cases are real cancer cases?". The positive predictive value is given by:

$$\text{ppv} = \text{precision} = \text{TP}/(\text{TP} + \text{FP}) \tag{5.19}$$

- Negative predictive value: the proportion of predicted negatives that are real negatives. For example: "What percentage of predicted healthy are in fact non-cancer cases?". The negative predictive value is given by:

$$\text{npv} = \text{TN}/(\text{TN} + \text{FN}) \tag{5.20}$$

- $F_1$ score: a measure combining recall (or sensitivity) and precision (or positive predictive value). The $F_1$ score is given by:

$$F_1 = 2\frac{\text{recall}\times\text{precision}}{\text{recall}+\text{precision}} \tag{5.21}$$

On the other hand, for multiple conditions the starting point is an $M \times M$ matrix as the one shown in Table 5.3, where each cell has the same meaning as in the $2 \times 2$ confusion matrix (Table 5.2).

The measures here are computed slightly differently. While the accuracy is also given by the proportion of diagonal cells, the other measures (sensitivity/recall, specificity, ppv/precision, npv and $F_1$) are condition-dependent, because they need to be defined in terms of "positives" and "negatives" in the following form. A positive case for condition $c$ is any case having condition $c$, whereas a negative case of condition $c$ is any case having any other condition. For example, to know the sensitivity for a condition $c$ (sensitivity$_c$), the

**Table 5.3** Confusion matrix for *M* conditions.[a]

| | | Real | | |
|---|---|---|---|---|
| | | Condition 1 | Condition *i* | Condition *M* |
| Predicted | Condition 1 | $\text{Count}_{11}$ | $\text{Count}_{1i}$ | $\text{Count}_{1M}$ |
| | Condition *i* | $\text{Count}_{i1}$ | $\text{Count}_{ii}$ | $\text{Count}_{iM}$ |
| | Condition *M* | $\text{Count}_{M1}$ | $\text{Count}_{Mi}$ | $\text{Count}_{MM}$ |

[a] *N*: Total count in the matrix

positives are samples of the condition *c*, and negatives are samples of any other condition. Here are the formulas for the same measures using an $M \times M$ confusion matrix:

$$\text{Accuracy} = \frac{\sum_{c}^{M} \text{Count}_{cc}}{N} \tag{5.22}$$

$$\text{Sensitivity}_c = \text{recall}_c = \frac{\text{Count}_{cc}}{\sum_{i}^{M} \text{Count}_{ic}} \tag{5.23}$$

$$\text{Specificity}_c = \frac{\sum_{i \neq c\, j \neq c}^{M} \text{Count}_{ij}}{\sum_{i\, j \neq c}^{M} \text{Count}_{ij}} \tag{5.24}$$

$$\text{ppv}_c = \text{precision}_c = \frac{\text{Count}_{cc}}{\sum_{j}^{M} \text{Count}_{cj}} \tag{5.25}$$

$$\text{npv}_c = \frac{\sum_{i \neq c\, j \neq c}^{M} \text{Count}_{ij}}{\sum_{i \neq c\, j}^{M} \text{Count}_{ij}} \tag{5.26}$$

$$F_{1c} = 2\frac{\text{recall}_c \times \text{precision}_c}{\text{recall}_c + \text{precision}_c} \tag{5.27}$$

Finally, another measure is the Cohen's kappa coefficient, which is a global statistic for evaluating the agreement between two classifications (in this case: predicted and real condition). The advantage of kappa is that it is more robust than the accuracy in cases of unbalanced datasets, where samples are not uniformly distributed across conditions. The kappa coefficient computes an expected agreement ($p_e$), or expected accuracy. Kappa ranges from 0 to 1: if the observed accuracy is not greater than the expected $p_e$, kappa is 0, whereas if the accuracy is 1, kappa is also 1. The kappa coefficient is given by:

$$\text{kappa} = \frac{\text{accuracy} - p_e}{1 - p_e} \quad (5.28)$$

Where $p_e$ is:

$$\frac{1}{N^2}\sum_{k}^{M}\left(\left(\sum_{i}^{M}\text{Count}_{ki}\right)\left(\sum_{j}^{M}\text{Count}_{jk}\right)\right) \quad (5.29)$$

### 5.5.4.1 ROC Analysis

ROC (receiver operating characteristics) is an intuitive analysis of the discriminative power of a candidate predictor measure (*e.g.* a biomarker or a trained model). Given a set of samples belonging to two different conditions and their value for the candidate predictor measure, this analysis computes the sensitivity and specificity for each possible threshold on the predictor measure.

For example, Table 5.4 contains a dataset with patients with recorded LDL cholesterol and whether they had a cardiovascular event within five years.

The potential of the LDL cholesterol as a CV risk predictor can be studied with ROC analysis by studying different thresholds. If a threshold of 120 is set, that is, values above or equal to 120 would be predicted as 'Yes', which sensitivity and specificity would this predictor get? All cases with a CV event within 5 years are above or equal to 120, so the sensitivity is 100%. And two of three cases without a CV event within 5 years are below 120, so the specificity is about 66.6%. The ROC analysis studies this result for all possible thresholds creating a ROC curve that will be shown in the following example. In R, the pROC package provides all tools that are needed to perform this kind of analysis.

Coming back to MS analysis, it can be used, for example, to study how biomarkers found previously in the Cancer Fiedler *et al.* 2009 dataset are able to discriminate between "cancer" and "control". Several potential biomarkers have been identified by using statistical tests (adjusted $p$-value $< 0.01$), for example, peak 1450.336 830 952 67.

**Table 5.4** Dataset with patients with recorded LDL cholesterol and whether they had a cardiovascular event within five years.

| LDL cholesterol | CV event within 5 years |
|---|---|
| 137 | Yes |
| 110 | No |
| 115 | No |
| 120 | Yes |
| 125 | No |

The `plot.roc` function does the analysis and shows the ROC curve. This function takes two arguments: the condition of the samples and the values of the predictor for those samples. The following code does this analysis for the peak 1450.336 830 952 67.

```
library("pROC")
source("load-maldiquant-cancer-fiedler.R")
binnedPeaksMatrix[is.na(binnedPeaksMatrix)] <- 0
rocobj <- plot.roc(binnedPeaksMatrix.conditions, binnedPeaksMatrix[,
"1450.33683095267"])
```

As can be seen from Figure 5.23, a ROC curve is a line graph with the $x$-axis for specificity (in descending order) and the $y$-axis for sensitivity. Each point corresponds to a possible threshold for the predictor value, in this example, a peak intensity to discriminate "cancer" from "control" cases.

Good predictors generate curves away from the diagonal, while bad predictors are near the diagonal. In this sense, one measure of the ROC analysis

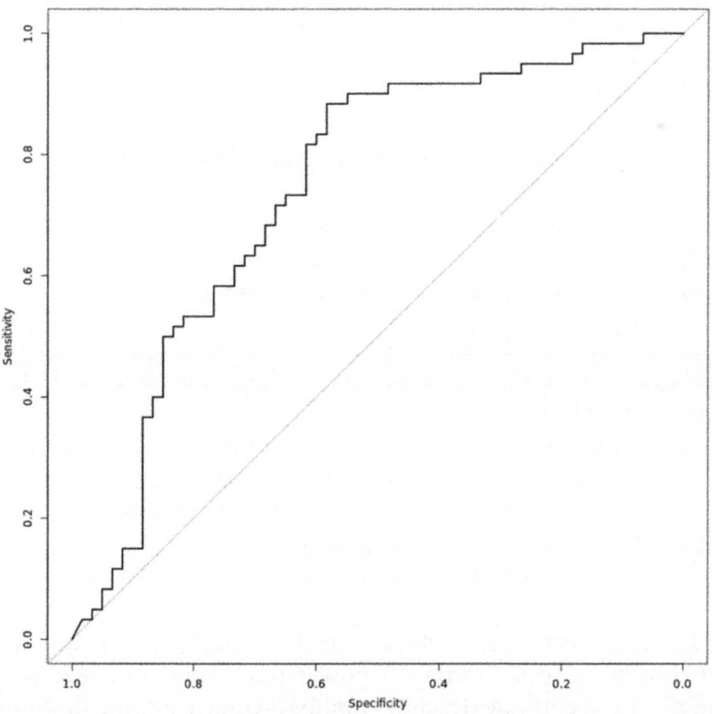


**Figure 5.23** ROC curve showing the discriminatory power for the peak 1450.336 830 952 67 in the Cancer Fiedler *et al.* 2009 dataset.

is AUC (area under the curve). This value is 0.5 for totally uncorrelated predictors and 1 for perfect predictors. AUC can be obtained by simply running:

```
rocobj$auc
Area under the curve: 0.7422
```

Another useful feature of ROC analysis is to obtain a good threshold in terms of sensitivity and specificity. To do so, the following code shows the best threshold.

```
coords(rocobj, "best", ret = c("threshold", "sensitivity", "specificity"))
   threshold  sensitivity  specificity
0.0005961044 0.8833333333 0.5833333333
```

The intensity of 0.000 596 104 4 is the best threshold. If "cancer" is predicted for samples having a peak intensity above this value, the sensitivity is 88.3% while specificity is 58.3%. The best threshold is a tradeoff of sensitivity and specificity. Concretely, pROC uses the Youden's index:

$$\text{Youden} = \text{sensitivity} + \text{specificity} - 1 \tag{5.30}$$

The ROC curve is also useful to compare several predictors in the same plot. The following code plots two candidate biomarkers (peak 1546.528 909 355 56 and peak 1546.528 909 355 56) on the same chart.

```
rocobj <- plot.roc(binnedPeaksMatrix.conditions, binnedPeaksMatrix[,
"1450.33683095267"], col = "red")
legend("bottomright", col = c("#ff0000", "#0000ff"), legend = c("peak
1450.33683095267", "peak 1546.52890935556"), lwd = 2)
rocobj2 <- plot.roc(binnedPeaksMatrix.conditions, binnedPeaksMatrix[,
"1546.52890935556"], col = "blue", add = TRUE)
```

As can be seen from Figure 5.24, the peak 1546.528 909 355 56 has a larger AUC and thus more discriminative power than peak 1450.336 830 952 67. Interestingly, the second peak has a smaller $p$-value than the first one.

ROC analysis can also be done over ML models. For example, the curve and the AUC for an ANN model can be computed. First, the predictions over a test dataset in the form of class probabilities are required. The "cancer" class probability produced by the neural network is considered the predictor where ROC analysis is performed.

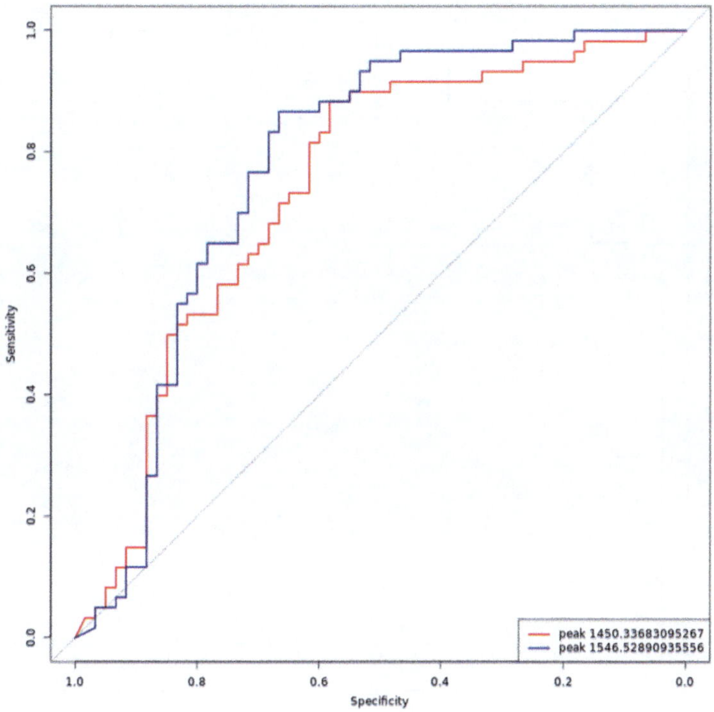


**Figure 5.24** ROC curve comparing two candidate biomarkers (peak 1546.528 909 355 56 and peak 1546.528 909 355 56) in the Cancer Fiedler *et al.* 2009 dataset. Peak 1546.528 909 355 56 has a larger AUC and thus more discriminative power than peak 1450.336 830 952 67, although this second peak has a smaller *p*-value.

```
nn <- train(condition ~ ., data = train.fixedColnames, method = "mlp", trControl =
trainControl(method = "none"), preProcess = c("center", "scale"))
nn.test.result <- predict(nn, type = "prob", newdata = test)

# probability of "cancer"
testSamples.cancerProb <- nn.test.result[, 1]

testSamples.class <- binnedPeaksMatrix.conditions[match(rownames(nn.test.result),
rownames(binnedPeaksMatrix))]
rocobj <- plot.roc(testSamples.class, testSamples.cancerProb)
coords(rocobj, "best", ret = c("threshold", "sensitivity", "specificity"))
rocobj$auc
Area under the curve: 0.929
```

As can be seen from Figure 5.25, this curve is farther from the diagonal as the previous individual biomarkers (AUC = 0.929) and thus it seems a better predictive model. However, it is more complex, since a neural network combines all input peaks in a single model.

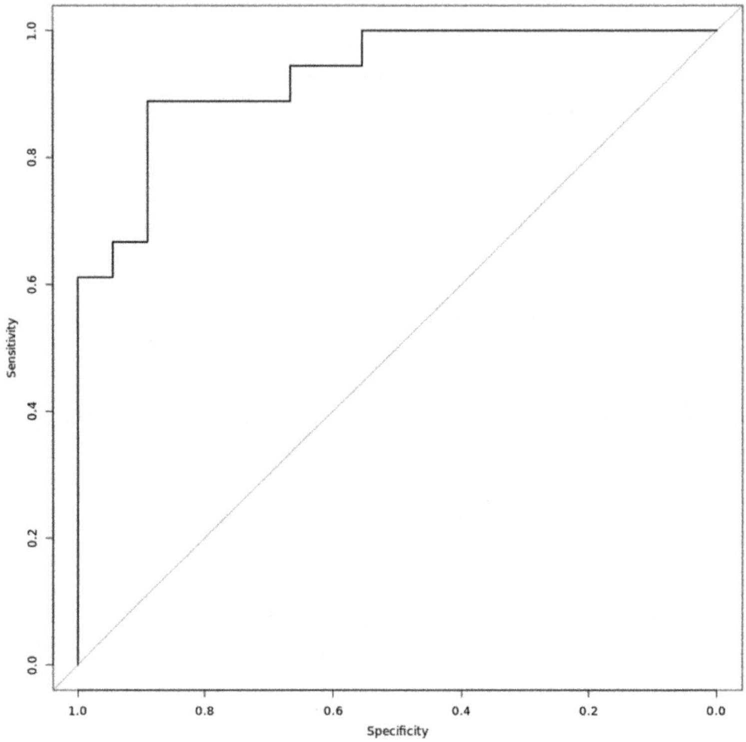


**Figure 5.25** ROC curve for an artificial neural network model fitted using the Cancer Fiedler *et al.* 2009 dataset (AUC = 0.929).

### 5.5.5 A Classification Case Study

In this subsection a classification case study is presented to show how the different concepts presented in this section are applied in a real world problem. Specifically, a classifier will be trained for the Cancer Fiedler *et al.* 2019 dataset, using only the Leipzig samples, in order to avoid an undesired batch effect. As in the examples of Section 5.5.2, the caret library is used as the base library to perform model evaluation.

Classification models, and almost any ML model, require every sample to have exactly the same features. Even when the value for a specific feature of a sample is unknown, this feature is still present in that sample with a N/A or missing value assigned. This is an important requirement that means that mass spectrometry samples, where the *m*/*z* value for the same peak in different samples may vary slightly, should be pre-processed to unify the peak *m*/*z* values, in a process known as peak binning.

When pre-processing the training and testing sets in mass spectrometry it is important to apply the steps that use all the information in the set, such as the peak alignment and peak binning steps, only to the training set. No preprocessing operation that uses information of several samples in the testing set should be applied to samples in this set, as when classifying a new single sample (*e.g.* diagnosing a patient) there will be no other samples to use. This kind of operation can be applied to testing samples, but the information of other samples should be taken from the training samples.

Implications of these rules can be seen in the `classification-case-study-load-cancer-fiedler.R` source code file, where the training and testing sets for this case study are prepared. As can be seen from this file, the original dataset is split just before the align spectra operation, as this operation uses the whole set to calculate the reference peaks used for alignment. Previous operations are applied independently to each sample, so they can be applied before splitting the original dataset. The peak binning operation also uses information from the whole dataset and, therefore, it should be applied carefully.

Pre-processing of the training set is done following the usual steps, with the only particularity that reference peaks are also calculated for further use. However, although the testing set is pre-processed with the same steps, peak alignment and peak binning are done in a different way to ensure that they are applied independently to each sample. On one hand, peak alignment is done using the reference peaks calculated for the training set, in order to avoid its calculation with the whole testing set. On the other hand, peak binning is done with a custom function that bins each peak in the testing samples to the closest peak bin of the training samples. This way, the pre-processing process for the testing set is applicable to any new sample that the final classifier should classify.

Once the training and the testing sets are prepared, the first step is to decide which partition method will be used to evaluate the models. In this case, a 10-fold cross-validation repeated five times is used as, in general, it is a good estimator method. The partition configuration in caret is configured with the `trainControl` function that generates an object that conducts the training process. This specific configuration can be created with the following code, where `repeatedcv` is the partition method to use, `number` is the number of folds and `repeats` the times that the cross-validation should be repeated.

```
ctrl <- trainControl(
  method = "repeatedcv",
  number = 10,
  repeats = 5
)
```

Some of the models used in this example, such as the MLP, need the variables to be at the same scale. Standardizing the variables is a way to achieve this requirement. Therefore, center and scale operations will be using to pre-process training subsets. These operations are stored in a variable that will be provided to the training function.

```
pp <- c("center", "scale")
```

In order to avoid selecting an inadequate classification model, five different models will be evaluated before selecting the best classifier. Specifically, the five models presented in Section 5.5.2. These models are stored in a variable that will be iterated to evaluate each model.

```
models <- c("rf", "glm", "mlp", "rpart1SE", "svmLinear2")
```

Now that all the parameters needed for model evaluation are ready, the training process can be started. In this case, the `lapply` function is used to apply the same evaluation to each model stored in models variable. As can be seen from the next code, the random seed is set to the same value before each evaluation, so that the same partitions are used to evaluate each model. Interestingly, the `train` function using the train control previously configured will evaluate several configurations of each classifier, changing the values of the main hyperparameters of each model. The configuration that yields the best performance results will be automatically selected to configure and train the candidate classifier for each model. This can be seen by printing the result object of each model.

The `resamples` function is very useful to compare the results of each model evaluated. This function receives a list of model fits and returns an object that can be visualized by printing it with the `summary` function or by creating plots, for example, with the `bwplot` or the `dotplot` functions. The following code shows how to create this object and the information printed by the summary function. In addition, Figure 5.26A shows the `bwplot`, where a boxplot is generated for the accuracy and kappa results of each model, while Figure 5.26B shows the `dotplot`, where the average accuracy and kappa with confidence intervals of each model is plotted.

```
modelFits <- lapply(models, function(model) {
    set.seed(2019)
    train(condition ~ ., data = trainingSet, method = model, preProcess = pp,
trControl = ctrl)
})

names(modelFits) <- models

modelFits$rf
Random Forest

 60 samples
150 predictors
  2 classes: 'cancer', 'control'

Pre-processing: centered (150), scaled (150)
Resampling: Cross-Validated (10 fold, repeated 5 times)
Summary of sample sizes: 54, 54, 54, 54, 54, 54, ...
Resampling results across tuning parameters:

  mtry  Accuracy   Kappa
    2   0.8233333  0.6466667
   76   0.8233333  0.6466667
  150   0.8266667  0.6533333

Accuracy was used to select the optimal model using the largest value.
The final value used for the model was mtry = 150.
```

Based on the mean accuracy and kappa of the models, the MLP model is finally selected, although Random Forest and SVM models also showed good performance. Once the best model is selected, it should be evaluated in the testing set reserved at the start of the process to get a hint of the performance of these model in a real environment by classifying new unseen samples. This can be done with the `predict` function, to classify the testing samples, and with the `confusionMatrix` function, to calculate the performance of the classifier.

As can be seen from this output, the performance of the selected classifier is very good in the testing set, an indication that the process followed to select the best model was effective. If the model performance in the test set was low, then the whole process to select a new model could be repeated with more models and/or with more data. However, in that case, the samples used for the testing set must be completely new, in order to avoid a bias in the process. The source code to reproduce this case study can be found in the `classification-case-study.R` and `classification-case-study-load-cancer-fiedler.R` files.

```
results <- resamples(modelFits)
summary(results)

Call:
summary.resamples(object = results)

Models: rf, glm, mlp, rpart1SE, svmLinear2
Number of resamples: 50

Accuracy
                Min.   1st Qu.    Median      Mean   3rd Qu.      Max. NA's
rf         0.5000000 0.6666667 0.8333333 0.8266667 1.0000000 1.0000000    0
glm        0.3333333 0.3750000 0.5000000 0.5500000 0.6666667 0.8333333    0
mlp        0.3333333 0.8333333 0.8333333 0.8466667 1.0000000 1.0000000    0
rpart1SE   0.5000000 0.6666667 0.8333333 0.7866667 0.8333333 1.0000000    0
svmLinear2 0.3333333 0.8333333 0.8333333 0.8433333 1.0000000 1.0000000    0

Kappa
                 Min.    1st Qu.    Median      Mean   3rd Qu.      Max. NA's
rf          0.0000000  0.3333333 0.6666667 0.6533333 1.0000000 1.0000000    0
glm        -0.3333333 -0.2500000 0.0000000 0.1000000 0.3333333 0.6666667    0
mlp        -0.3333333  0.6666667 0.6666667 0.6933333 1.0000000 1.0000000    0
rpart1SE    0.0000000  0.3333333 0.6666667 0.5733333 0.6666667 1.0000000    0
svmLinear2 -0.3333333  0.6666667 0.6666667 0.6866667 1.0000000 1.0000000    0
```

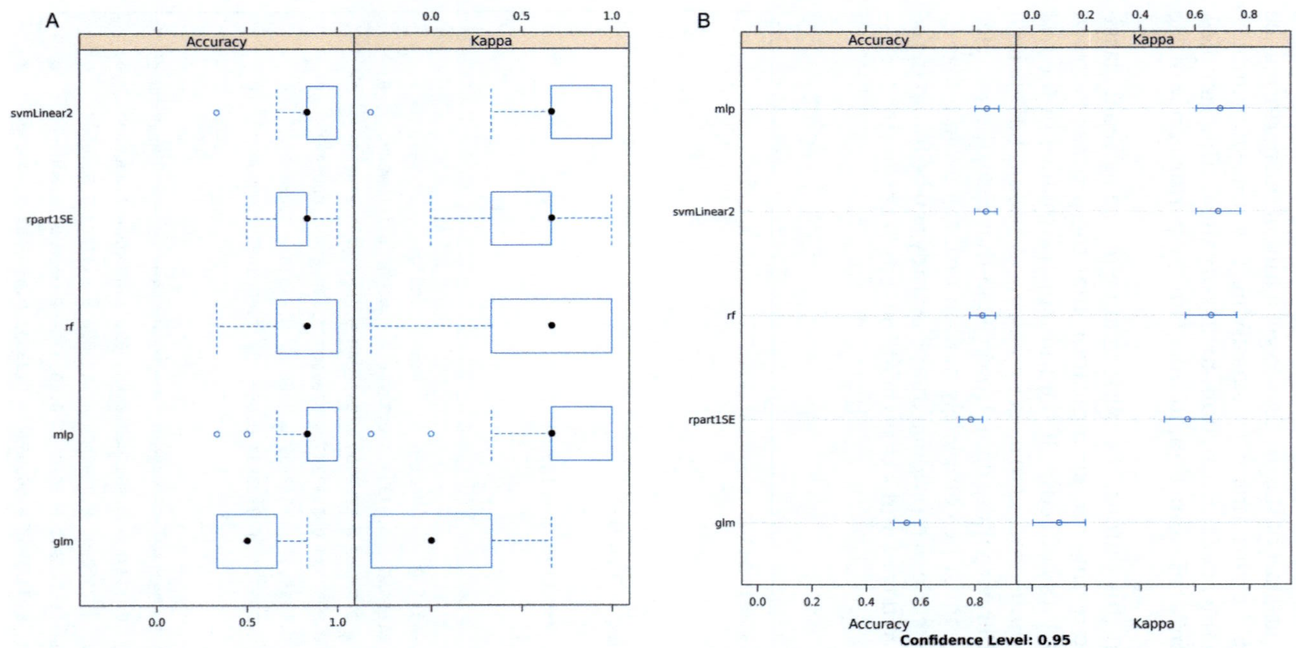


**Figure 5.26** Visualization of the model selection results: (A) plot generated with the `bwplot` function, where a boxplot is generated for the accuracy and kappa results of each model; (B) plot generated with the `dotplot` function, where the average accuracy and kappa with confidence intervals of each model is plotted.

```
cancerPrediction <- predict(modelFits$mlp, newdata=testingSet)

Multi-Layer Perceptron

 60 samples
150 predictors
  2 classes: 'cancer', 'control'

Pre-processing: centered (150), scaled (150)
Resampling: Cross-Validated (10 fold, repeated 5 times)
Summary of sample sizes: 54, 54, 54, 54, 54, 54, ...
Resampling results across tuning parameters:

  size  Accuracy   Kappa
  1     0.8466667  0.6933333
  3     0.8300000  0.6600000
  5     0.8400000  0.6800000

Accuracy was used to select the optimal model using the largest value.
The final value used for the model was size = 1.
confusionMatrix(data = cancerPrediction, testingSet$condition)

Confusion Matrix and Statistics

          Reference
Prediction cancer control
   cancer       8       0
   control      2      10

               Accuracy : 0.9
                 95% CI : (0.683, 0.9877)
    No Information Rate : 0.5
    P-Value [Acc > NIR] : 0.0002012

                  Kappa : 0.8
 Mcnemar's Test P-Value : 0.4795001

            Sensitivity : 0.8000
            Specificity : 1.0000
         Pos Pred Value : 1.0000
         Neg Pred Value : 0.8333
             Prevalence : 0.5000
         Detection Rate : 0.4000
   Detection Prevalence : 0.4000
      Balanced Accuracy : 0.9000

       'Positive' Class : cancer
```

# Acknowledgements

The SING group thanks the CITI (Centro de Investigación, Transferencia e Innovación) from the University of Vigo for hosting its IT infrastructure. This work was partially supported by the Consellería de Educación, Universidades e Formación Profesional (Xunta de Galicia) under the scope of the strategic funding ED431C2018/55-GRC Competitive Reference Group. H. López-Fernández is supported by a post-doctoral fellowship from Xunta de Galicia (ED481B 2016/068-0).

## References

1. R. López-Cortés, E. Oliveira, C. Núñez, C. Lodeiro, M. Páez de la Cadena, F. Fdez-Riverola, H. López-Fernández, M. Reboiro-Jato, D. Glez-Peña, J. Luis Capelo and H. M. Santos, *Talanta*, 2012, **100**, 239–245.
2. H. López-Fernández, H. M. Santos, J. L. Capelo, F. Fdez-Riverola, D. Glez-Peña and M. Reboiro-Jato, *BMC Bioinf.*, 2015, **16**, 318.
3. C. C. Aggarwal, A. Hinneburg and D. A. Keim, in *Database Theory — ICDT 2001*, ed. J. Van den Bussche and V. Vianu, Springer Berlin Heidelberg, 2001, pp. 420–434.
4. I. Eidhammer, K. Flikka, L. Martens and S.-O. Mikalsen, *MethodsforMassSpectrometryProteomics*,Wiley-Interscience,1stedn,2008.
5. F. E. Grubbs, *Technometrics*, 1969, **11**, 1–21.
6. H. Beyer, *Biom. J.*, 1981, **23**, 413–414.
7. T. Kohonen and T. Honkela, *Scholarpedia*, 2007, **2**, 1568.
8. R. Wehrens and J. Kruisselbrink, *J. Stat. Softw.*, 2018, **87**, 1–18.
9. J. Vesanto and E. Alhoniemi, *IEEE Trans. Neural Network.*, 2000, **11**, 586–600.
10. J. Tian, M. H. Azarian and M. Pecht, *Anomaly Detection Using Self-Organizing Maps-Based K-Nearest Neighbor Algorithm,* Nantes, France, 2014, available at https://www.phmsociety.org/node/1300.
11. M. Key, *BMC Bioinf.*, 2012, **13**, S10.
12. R. Suzuki and H. Shimodaira, *Bioinformatics*, 2006, **22**, 1540–1542.
13. S. Gibb and K. Strimmer, in *Statistical Analysis of Proteomics, Metabolomics, and Lipidomics Data Using Mass Spectrometry*, ed. S. Datta and B. J. A. Mertens, Springer International Publishing, Cham, 2017, pp. 101–124.
14. M. Ahdesmäki and K. Strimmer, *Ann. Appl. Stat.*, 2010, **4**, 503–519.
15. G. M. Fiedler, A. B. Leichtle, J. Kase, S. Baumann, U. Ceglarek, K. Felix, T. Conrad,H.Witzigmann,A.Weimann,C.Schutte,J.Hauss,M.BuchlerandJ. Thiery, *Clin. Cancer Res.*, 2009, **15**, 3812–3819.
16. W. Li, J. Freudenberg, Y. J. Suh and Y. Yang, *Comput. Biol. Chem.*, 2014, **48**, 77–83.
17. W. Li, *J. Bioinform. Comput. Biol.*, 2012, **10**, 1231003.
18. The MIT Press, Reinforcement Learning, https://mitpress.mit.edu/books/reinforcement-learning, accessed 23 April 2019.
19. M. Kuhn, *J. Stat. Softw.*, 2008, **28**, 1–26.
20. P. Bruce and A. Bruce, *Practical Statistics for Data Scientists*, 2017.
21. R. Kohavi, in *Proceedings of the 14th International Joint Conference on Artificial Intelligence*, Morgan Kaufmann Publishers Inc., San Francisco, CA, USA, 1995, vol. 2, pp. 1137–1143.
22. J.-H. Kim, *Comput. Stat. Data Anal.*, 2009, **53**, 3735–3745.
23. C. Beleites, R. Baumgartner, C. Bowman, R. Somorjai, G. Steiner, R. Salzer and M. G. Sowa, *Chemom. Intell. Lab. Syst.*, 2005, **79**, 91–100.
24. E. Dougherty, C. Sima, J. Hua, B. Hanczar and U. Braga-Neto, *Curr. Bioinf.*, 2010, **5**, 53–67.

# Part B

# Open MS Programs, Toolkits and Workflow Platforms

CHAPTER 6

# *OpenMS and KNIME for Mass Spectrometry Data Processing*

OLIVER ALKA[d], TIMO SACHSENBERG[d], LEON BICHMANN[d], JULIANUS PFEUFFER[d,j], HENDRIK WEISSER[k], SAMUEL WEIN[i], EUGEN NETZ[e], MARC RURIK[d], OLIVER KOHLBACHER[d,e,f,g,h] AND HANNES RÖST*[a,b,c]

[a]Donnelly Centre, University of Toronto, Toronto, Canada; [b]Department of Molecular Genetics, University of Toronto, Toronto, Canada; [c]Department of Computer Science, University of Toronto, Toronto, Canada; [d]Department for Computer Science, Applied Bioinformatics, University of Tübingen, Sand 14, 72076 Tübingen, Germany; [e]Biomolecular Interactions, Max Planck Institute for Developmental Biology, Max-Planck-Ring 5, 72076 Tübingen, Germany; [f]Institute for Translational Bioinformatics, University Hospital Tübingen, Hoppe-Seyler-Str. 9, 72076 Tübingen, Germany; [g]Institute for Biomedical Informatics, University of Tübingen, Sand 14, 72076 Tübingen, Germany; [h]Quantitative Biology Center, University of Tübingen, Auf der Morgenstelle 10, 72076 Tübingen, Germany; [i]Epigenetics Institute, Department of Cell and Developmental Biology, University of Pennsylvania, 9th Floor, Smilow Center for Translational Research 3400 Civic Center Blvd, Philadelphia, PA 19104, USA; [j]Department for Computer Science, Algorithmic Bioinformatics, Freie Universität Berlin, Takustr. 9, 14195 Berlin, Germany; [k]STORM Therapeutics Limited, Moneta Building, Babraham Research Campus, Cambridge CB22 3AT, UK
*E-mail: hannes.rost@utoronto.ca

New Developments in Mass Spectrometry No. 8
Processing Metabolomics and Proteomics Data with Open Software: A Practical Guide
Edited by Robert Winkler

Published by the Royal Society of Chemistry, www.rsc.org

## 6.1 Introduction

Computational mass spectrometry has seen exponential growth in recent years in data size and complexity, straining the existing infrastructure of many labs as they moved towards high-performance computing (HPC) and embraced big data paradigms. Transparent and reproducible data analysis has traditionally been challenging in the field due to a highly heterogeneous software environment, while algorithms and analysis workflows have grown increasingly complex. A multitude of competing and often incompatible file formats prevented objective algorithmic comparisons and, in some cases, access to specific software or file formats relied on a vendor license. Due to the fast technology progress in the field, many novel algorithms are proposed in the literature every year, but few are implemented with reusability, robustness, cross-platform compatibility and user-friendliness in mind, creating a highly challenging software and data storage environment that in some aspects is even opaque to experts.

The OpenMS software framework addresses these issues through a set of around 175 highly robust and transparent cross-platform tools with a focus on maximal flexibility.[1–3] Modern software engineering techniques ensure reproducibility between versions and minimize code duplication and putative errors in software. OpenMS is completely open-source, uses standardized data formats extensively and is available on all three major computing platforms (macOS, Windows, Linux). Multiple layers of access to the OpenMS algorithms exist for specialist, intermediate and novice users, providing low entrance barriers through sophisticated data visualization and graphical workflow managers.

The flexibility of OpenMS allows it to support a multitude of customizable and easily transmissible workflows in multi-omics data analysis, including metabolomics, lipidomics, and proteomics setups, supporting different quantitative approaches spanning label-free, isotopic, isobaric labeling techniques, as well as targeted proteomics. Its highly flexible structure and layered design allow different scientific groups to take full advantage of the software. Developers can fully exploit the sophisticated C++ library for tool and data structure development, while advanced Python bindings (pyOpenMS, see Chapter 16) wrap most of the classes,[4] providing an excellent solution for fast scripting and prototyping. Users, can either work on command line tool level or take advantage of industry-grade workflows systems, such as the KoNstanz Information MinEr (KNIME),[5,6] Galaxy,[7] Nextflow,[8] or Snakemake.[9] The framework is highly adaptable, allowing even novice users to generate complex workflows using the easy-to-learn graphical user interfaces of KNIME. Built-in support for most common workflow steps (such as popular proteomics search engines) ensures low entrance barriers while advanced users have high flexibility within the same framework. A modular and comprehensive codebase allows rapid development of novel methods as exemplified by the recent additions for metabolomics, SWATH-MS and cross-linking workflows.

In addition, a versatile visualization software (TOPPView) allows exploration of raw data as well as identification and quantification results.[10] The permissive BSD license encourages usage in commercial and academic projects, making the project especially suited for reference implementations of file formats and algorithms.

## 6.2 OpenMS for Developers

The OpenMS framework consists of different abstraction layers. The first layer consists of external libraries (Qt, Boost, Xerces, Seqan, Eigen, Wildmagic, Coin-Or, libSVM), which add additional data structures and functionality, simplifying complex tasks such as GUI-programming or XML parsing. The next layer consists of the OpenMS core library containing algorithms, data structures and input/output processing. The third layer encloses TOPP tools and utilities (~175), which allow various analysis tasks, such as signal processing, filtering, identification, quantification and visualization. The core library and the TOPP tools have Python bindings, which can be used for fast scripting and prototyping (pyOpenMS).[4] In the top layer, the tools are accessible from different workflow systems, for the construction of flexible, tool based workflows.

### 6.2.1 C++ Library

OpenMS has a multi-level architecture with an open source C++ library at its core which implements all data structures and algorithms required for MS data analyses. Its permissive license (3-clause BSD) allows using OpenMS both in academic as well as commercial projects without any licensing fees. Since its beginning, its aim has been to provide efficient data structures and algorithms for common MS data processing tasks. As such, the library targets computational biologists and algorithm developers with sound programming skills. Using the library, developers have direct access to the richest set of functionality and generally profit from highly-optimized implementations of core data structures and algorithms when developing novel methods. More than 1300 classes cover a broad range of functions in computational mass spectrometry, proteomics and metabolomics. These functionalities include

- file handling (mzXML, mzML, TraML, mzTab, fasta, pepxml, protxml, mzIdentML among others)
- chemistry (mass calculation, peptide fragmentation, isotopic abundances)
- signal processing and filtering (smoothing, filtering, de-isotoping, mass correction, retention time correction and peak picking)
- identification analysis (including peptide search, PTM analysis, cross-linked analytes, FDR control, RNA oligonucleotide search and small molecule search tools)
- quantitative analysis (including label-free, SILAC, iTRAQ, TMT, SWATH/DIA, and metabolomics analysis tools)

- chromatogram analysis (chromatographic peak picking, smoothing, elution profiles and peak scoring for targeted (SRM, MRM, PRM, SWATH, DIA) data)
- interaction with common tools in proteomics and metabolomics
  - search engines such as Comet, Crux, Mascot, MS-GF+, InsPecT, PepNovo, MSFragger, Myrimatch, OMSSA, Sequest, SpectraST, XTandem
  - post-processing tools such as Percolator, MSstats, Fido, EPIFANY
  - metabolomics tools such as SIRIUS

Ideally, newly developed methods, algorithms, or data structures of general interest are contributed by the community, and find their way back to the library to be used by other OpenMS developers. OpenMS itself builds on other open-source projects like Qt, Xerces, COIN-OR, libSVM, Eigen, WildMagic, or boost for tasks like GUI programming, XML parsing, non-linear optimizations, machine learning or fast linear algebra. To leverage the computing power of modern processors, the OpenMP library is used to parallelize many algorithms in OpenMS.

For a short example how a multi-threaded spectrum processing algorithm can be realized using the OpenMS library see Box 6.1.

### 6.2.2 Data Formats and Raw Data API

Using open data formats for storing, exchanging, and deposition of primary raw data and final analysis results in public data repository are prerequisites for reproducible science. OpenMS builds on open standards for reading and writing MS data (*e.g.*, mzML and mzXML format), transitions

**Box 6.1** Code example: Multi-threaded spectrum processing algorithm using the OpenMS library

```
// C++ example (excerpt):
// Load data, retain the 400 most intense peaks in a spectrum

MSExperiment exp;
MzMLFile().load("file.mzML", exp);
auto spectra = exp.getSpectra();

// construct a spectrum filter
NLargest nlargest_filter = NLargest (400);

// parallelize loop for concurrent execution using OpenMP


# pragma omp parallel for
for (auto it = spectra.begin(); it < spectra.end(); it++)
{
  // sort peaks by mass-to-charge position
  it->sortByPosition();

  // apply filter and keep only the 400 highest intensity peaks
  nlargest_filter.filterPeakSpectrum(*it);
}
```

(TraML), identifications (mzIdentML), or exporting final results (mzTab). In fact, members of the OpenMS community have been actively involved in the development of several HUPO-PSI standard formats in the past. For XML-based formats, OpenMS uses the powerful Xerces software library from the Apache project which implements both DOM and SAX parsing. For all data, in-memory data structures exist where data gets parsed from disk into these data structures, manipulated in memory and then written to disk again after transformation/manipulation. However, for raw MS data access in mzML files, OpenMS provides an advanced API that can handle access to spectra and chromatograms in different ways, thus optimizing the tradeoff between memory and CPU requirements of the algorithm:[11] (1) random access in memory, (2) random access on disk using indexedmzML, (3) random access on disk using a cached file format and (4) event-driven processing. While programmatically, it is easiest to program against interface (1) since all data is in memory and can be accessed very fast, this is not always possible for very large files that do not fit into the current computer's memory. For these cases, it is possible to access spectra from disk using either mechanism (2) or (3) which allows accessing raw data that is not held in memory. Finally, we have implemented an interface that uses a call-back mechanism similar to SAX parsing in XML parsing, which works on a per-spectra basis: the call-back function gets called as soon as a new spectrum is available from the reader, allowing a consumer to process spectra as they get read from disk. The processing function can either keep all results in memory or write them back to disk such that in an ideal case only a single spectrum is in memory at any given time.

### 6.2.3 Algorithms

The OpenMS library provides a multitude of algorithms ranging from low-level algorithms, like the generation of isotope patterns, processing and filtering of raw signals, to more complex algorithms like peptide database search. In the following we picked two examples to demonstrate how these algorithms are configured and executed on data. Other algorithms provide similar interfaces.

Some algorithms with few parameters provide a simple interface and can be directly called. For isotope pattern calculation see Box 6.2.

Note that multiple algorithms may be available through the same interface, here an isotopic distribution can be calculated using coarse (unit mass) resolution or using fine (hyperfine isotopic) resolution. A single line of code will switch between the two algorithms, but the rest of the code will work without change (note that the two algorithms take different parameter sets, either number of peaks or total isotopic probability covered). This allows OpenMS to implement new algorithms that improve performance or accuracy "under the hood" while the interface stays the same for the user and algorithms can be switched to the new interface with ease. In the above example, this may become important with high resolution instrumentation that can differentiate between the hyperfine isotopic peaks in an isotopic envelope.

**Box 6.2** Code example: Isotope pattern calculation

```
// excerpt: generate isotope pattern (max. 10 peaks or
// 99.9% of the isotopic probability) of Glucose

EmpiricalFormula molecule("C6H12O6");
IsotopeDistribution iso;
iso = molecule.getIsotopeDistribution(CoarseIsotopePatternGenerator(10));
iso = molecule.getIsotopeDistribution(FineIsotopePatternGenerator(1e-3));
```

**Box 6.3** Code example: Centroiding of profile spectra using the PeakPichkerHiRes class

```
// excerpt: centroiding profile spectra

// load profile spectra
MSExperiment profile_data, centroided_data;
MzMLFile().load("myData.mzML"), profile_data);

// get default parameters, change as desired and store changes
PeakPickerHiRes peak_picker;
Param param = peak_picker.getParam();
param.setValue("signal_to_noise", 1.0);
peak_picker.setParameters(param);

// run algorithm
peak_picker.pickExperiment(profile_data, centroided_data);
```

Other algorithms allow fine-tuning of many parameters *via* a so-called Param object. The signal processing algorithm PeakPickerHiRes for centroiding of profile data is one of those (see Box 6.3). Internally PeakPickerHiRes uses a spline interpolation of the raw data to determine the *m*/*z*, FWHM, and intensity of peak centroids. Several aspects, *e.g.*, the minimum signal-to-noise ratio to call a peak, can be configured *via* the Param object.

Available parameters are listed in the developer documentation of each algorithm and each parameter comes with information on allowed parameter ranges and a human-readable description of the parameter.

### 6.2.4 TOPP Tools (Developer Perspective)

TOPP tools are command line applications developed within the OpenMS framework that utilize the OpenMS C++ library to implement a specific function. OpenMS follows the UNIX philosophy which is based on providing individual tools with defined functionality that can be chained together to powerful workflows, allowing the greatest amount of flexibility to the user. Currently there are around 175 TOPP tools available which range from implementing a spectral noise filter to adaptors of third-party tools (*e.g.* the Comet search engine) and full implementations of quantitative analysis of SILAC, label-free or SWATH data. These tools are the preferred way to expose novel functionality in OpenMS and are available to user directly through the command line but are also automatically integrated into workflow engines such as KNIME.

**Box 6.4** Developer instructions

Follow the steps to start programming on the library:

Working with your own fork:
Fork the OpenMS repository (https://github.com/OpenMS/OpenMS)
Clone the respective fork locally (git clone https://github.com/username/OpenMS.git)
Compile OpenMS (build instructions below)
Have fun working with and on the library

Working on OpenMS/OpenMS:
Clone or download the source (https://github.com/OpenMS/OpenMS.git)
Compile OpenMS (build instructions below)
Have fun working with and on the library

Build instructions for Linux:

https://openms.de/documentation/install_linux.html

Build instructions for OS X:

https://openms.de/documentation/install_mac.html

Build instructions for Windows:

https://openms.de/documentation/install_win.html

For further instructions and information about coding conventions, please check out our WIKI:

https://github.com/OpenMS/OpenMS/wiki

OpenMS Developer C++ Guide:

https://openms.de/documentation/OpenMS_tutorial.html

OpenMS provides a base class and a template for the creation and integration of new tools. The base class handles logging, exception handling and command line argument parsing. The template has well-defined slots for adding tool parameters, input and output as well as algorithmic functionality which can be added in a straightforward manner. Further, before a tool can be merged into the OpenMS repository, it is required to provide its own regression tests, which allow for the automatic validation of its functionality in future releases of the OpenMS framework. Once added, the tool will become available as a command-line executable as well as a node in the KNIME workflow engine where users can integrate it into their workflows.

For further instruction on tool development in OpenMS please visit the tool section in the Developer C++ Quickstart Guide (see Box 6.4).

## 6.2.5 Visualization

MS data exhibits a large degree of variability and complex experimental setups may result in highly heterogeneous raw data. Visual inspection is regularly done in practice to assess data quality. TOPPView is a graphical application in OpenMS that allows users to inspect spectra, chromatograms, and identification results.[10] It uses the Qt 5 framework to implement advanced visualization which includes 1D (spectra, chromatogram), 2D and 3D (peak maps) plotting. Standard and re-useable Qt classes allow graphical plotting in multiple dimensions and OpenGL (Open Graphics Library) is used for high-performance interactive 3D plotting. Developers can build their own applications using the provided SpectrumWidget and SpectrumCanvas classes, which allows advanced plotting of mass spectrometric data without having to re-invent the wheel. Furthermore, TOPPView is tightly integrated with the available TOPP tools, which can be directly executed on the currently displayed data and the output of a transformation can then be loaded back into the same graphical view, providing direct feedback for a particular choice of tool and parameters. Fine-tuning these parameters sometimes may open the chance to analyze difficult data that fail to produce acceptable results using the default settings.

### 6.2.6 Code Quality and Community

OpenMS uses modern software engineering concepts more commonly found in industry settings than in academic environments. The project places great emphasis on modularity, reusability and extensive testing (using continuous integration), resulting in high code quality. The modular architecture of OpenMS tries to build upon existing standard libraries as much as possible, relying on them for sequence analysis, XML parsing, numerical computations and statistics. Modern object-oriented C++ is used exclusively throughout the code base, encapsulating raw data structures and discouraging manual heap-based memory management (when not provably crucial for efficiency), thus providing robust and error-tolerant code. Coding conventions are enforced and extensive English documentation is available for several thousand C++ functions as part of the public API. All development is performed in the open, using the public source code repository and ticketing system GitHub. Stringent code reviews and continuous integration, running a multitude of functional, unit and black-box tests, ensure continued support, robustness and correctness of the code. Automated tests verify correctness of functionality of both individual functions and whole units on all three supported platforms while also analyzing the code quality using automated tools such as cppcheck and cpplint. Additionally, the code is reviewed by other developers using the four-eyes principles and changes can be requested depending on the performance, accuracy, style and code quality.

The OpenMS development team is integrated into an international multi-site effort supported by leading labs in experimental and computational mass spectrometry across Europe and North America. It is unique in the field by providing industrial-strength high-performance algorithmic implementations for the majority of common tasks in computational proteomics as open-source software. Frequent physical meetings and training sessions educate users and transmit knowledge of established workflows to practitioners in the field, also providing opportunities for users and developers to meet and exchange ideas. The project sees high contributor activity and several downstream tools such as MSstats, aLFQ and Skyline have started to integrate their tools with OpenMS.[12–14]

### 6.2.7 Getting Started with the OpenMS Library for Developers

For getting started to develop a new tool or use specific classes from the OpenMS library please follow the steps in Box 6.4.

## 6.3 OpenMS for Users

### 6.3.1 TOPP Tools (User Perspective)

TOPP tools are individual tools, which usually perform one specific task. For example, the FeatureFinderCentroided can be used to detect two-dimensional features in centroided LC-MS data. A multitude of different tools exists ranging from simple format conversion, data filtering, to new data analysis, and data reduction algorithms. Additionally, wrappers exist and can be added upon request, which allow the usage of well-established third-party tools developed by non-OpenMS developers, such as MS-GF+[15] or SIRIUS[16] within the OpenMS Framework. All OpenMS tools provide a detailed choice of parameters that go beyond what classical software in the field offers and allow its function to be tailored to the specific needs of the user. The tools can be used individually or in an analysis pipeline either using the command line or a workflow engine. In the following, we would like to present a few examples of how to tackle common problem settings in mass spectrometry based multi-omics research with OpenMS tools and workflows.

### 6.3.2 Getting Started with OpenMS for Users

If you are interested in using OpenMS for your application and your research, have a look at the installation instructions and further tutorials on the usage of OpenMS (Box 6.5).

**Box 6.5** OpenMS installation instruction

Follow the steps to start using OpenMS TOPP and Command Line Tools:

Current release binary installer for MacOS, Windows and Linux (debian) can be obtained at https://www.openms.de/download/openms-binaries/

OpenMS Quickstart Guide:
https://openms.de/documentation/Quickstart.html

### 6.3.3 Workflows in MS

A bioinformatics workflow defines a series of computational steps which can be applied to single or multiple data sets. Therefore, the concept of a workflow clearly separates the computation from the data on which it operates, describing a reproducible set of steps with a well-defined input and output. In theory, if the same software is run on the same data in the same order with the same parameters then the user should obtain the same output. In practice, it is often difficult to exactly reproduce these conditions on a different computer or at a later date, leading to challenges with reproducible data analysis. This is the problem which a computational workflow solves.

The workflow describes the software, the order in which the software operates on the data and the chosen parameters, which allows the replication of the work with the same data. This will lead to consistent results after reanalysis, since the workflow contains all information on how to process the data – information that is often lost if only an input file and an output file is provided. It is therefore crucial to store and submit workflow files alongside any data output files for other users to have a machine-readable and reproducible description of how the scientific result was computed. Working without workflows can lead to irreproducible computational results when other researchers try to re-analyse data and potentially end up with different results because the computational methods were not well described.

#### *6.3.3.1 OpenMS in KNIME*

The workflow platform KNIME implements such workflows in a graphical user interface by connecting so called nodes (see Figure 6.1). The node is the smallest entity in a workflow, which represents a single operation. Nodes can be connected in the workflow using input and output ports. In KNIME different port types exists depending on the action to be performed. It distinguishes between tables (table ports – black triangle) and whole files (file ports – blue box).

KNIME supports a plugin system which allows the usage of a multitude of analysis software in synergy with OpenMS. These cover a wide area of applications, such as machine learning, hypothesis testing, data visualization and chemoinformatics methods. Here, for example RDKit can be used for the visualization of small molecule structures encoded in SMILES. Additionally, KNIME supports scripting nodes for R, Python and other languages, which can be used to run custom scripts within the workflow. Finished workflows can be saved and shared with the community using the KNIME Community Workflow Hub (https://hub.knime.com/). Further a large collection of plugins is provided by the user community, which can be integrated into the workflow.

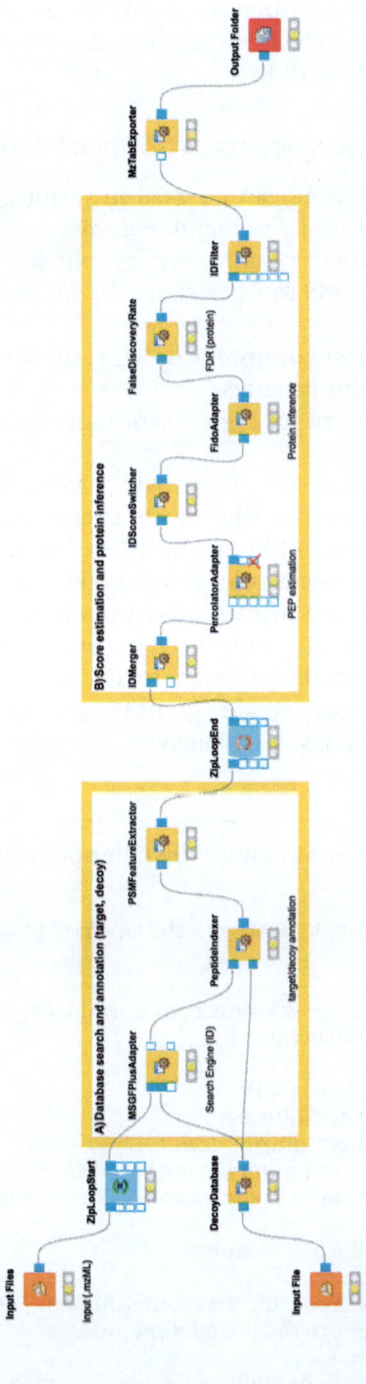


**Figure 6.1** Identification workflow using OpenMS and KNIME can be applied for peptide identification and protein inference.

### 6.3.3.2 Getting Started with OpenMS in KNIME

If you are interested in using OpenMS in combination with the workflow engine KNIME have a look at the installation instructions and further tutorials on the usage of OpenMS with KNIME (Box 6.6).

### 6.3.3.3 Other Workflow Systems with OpenMS Integrations

In addition to KNIME, OpenMS can be used in various other workflow systems due to the common easily wrappable command line interface of its tools. Other popular workflow systems with graphical user interfaces (GUI) for which wrappers of OpenMS tools exist are the following:

- Galaxy: One of the most commonly used server-based workflow editors and managers in bioinformatics.[7]
- gUSE: A workflow system for high-performance computing clusters.

Another workflow language that also provides software for automatic execution of OpenMS tools on various hardware backends (local, remote, HPC clusters or clouds) is nextflow.[8] Although nextflow does not provide a GUI yet, users can either script their own workflows or make use of community-made and well-maintained workflows available in its public workflow collection nf-core.[17] Lastly, OpenMS was successfully used in Pachyderm[18] and snakemake[9] pipelines as well. Easy installation even in restricted HPC environments can be achieved through ready-made OpenMS containers (see Chapter 19) or through the Bioconda[19] package manager.

**Box 6.6** KNIME with OpenMS plugin installation instructions and tutorials

Follow the steps to start using KNIME with the OpenMS plugin:

Installing KNIME
Download KNIME (https://www.knime.com/downloads)
Follow the installation instructions

Installing the OpenMS plugin in KNIME:
Go to "Help" -> "Install New Software..."
Select "KNIME Community Contributions (3.7) – http://update.knime.com/community-contributions/trusted/3.7" in "Work with"
Open the "KNIME Community Contributions – Bioinformatics & NGS" field
Select "OpenMS"
Click "Next" and follow the instructions.

In general, KNIME will automatically detect missing plugins upon opening of a workflow and directs the user to the installation process.

OpenMS/KNIME Tutorial (with handout, example data and workflows):
https://www.openms.de/tutorials/

## 6.3.4 Peptide Identification and Protein Inference

### *6.3.4.1 Search Engine Choice*

One task that is commonly needed for the analysis of shotgun proteomics data is the identification of peptides and inference of their proteins of origin. Identification of proteomics data can be performed in a workflow as depicted in Figure 6.1. Here, mass spectrometry input files (.mzML) are loaded in the "Input Files" node. All files are processed iteratively by all tools between the ZipLoopStart and ZipLoopEnd nodes (see Figure 6.1A). Here, the search engine MS-GF+[15] is applied using the MSGFPlusAdapter to identify MS2 spectra. Search parameters, such as mass error, fragmentation method, possible fixed and variable modifications, as well as charge range and peptide length can be specified. As an alternative to the MSGFPlusAdapter, OpenMS provides a multitude of different wrappers for classic proteomic search engines, such as Comet,[20] Crux,[21] InsPecT,[22] Mascot,[23] MSFragger,[24] MyriMatch,[25] OMSSA,[26] Sequest,[27] PepNovo,[28] and XTandem.[29] Hence, the search engine node within the workflow can be conveniently exchanged for other tools and thus one could test and find out the best performing method in a concise benchmark.

### *6.3.4.2 Sequence Database*

Database search engines make use of a provided sequence database fasta file ("Input File" node), which should contain all target proteins of the organism of choice and possible contaminants. The database and its size highly depend on the research question and the experimental design. In order to assess a false discovery rate in a later step, the database should additionally contain decoy proteins – shuffled or reversed sequences – of all provided targets. The OpenMS tool DecoyDatabase provides an option to concatenate multiple databases (*e.g.* Swiss-Prot human, cRAP) and generate decoys using different methods (*e.g.* protein- or peptide-based shuffling or pseudo-reversing) that will be appended to the provided database and tagged with a specified name prefix or suffix (*e.g.* "DECOY_").

### *6.3.4.3 Identification Post-processing*

After the database search (see Figure 6.1A), identified PSMs (peptide spectrum matches) undergo a series of consecutive post-processing steps to yield the FDR annotated lists of peptide and protein identifications. First target and decoy annotations are assigned to the respective PSM based on the fasta file decoy name tag entries (PeptideIndexer). Next, a number of descriptive features are computed and annotated in a standardized manner to each PSM in order to run a multivariate discrimination of target and decoy space at a later stage (PSMFeatureExtractor). Advanced users can also add specific or customized PSM features in this step. While each MS run is processed separately, it is commonly recommended to co-process the runs and compute a false discovery rate globally over the merged set of all identifications.[30] Hence, the ZipLoop ends here and IDMerger performs a merging of all identifications. Subsequently, the tool Percolator[31] is employed and computes a global FDR based on target and decoy PSM feature scores annotated previously. The Percolator version in OpenMS additionally supports basic protein inference capabilities which we will skip in favor of more advanced methods using ambiguous peptides.

#### 6.3.4.4 *Protein Inference*

Carrying out a robust protein inference and probabilistically distributing evidence of shared peptides in this workflow (see Figure 6.1B), requires the application of the FidoAdapter tool. However, as FidoAdapter was designed to work with OpenMS' own estimation tool for PEPs (IDPosteriorErrorProbability), it is necessary to pick and rename the right score from the PercolatorAdapter. Hence, IDScoreSwitcher, FidoAdapter and subsequently the FalseDiscoveryRate node is applied on protein-level to calculate a protein inference-based target-decoy FDR. As an alternative to Fido[32] one could instead employ the tool EPIFANY, which has recently been added to the OpenMS toolbox or PIA which is provided in a separate KNIME plugin.[33] Ultimately, the resulting peptide and protein identifications can be filtered by various criteria on both levels, including q-value, other metavalues or blacklists from fasta/text files (IDFilter) and exported in the community standard format (mzTab).

### 6.3.5 Further Peptide Identification Methods

#### 6.3.5.1 De Novo *Peptide Search*

Apart from database search, OpenMS also provides tools for *de novo* peptide identification for example the CompNovo or CompNovoCID[34] tool, as well as the NovorAdapter tool which supports *de novo* peptide search using Rapid Novor.[35]

#### 6.3.5.2 *Spectral Library Search*

Additionally, peptides can be identified *via* a spectral library search. Here, we provide a tool called SpecLibSearcher, which is able to identify MS2 spectra (.mzML) based on an input spectral library (.msp). Alternatively, we also support the SpectraST tool from the TPP through the SpectraSTSearchAdapter, which can be used for spectral library search.[36]

### 6.3.6 Additional Supported Methods

Additional tools are available which are able to use the identification data, for example for phosphosite localization and spectral clustering.

#### 6.3.6.1 Phosphosite Localization

The LuciphorAdapter uses the third-party tool LuciPHOr2[37] for the assessment of the phosphosite localization on a phosphopeptide with multiple possible sites, which can be critical in phosphorylation studies. It estimates a false localization rate based on a target decoy approach, which can be used for filtering later on.

#### 6.3.6.2 Spectral Clustering

OpenMS also provides an adapter to MaRaCluster[38] a third-party tool to cluster spectra into groups of similar spectra or to create consensus spectra. Applications range from yielding better and faster identification rates for database search to unsupervised clustering to identify spectra.

### 6.3.7 Peptide and Protein Quantification

The identification workflow above can be extended to perform identification and quantification. Here, depending on the experimental method (*e.g.* label free, SILAC, TMT) the respective nodes can be plugged into the workflow.

The workflow in see Figure 6.2 depicts how identification and label free quantification can be performed with OpenMS in KNIME.

#### 6.3.7.1 Feature Finding

The tools, which are able to perform label free quantification are named FeatureFinder in OpenMS. There are several different implementations available such as FeatureFinderCentroided[39] (see Figure 6.2A) or the FeatureFinderMultiplex. A FeatureFinder recognizes features in LC-MS maps, corresponding to peptides, by their characteristic isotope pattern. Quantification is carried out based on the sum of intensities within the feature region. Here, the position in *m*/*z* and *rt* as well as the charge of the analyte is computed. Resulting features are then represented with scores based on an isotope profile and retention time model. Data points which do not fit the model are removed from the feature region.

#### 6.3.7.2 Combining Post-processed Identification with Quantification Data (Peptide Level)

Afterwards, a score estimation and filtering of peptide identifications found in an individual file is performed (see Figure 6.2). For peptide level results, posterior error probability estimation is performed using Percolator[31] samplewise and a user-defined FDR filter is applied (*e.g.* 5% FDR). In order to combine quantitative and identification information per sample, peptide identifications are generated samplewise and mapped to their respective feature by the IDMapper.

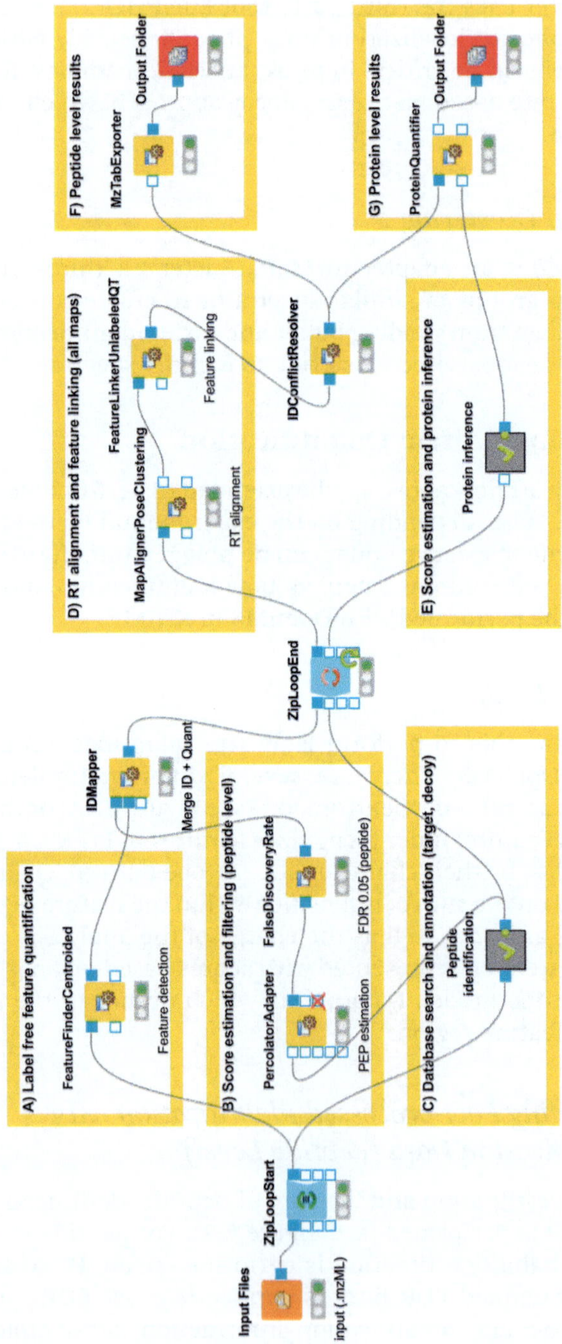


**Figure 6.2** Workflow using OpenMS and KNIME which can be applied for peptide, protein identification and label free quantification. The peptide identification (C) and protein inference (E) are described in the workflow in Figure 6.1.

#### *6.3.7.3 Retention Time Alignment, Feature Linking and Generation of Peptide Level Results*

The mapped information is further processed by the MapAlignerPoseClustering, which performs a linear retention time alignment of input maps to correct retention time shifts and distortions. This is based on a pose clustering algorithm, which uses affine transformation and later refinement based on feature grouping. The FeatureLinkerUnlabeledQT[39] uses QT-based clustering and linking to group corresponding features from multiple maps for label free data (see Figure 6.2). Additionally, OpenMS provides the FeatureLinkerUnlabeledKD, which uses a faster KD-tree based approach for linking. The KD-tree based algorithm has a speed advantage, which becomes apparent for larger datasets (hundreds of samples upwards). The IDConflictResolver is used to ensure that every feature is associated with one single identification based on its score. The peptide level results are then exported as MzTab (see Figure 6.2F).

#### *6.3.7.4 Generate Protein Level Results*

The peptide level information from the IDConflictResolver can also be used in conjunction with the protein inference information (see Figure 6.2C and E) to quantify on protein level (see Figure 6.1G). In this workflow the ProteinQuantifier[39] accumulates feature intensities to peptide abundances based on the identification. Afterwards it uses the protein inference information to average over the abundances the peptides referring to a protein.

### 6.3.8 Additional Supported Quantification Methods

#### *6.3.8.1 Label Free Quantification Based on Identification Data*

The quantification and retention time alignment can also be performed based on previous identifications. The FeatureFinderIdentification[40] can be used for label free quantification of the MS1 features, based on prior peptide identification. This is based on the intuition that a high confidence identification on MS2 level from a specific precursor will produce a corresponding feature at that position in all LC-MS maps of an experiment. Similarly, the MapAlignerIdentification performs retention time alignment based on previous peptide identification.

#### *6.3.8.2 Quantification Using Chemical or Isotopic Labeling*

Further, quantification can be performed for experiments using isotopically or chemically labelled peptides, such as dimethyl labeling or SILAC, by applying the FeatureFinderMultiplex. First, the algorithm finds pairs in a MS1 scan by using the mass difference based on the labeling, the charge and the intensity profile, its correlation and the averagine model. The assessed features are then filtered, and clusters are formed in the *rt* and *m*/*z* range. These clusters correspond to the monoisotopic trace referring to the lightest peptide of a pair (for example SILAC). Afterwards hierarchical clustering is used to assign the peaks to a specific cluster. Then linear regression is used to determine the relative amount of the peptide based on their labels.

#### *6.3.8.3 Quantification Using Isobaric Labeling*

In addition, quantification for isobaric labeling can be performed with the IsobaricAnalyzer, which is able to extract and normalize TMT and iTRAQ quantitative information. It is able to extract the data from centroided MS2 and MS3 spectra and performs an isotope correction based on the specified correction matrix (as provided by the manufacturer).

### 6.3.9 Targeted Analysis

Targeted proteomics is a field of proteomic analysis where accurate and reproducible quantification is required and is often used in clinical settings or laboratory experiments where quantification is of utmost importance. Multiple types of targeted proteomics workflows exist, traditional workflows on a triple quadrupole instrument (SRM or MRM) use a list of Q1 (precursor *m*/*z*) and Q3 (fragment *m*/*z*) for a set of peptides, through which the instrument will then cycle deterministically. These Q1-Q3 *m*/*z* pairs are called transitions and the instrument will typically measure 3–6 transitions per peptide over the course of an experiment, producing a chromatographic measurement for each transition. In more advanced setup, a set of transitions will not be measured during the whole LC-MS/MS experiment but only during a fixed amount of time centered around the putative elution of the target peptide. Similarly, PRM measurements use a list of target peptides, but acquire a high resolution full MS/MS scan for each target peptide (independent of whether a precursor signal was detected or not), resulting in a set of full MS/MS scans for a given precursor acquired deterministically over time. Software is then used to extract a set of $N$ (typically 3–6) transitions from these scans to determine the elution time-point of a peptide. One of the drawbacks of both SRM and PRM is that only a limited set of peptides can be targeted and no quantitative data is acquired for peptides that are not on the target list.

This limitation is addressed by DIA (or SWATH-MS) approaches, which partition the precursor *m*/*z* space into small windows of *ca.* 10 to 25 *m*/*z* and deterministically fragment all precursors in each window and record a full high-resolution MS/MS scan of the resulting fragment ions. Similar to PRM, software is used to extract fragment ion traces from these data, but unlike PRM, no target list of peptides is required since the whole mass range is targeted. In the original SWATH-MS implementation, 32 windows of 25 *m*/*z* width were used to target the mass range of 400–1200 Da, which covers most of the human tryptic peptides.

OpenMS supports targeted analysis through the OpenSWATH[41] module which allows automated analysis of targeted proteomics (SRM and PRM) as well as DIA/SWATH-MS data. For all targeted proteomics experiments, the analysis requires the raw MS data as well as an assay library that contains information (precursor *m*/*z*, fragment ion *m*/*z*, retention time, fragment intensity) about the target peptides. Like the rest of OpenMS, OpenSWATH works with standard file formats (mzML for raw MS data, TraML for the assay library) in order to provide interoperability with other software and standard compliance. The OpenSWATH module performs automatic retention time and *m*/*z* calibration using a set of anchor peptides (*e.g.* spiked in standards such as iRT peptides[42] or endogenous peptides[43]) using either a linear or non-linear function.

Next, OpenSWATH will perform chromatographic extraction, where extracted ion chromatograms are constructed for all peptides in the assay library and then performs chromatographic peak picking and scoring. For DIA data, it will perform chromatographic analysis of the data, but it will also consult the full scan data to obtain additional information such as mass accuracy and isotopic envelope information. Statistical analysis of the data using the target-decoy approach can then either be performed using Percolator[31] or the specifically designed pyProphet software,[44] which will compute false discovery rate (FDR) estimates. Each task described above can be performed by individual TOPP tools, but for convenience we offer an integrated tool OpenSwathWorkflow that performs all steps at once which speeds up execution.

As mentioned above, each targeted analysis requires an assay library. OpenMS provides multiple tools to generate such assays including OpenSwathAssayGenerator which can take spectral library input (*e.g.* in SpectraST format) and generate an assay library output using specific transition-level criteria. Similarly, OpenSwathDecoyGenerator offers several methods ('shuffle', 'pseudo-reverse', 'reverse', 'shift') for generating spectral decoys that are required for target-decoy approaches to FDR estimation.

Extensive documentation is provided at http://www.openswath.org/en/latest/ with detailed information on parameters, example data and extended tutorials on how to run OpenSWATH.

### 6.3.10 Metabolomics

OpenMS can be used for label-free LC-MS metabolomics data analysis. It supports quantification and different identification techniques, such as accurate mass search, *de novo* identification and spectral library search. Similar to label-free proteomics, feature detection and adduct grouping are first performed for each LC-MS map. Retention time alignment and grouping of features across multiple maps are then performed using MapAlignerPoseClustering and FeatureLinkerUnlabeledQT (see "Peptide and Protein quantification").

### 6.3.10.1 Metabolite Quantification

OpenMS provides several tools with the prefix "FeatureFinder" that are used to perform quantification on MS1-level. FeatureFinderMetabo[45] was specifically developed to detect small molecules in LC-MS samples with high sensitivity and specificity. As before, a feature contains all signals that are caused by the same metabolite in a certain charge state.

The algorithm first detects continuous mass traces by starting a new trace at the most intense peak and extending it in both retention time directions. Since low-intensity peaks are often less accurate, a heteroscedastic noise model is used to decide if an additional peak is added. Often, small molecules with the same mass elute at similar retention times and appear on a single continuous mass trace. FeatureFinderMetabo can detect these cases based on the elution profile and splits the mass trace at the local minimum of the elution profile. Finally, all mass traces that are caused by the same metabolite are assembled into a single feature, *i.e.* the monoisotopic trace and additional isotopic traces. To decide which mass traces are assembled, they need to co-elute at the correct *m*/*z* distance and exhibit the expected isotope abundance ratios. Due to the vast diversity of metabolites, it is not possible to use the averagine model. Instead, a support vector machine was trained to detect isotope abundance ratios that are likely to be observed for metabolites.

Each feature corresponds to a metabolite with a certain adduct. The same compound can be observed multiple times at a similar retention time with different adducts or neutral losses (*e.g.* sodium adduct or water loss). The OpenMS tool MetaboliteAdductDecharger can be used to group these features and annotate them with their adduct and charge state, which can be useful for subsequent analysis steps. For this, a list of potential adducts and their probabilities has to be provided. The MetaboliteAdductDecharger then builds a connected graph of co-eluting features, which is resolved using a corresponding Integer Linear Programming approach (ILP). With this, the solution that maximizes the overall probabilities is chosen in the end.

### 6.3.10.2 Metabolite Identification

Compound identification remains one of the major challenges in metabolomics. OpenMS supports commonly used approaches based on compound databases and spectral libraries. In addition, it integrates SIRIUS[16] for the *de novo* identification of metabolites.

**6.3.10.2.1 Compound Databases.** The tool AccurateMassSearch is the first step towards compound identification and can be used to annotate detected features with putative compound identifications using only their accurate mass. The tool considers arbitrary adducts for positive and negative polarity and a compound database providing access to the Human Metabolome Database (HMDB) by default.[46]

**6.3.10.2.2 Spectral Library Search.** Searching the accurate mass of unidentified metabolites against a compound database will provide insight into which compounds could be present in the sample, but the results are often ambiguous. To arrive at more confident identifications, MetaboliteSpectralMatcher can be used to search MS/MS spectra against spectral libraries containing experimentally acquired reference spectra. Spectrum matches are scored using a modified version of the hyperscore introduced by Fenyö and Beavis.[47] Any spectral library in the mzML file format can be used, by default MassBank is used.[48]

**6.3.10.2.3 De Novo.** *De novo* identification has the advantage that it does not rely on a spectral library of previously measured compounds. SIRIUS reported identification rates of more than 70%.[16] *De novo* approaches are computationally expensive making their application in large high-throughput studies cumbersome. To this end SIRIUS has been integrated into OpenMS (SiriusAdapter), which allows the pre-processing and complexity reduction of mass spectrometry data, by providing feature, charge and adduct information.

### 6.3.11 Metaproteomics

The field of Metaproteomics studies communities of (micro-) organisms like the gut microbiome at the proteome level. Besides host–pathogen interaction, central topics are the degradation of substrates and nutrients – including feeding on other organisms. The characterization of organisms helps to understand clinical relevant processes in microbiomes and potentially associated diseases. In OpenMS, we provide the tool MetaProSIP to perform stable isotope probing of metaproteomic communities. It determines to what extent isotopes from the labeled substrate were incorporated into newly synthesized proteins and the labeling ratio to characterize the speed of protein biosynthesis (protein turnover). Carefully designed experiments and time-series analysis of MetaProSIP results allows reconstructing the elemental flow between functional groups of organisms in a complex community.[49]

### 6.3.12 Cross-linking MS

Structural proteomics is an emerging field combining different experimental methodologies with mass spectrometry analysis to gain insights into the structures of biomolecular complexes. Cross-linking is one of these methods and involves inducing non-native covalent bonds between different molecules or different moieties within the same molecule using either chemical reagents, UV light or both. In protein–protein cross-linking usually side chains of protein residues are bound using a chemical cross-linker. The linker has a specific length that it can span, so the identification of the two linked residues gives us an upper bound for the distance between these residues. They can be part of the same protein or two separate proteins interacting with each other, therefore cross-linking MS yields information about the structures of single proteins as well as protein complexes of any size.[50] In nucleic acid to protein cross-linking experiments, covalent bonds are induced between proteins and RNA or DNA strands in close proximity. These bonds are induced while the molecules are as close as possible to their native state or a specific state of interest. Afterwards the proteins are digested with enzymes and the linked peptide pairs or peptides linked with nucleotide oligos are analyzed by mass spectrometry. Because of the increased search space and more complex MS2 fragmentation patterns, linear peptide search algorithms cannot be effectively used to analyze these types of data.

OpenMS enables the analysis of both of these types of data through the dedicated search algorithms OpenPepXL for protein–protein cross-linking and RNP$^{XL}$ for protein-RNA cross-linking.

OpenPepXL requires the centroided MS data from a protein–protein cross-linking experiment and a fasta database of the targeted proteins and decoys. If an isotopically labeled cross-linking reagent was used, it is possible to combine the information from two MS2 spectra (the same peptide pair linked by the light and the heavy linker) to reduce search time and increase the specificity of the search. In this case an additional consensusXML file produced by the tool FeatureFinderMultiplex is necessary to link together MS1 features from the light and heavy cross-linkers. OpenPepXL will then digest the fasta database and preprocess the spectra or spectra pairs by deisotoping and filtering. For each spectrum a list of candidate peptide pairs is generated according to its precursor mass and the given precursor tolerance. Theoretical spectra are generated by considering both peptides and the cross-linker as a single molecule to accurately model the fragment masses of cross-linked peptides. The experimental spectrum and the theoretical spectrum are matched and scored against each other using the OpenPepXL scoring function. The hits for all MS2 spectra in the input mzML file are then written out in idXML or mzIdentML. These can be further processed by XFDR, the dedicated false discovery rate estimation tool for protein–protein cross-linking based on xProphet.[51]

RNP$^{XL}$ identifies protein-RNA cross-links in UV-induced cross-linking experiments. Input files are centroided spectra and a fasta database of the targeted proteins and decoys. A detailed description on data analysis is given in the Supplementary material of Kramer *et al.*[52]

### 6.3.13 RNA (Modification) Analysis

Besides proteomics and metabolomics, another important application of biological mass spectrometry is the sequence analysis of nucleic acids. Before the advent of "next generation" sequencing approaches, mass spectrometry was being pursued as a potential tool for high-throughput DNA sequencing.[53] Recently, the nascent field of epitranscriptomics (RNA epigenetics) has spurred a growing interest in chemical modifications on RNA, and an appreciation of their varied biological roles. Mass spectrometric analysis of intact RNA oligonucleotides has become an important method in this area; it has the unique advantages of allowing the detection and localization of multiple different modifications at the same time, with single-nucleotide resolution. The experimental and computational analysis workflows are largely analogous to shotgun proteomics (see Chapter 4): Purified RNA samples are enzymatically digested (typically with RNase T1, which cuts after guanosines), the oligonucleotides separated by liquid chromatography (typically ion-pair HPLC) and characterized by tandem mass spectrometry in negative ion mode.

OpenMS now includes tools to analyze data from such experiments. The NucleicAcidSearchEngine (NASE) provides functionality for the identification of RNA oligonucleotides based on tandem mass spectra (mzML) and a sequence database (fasta).[54] Analogously to database search engines for shotgun proteomics, NASE has a variety of options that can be adjusted by users, including support for a plethora of ribonucleotide modifications and different digestion enzymes. During development NASE has been tested on tRNA, rRNA, and miRNA samples.

Beyond RNA identification, OpenMS offers basic capabilities for label-free quantification of RNA analytes. To this end, NASE can produce a file with "target coordinates" for its search results, which the FeatureFinderMetaboIdent tool can use to perform targeted quantification of the identified oligonucleotides.

### 6.3.14 Visualization Capabilities (User Perspective)

The graphical application TOPPView provides advanced visualization of mass spectrometric data. It allows examining raw spectra and chromatograms, the effects of data processing steps as well as identification and quantification results in a graphical user interface (GUI). TOPPView offers one dimensional visualization of spectra (*m*/*z* *versus* intensities) and chromatograms (RT *versus* intensities). Additionally, whole experimental maps can be visualized in 2D (RT *versus* *m*/*z*) and 3D (RT *versus* *m*/*z* *versus* intensity) allowing visual quality inspection of the current experiment. From the 2D view, projections on either the RT axis (extracted ion chromatograms, XIC) or projections on the *m*/*z* axis (integrated spectra) can be computed (see Figure 6.3).

Furthermore, chromatographic data as acquired in SRM or extracted ion chromatograms (from PRM/DIA or SWATH-MS data) can be visualized using TOPPView's chromatographic visualization module (RT *versus* intensities). DIA or SWATH-MS data can be visualized using full high-resolution MS/MS spectra by displaying SWATH maps individually in 2D or 3D (fragment ion *m*/*z* *versus* RT). TOPPView is also capable of displaying ion mobility (IM) data is individual spectra ("frames") contain additional ion mobility data, either annotated as meta-data or in 2D or 3D (*m*/*z* *versus* IM) by right-clicking on a spectrum and selecting "Switch to ion mobility view".

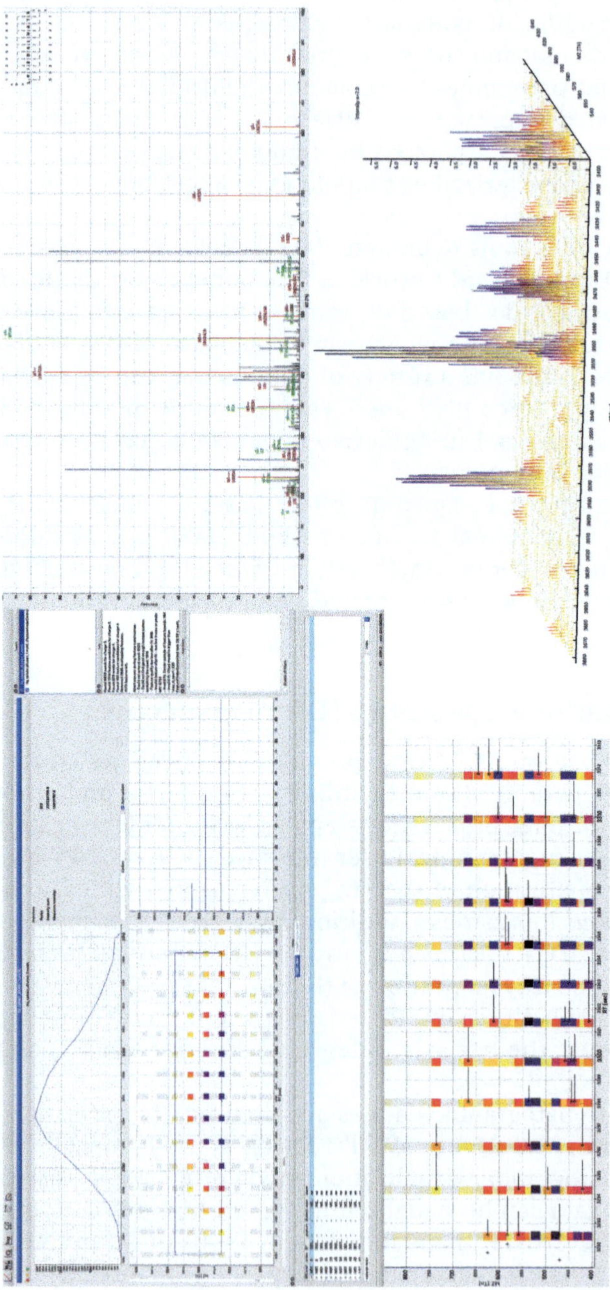

**Figure 6.3** TOPPView – Tool for data visualization. (Top-left) Extracted ion chromatogram view of a specific feature in a segment of the current 2D map. (Bottom-left) Time points of fragmentation events in the 2D view. (Top-right) MS2 spectrum in a 1D view overlaid with the identification based on database search (Identification View). (Bottom-right) 3D representation of a segment from a mass spectrometry map.

Identification data from search engines can be superimposed on individual spectra to annotate fragment ion peaks and display the highest-scoring peptide identification in the same graphical frameworks. These peptide-spectrum matches (PSMs) can be visualized and manually curated.

The TOPPView application is tightly integrated with the TOPP tools provided by OpenMS, offering graphical dialogues to conveniently configure and run TOPP tools without resorting to executing the tool on the command line. Possible applications are optimizing tool configurations to find the best parameters for a particular type of instrument or data.

TOPPView is highly configurable, where the user can select the colors of the display, the position of the axes and the scaling of the data (log, relative, absolute). Furthermore, the user can choose to not load all data into memory at once, which can be suitable for memory-constrained situations or in the case of very large files. In this case, TOPPView is capable of only loading the requested spectra into memory using the indexedmzML data standard that allows random-access to individual spectra even in large XML files.[11]

### 6.3.15 Containerization and Reproducibility

Containerization of software is usually defined as a method to bundle code, configurations and dependencies into one object that is quickly and reliably deployable on different computing environments. Leading providers of software that is able to create and run such containers include Docker and Singularity. OpenMS regularly provides updated containers and recipes (see Box 6.7) to create such containers for different configurations and scenarios. Those scenarios include but are not limited to development, usage in container-enabled workflow-systems (*e.g.* nextflow) or spawning workers in cloud or HPC environments to scale up analyses. Even full workflows can be containerized (*e.g.* with the deNBI-CIBI plugin from KNIME) to create a snapshot of the environment with which a certain analysis was performed. This is useful *e.g.* for generating referenceable identifiers in scientific journals and enabling reproducible research.

**Box 6.7** Information about containerization

Further information about OpenMS containers can be found here:

OpenMS DockerHub pages:
https://hub.docker.com/u/openms and https://hub.docker.com/u/hroest

Biocontainers collection:
https://biocontainers.pro

Containerization of KNIME workflows:
https://www.knime.com/denbicibi-contributions
https://github.com/OpenMS/OpenMS/wiki/Exporting-a-KNIME-workflow-as-a-Docker-container

## Acknowledgements

O.K., L.B., J.P. and T.S. were supported by a grant from the German Federal Ministry of Education and Research (BMBF) under grant no. 031A535A (German Network for Bioinformatics, de.NBI/CIBI).

This Project was funded by the Government of Canada through Genome Canada and the Ontario Genomics Institute (OGI-164). This work was supported by the Canadian Institutes for Health Research, the Natural Sciences and Engineering Research Council of Canada, and the Canada Research Coordinating Committee. H.L.R. is supported by the Canadian Foundation for Innovation and the John R. Evans Leaders Fund and is the Canada Research Chair in Mass Spectrometry-based Personalized Medicine.


## References


1. J. Pfeuffer, *et al.*, OpenMS – A Platform for Reproducible Analysis of Mass Spectrometry Data, *J. Biotechnol.*, 2017, **261**(February), 142–148.
2. H. L. Röst, *et al.*, OpenMS: A Flexible Open-Source Software Platform for Mass Spectrometry Data Analysis, *Nat. Methods*, 2016, **13**(9), 741–748.
3. M. Sturm, *et al.*, OpenMS - an Open-Source Software Framework for Mass Spectrometry, *BMC Bioinf.*, 2008, **9**, 163.
4. H. L. Röst, U. Schmitt, R. Aebersold and L. Malmström, PyOpenMS: A Python-Based Interface to the OpenMS Mass-Spectrometry Algorithm Library, *Proteomics*, 2014, **14**(1), 74–77.
5. M. R. Berthold *et al.*, Studies in Classification, Data Analysis, and Knowledge Organization (GfKL 2007), *KNIME: The Konstanz Information Miner*. Springer, 2007.
6. A. Fillbrunn, *et al.*, KNIME for Reproducible Cross-Domain Analysis of Life Science Data, *J. Biotechnol.*, 2017, **261**(February), 149–156.
7. E. Afgan, *et al.*, The Galaxy Platform for Accessible, Reproducible and Collaborative Biomedical Analyses: 2018 Update, *Nucleic Acids Res.*, 2018, **46**(W1), W537–W544.
8. P. Di Tommaso, *et al.*, Nextflow Enables Reproducible Computational Workflows, *Nat. Biotechnol.*, 2017, **35**(4), 316–319.
9. J. Koster and S. Rahmann, Snakemake–a Scalable Bioinformatics Workflow Engine, *Bioinformatics*, 2012, **28**(19), 2520–2522.
10. M. Sturm and O. Kohlbacher, TOPPView: An Open-Source Viewer for Mass Spectrometry Data, *J. Proteome Res.*, 2009, **8**(7), 3760–3763.
11. H. L. Röst, U. Schmitt, R. Aebersold and L. Malmström, Fast and Efficient XML Data Access for Next-Generation Mass Spectrometry, *PLoS One*, 2015, **10**(4), e0125108.
12. M. Choi, *et al.*, MSstats: An R Package for Statistical Analysis of Quantitative Mass Spectrometry-Based Proteomic Experiments, *Bioinformatics*, 2014, **30**(17), 2524–2526.

13. B. MacLean, *et al.*, Skyline: An Open Source Document Editor for Creating and Analyzing Targeted Proteomics Experiments, *Bioinformatics*, 2010, **26**(7), 966–968.
14. G. Rosenberger, *et al.*, ALFQ: An R-Package for Estimating Absolute Protein Quantities from Label-Free LC-MS/MS Proteomics Data, *Bioinformatics*, 2014, **30**(17), 2511–2513.
15. S. Kim and P. A. Pevzner, Database Search Tool for Proteomics, *Nat. Commun.*, 2014, **5**, 1–10.
16. K. Dührkop, *et al.*, SIRIUS 4: A Rapid Tool for Turning Tandem Mass Spectra into Metabolite Structure Information, *Nat. Methods*, 2019, **16**(4), 299–302.
17. P. A. Ewels, *et al.*, Nf-Core: Community Curated Bioinformatics Pipelines, *bioRxiv*, 2019, 610741.
18. J. A. Novella, *et al.*, Container-Based Bioinformatics with Pachyderm" ed. Jonathan Wren, *Bioinformatics*, 2019, **35**(5), 839–846.
19. B. Grüning, *et al.*, Bioconda: Sustainable and Comprehensive Software Distribution for the Life Sciences, *Nat. Methods*, 2018, **15**(7), 475–476.
20. J. K. Eng, T. A. Jahan and M. R. Hoopmann, Comet: An Open-Source MS/MS Sequence Database Search Tool, *Proteomics*, 2013, **13**(1), 22–24.
21. C. Y. Park, *et al.*, Rapid and Accurate Peptide Identification from Tandem Mass Spectra, *J. Proteome Res.*, 2008, **7**(7), 3022–3027.
22. S. Tanner, *et al.*, InsPecT: Identification of Posttranslationally Modified Peptides from Tandem Mass Spectra, *Anal. Chem.*, 2005, **77**(14), 4626–4639.
23. D. N. Perkins, D. J. C. Pappin, D. M. Creasy and J. S. Cottrell, Probability-Based Protein Identification by Searching Sequence Databases Using Mass Spectrometry Data, *Electrophoresis*, 1999, **20**(18), 3551–3567.
24. A. T. Kong, *et al.*, MSFragger: Ultrafast and Comprehensive Peptide Identification in Mass Spectrometry–Based Proteomics, *Nat. Methods*, 2017, **14**(5), 513–520.
25. D. L. Tabb, C. G. Fernando and M. C. Chambers, MyriMatch: Highly Accurate Tandem Mass Spectral Peptide Identification by Multivariate Hypergeometric Analysis, *J. Proteome Res.*, 2007, **6**(2), 654–661.
26. L. Y. Geer, *et al.*, Open Mass Spectrometry Search Algorithm, *J. Proteome Res.*, 2004, 958–964.
27. J. K. Eng, A. L. McCormack and J. R. Yates, An Approach to Correlate Tandem Mass Spectral Data of Peptides with Amino Acid Sequences in a Protein Database, *J. Am. Soc. Mass Spectrom.*, 1994, **5**(11), 976–989.
28. A. Frank and P. Pevzner, PepNovo: *De Novo* Peptide Sequencing *via* Probabilistic Network Modeling, *Anal. Chem.*, 2005, **77**(4), 964–973.
29. R. Craig and R. C. Beavis, TANDEM: Matching Proteins with Tandem Mass Spectra, *Bioinformatics*, 2004, **20**(9), 1466–1467.
30. O. Serang and L. Käll, Solution to Statistical Challenges in Proteomics Is More Statistics, Not Less, *J. Proteome Res.*, 2015, **14**(10), 4099–4103.

31. M. The, M. J. MacCoss, W. S. Noble and L. Käll, Fast and Accurate Protein False Discovery Rates on Large-Scale Proteomics Data Sets with Percolator 3.0, *J. Am. Soc. Mass Spectrom.*, 2016, **27**(11), 1719–1727.
32. O. Serang, M. J. MacCoss and W. S. Noble, Efficient Marginalization to Compute Protein Posterior Probabilities from Shotgun Mass Spectrometry Data, *J. Proteome Res.*, 2010, **9**(10), 5346–5357.
33. J. Uszkoreit, *et al.*, PIA: An Intuitive Protein Inference Engine with a Web-Based User Interface, *J. Proteome Res.*, 2015, **14**(7), 2988–2997.
34. A. Bertsch, *et al.*, *De Novo* Peptide Sequencing by Tandem MS Using Complementary CID and Electron Transfer Dissociation, *Electrophoresis*, 2009, **30**(21), 3736–3747.
35. B. Ma, Novor: Real-Time Peptide *De Novo* Sequencing Software, *J. Am. Soc. Mass Spectrom.*, 2015, **26**(11), 1885–1894.
36. H. Lam, *et al.*, Development and Validation of a Spectral Library Searching Method for Peptide Identification from MS/MS, *Proteomics*, 2007, **7**(5), 655–667.
37. D. Fermin, D. Avtonomov, H. Choi and A. I. Nesvizhskii, LuciPHOr2: Site Localization of Generic Post-Translational Modifications from Tandem Mass Spectrometry Data, *Bioinformatics*, 2015, **31**(7), 1141–1143.
38. M. The and L. Käll, MaRaCluster: A Fragment Rarity Metric for Clustering Fragment Spectra in Shotgun Proteomics, *J. Proteome Res.*, 2016, **15**(3), 713–720.
39. H. Weisser, *et al.*, An Automated Pipeline for High-Throughput Label-Free Quantitative Proteomics, *J. Proteome Res.*, 2013, **12**(4), 1628–1644.
40. H. Weisser and J. S. Choudhary, Targeted Feature Detection for Data-Dependent Shotgun Proteomics, *J. Proteome Res.*, 2017, **16**(8), 2964–2974.
41. H. L. Röst and G. Rosenberger, *et al.*, OpenSWATH Enables Automated, Targeted Analysis of Data-Independent Acquisition MS Data, *Nat. Biotechnol.*, 2014, **32**(3), 219–223.
42. C. Escher, *et al.*, Using IRT, a Normalized Retention Time for More Targeted Measurement of Peptides, *Proteomics*, 2012, **12**(8), 1111–1121.
43. S. J. Parker, *et al.*, Identification of a Set of Conserved Eukaryotic Internal Retention Time Standards for Data-Independent Acquisition Mass Spectrometry, *Mol. Cell. Proteomics*, 2015, **14**(10), 2800–2813.
44. J. Teleman, *et al.*, DIANA-Algorithmic Improvements for Analysis of Data-Independent Acquisition MS Data, *Bioinformatics*, 2015, **31**(4), 555–562.
45. E. Kenar, *et al.*, Automated Label-Free Quantification of Metabolites from Liquid Chromatography-Mass Spectrometry Data, *Mol. Cell. Proteomics*, 2014, **13**(1), 348–359.
46. D. S. Wishart, *et al.*, HMDB 4.0: The Human Metabolome Database for 2018, *Nucleic Acids Res.*, 2018, **46**(D1), D608–D617.
47. D. Fenyö and R. C. Beavis, A Method for Assessing the Statistical Significance of Mass Spectrometry-Based Protein Identifications Using General Scoring Schemes, *Anal. Chem.*, 2003, **75**(4), 768–774.
48. H. Horai, *et al.*, MassBank: A Public Repository for Sharing Mass Spectral Data for Life Sciences, *J. Mass Spectrom.*, 2010, **45**(7), 703–714.

49. T. Sachsenberg, *et al.*, MetaProSIP: Automated Inference of Stable Isotope Incorporation Rates in Proteins for Functional Metaproteomics, *J. Proteome Res.*, 2015, **14**(2), 619–627.
50. A. Leitner, M. Faini, F. Stengel and R. Aebersold, Crosslinking and Mass Spectrometry: An Integrated Technology to Understand the Structure and Function of Molecular Machines, *Trends Biochem. Sci.*, 2016, **41**(1), 20–32.
51. T. Walzthoeni, *et al.*, False Discovery Rate Estimation for Cross-Linked Peptides Identified by Mass Spectrometry, *Nat. Methods*, 2012, **9**(9), 901–903.
52. K. Kramer, *et al.*, Photo-Cross-Linking and High-Resolution Mass Spectrometry for Assignment of RNA-Binding Sites in RNA-Binding Proteins, *Nat. Methods*, 2014, **11**(10), 1064–1070.
53. A. Apffel, *et al.*, Analysis of Oligonucleotides by HPLC–Electrospray Ionization Mass Spectrometry, *Anal. Chem.*, 1997, **69**(7), 1320–1325.
54. S. Wein, *et al.*, A Computational Platform for High-Throughput Analysis of RNA Sequences and Modifications by Mass Spectrometry, *bioRxiv*, 2018, 501668.

CHAPTER 7

# Metabolomics Data Analysis Using MZmine

TOMÁŠ PLUSKAL*[a], ANSGAR KORF[b], ALEKSANDR SMIRNOV[c], ROBIN SCHMID[b], TIMOTHY R. FALLON[a,d], XIUXIA DU[c] AND JING-KE WENG*[a,d]

[a]Whitehead Institute for Biomedical Research, 455 Main Street, Cambridge, MA 02142, USA; [b]University of Münster, Institute of Inorganic and Analytical Chemistry, Department of Analytical Chemistry, Corrensstraße 28/30, Münster, 48149, Germany; [c]University of North Carolina at Charlotte, Department of Bioinformatics and Genomics, 9331 Robert D. Snyder Rd, Charlotte, NC 28223, USA; [d]Massachusetts Institute of Technology, Department of Biology, 77 Massachusetts Ave, Cambridge, MA 02139, USA
*E-mail: pluskal@wi.mit.edu, wengj@wi.mit.edu

## 7.1 Introduction

Rapid improvements in high-resolution mass spectrometry (HRMS) instrumentation since the early 2000s have led to equally dramatic developments in the fields of targeted and untargeted metabolomics.[1] However, the MS instrument vendors initially lagged behind in software development, and the gap in raw MS data processing tools for metabolomics has been primarily filled by efforts from academia. The MZmine project was originally started in 2005 by Matej Orešic's group at VTT Biotechnology in Finland.[2] It received a major overhaul towards modularity, spearheaded mainly by Tomáš Pluskal at the Okinawa Institute of Science and technology in Japan, and its second version, MZmine 2, was introduced in

New Developments in Mass Spectrometry No. 8
Processing Metabolomics and Proteomics Data with Open Software: A Practical Guide
Edited by Robert Winkler

Published by the Royal Society of Chemistry, www.rsc.org

2010.[3] Since then, MZmine has grown into a worldwide collaborative project with many research labs and companies having contributed new code and data-processing modules. As of 2019, the project contains over 180 000 lines of Java source code. GitHub statistics indicate that the software has been downloaded over 56 000 times over the last four years, and our internal tracking system shows that over 4.3 million individual module runs have been performed in 2018 alone. In the last three years, MZmine has also been participating in the "Google Summer of Code" program, offering opportunities to computer science students to receive funding from Google for their contributions to the development of MZmine.[4]

MZmine is implemented in Java and can, therefore, be readily used on many different computer platforms. It has been designed as a modular system and a particular emphasis has been given to its powerful visualization modules (Figure 7.1a), which distinguish MZmine from other MS data-processing tools such as XCMS or OpenMS.[5,6] Raw mass spectra can be imported into MZmine in common file formats, including netcdf, mzML, mzXML, and mzData.[7] When running on Microsoft Windows, MZmine can also directly import the native *.raw* files of Thermo and Waters instruments using vendor libraries. MZmine assumes the input data comes from MS experiments coupled to liquid chromatography (LC-MS) or gas chromatography (GC-MS). Although there are no specific modules in MZmine for processing direct infusion data, the existing modules can perform feature detection, deisotoping, and metabolite identification based on such data simply by ignoring the retention time values.

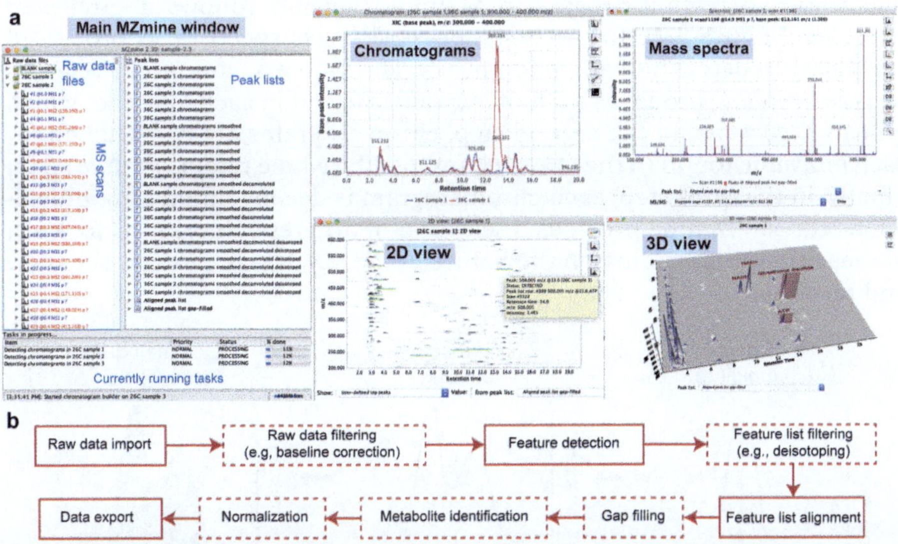


**Figure 7.1** (a), Main visualization modules in MZmine for viewing MS data. (b), A schema of the general data processing workflow in MZmine. Optional steps are indicated with dashed borders.

The general data-processing workflow in MZmine (Figure 7.1b) starts with raw data filtering (*e.g.*, cropping, baseline correction, or smoothing) followed by feature detection, which is the cornerstone of the process. Feature detection identifies *m*/*z* and retention time pairs to call features in the 3D space defined by retention time (*x*-axis), *m*/*z* value (*y*-axis) and signal intensity (*z*-axis). We use the term 'feature' to emphasize the 3D nature of the signal, as opposed to the term 'peak', which is typically used for 2D datasets (*e.g.*, single ions in a mass spectrum can be called peaks). Detected features in each file are listed in *feature lists*, which are then further processed (*e.g.*, to remove features produced by natural isotopes) and aligned to connect corresponding features across all samples. Secondary feature detection (gap filling) can then be performed on the aligned feature lists to cope with missing features that might be artifacts of the feature-detection process. The detected features can further be identified by searching compound or spectral databases and their peak areas can be normalized (*e.g.*, using internal standards). Finally, the results are exported for downstream statistical or multivariate analysis (*e.g.*, using MetaboAnalyst)[8]. In this chapter, we will mainly discuss new data-processing methods that have been added to MZmine since the introduction of MZmine 2 in 2010.[3]

## 7.2 Feature Detection

Feature detection is the cornerstone of each MS data-processing software. A number of algorithmic approaches have been applied for this purpose, including wavelet transform,[9] Kalman filters,[10] or *k*-means clustering.[11] The feature-detection process in MZmine typically follows a three-step approach (Figure 7.2). In the first step, each mass spectrum is processed separately to detect individual ion peaks. This process, commonly referred to as *centroiding*, produces a list of *m*/*z* values found in each MS scan, which we call a *mass list*. In the second step, chromatograms are constructed for each *m*/*z* value found in the mass lists across the whole retention time span. Finally, in the third step, each chromatogram is deconvoluted into individual features. MZmine provides a selection of different algorithms for each of these steps, depending on the nature of the MS data (*e.g.*, mass accuracy and resolution).

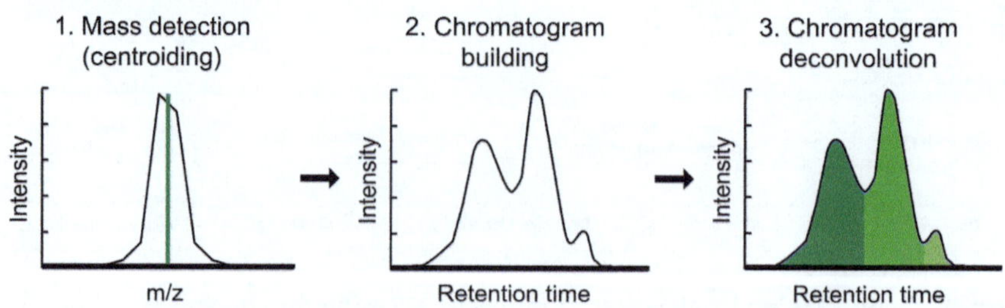


**Figure 7.2** Typical feature detection workflow in MZmine.

### 7.2.1 ADAP Feature Detection Methods

Development of automated data analysis pipeline (ADAP) feature detection started in 2016 to address the issue of false features that were detected by many software tools and reported in software evaluation publications.[12–14] ADAP feature detection starts with building extracted ion chromatograms (EICs, Figure 7.3a). Unlike the EIC builder in other open-source software tools such as XCMS that builds EICs chronologically in retention time, the ADAP algorithm works from the largest intensity point in a data file down to the smallest. This method allows ADAP to start each EIC at the highest intensity point that also has the highest mass measurement accuracy among all of the data points that belong to this EIC. This way of EIC building is especially important for mass spectra that are acquired by time-of-flight mass (TOF) mass analyzers. TOF mass spectra exhibit stronger association between mass measurement accuracy and signal intensity in comparison to other types of mass analyzers such as Orbitrap.

After all of the data points in a data file have been examined and each data point has been either allocated to a specific EIC or considered non-EIC-forming, ADAP detects chromatographic features from each EIC using continuous wavelet transform (CWT) and ridgeline detection (Figure 7.3b). Wavelet transform is a widely used signal-processing technique that can represent a 1D temporal signal in a 2D time-scale space. This redundant way of representing the 1D temporal signal in a 2D space facilitates the detection of not only the different frequencies that the signal contains but also the temporal location of the frequency components. As a result, wavelet transform has been applied widely in the analysis of non-stationary signals (*i.e.*, the frequency content of the signal changes with respect to time). EICs are typical non-stationary signals. As such, results from the wavelet transform automatically provide information for locating the time interval where a chromatographic peak appears, regardless of the width of the chromatographic peak. This level of robustness is desired for any feature detection method.[15]

The centWave algorithm that XCMS uses for detecting chromatographic features is also CWT-based. However, there are significant differences between the ADAP feature detection and centWave in terms of filtering false features based on ridgeline length and signal-to-noise ratio (SNR) of a feature. In particular, ADAP uses a more streamlined approach to estimate SNR compared to what is implemented in centWave.[14] Furthermore, ADAP adjusts the left and right boundaries of each feature using a minimum-intensity search around the initial estimate of feature boundaries derived from ridgeline detection. This adjustment is necessary because the left and right boundaries estimated from ridgeline detection results are symmetric, *i.e.*, equal distances from the feature apex. But chromatographic feature shapes are usually non-symmetric and are affected by chromatography.

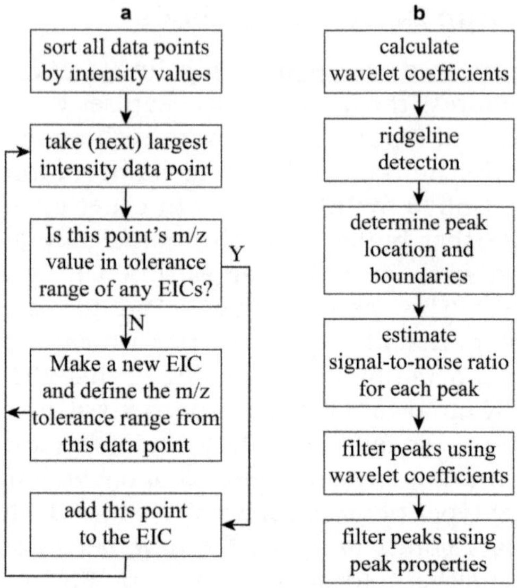


**Figure 7.3** Simplified flow diagram of the ADAP EIC construction and peak picking process. (a) EIC construction. (b) Peak picking. Reproduced from ref. 14 with permission from American Chemical Society, Copyright 2017.

### 7.2.2 GridMass – 2D Feature Detection

The GridMass algorithm was introduced into MZmine by Victor Treviño in 2015.[16] Unlike the typical workflow described above, this method requires the input of mass spectra acquired in profile mode. GridMass takes advantage of the continuous nature of profile-mode spectra by placing a large number of probes across the whole dataset, and then converges the probes towards local maxima (Figure 7.4). The initial locations of all probes that converge to the same local maximum are then used to define the boundaries of the detected feature.

### 7.2.3 Evaluation of Feature Detection Methods

Unbiased evaluation of feature detection algorithms is a difficult task, because no ground truth is defined for experimental LC-MS or GC-MS datasets, and the algorithms must balance sensitivity *versus* specificity. Furthermore, the results obtained by each algorithm strongly depend on the parameter settings, which are non-trivial to optimize.[17] Coble and Fraga compared the results obtained with SpectConnect, MetAlign, XCMS, and MZmine, and concluded that while each software tool generated a large number of false positive signals, combining the results of multiple preprocessing tools might be a suitable strategy to maximize the chance of detecting low-abundance

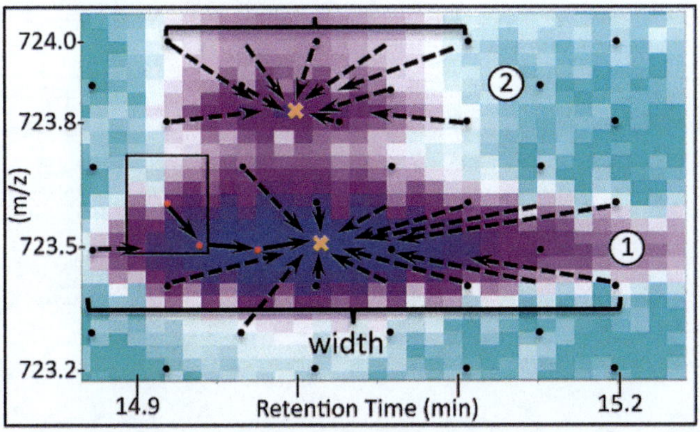


**Figure 7.4** The principle of the GridMass algorithm. Black dots represent individual probes, and orange crosses represent local maxima. Two detected features are annotated as ① and ②. Reproduced from ref. 16 with permission from John Wiley and Sons, © 2015 John Wiley & Sons, Ltd.

components.[12] Myers *et al.* performed a thorough comparison of the ADAP feature detection, the original MZmine feature detection, and the XCMS centWave algorithm by manually evaluating the peak shapes of features sampled randomly from the sets of all detected features.[14] In this evaluation, the ADAP algorithms provided the most good-quality peak shapes detected across all tested files. Recently, Li *et al.* compared the quantification accuracy of five commercial and open-source MS data-processing tools by analyzing standard mixtures consisting of 1100 compounds. The authors concluded that MZmine provides the best performance in terms of quantification accuracy and reports the most true sample-discriminating markers together with the fewest false markers.[18]

## 7.3 Spectral Deconvolution

In GC-MS experiments, each compound produces multiple fragments that appear in the raw data as features with similar retention times and different *m*/*z* values. The spectral deconvolution procedure is intended to estimate the number and location of compounds that produced those features and to construct their pure fragmentation mass spectra. However, the latter task can be difficult due to co-eluting compounds, where features from these compounds may be mixed together. The retention-time resolution of GC is often not sufficient to completely separate all features in a complex sample. Thus, spectral deconvolution is a necessary step in GC-MS data processing.

In a typical workflow of GC-MS data processing, spectral deconvolution is applied after all the features have been detected (Figure 7.5a and b). Its function is two-fold: (1) estimation of the number and retention time of

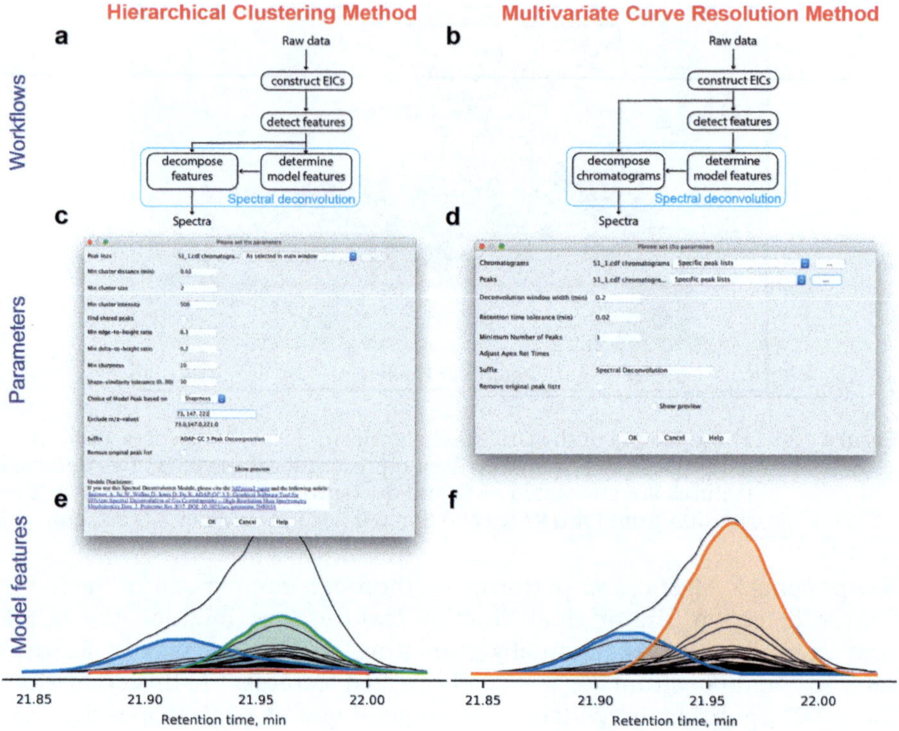


**Figure 7.5** Spectral deconvolution modules in MZmine – Hierarchical Clustering Method and MCR Method. (a) and (b) Data processing workflows for the two spectral deconvolution methods. (c) and (d) MZmine parameter windows for these methods. e and f, Model features (displayed as colored areas) constructed by Hierarchical Clustering and MCR. EIC stands for extracted ion chromatogram.

compounds that produced the detected features, and (2) construction of pure fragmentation spectra of those compounds. The constructed spectra later can be used for identification and relative quantitation of specific compounds in data samples.

Spectral deconvolution can be viewed as a mathematical problem of decomposing matrix $X$ containing the elution profiles of detected features, into the product of two matrices $C$ and $S$ representing the elution profiles and pure fragmentation spectra of compounds respectively:

$$X = CS^T + E \tag{7.1}$$

where $E$ is an error matrix. There are multiple reported approaches to perform spectral deconvolution, and each has some strengths and weaknesses. However, all approaches can be classified into two large categories: (1) traditional two-step approach that first constructs matrix $C$ and then solves an optimization problem with respect to matrix $S$, and (2) multivariate curve resolution (MCR) approach that constructs matrices *C* and *S* simultaneously.

MZmine users can choose between the traditional and MCR approaches by using one of the two spectral deconvolution modules: Hierarchical Clustering and MCR, respectively.

### 7.3.1 Hierarchical Clustering Method

The hierarchical clustering spectral deconvolution method was developed by Yan Ni *et al.*[15] and further modified by Smirnov *et al.*[19] It follows the traditional two-step approach, where the elution profiles of perceived compounds (matrix *C*) are determined first, followed by the fragmentation mass spectra of those compounds (matrix *S*). The identification and quantitation performance of the hierarchical clustering method was evaluated on both unit-mass-resolution and high-mass-resolution data from standard-mixture and urine samples, and outperformed several other available softwares.[19] The MZmine parameter window of the hierarchical clustering method and produced elution profiles for two co-eluting compounds are shown in Figure 7.5c and e, respectively.

The hierarchical clustering method infers the presence of compounds and constructs their mass fragmentation spectra in several steps. First, DBSCAN clustering[20] is used to find groups of features that overlap in the retention time domain. Second, a filter is applied to these features so that only features with high sharpness,[15] a single local maximum, and low edge-to-apex intensity ratios[15] are retained in each group. Third, the hierarchical clustering of features is used to infer the number of compounds in each group and select the model feature for each compound, where the similarity between the elution profiles of detected features is used as a distance measure in the hierarchical clustering. Fourth, each feature is decomposed into a linear combination of the model features to form the fragmentation spectrum of each inferred compound.

Although the hierarchical clustering spectral deconvolution method is computationally efficient and can produce superior identification and quantitation results, it has several drawbacks. First, hierarchical clustering involves several steps and each step requires the user to specify certain parameters. As a result, the total number of user parameters for hierarchical clustering is rather high (Figure 7.5c). Second, the produced spectral deconvolution results heavily depend on a choice of model features selected by the hierarchical clustering method. Specifically, if data contain co-eluting compounds, the user has to make sure that no composite features produced by co-eluting compounds are selected as model features. Otherwise, selecting a composite feature would result in incorrect fragmentation mass spectra and omission of at least some of co-eluting compounds. Thus, this algorithm requires the user to go through a trial-and-error procedure to choose the correct parameters and eventually arrive at appropriate spectral deconvolution results.

### 7.3.2 MCR Method

The MCR spectral deconvolution method is designed to mitigate the aforementioned drawbacks of the hierarchical clustering by following two principles: (1) including only the minimum number of user-specified parameters, and (2) avoiding the selection of model features. MCR employs non-negative matrix factorization,[21] which involves iteratively updating of matrices *C* and *S* with the purpose to minimize the error matrix (eqn (7.1)).

MCR-based methods have demonstrated their ability to computationally separate features in complex samples.[22] However, solution of the MCR problem may be ambiguous, so its application to spectral deconvolution requires imposing additional constraints such as the unimodality and smoothness of the constructed elution profiles, sparse mass fragmentation spectra, robust initialization, *etc.*[23] Moreover, the MCR-based spectral deconvolution is more time intensive than the traditional two-step spectral deconvolution approach.

The MCR method in MZmine is a new implementation of MCR-based spectral deconvolution different from other implementations in several aspects. First, the entire retention time range of a file is split into deconvolution windows, and MCR is applied to each window separately. Using these deconvolution windows helps speed up the overall spectral deconvolution process. Second, the number of compounds is inferred based on clustering the retention times of detected features, where the retention time of a feature is adjusted by fitting a parabola in the top half of that feature. Third, after MCR is completed, the pure fragmentation mass spectra are determined by decomposing extracted-ion chromatograms (EICs) instead of features. The latter helps to recover features that were missed by the chromatogram deconvolution step.

The MZmine parameter window of the MCR method, and produced elution profiles for two co-eluting compounds are shown in Figure 7.5d and f, respectively.

## 7.4 Compound Identification

Compound identification has long been recognized as the principal bottleneck in mass-spectrometry-based metabolomics.[24] Consequently, this area has received a lot of attention in recent MZmine developments. MZmine currently supports annotation of features with chemical formulas, compound structures from chemical and biological databases, and *in silico* predicted chemical structures (Figure 7.6). In addition, MZmine allows matching of spectra to records from mass spectral databases, and provides specific visualization tools for the identification of lipids.

### 7.4.1 Chemical Formula Prediction

The measured mass information ($m/z$ value) of an ion is not sufficient to determine the molecular formula of the ion even with the most accurate mass spectrometers, due to a large number of potential candidate formulas

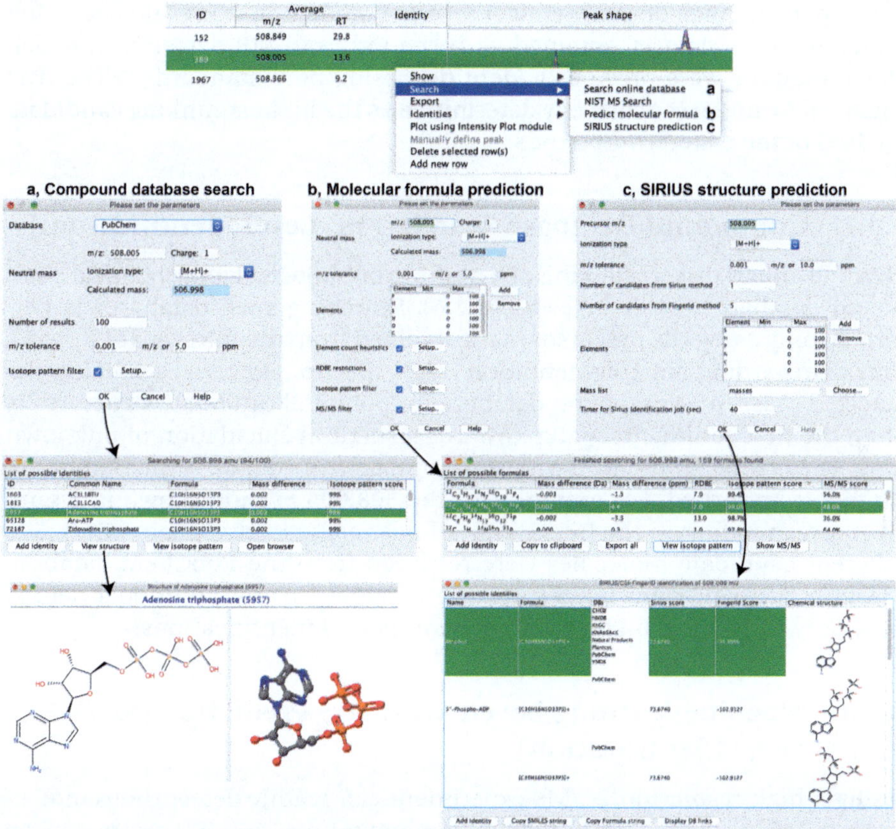


**Figure 7.6** Main compound identification tools in MZmine. The selected feature of 508.005 *m*/*z*, corresponding to an $[M+H]^+$ ion of adenosine triphosphate (ATP) is assigned a tentative identity by searching a public compound database (a), by predicting its chemical formula (b), or using machine learning-based SIRIUS structure prediction (c).

even for relatively small molecules (*e.g.*, above 300 Da).[25] MZmine contains a chemical formula prediction tool that applies a combinatorial approach to rank candidate formulas for each ion (Figure 7.6). The tool calculates all possible formulas within the mass window of each ion, constrained by selected chemical elements, and uses heuristic rules known as "seven golden rules" to discard formulas that are unreasonable in the context of organic chemistry.[26] Next, each candidate formula is scored based on how the natural distribution of isotopes for that formula matches the isotope pattern detected in the MS data. In addition to isotope pattern scoring, MZmine also includes an MS/MS fragmentation filter, which examines the high-resolution MS/MS spectra of the ions (if available) and checks whether the observed fragments can be interpreted using a subset of each candidate formula. This filter can improve the final scoring in cases where the isotope distribution is ambiguous.

The performance of the chemical formula prediction was evaluated using a metabolomic dataset obtained with the Orbitrap MS detector, in which 48 compounds were previously identified using pure standards.[27] The true chemical formula was correctly determined as the highest-ranking candidate for 79% of the tested compounds.

### 7.4.2 Compound Database Search ($MS^1$ Level Identification)

MZmine allows direct querying of a number of biological and chemical compound databases (Table 7.1; Figure 7.6). Searching such databases is performed only using the precursor mass obtained from the full scan ($MS^1$ scan), thus disregarding any fragmentation ($MS^n$) spectra. However, searching the detected mass in a compound database for potential candidate structures is often the first rudimentary step towards structural elucidation of unknown ions. The obvious limitation with this approach, of course, is the number of candidates returned. For example, for the 508.005 *m*/*z* ion shown in Figure 7.6, corresponding to the $[M+H]^+$ ion of adenosine triphosphate (ATP), 577 different candidate molecules were retrieved from the PubChem database within a narrow 5-ppm mass tolerance window. Clearly, additional data is necessary to produce high-confidence compound identifications.

### 7.4.3 Machine-learning-based Structure Prediction (MS/MS Level Identification)

A single high-resolution LC/MS experiment can readily detect thousands of distinct $MS^1$ features, while further fragmentation MS/MS spectra can be collected for many hundreds of these features. For certain classes of compounds, such as lipids or peptides, simple fragmentation rules allow for identification of these features from their MS/MS spectra, through comparison of these experimental fragmentation patterns to *in silico* fragmentation libraries produced from chemical structure databases. But for other classes of compounds, including most small molecules, simple fragmentation rules do not exist, making *in silico* prediction of fragmentation spectra rather challenging.

**Table 7.1** Compound databases that can be queried directly from MZmine.

| Database | Purpose | # Compounds (May 2019) |
|---|---|---|
| KEGG[57] | Metabolic pathways | 18 532 |
| PubChem[58] | General chemicals | 97 915 204 |
| HMDB[59] | Human metabolites | 114 100 |
| YMDB[60] | Yeast metabolites | 16 042 |
| LIPID MAPS[61] | Lipids | 43 403 |
| MassBank.eu[62] | Compounds with experimental spectra | 5923 |
| ChemSpider[63] | General chemicals | 67 000 000+ |
| MetaCyc[64] | Metabolic pathways | 15 655 |

For example, the Mass Frontier software (HighChem) uses a curated library of tens of thousands of fragmentation mechanisms published in scientific literature to predict the molecular transformations that occur in the collision chamber of the mass spectrometer. The latest trend in the metabolomics field is to combine the use of fragmentation rules with machine learning methods such as support vector machines or Markov chains that can learn patterns of molecular fragmentation from large collections of MS/MS spectra contained in public databases.[24] Such learned patterns can then be used to predict fragmentation spectra from chemical structures (CFM-ID)[28] or to associate unknown spectra with most probable molecular structures from non-specific chemical databases such as PubChem or ChemSpider (SIRIUS/CSI:FingerID, MetFrag, MS-FINDER, MAGMa, and others)[29–32].

Among the various algorithms developed for compound identification in recent years, the SIRIUS/CSI:FingerID approach is arguably one of the most sophisticated, achieving ~70% prediction accuracy.[33] The algorithm works in three stages. In the first stage, it generates all possible candidate formulas for the precursor $m/z$ value and constructs fragmentation trees that interpret fragment ions observed in the MS/MS spectra. The best tree is selected using multiple heuristic rules such as isotope pattern matching and the proportion of fragments that could be interpreted. In the second stage, the algorithm uses previously trained predictors to estimate the most likely chemical fingerprint (a binary descriptor of a molecule) of the unknown compound that generated the spectra, using the spectra and the fragmentation tree as inputs. Finally, in the third stage the algorithm scores molecules from a chemical database based on how well they fit the estimated fingerprint, and outputs a list of scored candidate structures. MZmine can export the MS/MS spectra, isotope pattern and MS scans of selected features or whole feature lists into an MGF file format that can be imported into the stand-alone SIRIUS application.[33] Additionally, MZmine provides a module to perform the structure prediction directly from the MZmine interface (Figure 7.6).

### 7.4.4 Spectral Similarity

Although MS/MS spectra can be used for structure prediction as described in the last section, a direct comparison to previously acquired spectra might add further confidence to the tentative identification, and, in some cases, help to identify common substructures or similarities among compounds that are completely unknown. There have been significant advances in making fully public or semi-public MS/MS spectral datasets available to assist with compound identification, such as the MassBank of North America (MoNA) database,[34] MassBank of Europe,[35] the Global Natural Products Social Molecular Networking (GNPS) database,[36] METLIN,[37] and the mzCloud database (Thermo Fisher Scientific).[38] Unfortunately, MS/MS spectral databases are still very fragmented, with a relatively small overlap of contained compounds[39] as well as a lack of data sharing. Unlike the sequencing field, where the nearly four-decade-old International Nucleotide Sequence Database

Collaboration (INSDC)[40] continues to produce a single synchronized and internationally accepted nucleotide reference database that any researcher can contribute data to, mass spectrometry databases have instead angled towards closed approaches, where full datasets are in some cases only available through commercial software or subscriptions, and only select trusted members are able to contribute to the database. There is, however, ongoing development towards data sharing among GNPS, MoNA, and MassBank EU (personal communication).

The local spectral library search in MZmine enables users to match a single spectrum or a whole feature list against a locally saved spectral library of any size. Parsers are provided for the major database formats, which are used by NIST (.msp), MoNA (.json, .msp), GNPS (.mgf, .json), and JCAMP-DX (.jdx). Many open databases allow users to download complete database contents as spectral libraries in at least one of these file formats. Furthermore, MZmine's spectral library creation module facilitates the submission of new entries to local libraries and the GNPS database. This significantly reduces the invested time and work to share new library spectra with the GNPS community and to create specific local libraries, while giving a high level of support and control for filtering and sorting the spectra by quality, selecting the best spectra, and providing metadata. When creating MS/MS spectral entries, multiple different ions of the same molecule can be selected at once, leading to a higher library coverage of ion types, such as *in-source* fragments, adducts, and multimeric species (*e.g.*, $[M-H_2O+H]^+$, $[M+Na]^+$, and $[2M+H]^+$, respectively). MZmine implements multiple similarity functions to match experimental spectra against any local spectral library. First, experimental spectra are extracted from the spectra visualizer, a feature list, or multiple selected feature list rows. The spectrum type is then specified as (1) an MS/MS spectrum with a precursor *m*/*z*, which is often recorded in LC-MS experiments with data-dependent acquisition (DDA), or (2) a spectrum without precursor *m*/*z*, *e.g.*, acquired with GC-EI-MS, all-ion-fragmentation (AIF), or elevated in-source fragmentation. Finally, all experimental spectra are searched against all library spectra. The results can be visualized as spectra mirror charts (Figure 7.7a). To increase the similarity score of spectra which were acquired on different instruments or with modified methods, optional filter steps are implemented to run before spectral similarity calculation. This includes a $^{13}C$-isotope filter, which is applied to the query and library spectrum, and a limitation to signals that fall within the intersecting *m*/*z* range of both spectra.

Apart from providing a spectral database, the GNPS web server enables the analysis of large-scale untargeted mass spectrometry studies and links different studies, results, and annotations in a community curated knowledge base. Molecular networking, the main workflow in GNPS, has emerged as an essential tool to interpret LC-MS data by matching all MS/MS spectra against the spectral library and by creating MS/MS similarity networks, where molecular/spectral families often cluster in sub networks. Feature-based molecular networking (FBMN) was introduced to combine the capabilities

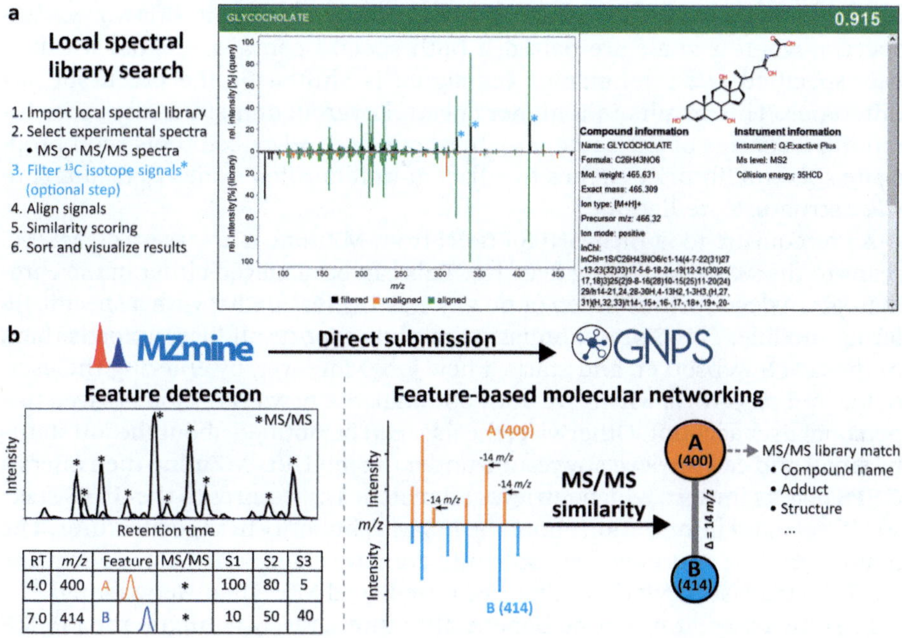


**Figure 7.7** (a) A simplified workflow of the local spectral library search in MZmine, which matches experimental MS or MS/MS spectra against spectral library entries in different common file formats. The results pane depicts the match with metadata and as a spectral mirror chart, highlighting all filtered (black), unaligned (orange), and aligned (green) signals. The query and library spectra of glycocholic acid were acquired on a time-of-flight (TOF) and an orbital Fourier-transform (FT)-based instrument, respectively. Due to a smaller precursor *m/z* isolation width for the library spectrum, the match score was increased by filtering out all $^{13}C$-isotope signals in the query spectrum. (b) Feature-based molecular networking by direct submission of MZmine feature detection results to the GNPS webserver. Network creation with structure modification tolerant MS/MS similarity scoring is illustrated for two features, with B being a putative methylated derivative of A with a precursor *m/z* delta of 14. All MS/MS spectra are searched against a spectral library and the matching structures, visualized for A by a larger node, can be propagated to adjacent nodes using the spectral similarity edges.

and mainly the feature detection workflows of different mass spectrometry processing tools with molecular networking on GNPS. Therefore, specific workflows and export modules were developed in MZmine, XCMS, OpenMS, MS-DIAL, and MetaboScape.[41] Currently, MZmine provides the function to submit all needed data and metadata directly to GNPS to start a new FBMN job (Figure 7.7b). This includes the feature list as a quantification table, the MS/MS spectra of all features in an MGF file format, and an optional sample metadata sheet. The FBMN result is a network of nodes (features with MS/MS scans) which are linked by edges based on a modified MS/MS spectral cosine similarity score, ranging from 0 (dissimilar) to 1 (identical). The scoring is preceded by a structure-modification-tolerant alignment of two MS/MS spectra, where signals are paired if both spectra contain a signal within a user-specified *m*/*z* tolerance or the signal is shifted by the precursor *m*/*z* difference. This results in a higher spectral overlap and similarity score for modified species of the same structural family.[36] Advanced tools then propagate spectral library matches to adjacent unidentified nodes to facilitate *in silico* structure prediction.[42]

A prerequisite to launch GNPS FBMN from MZmine is to assign all MS/MS scans to their corresponding features. This can be achieved either in the chromatogram deconvolution step or on any existing feature list with a specific filtering module. The GNPS submission module exports all files, uploads them to the GNPS webserver, and starts a new job. Moreover, by entering the username and password, which are both optional, the new FBMN job is saved to a personal user account. Otherwise, the user can be notified about the job status by email and can retrieve any results under the job URL. MZmine then offers a GNPS results import, which retrieves all matches of features to the GNPS spectral library and information about the MS/MS similarity between features. The main workflow and new developments are covered as video tutorials in the YouTube playlist "GNPS/MZmine – Feature-Based Molecular Networking".[43]

In some cases, it might be beneficial to interactively compare the MS/MS spectral similarity in a single experiment to identify ions that share structural similarities. With this in mind, we developed an MS/MS similarity searching module for MZmine, which allows for simple visualization of fragmentation pattern similarity of all detected features within a dataset or between two datasets. This module requires preprocessed feature lists with associated MS/MS fragmentation spectra. The user can choose to compare MS/MS fragmentation spectra within a single feature list, typically representing a single chromatographic run, or between two feature lists, ideally experimental runs produced at similar times with similarly calibrated *m*/*z* values. The module performs an all-to-all comparison of the centroided ion *m*/*z* values across all the MS/MS spectra in the feature list. The similarity calculation is simple: the ions are considered to be "matched" across spectra if their *m*/*z* values are within user configurable parameters, while the overall matching score is the sum of the product of intensities of all matched ions. It is possible to set the *m*/*z* window where ions are considered "matched" to a range of only a few ppm, which is well suited to high-mass-accuracy LC/MS instruments. The module records the calculated MS/MS similarity results into the "Identity" column of a given feature.

### 7.4.5 Lipid Identification

Lipids play important roles in basic cell function and organismal physiology.[44] This group of biomolecules possess a broad and complex variety of chemical structures, defined mainly by the length of the acyl and alkyl chains, the degree of unsaturation, double bond positions, and stereochemistry (for in-chain modified chiral carbons).

HRMS has emerged as the gold standard for the identification of lipids in complex biological samples. In particular, LC-HRMS enables accurate and sensitive detection of a great number of lipid species in a single analytical run. Data-dependent tandem mass spectrometry (MS/MS) methods enable structural elucidation of lipid species to some extent. While HRMS alone enables the prediction of a lipid's molecular formula (Figure 7.8a, L1), collision induced dissociation (CID) MS/MS experiments allow elucidation up to the chain composition with identification or verification of the lipid class based on headgroup fragments (Figure 7.8a, L2), revealing the lipid class, the acyl chain length and the degree of unsaturation of the analyte. However, the possible presence of constitutional isomers, such as phosphatidylglycerol (PG) and bis(monoacylglycero)phosphate (BMP), need to be ruled out. This can be achieved by chromatographic lipid separation prior to mass spectrometric detection (Figure 7.8a, L3.1).[45] LC-MS does not provide any information on the position of the acyl chain at the glycerol backbone (*sn*-position; Figure 7.8a, L3.2). A promising instrumental solution was recently published by Maccarone *et al.*, separating unsaturated phosphatidylcholine (PC) constitutional isomers with ion-mobility (IM)-MS.[46] It is noted that the separation was only possible after adding $Ag^+$ to the solution, which resulted in the formation of PC-$Ag^+$ adducts. IM-MS can also be an alternative to differentiate between *cis*/*trans* isomers (Figure 7.8a, L4). The determination of acyl chains double bonds was recently addressed based on various double-bond functionalizations, such as ozone-induced dissociation or the Paternò-Büchi (PB) reaction, which allows the use of conventional CID for its elucidation (Figure 7.8a, L3.3).[47,48]

MZmine enables the identification of lipids from molecular formula prediction (Figure 7.8a, L1) to double bond position prediction (Figure 7.8a, L3.3). Currently, the differentiation of *sn*-positional and *cis*/*trans* isomers is not supported. The annotations are carried out according to the standardized notations for lipids proposed by Liebisch *et al.* to avoid misinterpretation.[49] For the untargeted lipid analysis in LC-HRMS datasets, a novel 3D adaptation of the Kendrick mass defect (KMD) analysis was implemented as an interactive visualization module in MZmine.[50] The module allows visualization of feature lists as Kendrick mass plots. KMD analysis was first introduced in 1963.[51] KMD analysis reduces complex spectra of organic compounds by introducing a new mass scale based on $CH_2$ = 14.0000 u ($KM_{base}$). The Kendrick mass scale (KM) can be calculated by multiplying any IUPAC mass ($m_{IUPAC}$) by the Kendrick mass factor, which can be calculated by dividing the nominal mass of $CH_2$ by the IUPAC mass of $CH_2$ (eqn (7.2)). The $KM_{base}$ $CH_2$ is replaceable by any other molecular formula.

$$KM = \text{nominal mass of } CH_2/\text{exact mass of } CH_2 \cdot m_{IUPAC} \tag{7.2}$$

The KMD is defined as the delta of the nominal KM ($KM_{nom}$) and the KM (eqn (7.3)).

$$KMD = KM_{nom} - KM \tag{7.3}$$

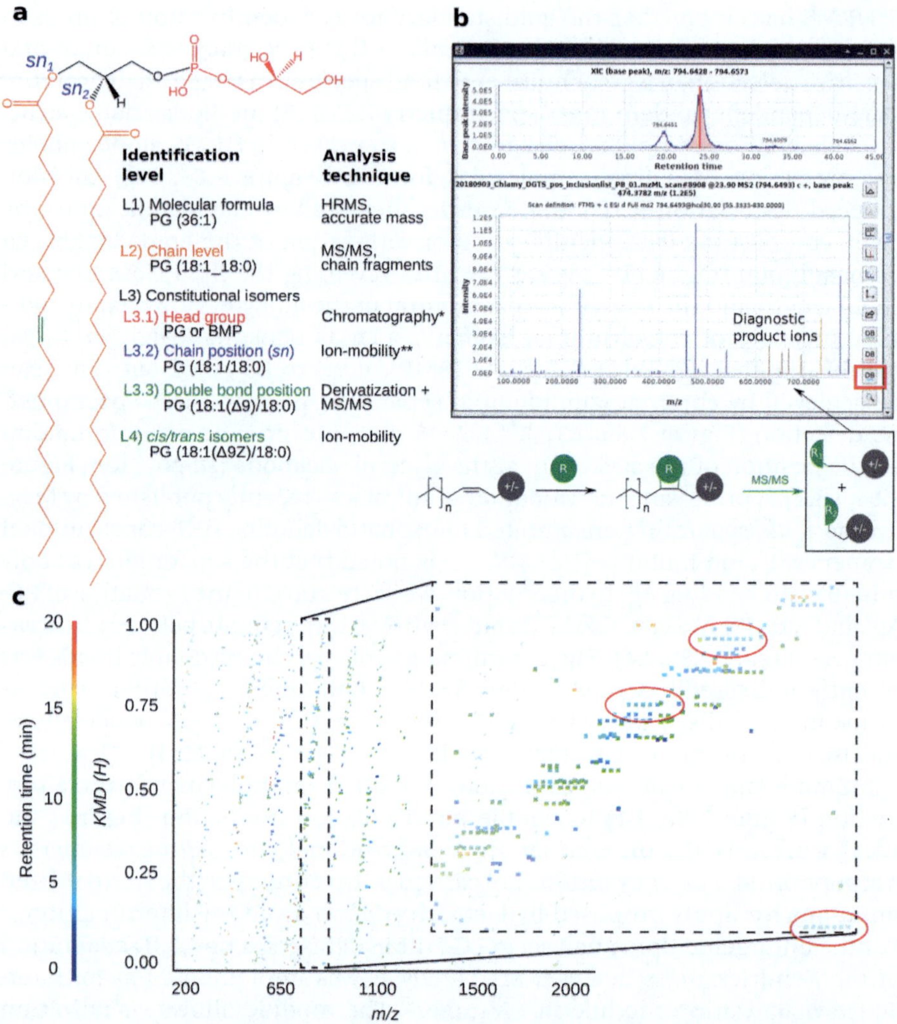


**Figure 7.8** (a) Identification levels of lipids and MS-based techniques to potentially achieve structural elucidation exemplified on PG (18:1(Δ9Z)/18:0). Further techniques, *e.g.*, using enzymatic reactions prior to analysis, are not mentioned. *Chromatography is one possible solution and has been shown for the example of PG and BMP. **Only a possible solution, which has yet solely been shown for PC species as $Ag^+$ adducts. (b) All MS/MS scans summary frame with an extracted ion chromatogram (top) including a red marker for the MS/MS scan recording time. The signals of the diagnostic product ions are highlighted in orange. Highlighted with a red rectangle is the Lipid Search module to annotated signals directly in the spectrum. A general double bond functionalization reaction prior to CID is displayed as a scheme at the bottom (for the PB-reaction R is acetone). (c) 3D Kendrick mass plot of a green alga lipid extract. Hydrogen is used as the $KM_{base}$ to analyze differences in the lipid species' saturation. The retention time is plotted in a color-coded third dimension to group coeluting lipid species by their lipid class. Exemplarily, the red ellipses mark coeluting lipid species of the same lipid class.[50]

Traditionally, KMD analysis was carried out on spectral data. Using chromatographically separated features instead of *m*/*z* signals of a selected spectrum enables the addition of chromatographic characteristics, such as the retention time, in a third-dimension (Figure 7.8c). Figure 7.8c shows all detected features in a green alga lipid extract, which was separated by means of hydrophilic interaction liquid chromatography (HILIC) and detected with an Orbitrap mass analyzer. Using hydrogen as $KM_{base}$ instead of $CH_2$ results in an order in which features that only differ in their number of hydrogen atoms appear in a horizontal line. This characteristic allows the grouping of lipid species of the same class that only differ in their saturation but have the same acyl chain length. HILIC enables the separation of lipids by class due to their polar head group. As a result, lipid species that belong to the same lipid class have very similar retention times and therefore exhibit the same color in the 3D Kendrick mass plot (see the example in Figure 7.8c with red ellipses). This allows a fast graphical analysis of a complex lipid extract to reduce the size of the target of potential lipid species.

The MZmine Lipid Search module allows annotation of the graphically spotted features as potential lipids at the molecular formula and chain levels.[52] The module compares the accurate *m*/*z* of all features with a custom lipid database, which is generated based on selected user parameters, such as lipid class, chain length, and unsaturation status. Furthermore, every generated lipid database entry can be rapidly modified by the “lipid modification parameter”, which allows the addition and/or subtraction of any molecular formula. This enables the simultaneous search for adducts, in-source fragments, and oxidation products. Furthermore, the algorithm automatically searches MS/MS scans of each feature for specific chain and head group fragments to reconstruct possible lipid species identities at the chain level.

The Lipid Search module can also be applied directly to a single mass spectrum. This feature becomes more useful when combined with the “lipid modification parameter” to search for product ions in MS/MS spectra. MZmine has a summary frame of all recorded MS/MS scans of a selected feature list row (Figure 7.8b, top panel). For each scan, an EIC is shown above the MS/MS scan, including a red marker to display the retention time when the MS/MS scan was recorded.

Located on the right-hand side of each scan is a toolbar, which provides methods to rapidly annotate the spectrum. Custom feature database search, spectral database search, online compound database search, molecular formula prediction, and the Lipid Search module are included. The Lipid Search module allows the annotation of diagnostic product ions of derivatization products, which is mandatory for the annotation of lipid species on double bond position level, using conventional CID (Figure 7.8a, c3). Figure 7.8b displays an MS/MS scan of a lipid species PB-product. The diagnostic product ions for the localization of the double bond position are highlighted in orange. The data was recorded with an LC post-column derivatization set up based on a protocol developed by Jeck *et al.*[53] The "lipid modification parameter" of the Lipid Search module can be used to create all possible diagnostic product ions of any lipid species, without limiting the module to specific derivatization reactions.

## 7.5 Batch Mode

The data-processing steps in MZmine can be executed not only through the interactive graphical user interface, but also through a "batch execution" mode. For the batch mode, a sequence of data-processing steps can be defined together with their parameter values and saved as a batch script file. The batch script can then be executed from the graphical user interface or using the command line. This feature enables relatively simple creation of well-defined workflows for reproducible processing of multiple experiments, as well as for execution of large-scale data processing tasks on computing clusters.

## 7.6 Conclusions

MZmine is a comprehensive data-processing and visualization platform with over 15 years of development history. Over this period, the MZmine user base among academic researchers conducting metabolomics experiments has also grown significantly. For new users, the MZmine website provides both text- and video-based tutorials, as well as sample datasets that demonstrate the function of individual modules.[54] A development tutorial is also available for researchers interested in contributing new modules for MS data-processing or visualization.

Development of MZmine is ongoing. Among the planned features are support for imaging mass spectrometry and the corresponding imzML data file format,[55] import and export of processed metabolomics datasets into the recently introduced mzTab-M format,[56] spectral deconvolution for LC-MS datasets acquired using data-independent fragmentation, support for ion mobility datasets, and integration of additional compound identification algorithms such as MetFrag[30] and CFM-ID.[28]

## Acknowledgements

T.P. is a Simons Foundation Fellow of the Helen Hay Whitney Foundation. This work is in part supported by the National Science Foundation (CHE-1709616 and MCB-1818132) and the Richard and Susan Smith Family Foundation. We are grateful to many individual developers worldwide who contributed both small and large pieces of MZmine source code. We acknowledge the generous support of the Google Summer of Code program, which has funded the development of several MZmine modules through student projects.

## References


1. G. J. Patti, O. Yanes and G. Siuzdak, *Nat. Rev. Mol. Cell Biol.*, 2012, **13**, 263–269.
2. M. Katajamaa and M. Oresic, *BMC Bioinf.*, 2005, **6**, 179.
3. T. Pluskal, S. Castillo, A. Villar-Briones and M. Oresic, *BMC Bioinf.*, 2010, **11**, 395.
4. Google Summer of Code, https://summerofcode.withgoogle.com (accessed 7 May 2019).
5. C. A. Smith, E. J. Want, G. O'Maille, R. Abagyan and G. Siuzdak, *Anal. Chem.*, 2006, **78**, 779–787.
6. H. L. Röst, T. Sachsenberg, S. Aiche, C. Bielow, H. Weisser, F. Aicheler, S. Andreotti, H.-C. Ehrlich, P. Gutenbrunner, E. Kenar, X. Liang, S. Nahnsen, L. Nilse, J. Pfeuffer, G. Rosenberger, M. Rurik, U. Schmitt, J. Veit, M. Walzer, D. Wojnar, W. E. Wolski, O. Schilling, J. S. Choudhary, L. Malmström, R. Aebersold, K. Reinert and O. Kohlbacher, *Nat. Methods*, 2016, **13**, 741–748.
7. E. W. Deutsch, *Mol. Cell. Proteomics*, 2012, **11**, 1612–1621.
8. J. Chong, O. Soufan, C. Li, I. Caraus, S. Li, G. Bourque, D. S. Wishart and J. Xia, *Nucleic Acids Res.*, 2018, **46**, W486–W494.
9. R. Tautenhahn, C. Böttcher and S. Neumann, *BMC Bioinf.*, 2008, **9**, 504.
10. C. J. Conley, R. Smith, R. J. O. Torgrip, R. M. Taylor, R. Tautenhahn and J. T. Prince, *Bioinformatics*, 2014, **30**, 2636–2643.
11. H. Ji, F. Zeng, Y. Xu, H. Lu and Z. Zhang, *Anal. Chem.*, 2017, **89**, 7631–7640.
12. J. B. Coble and C. G. Fraga, *J. Chromatogr. A*, 2014, **1358**, 155–164.
13. O. D. Myers, S. J. Sumner, S. Li, S. Barnes and X. Du, *Anal. Chem.*, 2017, **89**, 8689–8695.
14. O. D. Myers, S. J. Sumner, S. Li, S. Barnes and X. Du, *Anal. Chem.*, 2017, **89**, 8696–8703.
15. Y. Ni, M. Su, Y. Qiu, W. Jia and X. Du, *Anal. Chem.*, 2016, **88**, 8802–8811.
16. V. Treviño, I.-L. Yañez-Garza, C. E. Rodriguez-López, R. Urrea-López, M.-L. Garza-Rodriguez, H.-A. Barrera-Saldaña, J. G. Tamez-Peña, R. Winkler and R.-I. Díaz de-la-Garza, *J. Mass Spectrom.*, 2015, **50**, 165–174.
17. M. Hu, M. Krauss, W. Brack and T. Schulze, *Anal. Bioanal. Chem.*, 2016, **408**, 7905–7915.
18. Z. Li, Y. Lu, Y. Guo, H. Cao, Q. Wang and W. Shui, *Anal. Chim. Acta*, 2018, **1029**, 50–57.
19. A. Smirnov, W. Jia, D. I. Walker, D. P. Jones and X. Du, *J. Proteome Res.*, 2018, **17**, 470–478.
20. M. Ester, H.-P. Kriegel, J. Sander, X. Xu, *et al.*, in *KDD-96*, 1996, vol. 96, pp. 226–231.
21. D. D. Lee and H. S. Seung, in *Advances in Neural Information Processing Systems 13*, ed. T. K. Leen, T. G. Dietterich and V. Tresp, MIT Press, 2001, pp. 556–562.
22. L. W. Hantao, H. G. Aleme, M. P. Pedroso, G. P. Sabin, R. J. Poppi and F. Augusto, *Anal. Chim. Acta*, 2012, **731**, 11–23.

23. H.-T. Gao, T.-H. Li, K. Chen, W.-G. Li and X. Bi, *Talanta*, 2005, **66**, 65–73.
24. I. Blaženović, T. Kind, J. Ji and O. Fiehn, *Metabolites*, 2018, 31.
25. T. Kind and O. Fiehn, *BMC Bioinf.*, 2006, **7**, 234.
26. T. Kind and O. Fiehn, *BMC Bioinf.*, 2007, **8**, 105.
27. T. Pluskal, T. Uehara and M. Yanagida, *Anal. Chem.*, 2012, **84**, 4396–4403.
28. Y. Djoumbou-Feunang, A. Pon, N. Karu, J. Zheng, C. Li, D. Arndt, M. Gautam, F. Allen and D. S. Wishart, *Metabolites*, 2019, 72.
29. K. Dührkop, H. Shen, M. Meusel, J. Rousu and S. Böcker, *Proc. Natl. Acad. Sci. U. S. A.*, 2015, **112**, 12580–12585.
30. C. Ruttkies, E. L. Schymanski, S. Wolf, J. Hollender and S. Neumann, *J. Cheminf.*, 2016, **8**, 3.
31. H. Tsugawa, T. Kind, R. Nakabayashi, D. Yukihira, W. Tanaka, T. Cajka, K. Saito, O. Fiehn and M. Arita, *Anal. Chem.*, 2016, **88**, 7946–7958.
32. L. Ridder, J. J. J. van der Hooft and S. Verhoeven, *Mass Spectrom.*, 2014, **3**, S0033.
33. K. Dührkop, M. Fleischauer, M. Ludwig, A. A. Aksenov, A. V. Melnik, M. Meusel, P. C. Dorrestein, J. Rousu and S. Böcker, *Nat. Methods*, 2019, **16**, 299–302.
34. MassBank of North America (MoNA), http://mona.fiehnlab.ucdavis.edu/ (accessed 29 April 2019).
35. MassBank, European MassBank, https://massbank.eu (accessed 29 April 2019).
36. M. Wang, J. J. Carver, V. V. Phelan, L. M. Sanchez, N. Garg, Y. Peng, D. D. Nguyen, J. Watrous, C. A. Kapono, T. Luzzatto-Knaan, C. Porto, A. Bouslimani, A. V. Melnik, M. J. Meehan, W.-T. Liu, M. Crüsemann, P. D. Boudreau, E. Esquenazi, M. Sandoval-Calderón, R. D. Kersten, L. A. Pace, R. A. Quinn, K. R. Duncan, C.-C. Hsu, D. J. Floros, R. G. Gavilan, K. Kleigrewe, T. Northen, R. J. Dutton, D. Parrot, E. E. Carlson, B. Aigle, C. F. Michelsen, L. Jelsbak, C. Sohlenkamp, P. Pevzner, A. Edlund, J. McLean, J. Piel, B. T. Murphy, L. Gerwick, C.-C. Liaw, Y.-L. Yang, H.-U. Humpf, M. Maansson, R. A. Keyzers, A. C. Sims, A. R. Johnson, A. M. Sidebottom, B. E. Sedio, A. Klitgaard, C. B. Larson, C. A. B. P, D. Torres-Mendoza, D. J. Gonzalez, D. B. Silva, L. M. Marques, D. P. Demarque, E. Pociute, E. C. O'Neill, E. Briand, E. J. N. Helfrich, E. A. Granatosky, E. Glukhov, F. Ryffel, H. Houson, H. Mohimani, J. J. Kharbush, Y. Zeng, J. A. Vorholt, K. L. Kurita, P. Charusanti, K. L. McPhail, K. F. Nielsen, L. Vuong, M. Elfeki, M. F. Traxler, N. Engene, N. Koyama, O. B. Vining, R. Baric, R. R. Silva, S. J. Mascuch, S. Tomasi, S. Jenkins, V. Macherla, T. Hoffman, V. Agarwal, P. G. Williams, J. Dai, R. Neupane, J. Gurr, A. M. C. Rodríguez, A. Lamsa, C. Zhang, K. Dorrestein, B. M. Duggan, J. Almaliti, P.-M. Allard, P. Phapale, L.-F. Nothias, T. Alexandrov, M. Litaudon, J.-L. Wolfender, J. E. Kyle, T. O. Metz, T. Peryea, D.-T. Nguyen, D. VanLeer, P. Shinn, A. Jadhav, R. Müller, K. M. Waters, W. Shi, X. Liu, L. Zhang, R. Knight, P. R. Jensen, B. O. Palsson, K. Pogliano, R. G. Linington, M. Gutiérrez, N. P. Lopes, W. H. Gerwick, B. S. Moore, P. C. Dorrestein and N. Bandeira, *Nat. Biotechnol.*, 2016, **34**, 828–837.

37. C. Guijas, J. Rafael Montenegro-Burke, X. Domingo-Almenara, A. Palermo, B. Warth, G. Hermann, G. Koellensperger, T. Huan, W. Uritboonthai, A. E. Aisporna, D. W. Wolan, M. E. Spilker, H. Paul Benton and G. Siuzdak, *Anal. Chem.*, 2018, **90**, 3156–3164.
38. mzCloud – Advanced Mass Spectral Database, https://www.mzcloud.org (accessed 29 April 2019).
39. M. Vinaixa, E. L. Schymanski, S. Neumann, M. Navarro, R. M. Salek and O. Yanes, *TrAC, Trends Anal. Chem.*, 2016, **78**, 23–35.
40. Y. Nakamura, G. Cochrane, I. Karsch-Mizrachi on behalf of the International Nucleotide Sequence Database Collaboration, *Nucleic Acids Res.*, 2012, **41**, D21–D24.
41. FBMN Workflow – GNPS Documentation, https://ccms-ucsd.github.io/GNPSDocumentation/featurebasedmolecularnetworking/ (accessed 16 May 2019).
42. R. R. da Silva, M. Wang, L.-F. Nothias, J. J. J. van der Hooft, A. M. Caraballo-Rodríguez, E. Fox, M. J. Balunas, J. L. Klassen, N. P. Lopes and P. C. Dorrestein, *PLoS Comput. Biol.*, 2018, **14**, e1006089.
43. *GNPS/MZmine – Feature-Based Molecular Networking*, Youtube.
44. M. R. Wenk, *Nat. Rev. Drug Discovery*, 2005, **4**, 594–610.
45. C. Vosse, C. Wienken, C. Cadenas and H. Hayen, *J. Chromatogr. A*, 2018, **1565**, 105–113.
46. A. T. Maccarone, J. Duldig, T. W. Mitchell, S. J. Blanksby, E. Duchoslav and J. L. Campbell, *J. Lipid Res.*, 2014, **55**, 1668–1677.
47. M. C. Thomas, T. W. Mitchell, D. G. Harman, J. M. Deeley, J. R. Nealon and S. J. Blanksby, *Anal. Chem.*, 2008, **80**, 303–311.
48. X. Ma and Y. Xia, *Angew. Chem., Int. Ed.*, 2014, **53**, 2592–2596.
49. G. Liebisch, J. A. Vizcaíno, H. Köfeler, M. Trötzmüller, W. J. Griffiths, G. Schmitz, F. Spener and M. J. O. Wakelam, *J. Lipid Res.*, 2013, **54**, 1523–1530.
50. A. Korf, C. Vosse, R. Schmid, P. O. Helmer, V. Jeck and H. Hayen, *Rapid Commun. Mass Spectrom.*, 2018, **32**, 981–991.
51. E. Kendrick, *Anal. Chem.*, 1963, **35**, 2146–2154.
52. A. Korf, V. Jeck, R. Schmid, P. O. Helmer and H. Hayen, *Anal. Chem.*, 2019, **91**, 5098–5105.
53. V. Jeck, A. Korf, C. Vosse and H. Hayen, *Rapid Commun. Mass Spectrom.*, 2019, **33**, 86–94.
54. MZmine 2, https://mzmine.github.io (accessed 12 May 2019).
55. A. Römpp, T. Schramm, A. Hester, I. Klinkert, J.-P. Both, R. M. A. Heeren, M. Stöckli and B. Spengler, *Methods Mol. Biol.*, 2011, **696**, 205–224.
56. N. Hoffmann, J. Rein, T. Sachsenberg, J. Hartler, K. Haug, G. Mayer, O. Alka, S. Dayalan, J. T. M. Pearce, P. Rocca-Serra, D. Qi, M. Eisenacher, Y. Perez-Riverol, J. A. Vizcaíno, R. M. Salek, S. Neumann and A. R. Jones, *Anal. Chem.*, 2019, **91**, 3302–3310.
57. M. Kanehisa, M. Furumichi, M. Tanabe, Y. Sato and K. Morishima, *Nucleic Acids Res.*, 2017, **45**, D353–D361.
58. S. Kim, J. Chen, T. Cheng, A. Gindulyte, J. He, S. He, Q. Li, B. A. Shoemaker, P. A. Thiessen, B. Yu, L. Zaslavsky, J. Zhang and E. E. Bolton, *Nucleic Acids Res.*, 2019, **47**, D1102–D1109.

59. D. S. Wishart, Y. D. Feunang, A. Marcu, A. C. Guo, K. Liang, R. Vázquez-Fresno, T. Sajed, D. Johnson, C. Li, N. Karu, Z. Sayeeda, E. Lo, N. Assempour, M. Berjanskii, S. Singhal, D. Arndt, Y. Liang, H. Badran, J. Grant, A. Serra-Cayuela, Y. Liu, R. Mandal, V. Neveu, A. Pon, C. Knox, M. Wilson, C. Manach and A. Scalbert, *Nucleic Acids Res.*, 2018, **46**, D608–D617.
60. M. Ramirez-Gaona, A. Marcu, A. Pon, A. C. Guo, T. Sajed, N. A. Wishart, N. Karu, Y. Djoumbou Feunang, D. Arndt and D. S. Wishart, *Nucleic Acids Res.*, 2017, **45**, D440–D445.
61. M. Sud, E. Fahy, D. Cotter, A. Brown, E. A. Dennis, C. K. Glass, A. H. Merrill Jr, R. C. Murphy, C. R. H. Raetz, D. W. Russell and S. Subramaniam, *Nucleic Acids Res.*, 2007, **35**, D527–D532.
62. MassBank, MassBank | European MassBank (NORMAN MassBank) Mass Spectral DataBase, https://massbank.eu/MassBank/ (accessed 7 May 2019).
63. H. E. Pence and A. Williams, *J. Chem. Educ.*, 2010, **87**, 1123–1124.
64. R. Caspi, R. Billington, C. A. Fulcher, I. M. Keseler, A. Kothari, M. Krummenacker, M. Latendresse, P. E. Midford, Q. Ong, W. K. Ong, S. Paley, P. Subhraveti and P. D. Karp, *Nucleic Acids Res.*, 2018, **46**, D633–D639.

CHAPTER 8

# *Pre-processing and Analysis of Metabolomics Data with XCMS/R and XCMS Online*†

LAILA PAMELA PARTIDA-MARTÍNEZ[a] AND ROBERT WINKLER*[b,c]

[a]Department of Genetic Engineering, Center for Research and Advanced Studies (CINVESTAV) Irapuato, Km. 9.6 Libramiento Norte Carr. Irapuato-León, 36824 Irapuato, Guanajuato, Mexico; [b]Department of Biochemistry and Biotechnology, Center for Research and Advanced Studies (CINVESTAV) Irapuato, Km. 9.6 Libramiento Norte Carr. Irapuato-León, 36824 Irapuato, Gto., Mexico; [c]Mass Spectrometry Group, Max Planck Institute for Chemical Ecology, Hans-Knöll-Straße 8, 07745 Jena, Germany
*E-mail: robert.winkler@cinvestav.mx

## 8.1 Introduction

XCMS was originally developed to process liquid-chromatography/mass spectrometry (LC-MS) data in metabolomic studies with the R statistics and graphics language (http://www.r-project.org).[1–3]

†To facilitate the training of new metabolomics users, the example data and the XCMS/R script are freely available from http://doi.org/10.5281/zenodo.2635400

New Developments in Mass Spectrometry No. 8
Processing Metabolomics and Proteomics Data with Open Software: A Practical Guide
Edited by Robert Winkler

Published by the Royal Society of Chemistry, www.rsc.org

The main focus of XCMS is the processing of raw data and statistical analysis, but it also has built connections to the METLIN database of the Scripps Institute (http://metlin.scripps.edu), which provides reference spectra for numerous metabolites.[4,5] In addition, XCMS2 enables mass fragmentation (MS/MS) data to be matched against the METLIN reference library.[6]

The first releases of XCMS focused on comparing two sample groups, such as a diseased *versus* a control group. Subsequently, metaXCMS allowed multiple untargeted metabolics datasets to be compared.[7,8]

Data processing with the XCMS/R package requires some manual programming, and XCMS Online (https://xcmsonline.scripps.edu/) was developed to facilitate the use of XCMS by casual users.[9–12] This platform also provides 'cloud plots' for visualizing metabolic differences.[13]

Both versions (XCMS/R and XCMS Online) are open source and free to use. However, for professional environments, SCIEX also offers a commercial solution, XCMS$^{plus}$ (https://sciex.com/products/software/xcms-plus-software).

In this section, we describe the processing of an LC-MS metabolomics dataset using XCMS/R and XCMS Online.

## 8.2 Example Project: A Biological Background and Analytical Question

*Arabidopsis thaliana* is an important model in plant sciences. This small plant completes its life cycle in only six weeks.[14] *Arabidopsis* was also the first plant to have its full genome published,[15] which stimulated numerous genetic studies.

Investigating naturally occurring variants of *Arabidopsis thaliana* is a forward genetics approach,[16] and wild *Arabidopsis* plants have been collected from different locations and were initially referred to as 'ecotypes';[17] the term 'accessions' is now more commonly used in *Arabidopsis* genetics.[18]

Two frequently studied accessions are 'Col-0', which originates from Columbia (USA), and 'Ws-3' from Wassilewskija (Russia) (https://www.arabidopsis.org/). Floral tissues from these two accessions were subjected to metabolic fingerprinting using liquid chromatography coupled with high-resolution mass spectrometry (LC-MS).[19]

Based on the data generated in this study, we wanted to test whether or not the inflorescences of the two *Arabidopsis* accessions (Col and Ws) exhibit distinct metabolic phenotypes.

## 8.3 Raw Data Preparation and Preview

### 8.3.1 Data and Code Availability

Herein, we evaluate the published LC-MS metabolomic-fingerprinting data of the Col-0 and Ws-3 accessions of *Arabidopsis* using XCMS/R and XCMS Online.

The pre-processed raw .mzML data and an R script `xcms_mzml.R` for processing these data using XCMS/R have been deposited at Zenodo: Robert Winkler (2019). RSC book example dataset: Metabolomics data and R script for XCMS/R preprocessing [Data set]. Zenodo. http://doi.org/10.5281/zenodo.2635400.

## 8.3.2 Converting Raw Files in Vendor Format

The complete methodology used herein has been described previously.[19] High-resolution LC-MS data were acquired on a UPLC-Q-ToF instrument (Waters Corps., Mexico), after which the data were converted into .mzML format in centroided mode using ProteoWizard.[20]

TRaw data preparation is crucial for any subsequent processing. Typical steps include the conversion of binary vendor files into community formats, peak-picking (centroiding), and the application of filters (noise filter, extraction of relevant mass/retention-time ranges). The ProteoWizard project (http://proteowizard.sourceforge.net/) provides different tools and interfaces for these operations.[20] For Windows, a graphical user interface (GUI) for file conversion and filtering is provided, in addition to a command-line version `msconvert`. Most of the ProteoWizard programs and libraries, such as the extremely versatile `msconvert` command, are cross-platform compatible. However, some functions, such as the direct conversion from binary formats, require vendor licenses that are only available on Windows. Thus, the raw files are ideally converted directly after acquisition on the MS conversion of msconvert can be used on supported platforms (Windows, MacOS, Linux). For example, the conversion of raw data from a Linux platform is possible by:

```
docker run -it --privileged=true -v /home/user/DATA/xcms_data/:/data
chambm/pwiz-skyline-i-agree-to-the-vendor-licenses wine msconvert
*.raw
--filter "scanTime [30,1500]"
```

The files can be compressed using the zlib (https://www.zlib.net/) algorithm by setting the msconvert `-z` flag. Furthermore, the files can be packed with GNU Gzip (https://www.gnu.org/software/gzip/), adding the `-g` flag. These compression protocols considerably reduce file size, but may lead to incompatibility issues. A substantial reduction in file size (one-to-two orders of magnitude) is already possible by converting profile data to noise-filtered centroided data. High-resolution LC-MS metabolomics profile data files are usually relatively large (*e.g.*, 1 Gb/sample). Therefore, peak picking/centroiding greatly facilitates downstream data handling and processing. XCMS accepts centroid data, which, therefore, should be used by default. The original profile data files can be retained as backups, for example, if different data processing software requires profile information. The LC-MS mzML data deposited at Zenodo are already prepared for XCMS (mzML, centroid, noise, and scan time filtered).

## 8.3.3 mzML File Preview

Before commencing MS data processing, the experimental data in the representative raw data files should be visualized, for example with MZmine 2 (http://mzmine.github.io/). Important criteria used to judge the quality of data are noise level, chromatographic peak width and separation, variations in retention times and mass-to-charge (*m*/*z*) values between runs, and the signal intensity. Data quality knowledge is indispensable when optimizing parameters during data processing.

## 8.4 XCMS/R

### 8.4.1 Directory Structure of Data

The example script uses the following directory structure, in which the LC-MS data of different samples are in subdirectories of `/mzML` and the R script `xcms_mzml.R` are the directory `xcms_R`:

```
/mzML/Col
/mzML/Ws
xcms_R
```

### 8.4.2 RStudio Editor for R

To facilitate the edition of R and the running of R scripts, the use of an integrated development environment (IDE) is recommended. In this tutorial, we use RStudio (https://www.rstudio.com/), which is available for different platforms (Linux, Mac OS, and Windows).

### 8.4.3 Installing the R Packages

The current version of XCMS is installed in the R/RStudio console by:

```
if (!requireNamespace("BiocManager", quietly = TRUE))
    install.packages("BiocManager")
BiocManager::install("xcms")
```

Other R packages that are required by the example script can be installed by:

```
install.packages(c('RColorBrewer', 'pander', 'magrittr', 'gplots',
                   'cluster', 'factoextra'))
```

### 8.4.4 Loading and Running the R Script

Please load the R script file `xcms_mzml.R` in the RStudio IDE. To save the output of the script (figures and tables) to the same directory, choose: Session – Set Working Directory – To Source File Location. This example is based on the documentation provided on the R package for 'LCMS data pre-processing and analysis with xcms' by Johannes Rainer (https://bioconductor.org/packages/release/bioc/vignettes/xcms/inst/doc/xcms.html, 4 January 2019).

To run the complete script, which is displayed in the editor, choose 'Source'. For running single lines or sections of code, the required region can be selected and executed using 'Run' (or the 'Shift-Enter' keys).

The R script is explained in the following sections; further information can be found in the form of comments in the script itself.

### 8.4.5 Loading Required R Libraries

R libraries provide additional functions. Depending on the purpose of a script, less or more libraries may be included.

```
library(xcms)
library(RColorBrewer)
library(pander)
library(magrittr)
library(gplots)
library(cluster)
library(factoextra)
```

### 8.4.6 Reading and Annotating Raw Data

The script has to be provided with the data localized. Furthermore, the individual files are annotated within a group. The first six files belong to the group 'Col-0', the next six files to 'Ws-3'.

```
msdataDir <- "../mzML"
msdataFiles <- list.files(msdataDir, pattern = "*.mzML", full.names
= TRUE, recursive = TRUE)
pd <- data.frame(sample_name = sub(basename(msdataFiles), pattern =
".mzML", replacement = "", fixed = TRUE),
                 sample_group = c(rep("Col-0", 6), rep("Ws-3", 6)),
                 stringsAsFactors = FALSE)
```

The correct correlation between samples and groups can be verified by:

```
> print(pd)
   sample_name sample_group
1        Col_1        Col-0
2        Col_2        Col-0
3        Col_3        Col-0
4        Col_4        Col-0
5        Col_5        Col-0
6        Col_6        Col-0
7         Ws_1         Ws-3
8         Ws_2         Ws-3
9         Ws_3         Ws-3
10        Ws_4         Ws-3
11        Ws_5         Ws-3
12        Ws_6         Ws-3
```

### 8.4.7 Defining Colours

For visualization purposes, the same colours should be always used and named according to their sample group:

```
group_colors<- paste0(brewer.pal(3, "Dark2")[1:2])
names(group_colors) <- c("Col-0", "Ws-3")
```

### 8.4.8 Reading the Raw Data

The raw data can now be read. onDisk mode is memory efficient because it only imports general information from a file and reads the full data from the disk when necessary:

```
raw_data<- readMSData(files = msdataFiles, pdata =
new("NAnnotatedDataFrame", pd),
                     mode = "onDisk")
```

### 8.4.9 Plotting Base Peak Chromatograms

To provide a general overview of the chromatographic runs, the maximum intensity of each spectrum can be extracted, and all base peak chromatograms (BPC) can be plotted in a single pdf. The plot lines are coloured according to their sample group:

```
bpis <- chromatogram(raw_data, aggregationFun = "max")
pdf("1_all_base_peak_chromatograms.pdf")
plot(bpis, col = group_colors[raw_data$sample_group])
legend("topright", inset = .02,lty=c(1,1),legend=names(group_colors),
       col = group_colors)
dev.off()
```

### 8.4.10 Total Ion Current Box Plot

The total ion current (TIC) for all samples in a box plot can also be provided:

```
tc <- split(tic(raw_data), f = fromFile(raw_data))
pdf("2_all_TICs_boxplots.pdf")
boxplot(tc, col = group_colors[raw_data$sample_group],
        ylab = "intensity", main = "Total ion current")
legend("topleft", inset = .02, legend=names(group_colors), fill =
group_colors)
dev.off()
```

### 8.4.11 Test a Single Feature

A feature is characterized by a retention time and a mass-to-charge ratio ($m/z$). To evaluate the consistency of data between runs and to determine reasonable tolerances, a feature present in all samples is plotted. As an example, MZmine (http://mzmine.github.io/) can be used to visually inspect mzML files and to determine suitable features.

```
## Define the rt and m/z range of the peak area
rtr <- c(400, 600)
mzr <- c(735, 737)

## extract the chromatogram for that feature region
chr_raw <- chromatogram(raw_data, mz = mzr, rt = rtr)
pdf("3_plot_test_feature.pdf")
plot(chr_raw, col = group_colors[chr_raw$sample_group])
legend("topright", inset = .02,lty=c(1,1),legend=names(group_colors),
      col = group_colors)
dev.off()

## plot extracted ion chromatograms
## WORKS ONLY FOR FEW SAMPLES E.G. 12 - OTHERWISE THE PDF GETS TOO
LARGE
pdf("4_extracted_ion_chromatograms_test_feature.pdf")
raw_data %>%
  filterRt(rt = rtr) %>%
  filterMz(mz = mzr) %>%
  plot(type = "XIC")
dev.off()
```

## 8.4.12 Feature Detection

We can now search for the features in all LC-MS datasets. The **parameter choice is highly critical for further interpretation.** To provide information on possible settings, we can use the `?CentWaveParam` command on the R console; the feature detection parameters need to be optimized and depend on data quality and the evaluation results below. To begin, we can use settings that are typical for high resolution LC-MS experiments and congruent with our initial data revision.

```
pickingParam <- CentWaveParam(ppm = 15, peakwidth = c(5, 20),
integrate = 1, fitgauss = TRUE, snthresh = 10, mzdiff = -0.001)
xdata<- findChromPeaks(raw_data, param = pickingParam)
```

In the next steps, we use `xdata`, which is an `XCMSnExp` object. To evaluate our detected features, we create a summary statistics table:

```
summary_fun <- function(z) {
  c(peak_count = nrow(z), rt = quantile(z[, "rtmax"] - z[,"rtmin"]))
}
T <- lapply(split.data.frame(chromPeaks(xdata),
            f = chromPeaks(xdata)[, "sample"]),
            FUN = summary_fun)
T <- do.call(rbind, T)
rownames(T) <- basename(fileNames(xdata))
pandoc.table(T,
    caption = paste0("Summary statistics on identified
chromatographic", " peaks. Shown are number of identified peaks per",
                    " sample and widths/duration of chromatographic ",
                    "peaks."))
```

The R console presents a pandoc table with information on the detected features:

```
-------------------------------------------------------------------------------
               peak_count   rt.0%   rt.25%   rt.50%   rt.75%   rt.100%
----------------- ------------ ------- -------- -------- -------- ---------
**Col_1.mzML**        1976      2.028     10.4    14.46    18.74     66.36
**Col_2.mzML**        2153      2.016     10.4    14.47    18.74     81.07
**Col_3.mzML**        2196      2.02      10.4    14.69    18.74     74.67
**Col_4.mzML**        2163      2.021     10.4    14.45    18.73      133
**Col_5.mzML**        2143      2.017     10.4    14.44    18.73     97.75
**Col_6.mzML**        2149      2.018    10.39    14.46    18.74     132.9
**Ws_1.mzML**         2610      2.018    10.39    14.47    18.76     72.77
**Ws_2.mzML**         2670      2.021    10.39    14.69    18.76     83.14
**Ws_3.mzML**         2486      2.019     10.4    14.69    18.76     93.54
**Ws_4.mzML**         2426      2.019    10.39    14.46    18.76     87.4
**Ws_5.mzML**         2294      2.018    10.39    14.69    18.76     83.06
**Ws_6.mzML**         2353      2.017    10.39    14.44    18.75     91.43
-------------------------------------------------------------------------------
```

Between 1976 and 2670 features were detected in the various samples, which is reasonable for a plant-tissue extract. The retention time windows of 75% of the features are less than ~19 seconds, which is consistent with observations made through manual data inspection. Therefore, the chosen parameters for feature detection are acceptable.

### 8.4.13 Retention Time Correction

Before we bin and compare features, possible shifts and deviations in retention times must be corrected for. These variations are caused by changes in solvent composition and temperature, among others, during chromatography. To avoid artefacts during data analysis, LC-MS should be conducted in random sample order.

Different approaches can be used to align chromatograms. In the R script example, two options are provided, of which one needs to be chosen. In the following, we chose the Orbiwarp algorithm:

```
xdata <- adjustRtime(xdata, param = ObiwarpParam(binSize = 0.6))
```

The binSize refers to *m*/*z* bins. Depending on the plot produced, the value of this parameter can be optimized. To evaluate the adjustment, the aligned BPCs and retention time corrections can be plotted:

```
bpis_adj <- chromatogram(xdata, aggregationFun = "max")

pdf("5_retention_time_correction.pdf")
par(mfrow = c(2, 1), mar = c(4.5, 4.2, 1, 0.5))
plot(bpis_adj, col = group_colors[bpis_adj$sample_group])

## Plot also the difference of adjusted to raw retention time.
plotAdjustedRtime(xdata, col = group_colors[xdata$sample_group])
dev.off()
```

To verify that the `xdata` object contains the corrected retention times we use:

```
hasAdjustedRtime(xdata)
```

### 8.4.14 Grouping/Binning Features

For statistical analysis, group features of different samples, which most likely represent the same compound, need to be grouped. To determine the values of suitable parameters, we define a test region that contains various features ($m/z$ range: 300–330; retention time: 800–1000 s):

```
mzr <- c(300, 330)
rtt <- c(800, 1000)
```

The data in this region and the detected features can now be plotted using:

```
chr_mzr <- chromatogram(xdata, mz = mzr, rt = rtt)

pdf("6_test_peak_picking_and_binning.pdf")
par(mfrow = c(3, 1), mar = c(1, 4, 1, 0.5))
cols <- group_colors[chr_mzr$sample_group]
plot(chr_mzr, col = cols, xaxt = "n", xlab = "")
legend("topright", inset = .02,lty=c(1,1),legend=names(group_colors),
      col = group_colors)
highlightChromPeaks(xdata, mz = mzr, col = cols, type = "point", pch
= 16)
```

Note: the pdf is not yet closed, since two more plots will be saved in it. We define the first set of parameters for grouping by the peak density method and plot a second chromatogram:

```
pdp <- PeakDensityParam(sampleGroups = xdata$sample_group,
                        minFraction = 0.2, bw = 2)
par(mar = c(4, 4, 1, 0.5))
plotChromPeakDensity(xdata, mz = mzr, col = cols, param = pdp,
                     pch = 16, xlim = rtt)
```

The bandwidth parameter (`bw`) is now changed from `2` to `0.5`, the result plotted, and the pdf closed:

```
pdp <- PeakDensityParam(sampleGroups = xdata$sample_group,
                        minFraction = 0.2, bw = 0.5)
plotChromPeakDensity(xdata, mz = mzr, col = cols, param = pdp,
                     pch = 16, xlim = rtt)
dev.off()
```

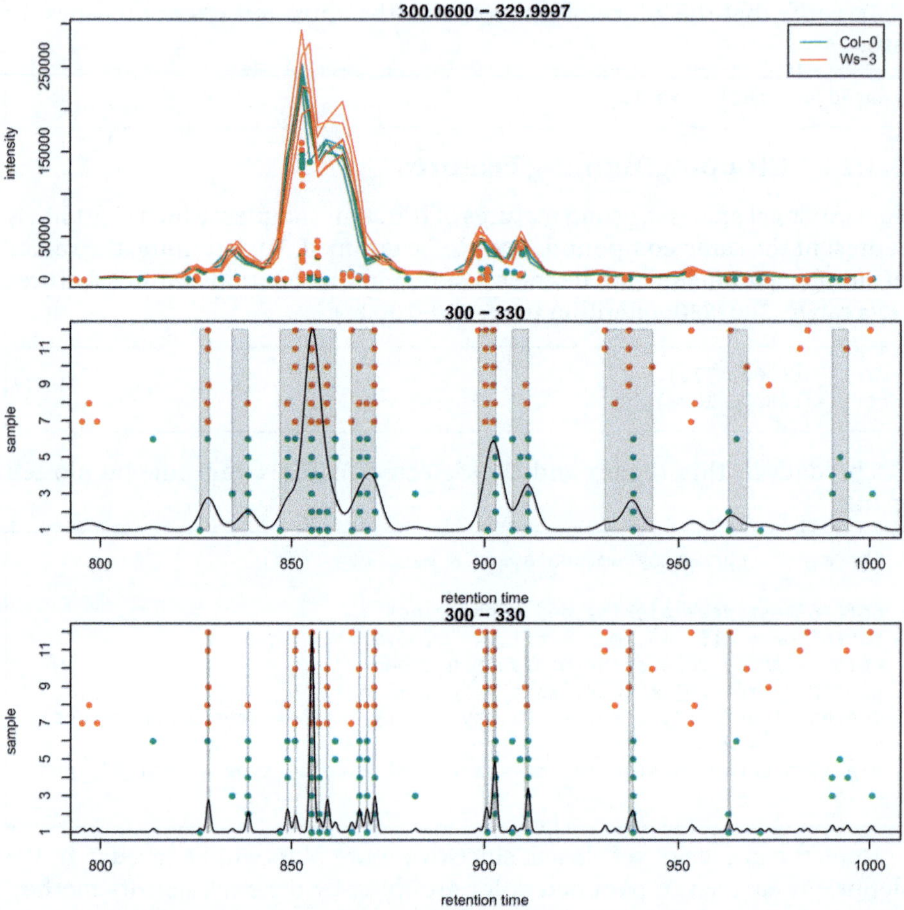


**Figure 8.1** Evaluating retention time alignment, feature detection, and grouping.

Figure 8.1 reveals that a bandwidth parameter value of `0.5` provides the correct grouping of features; this parameter value is then applied to the complete `XCMSnExp` object `xdata`:

```
pdp <- PeakDensityParam(sampleGroups = xdata$sample_group,
                        minFraction = 0.2, bw = 0.5)
xdata <- groupChromPeaks(xdata, param = pdp)
```

The feature definitions and their values can be reviewed by:

```
print(featureDefinitions(xdata))
print(head(featureValues(xdata, value = "into")))
```

### 8.4.15 Filling Data for Missing Peaks

The expected features for some samples might be missing; they can be filled with the measured values in the raw data:

```
xdata <- fillChromPeaks(xdata)
```

Successful execution can be verified by:

```
print(head(featureValues(xdata, value = "into")))
```

The following commands can be used to compare the number of missing values before and after peak filling:

```
apply(featureValues(xdata, filled = FALSE), MARGIN = 2, FUN =
function(z)
      sum(is.na(z)))
apply(featureValues(xdata), MARGIN = 2, FUN = function(z)
sum(is.na(z)))
```

### 8.4.16 Creating a Feature Summary

To create a matrix that contains the feature definitions as well as the summary information for each samples, we perform:

```
featDefframe <- featureDefinitions(xdata)
featDef <- as.matrix(featDefframe[,1:6])
featSum <- featureSummary(xdata)
featData <-  cbind(featDef, featSum)
print(head(featData))
```

### 8.4.17 Histogram of Features

As a first statistical evaluation, we plot a histogram that shows the occurrence of features in the various samples:

```
pdf("7_histogram_of_features.pdf")
hist(featSum[,2], main = "summary of features",
      xlab = "occurrence of feature in samples [%]",col="darkgreen")
dev.off()
```

The histogram reveals that more than 2000 of the 4267 features are only found in up to 25% of the samples. Approximately 300 features are found in at least 80% of the samples; these are referred to as 'high-quality' (HQ) features. However, compounds that only occur in one sample group, such as a disease biomarker, are lost if we limit our feature matrix to only such HQ features. Therefore, different subsets are created for these results.

### 8.4.18 Sub-setting and Exporting Features

First, we combine all feature information and export the complete matrix as a comma-separated value (CSV) file. Such files can be read as tables in spreadsheet programs (EXCEL, LibreOffice), or imported into databases (SQL, SQLite):

```
featVal <- featureValues(xdata, value = "into", filled = TRUE)
featAll <- cbind(featData, featVal)
write.csv(featAll, file = "xcms_all_features.csv")
```

As a first subset, we only consider features that occur in more than 80% of the samples:

```
featSubSet80 <- featAll[featAll[,8]>80,]
write.csv(featSubSet80, file = "xcms_subset_80.csv")
```

To take into account features that are only present in a particular group, we also generate a subset of features that are detected in more than 40% of the samples:

```
featSubSet40 <- featAll[featAll[,8]>40,]
write.csv(featSubSet40, file = "xcms_subset_40.csv")
```

For further statistical analysis, we also create a data matrix of samples and their values:

```
featSubValues <- featSubSet80[,12:23]
```

### 8.4.19 Principal Component Analysis (PCA)

Principal component analysis (PCA) was used as a first unsupervized statistical evaluation method. Features with missing values (NA) need to be removed from the dataset:

```
## featurs with NA are filtered out
featSubValues <- na.omit(featSubValues)
```

To include only the 100 strongest signals, we sort the data by feature intensity, taking one of the samples as reference (Col_4.mzML in this example):

```
   heatmapdata <- featSubValues[order(featSubValues[,'Col_4.mzML'],
                                     decreasing = TRUE),]
   #heatmapdata <- featSubValues
```

PCA and result plotting is performed by:

```
pc <- prcomp(t(heatmapdata[1:100,]))

pdf("8_PCA.pdf")
cols <- group_colors[xdata$sample_group]
pcSummary <- summary(pc)
plot(pc$x[,1], pc$x[,2], pch = 21, main = "",
     xlab = paste0("PC1: ", format(pcSummary$importance[2, 1] * 100,
     digits = 3), " % variance"),
     ylab = paste0("PC2: ", format(pcSummary$importance[2, 2] * 100,
     digits = 3), " % variance"),col = "darkgrey", bg = cols, cex = 2)
grid()
text(pc$x[, 1], pc$x[,2], labels = xdata$sample_name, col =
"darkgrey",
    pos = 3, cex = 1)
legend("topright", inset = .02, legend=names(group_colors), fill =
group_colors)
dev.off()
```

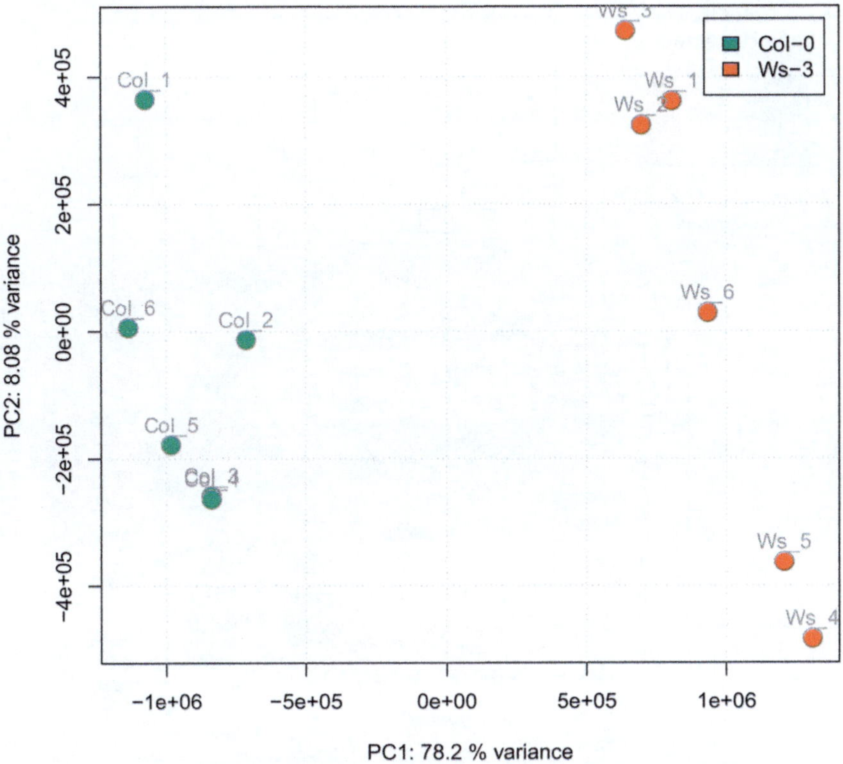


**Figure 8.2** Principal component analysis (PCA) of the 100 highest-intensity features separates the two sample classes.

The resulting PCA plot (Figure 8.2) indicates two clearly distinct metabolic phenotypes for the 'Col-0' and 'Ws-3' samples.

## 8.4.20 Hierarchical Cluster Analysis (HCA)

Hierarchical cluster analysis (HCA) is another important strategy for evaluating the metabolic relationships between samples:[21,22]

```
pdf("9_heatmap-top50.pdf")
heatmap.2(heatmapdata[1:50,], cexCol = 0.5, cexRow = 0.5,
          col = "bluered", scale = "row")
dev.off()
```

Figure 8.3 also reveals that the two groups are separated into two main branches, with the dendrograms indicating further relationships between samples or features.

## 8.4.21 PAM Clustering

The data are finally divided into a predefined number of *k* clusters, and for this task we partition the data into *k* clusters 'around medoids'. The PAM clustering approach is more robust than the *K*-means algorithm

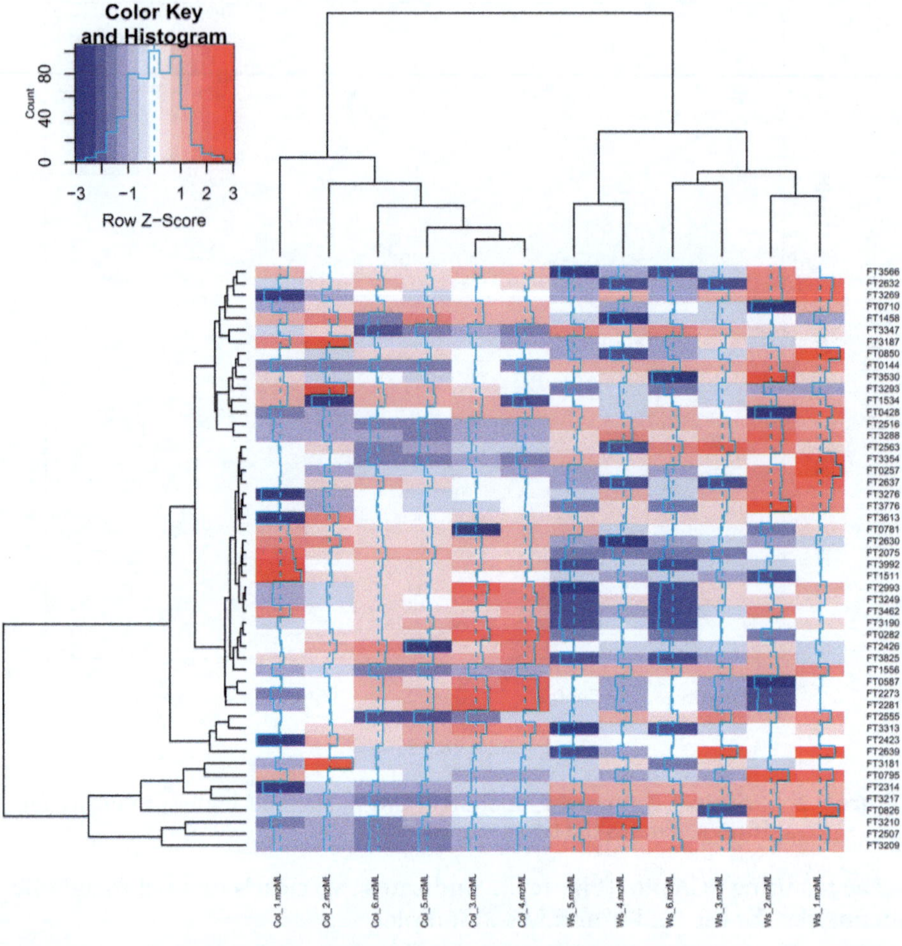


**Figure 8.3** Hierarchical cluster analysis (HCA) of the 50 highest-intensity features, which shows the two major sample branches, as well as the relationships between features.

(<https://stat.ethz.ch/R-manual/R-patched/library/cluster/html/pam.html>).[23] The 100 most prominent features are evaluated:

```
pam.res<- pam(t(heatmapdata), 2)
pdf("10_PAM_cluster.pdf")
fviz_cluster(pam.res)
dev.off()
```

PAM clustering corretly classified all samples (Figure 8.4), lending support to the defined metabolic identities of the two *Arabidopsis* accessions/ecotypes, 'Col-0' and 'Ws-3'.

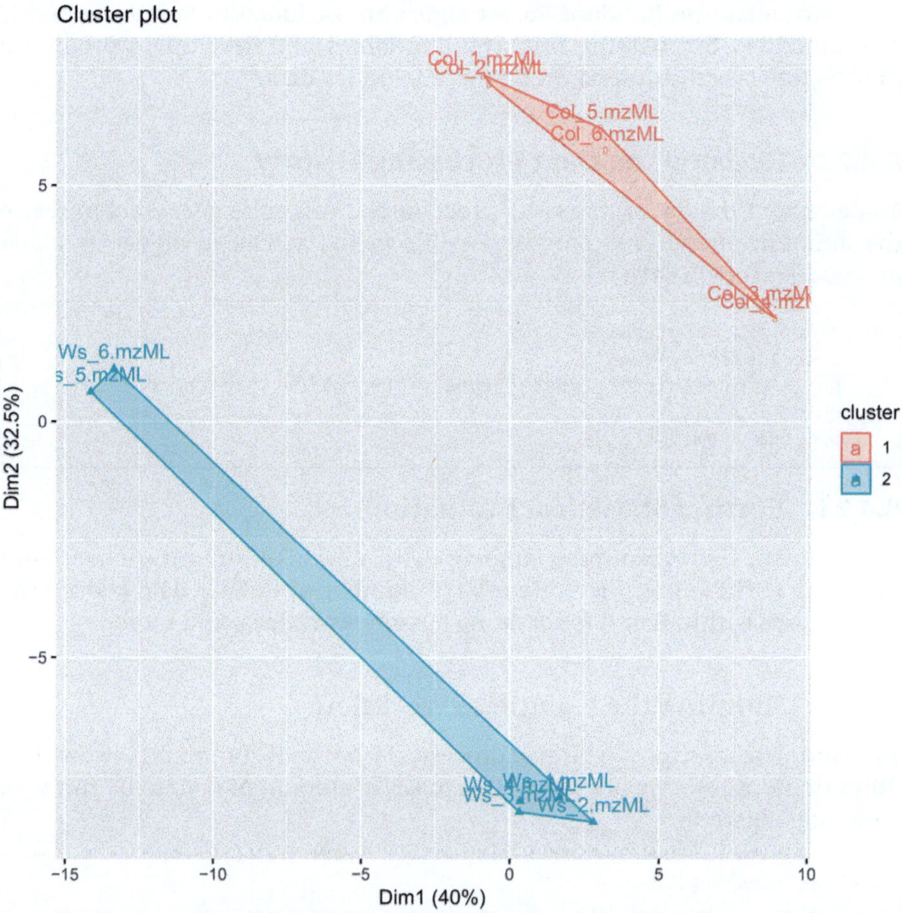


**Figure 8.4** PAM clustering, using the 100 most intense features, correctly predicts the sample groups on the basis of LC-MS metabolic data.

## 8.4.22 Additional Data Mining with Rattle

For more complex data mining operations, the `Rattle` R package provides a convenient GUI for multiple statistical operations and algorithms.[24–26] To create a data table suitable for importing into `Rattle`, features can be exported as follows:

```
featRattle <- featAll[,12:23]
featRattle <- t(featRattle)
featRattleExp <- cbind(pd$sample_group,featRattle)
write.csv(featRattleExp, file = "xcms_all_features_Rattle.csv")
```

In particular, the Random Forest algorithm included in `Rattle` is highly recommended for locating important variables and developing predictive models for classification based on metabolomics data.[26]

### 8.4.23 Revision the Data Processing History

To document the data processing procedure, it might be necessary to revise the different operations. This is possible in the `XCMSnExp` object by using `processHistory(xdata)`.

```
processHistory(xdata)
processHistoryTypes()
ph <- processHistory(xdata, type = "Retention time correction")
ph
processParam(ph[[1]])
```

### 8.4.24 Further Statistical Evaluation

R is a statistical programming language with a lively community of users and developers. Therefore, there are a large number of further data-processing possibilities within R and for creating custom workflows and tools.

### 8.4.25 Running the Complete Workflow

The complete R script can be also processed from an R command line, which, for example, allows a workflow with a massive amount of data to be run on a computer cluster:

```
source('xcms_mzml.R')
```

The example dataset requires about 10 minutes to process on a standard Microsoft Windows 10 desktop PC.

### 8.4.26 Metabolite Identification

Identifying metabolites that correspond to features of interest is outside of the scope of XCMS/R. However, putative identifications can be made with external tools such as MetaboQuest (http://www.omicscraft.com/MetaboQuest/). More metabolite identification tools are presented below in this chapter.

## 8.5 XCMS Online

XCMS Online (https://xcmsonline.scripps.edu/) is a platform that makes metabolomic analyses accessible to any user, irrespective of their expertise in programming and/or in statistical/analytical tools. XCMS Online has constantly expanded its functionality, from the processing and analysis of raw MS data to the integration of diverse omics data. The inclusion of genomic, transcriptomic, and proteomic data render it a useful tool for systems biology.[27] Until now, the use of XCMS Online has been free and supported by experts.

Herein, we use XCMS Online to determine putative differential metabolites and pathways in the 'Col' and 'Ws' accessions of *A. thaliana.*

### 8.5.1 Registration as a User

The first step towards using XCMS Online is to create an account. With a free basic-user account, one can upload and store up to 25.04 GB of data and to run one job at a time.

### 8.5.2 Dataset Specification

Prior to any metabolomic analysis, it is advisable to know in detail the characteristics of the **MS device** used to acquire data. Nineteen parameter sets for commercial instruments are available in version 3.7.1 of XCMS Online.

After choosing the appropriate instrument, the selected **parameter set** can be edited and changed. These adjustments can be made as the instrument for a job is selected, and are mainly related to the instrument (equipment, polarization mode), the run (statistics, database used for match identification, annotation, *etc.*), and the identification of features (deviations in ppm of peaks, signal/noise threshold, *etc.*). Importantly, the **reference organism** for which the metabolites and pathways are being identified should be specified here. Currently, more than 7600 organisms are available based on BioCyc (https://biocyc.org/), including *A. thaliana* (ARA). By default, XCMS Online will compare data against the human database.

### 8.5.3 Preparing Data for Uploading

The MS data files need to be transformed into a compatible format. Many formats are already compatible with the platform, including mzXML, mzDATA, .cdf NetCDF (AIA/ANDI), .d folders (Agilent; Bruker), and .wiff files (SCIEX), among others. It is not possible to upload .zip files. Compatible formats are listed at https://xcmsonline.scripps.edu/docs/fileformats.html.

### 8.5.4 Uploading Datasets

Files are uploaded according to sample group or treatment (see Figure 8.5). Datasets should be separately uploaded and assigned unique names; only these defined datasets can be subsequently compared. Uploading can take several minutes, depending on the size of the dataset and the speed of the internet connection.

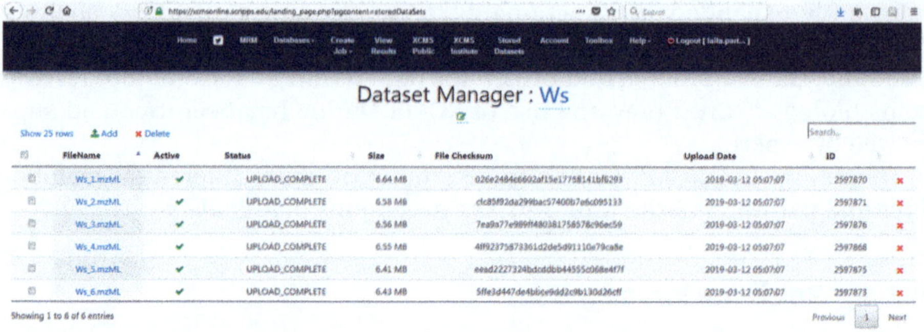


**Figure 8.5** Uploading an LC-MS dataset to XCMS Online.

### 8.5.5 Available Analysis Types

The analysis strategy used to compare multiple datasets can be chosen from the following options:

- *Single*: Annotates and identifies chemical features (unique *m*/*z* ions and retention times) in one dataset.
- *Pairwise*: Compares two datasets and identifies differential metabolites present in each group. Automatic predictive pathway analysis is only available for this type of comparison.
- *Meta XCMS*: Compares different perturbations to a single control. Pairwise comparisons of each perturbation *vs.* the control should be performed first in this analysis method.
- *Multigroup*: Compares large datasets with multiple conditions and/or time points, including at least one control.
- *Multi modal*: Compares multiple jobs.

### 8.5.6 Creating and Submitting a Job

To create and submit a job, 'Create job/Pairwise' is selected, and then the datasets to be compared, and the instrument parameters are set for analysis. In this example, 'HPLC/Waters TOF (3237)' is the most suitable instrument. XCMS Online confirms by email that the job has been submitted and when it is complete.

To create and submit the job, we select 'Create job/Pairwise', then the datasets to be compared, and the instrument parameter set for the analysis. In this example, 'HPLC/Waters TOF (3237)' represents best the instrument used by Sotelo-Silveira *et al.*[19] XCMS Online confirms by email that the job has been submitted and when it has been completed.

### 8.5.7 Visualizing the Results

The results can be visualized in the 'View Results' menu. The main results page displays the number of aligned features, analysis parameters, several graphics, and links to other analytical modules, as well as a button to

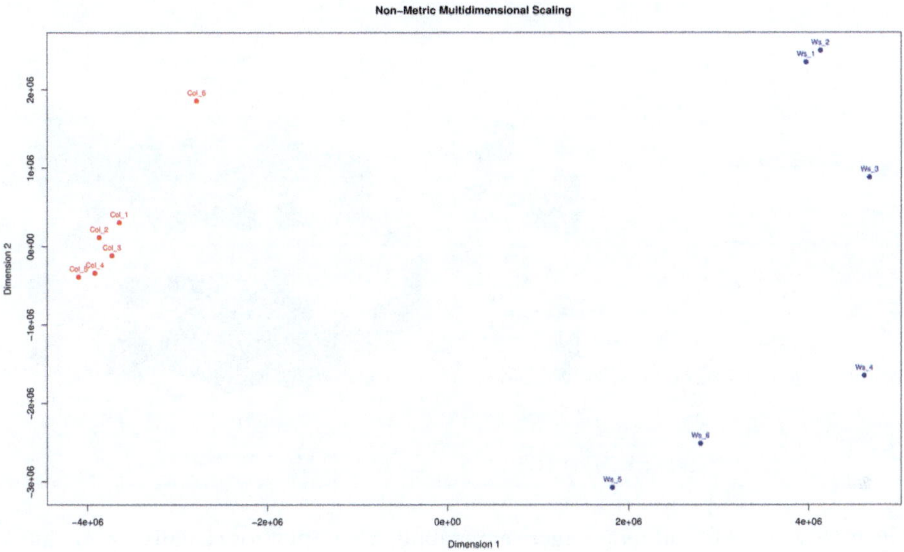


**Figure 8.6** Non-multi-dimensional scaling (NMDS) analysis after pairwise comparing the 'Col' and 'Ws' datasets in XCMS Online.

download a zip file with your results. Graphical comparisons across datasets are presented in several forms, namely non-multi-dimensional scaling (NMDS) analysis, principal component (PCA) analysis, metabolic cloud plot, and pathway cloud plot.

### 8.5.8 Non-multi-dimensional Scaling (NMDS) Analysis

The NMDS plot indicates significant distinct metabolic profiles for the inflorescences of the *A. thaliana* 'Col' and 'Ws' accessions (Figure 8.6).

### 8.5.9 Metabolic Cloud Plot

The interactive metabolic cloud plot (Figure 8.7) helps to visualize differential features/metabolites along the chromatogram, likely metabolite matches, and the statistical significance of the differences. In our example, 5146 of a total of 8661 features are found to be differential, which means that approximately 60% of the features exhibit a fold change of ≥ 1.5 and a *p*-value ≤ 0.01. The interactive metabolic cloud allows the desired fold change and/or *p*-value to be dynamically modified and to see the *m*/*z* ions that match the new criteria. Navigating the different bubbles informs about the feature's *m*/*z*, its retention time, fold change, *p*-value, and putative metabolite(s) (METLIN). The results are also listed in a table.

In this example, 787 features are found to have a fold change ≥ 30 and a *p*-value ≤ 0.001, as depicted in Figure 8.7. The high number of differential features across these two accessions demonstrates the power of high-resolution mass spectrometry for characterizing biological samples.

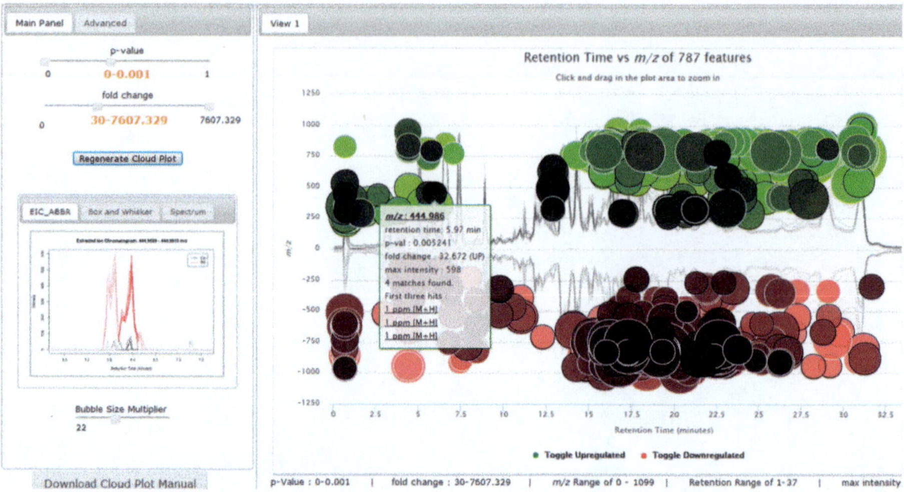


**Figure 8.7** XCMS Online interactive metabolic cloud plot for visualizing and analyzing differential features.

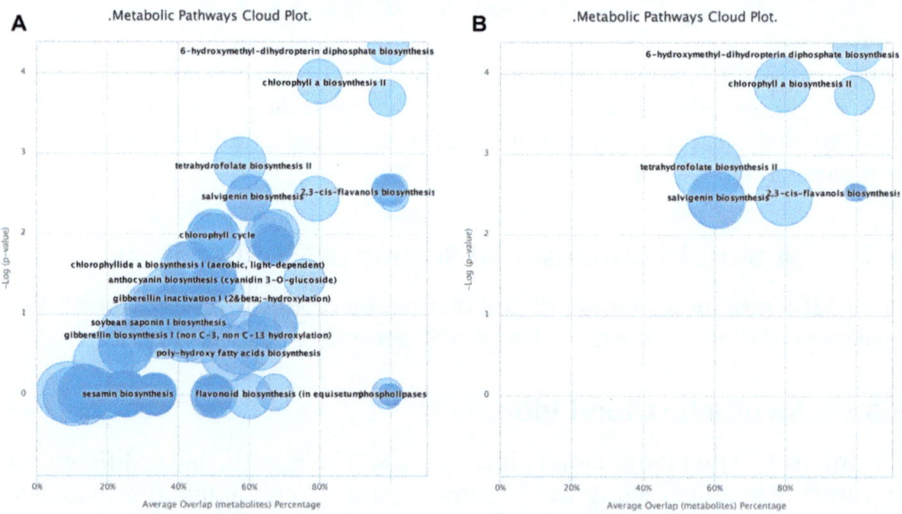


**Figure 8.8** (A) Metabolic pathways with representative metabolites in the datasets. (B) Metabolic pathways with differential metabolites in the datasets ($p$-value $\leq 0.01$).

## 8.5.10 Pathway Cloud Plot and Systems Biology Results

Interpreting the metabolic differences is not trivial. Hence, the platform first enables the *in situ* annotation of the represented and differential pathways, which can be graphically viewed using the 'pathway cloud plot' button. In this module, a $p$-value can be modified to match the pathway criteria. If the $p$-value is set to 1, all of the identified pathways with representative metabolites are shown (see Figure 8.8A). In our example, by using a $p$-value of 0.01, we can infer

| Pathway | Overlapping genes | Overlapping genes | All genes | Overlapping proteins | Overlapping proteins | All proteins | Overlapping putative metabolites1 | Overlapping putative metabolites1 | All metabolites2* | p-values |
|---|---|---|---|---|---|---|---|---|---|---|
| 6-hydroxymethyl-dihydropterin diphosphate biosynthesis I | 0 | 0% | 0 | 0 | 0% | 6 | 4 | 100.0% | 4 | 4.2e-5 |
| chlorophyll a biosynthesis II | 0 | 0% | 2 | 0 | 0% | 2 | 4 | 80.0% | 5 | 1.4e-4 |
| vitamin E biosynthesis (tocotrienols) | 0 | 0% | 0 | 0 | 0% | 4 | 3 | 100.0% | 3 | 1.9e-4 |
| tetrahydrofolate biosynthesis II | 0 | 0% | 0 | 0 | 0% | 23 | 4 | 57.1% | 7 | 1.5e-3 |
| 2,3-cis-flavanols biosynthesis | 0 | 0% | 0 | 0 | 0% | 1 | 2 | 100.0% | 2 | 3.1e-3 |
| ent-kaurene biosynthesis I | 0 | 0% | 0 | 0 | 0% | 3 | 2 | 100.0% | 2 | 3.1e-3 |
| chlorophyll a biosynthesis I | 0 | 0% | 0 | 0 | 0% | 1 | 2 | 100.0% | 2 | 3.1e-3 |
| phytol salvage pathway | 0 | 0% | 0 | 0 | 0% | 2 | 2 | 100.0% | 2 | 3.1e-3 |
| leucopelargonidin and leucocyanidin biosynthesis | 0 | 0% | 6 | 0 | 0% | 15 | 2 | 100.0% | 2 | 3.1e-3 |
| chlorophyll a degradation II | 0 | 0% | 11 | 0 | 0% | 11 | 2 | 100.0% | 2 | 3.1e-3 |
| cutin biosynthesis | 0 | 0% | 0 | 0 | 0% | 16 | 4 | 80.0% | 5 | 3.6e-3 |
| salvigenin biosynthesis | 0 | 0% | 0 | 0 | 0% | 2 | 3 | 60.0% | 5 | 3.6e-3 |
| carotenoid cleavage | 0 | 0% | 4 | 0 | 0% | 4 | 3 | 60.0% | 5 | 3.6e-3 |
| chlorophyll cycle | 0 | 0% | 0 | 0 | 0% | 6 | 3 | 50.0% | 6 | 1.1e-2 |
| gibberellin biosynthesis II (early C-3 hydroxylation) | 0 | 0% | 0 | 0 | 0% | 2 | 4 | 66.7% | 6 | 1.1e-2 |

**Figure 8.9** Pathways associated with putative metabolites.

the major affected 'Col' and 'Ws' pathways to be 6-hydroxymethyldihydropterin diphosphate biosynthesis I, chlorophyll A biosynthesis II, and vitamin E biosynthesis (tocotrienols), among others (see Figure 8.8B).

By navigating into 'Systems Biology Results', the same results can be shown in table form (Figure 8.9). In this table, (1) dysregulated putative metabolites are displayed, as well as (2) the total metabolites identified in association to the said pathway. For this run, a total of 545 pathways were found, with 35 of them (6.4%) exhibiting a $p$-value $\leq 0.05$.

From this table, the 'Predictive Metabolites Results' links pathways to specific metabolites and associated features. Figure 8.10 shows the metabolites associated with 'vitamin E biosynthesis (tocotrienols)'. Several features are associated with three defined metabolites (similar $m/z$, different retention times). Interestingly, 216 features (2% of all detected) are found to be predictive among these two accessions, while none is classified as predictive when the human BioSource database is used (data not shown). These results highlight the importance of carefully selecting the parameters that best apply to the investigated samples and the instrument used for metabolic analysis.

Closer inspection of two of these features clearly confirms the differences in the chromatograms as well as box plots of these *A. thaliana* accessions. The LC-MS data for individual features can be evaluated from the extracted ion chromatograms and box plots (see Figure 8.11). For example, MSn fragmentation analyses can be performed to confirm the identities of predicted metabolites.

## 8.5.11 Interpreting Results

XCMS Online can process metabolomics data sets and integrate other omics data. Nevertheless, the results still depend strongly on the experimental design, analytical methods, and the settings of the data-processing parameters.

Overall, XCMS Online is a valuable tool for metabolomics data analysis, as demonstrated by the identification of plant compounds with potential antiviral activities, as an example.[28]

Metabolite Overlap with Pathway:

Search:

| Metabolites | METLIN ID | KEGG ID | Dysregulation | Fold Change | p-value | m/z | Retention Time | Select Job | Feature Details |
|---|---|---|---|---|---|---|---|---|---|
| &delta-tocotrienol | | | | | | | | | |
| | 6134 | C00353 | "UP" | 5.5 | 5.2e-7 | 419.2912 | 16.33 | | 994 |
| | 6134 | C00353 | "UP" | 3.3 | 2.8e-6 | 419.2908 | 20.31 | | 1499 |
| | 6134 | C00353 | "UP" | 2.2 | 7.0e-8 | 419.2902 | 20.73 | | 596 |
| | 6134 | C00353 | "UP" | 3.2 | 5.0e-6 | 419.2902 | 24.97 | | 1689 |
| | 6134 | C00353 | "UP" | 1.9 | 5.3e-7 | 419.2911 | 31.04 | | 1000 |
| | 6134 | C00353 | "UP" | 2.8 | 5.7e-6 | 419.2906 | 28.47 | | 1739 |
| | 6134 | C00353 | "UP" | 3.5 | 9.9e-8 | 419.2903 | 15.53 | | 658 |
| | 6134 | C00353 | "UP" | 3.8 | 2.3e-7 | 419.2910 | 15.88 | | 816 |
| 2-methyl-6-geranylgeranyl-1,4-benzoquinol | | | | | | | | | |
| | 6134 | C00353 | "UP" | 5.5 | 5.2e-7 | 419.2912 | 16.33 | | 994 |
| | 6134 | C00353 | "UP" | 3.3 | 2.8e-6 | 419.2908 | 20.31 | | 1499 |
| | 6134 | C00353 | "UP" | 2.2 | 7.0e-8 | 419.2902 | 20.73 | | 596 |
| | 6134 | C00353 | "UP" | 3.2 | 5.0e-6 | 419.2902 | 24.97 | | 1689 |
| | 6134 | C00353 | "UP" | 1.9 | 5.3e-7 | 419.2911 | 31.04 | | 1000 |
| | 6134 | C00353 | "UP" | 2.8 | 5.7e-6 | 419.2906 | 28.47 | | 1739 |
| | 6134 | C00353 | "UP" | 3.5 | 9.9e-8 | 419.2903 | 15.53 | | 658 |
| | 6134 | C00353 | "UP" | 3.8 | 2.3e-7 | 419.2910 | 15.88 | | 816 |
| geranylgeranyl diphosphate | | | | | | | | | |
| | 6134 | C00353 | "DOWN" | 1.9 | 1.4e-5 | 468.2297 | 6.52 | | 2081 |
| | 6134 | C00353 | "DOWN" | 2.0 | 9.6e-6 | 468.2287 | 6.87 | | 1932 |

Previous 1 Next

**Figure 8.10** Features associated with vitamin E biosynthesis.

### 8.5.12 Data Sharing

XCMS Online results can be shared with specific individuals *via* email, or made publicly available, with the authors and publication details specified.

## 8.6 Comparing XCMS/R and XCMS Online

To compare the results from the XCMS/R and XCMS Online workflows, we counted the features that were identified in all twelve samples by each strategy, with differences in *m*/*z* of less than 0.005, and in retention time of less than 5 s. In total, 128 features were found in each sample using the XCMS/R workflow, but 1157 features were found using XCMS Online; *i.e.*, about ten times as many features were identified with the latter. Fifteen features were only detected in the XCMS/R workflow. All remaining features, which corresponds to 88% of the XCMS/R features, were found using both strategies (see Figure 8.12). This does not necessarily mean that the XCMS

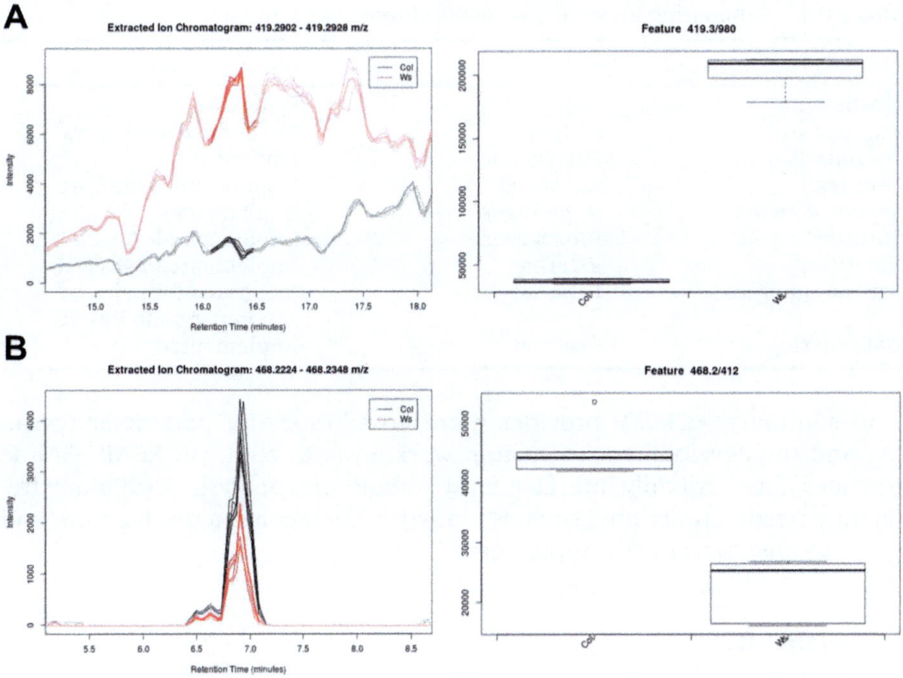


**Figure 8.11** Extracted ion chromatograms (XICs) and box plots of (A) up-regulated delta-tocotrienol, and (B) down-regulated geranylgeranyl diphosphate predictive features.

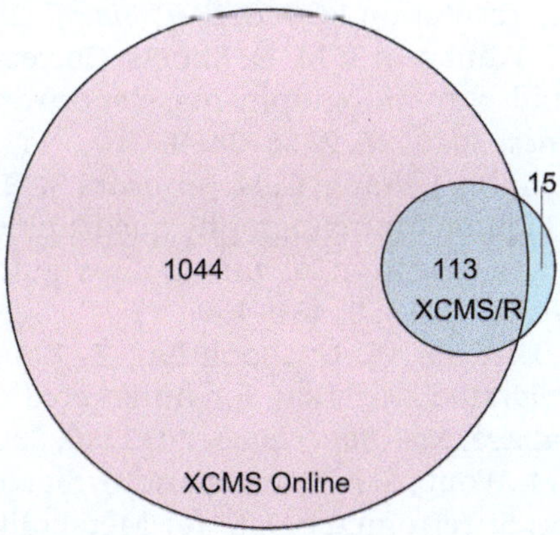


**Figure 8.12** Features identified in all twelve samples by XCMS/R and XCMS Online.

Online algorithms are more sensitive, since the parameter settings used in XCMS/R strongly affect the outcome. Depending on the research question, the XCMS/R settings could can adjusted, for example, to allow lower signal/noise ratio features. Table 8.1 compares the different characteristics of the XCMS/R and XCMS Online workflows.

**Table 8.1** Comparing XCMS/R and XCMS Online.

| Property | XCMS/R | XCMS online |
|---|---|---|
| Hosting | Local | Web server |
| Automation | Scripts | Pre-defined workflows |
| Optimization | Full access | Limited |
| User level | Advanced | Beginner/intermediate |
| Export of results | R/packages | Implemented |
| Statistics | R/packages | Implemented |
| Identification | External | Implemented/METLIN |
| Pathway analysis | External | Integrated for pairwise comparisons/Biocyc |
| Data sharing | External | Implemented |

In summary, XCMS/R provides more possibilities for parameter tweaking and the development of custom workflows. In contrast, XCMS Online provides a user-friendly interface for metabolomics analysis, with many frequently needed functions, such as statistics, metabolite identification, and pathway analysis, already implemented.

## References


1. C. A. Smith, E. J. Want, G. O'Maille, R. Abagyan and G. Siuzdak, XCMS: Processing mass spectrometry data for metabolite profiling using nonlinear peak alignment, matching, and identification, *Anal. Chem.*, 2006, **78**, 779–787.
2. R. Tautenhahn, C. Böttcher and S. Neumann, Highly sensitive feature detection for high resolution LC/MS, *BMC Bioinf.*, 2008, **9**, 504.
3. H. P. Benton, E. J. Want and T. M. D. Ebbels, Correction of mass calibration gaps in liquid chromatography-mass spectrometry metabolomics data, *Bioinformatics*, 2010, **26**, 2488–2489.
4. Z.-J. Zhu, A. W. Schultz, J. Wang, C. H. Johnson, S. M. Yannone, G. J. Patti and G. Siuzdak, Liquid chromatography quadrupole time-of-flight mass spectrometry characterization of metabolites guided by the METLIN database, *Nat. Protoc.*, 2013, **8**, 451–460.
5. R. Tautenhahn, K. Cho, W. Uritboonthai, Z. Zhu, G. J. Patti and G. Siuzdak, An accelerated workflow for untargeted metabolomics using the METLIN database, *Nat. Biotechnol.*, 2012, **30**, 826–828.
6. H. P. Benton, D. M. Wong, S. A. Trauger and G. Siuzdak, XCMS2: Processing Tandem Mass Spectrometry Data for Metabolite Identification and Structural Characterization, *Anal. Chem.*, 2008, **80**, 6382–6389.
7. R. Tautenhahn, G. J. Patti, E. Kalisiak, T. Miyamoto, M. Schmidt, F. Y. Lo, J. McBee, N. S. Baliga and G. Siuzdak, metaXCMS: Second-Order Analysis of Untargeted Metabolomics Data, *Anal. Chem.*, 2011, **83**, 696–700.
8. G. J. Patti, R. Tautenhahn and G. Siuzdak, Meta-analysis of untargeted metabolomic data from multiple profiling experiments, *Nat. Protoc.*, 2012, **7**, 508–516.

9. R. Tautenhahn, G. J. Patti, D. Rinehart and G. Siuzdak, XCMS Online: A web-based platform to process untargeted metabolomic data, *Anal. Chem.*, 2012, **84**, 5035–5039.
10. D. Rinehart, C. H. Johnson, T. Nguyen, J. Ivanisevic, H. P. Benton, J. Lloyd, A. P. Arkin, A. M. Deutschbauer, G. J. Patti and G. Siuzdak, Metabolomic data streaming for biology-dependent data acquisition, *Nat. Biotechnol.*, 2014, **32**, 524–527.
11. H. Gowda, J. Ivanisevic, C. H. Johnson, M. E. Kurczy, H. P. Benton, D. Rinehart, T. Nguyen, J. Ray, J. Kuehl, B. Arevalo, P. D. Westenskow, J. Wang, A. P. Arkin, A. M. Deutschbauer, G. J. Patti and G. Siuzdak, Interactive XCMS Online: Simplifying Advanced Metabolomic Data Processing and Subsequent Statistical Analyses, *Anal. Chem.*, 2014, **86**, 6931–6939.
12. H. P. Benton, J. Ivanisevic, N. G. Mahieu, M. E. Kurczy, C. H. Johnson, L. Franco, D. Rinehart, E. Valentine, H. Gowda, B. K. Ubhi, R. Tautenhahn, A. Gieschen, M. W. Fields, G. J. Patti and G. Siuzdak, Autonomous Metabolomics for Rapid Metabolite Identification in Global Profiling, *Anal. Chem.*, 2015, **87**, 884–891.
13. G. J. Patti, R. Tautenhahn, D. Rinehart, K. Cho, L. P. Shriver, M. Manchester, I. Nikolskiy, C. H. Johnson, N. G. Mahieu and G. Siuzdak, A View from Above: Cloud Plots to Visualize Global Metabolomic Data, *Anal. Chem.*, 2013, **85**, 798–804.
14. D. W. Meinke, J. M. Cherry, C. Dean, S. D. Rounsley and M. Koornneef, Arabidopsis thaliana: A Model Plant for Genome Analysis, *Science*, 1998, **282**, 662–682.
15. Arabidopsis Genome Initiative, Analysis of the genome sequence of the flowering plant Arabidopsis thaliana, *Nature*, 2000, **408**, 796–815.
16. C. Alonso-Blanco and M. Koornneef, Naturally occurring variation in Arabidopsis: An underexploited resource for plant genetics, *Trends Plant Sci.*, 2000, **5**, 22–29.
17. F. Laibach, Über sommer- und winterannuelle Rassen von Arabidopsis thaliana (L.) Heynh. Ein Beitrag zur ätiologie der Blütenbildung, *Beitr. Biol. Pflanz.*, 1951, **28**, 173–210.
18. M. Pigliucci, Ecological and evolutionary genetics of Arabidopsis, *Trends Plant Sci.*, 1998, **3**, 485–489.
19. M. Sotelo-Silveira, A.-L. Chauvin, N. Marsch-Martínez, R. Winkler and S. De Folter, Metabolic fingerprinting of Arabidopsis thaliana accessions, *Front. Plant Sci.*, 2015, **6**, 1–13.
20. D. Kessner, M. Chambers, R. Burke, D. Agus and P. Mallick, ProteoWizard: Open source software for rapid proteomics tools development, *Bioinformatics*, 2008, **24**, 2534–2536.
21. M. Rendón-Anaya, J. M. Montero-Vargas, S. Saburido-Álvarez, A. Vlasova, S. Capella-Gutierrez, J. J. Ordaz-Ortiz, O. M. Aguilar, R. P. Vianello-Brondani, M. Santalla, L. Delaye, T. Gabaldón, P. Gepts, R. Winkler, R. Guigó, A. Delgado-Salinas and A. Herrera-Estrella, Genomic history of the origin and domestication of common bean unveils its closest sister species, *Genome Biol.*, 2017, **18**, 60.

22. A. Tiessen, E. A. Cubedo-Ruiz and R. Winkler, Improved Representation of Biological Information by Using Correlation as Distance Function for Heatmap Cluster Analysis, *Am. J. Plant Sci.*, 2017, **08**, 502.
23. A. P. Reynolds, G. Richards, B. de la Iglesia and V. J. Rayward-Smith, Clustering Rules: A Comparison of Partitioning and Hierarchical Clustering Algorithms, *J. Math. Model. Algorithms*, 2006, **5**, 475–504.
24. G. J. Williams, Rattle: A Data Mining GUI for R, *R J.*, 2009, **1**, 45–55.
25. G. Williams, *Data Mining with Rattle and R: The Art of Excavating Data for Knowledge Discovery (Use R!)*, Springer, 1st edn, 2011.
26. R. Winkler, An evolving computational platform for biological mass spectrometry: Workflows, statistics and data mining with MASSyPup64, *PeerJ*, 2015, **3**, 1–34.
27. E. M. Forsberg, T. Huan, D. Rinehart, H. P. Benton, B. Warth, B. Hilmers and G. Siuzdak, Data processing, multi-omic pathway mapping, and metabolite activity analysis using XCMS Online, *Nat. Protoc.*, 2018, **13**, 633–651.
28. G. Prinsloo and J. Vervoort, Identifying anti-HSV compounds from unrelated plants using NMR and LC–MS metabolomic analysis, *Metabolomics*, 2018, **14**, 134.

CHAPTER 9

# Statistical Evaluation and Integration of Multi-omics Data with MetaboAnalyst

DAVID S. WISHART*[a,b]

[a]Department of Computing Science, University of Alberta, Edmonton, AB, Canada T6G 2E8; [b]Department of Biological Sciences, University of Alberta, Edmonton, AB, Canada T6G 2E8
*E-mail: dwishart@ualberta.ca

## 9.1 Introduction to MetaboAnalyst

MetaboAnalyst is a comprehensive, web-based tool for metabolomic data processing, data visualization, data interpretation, and data integration with other omics data. It has been carefully designed to enable researchers, with little statistical or computational background, to perform complex data analysis procedures commonly used in metabolomic data analysis. It has also been structured to support the analysis of data generated from both targeted and untargeted metabolomics approaches. Because of its easy accessibility, interactivity, high quality graphics and broad scope, MetaboAnalyst has become one of the most popular software tools in the field of metabolomics. Indeed, based on the most recent citation count statistics, approximately one third of all published papers in the area of metabolomics, make use of MetaboAnalyst.

New Developments in Mass Spectrometry No. 8
Processing Metabolomics and Proteomics Data with Open Software: A Practical Guide
Edited by Robert Winkler

Published by the Royal Society of Chemistry, www.rsc.org

To keep up with the rapid growth and development of metabolomics, MetaboAnalyst has continuously evolved, with four major releases over the past ten years. The first version of the web server, MetaboAnalyst 1.0[1] was designed primarily for the processing and statistical analysis of targeted metabolomic data. The second version, MetaboAnalyst 2.0,[2] included tools for handling untargeted metabolomic data and incorporated additional components for metabolomic data interpretation such as metabolic pathway analysis or MetPA[3] and metabolite set enrichment analysis or MSEA.[4] The third version of the server, MetaboAnalyst 3.0[5] added four new modules including: (1) two-factor and time-series data analysis; (2) biomarker discovery/analysis; (3) sample size estimation *via* power analysis and (4) integrative pathway analysis. Finally, the latest version of the web-server, MetaboAnalyst 4.0[6] involved a complete overhaul of the user interface. This gave MetaboAnalyst a more modern "look and feel". The latest version also included a new window to display a user's R command history in real-time, which can be used in conjunction with the newly released MetaboAnalystR package,[7] a downloadable, command-line version of MetaboAnalyst. Additionally, MetaboAnalyst 4.0 incorporates three new modules for: (1) predicting pathway activity from untargeted mass spectrometry data *via* mummichog;[8] (2) identifying biomarkers across multiple metabolomics datasets, and (3) integrating multi-omics (metabolomics, metagenomics, and/or transcriptomics) data together.

The focus of this chapter will be on the latest version of MetaboAnalyst (version 4.0) with a particular emphasis on those modules most relevant to MS-based metabolomics. This chapter will be divided into seven parts. The first is the introduction to MetaboAnalyst. The second part provides a broad overview of MetaboAnalyst, its various modules, its strengths and limitations. The third section provides a description of what kinds of data MetaboAnalyst accepts and how these data should be properly formatted. The fourth section describes how data are processed and handled through MetaboAnalyst's Statistical Analysis module (its most popular module). The fifth section describes MetaboAnalyst's Enrichment Analysis and Pathway Analysis modules (the next most popular modules). The sixth section describes the MS Peaks-to-Pathways (or mummichog) module, which is rapidly growing in popularity, while the final section provides a short summary.

Because of MetaboAnalyst's very broad scope and large number of functions or modules, it is simply not possible to cover all of its many features or to create a step-by-step tutorial. Readers who are interested in learning more about MetaboAnalyst's other modules or who would prefer to access a comprehensive tutorial are encouraged to look at the MetaboAnalyst tutorials published in Current Protocols in Bioinformatics[9] or Nature Protocols,[10] to download tutorials that are available on the MetaboAnalyst website or to view some of the many videos on MetaboAnalyst that are posted on YouTube.

## 9.2 MetaboAnalyst Overview

Many tasks in metabolomic data analysis can be divided into modules, steps or sections. These may include: (1) spectral alignment and peak selection; (2) metabolite annotation; (3) batch or outlier correction; (4) data normalization and scaling; (5) univariate or multivariate statistical analysis; (6) pathway and biological interpretation; (7) biomarker identification or (8) integration with other omics data. Some steps in the data analysis workflow are interdependent and need to be performed sequentially, while other steps can be done independently or are only used in specialized situations. Because of the many different data analysis scenarios available in any given metabolomics workflow MetaboAnalyst has evolved to become a modular web service. That is, users are able to select different MetaboAnalyst modules to perform different analytical tasks at different times or with different data sets. Furthermore, to reduce the likelihood of errors, each MetaboAnalyst module is self-contained so that the sub-tasks or steps needed to properly conduct each type of analysis are contained entirely within that module.

As shown in Figure 9.1, MetaboAnalyst has 12 modules that are selectable from its "hour clock" panel. Clicking on any of the 12 circles will launch the selected module. As can be seen in Figure 9.1, the modules are color coded and strategically placed around the circle. The most popular modules are located at the top of the clock (Statistical Analysis, Enrichment Analysis, Pathway Analysis, Biomarker Analysis). The solid blue modules are compatible with both targeted and untargeted metabolomic data sets, while the solid green modules require annotated metabolomic data (preferably with absolute concentrations). The modules with hashed lines are designed for multi-omics integration, the cyan module (Peaks-to-Pathways) is specifically for analyzing untargeted high-resolution MS metabolomic and the grey modules at the bottom of the clock provide spectral analysis options and other general utilities for NMR, GC-MS, LC-MS and lipidomics analysis. Most of MetaboAnalyst's modules have a similar layout or design. Upon launching a given module users will typically see a three panel view with the left panel or left column listing the data processing steps (which are all hyperlinked), a large central panel or window where users upload/view their data or select their data processing options and a right panel or column which lists a user's R command history (Figure 9.2). Each module is designed to be self-contained, self-guiding, step-wise in structure and fully interactive. Users cannot proceed from one step to the other without first completing the task or checking their data. This interface design is intended to ensure that new or naïve users do not improperly use the very powerful statistical or analytical tools in MetaboAnalyst.

The fact that MetaboAnalyst uses a step-wise data processing design to guide users through all the major data processing steps within a given module is one of the reasons why it has gained such broad user appeal. Another reason for its popularity lies in the fact that MetaboAnalyst is

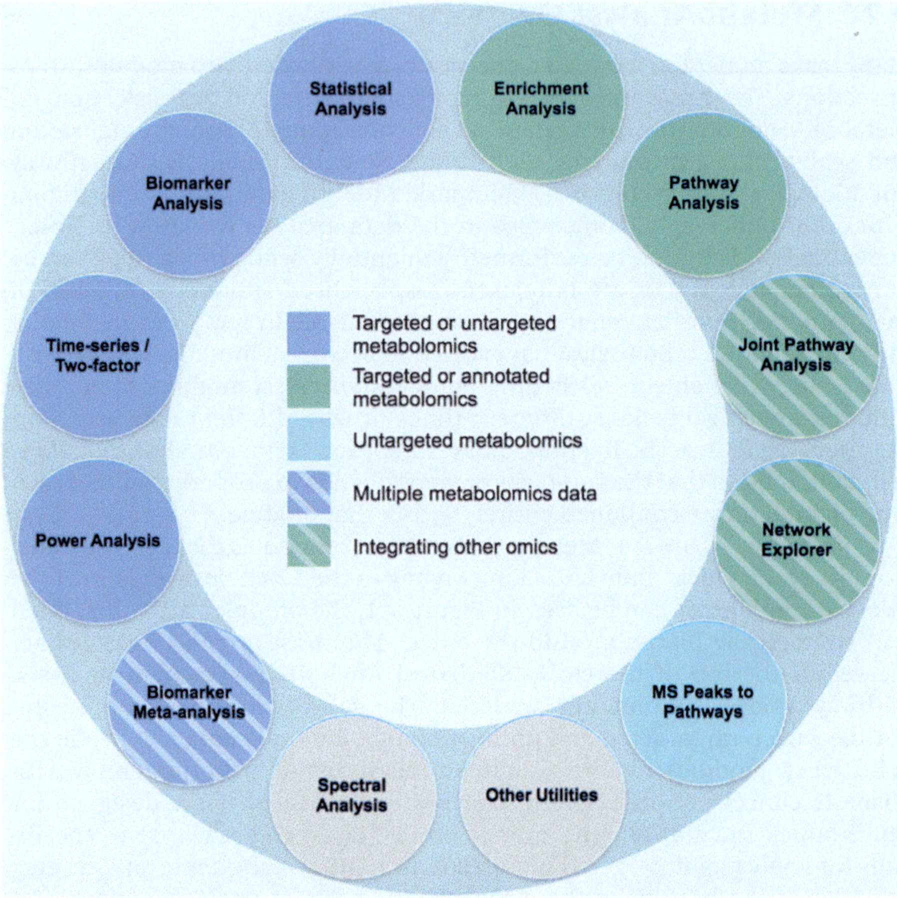


**Figure 9.1** A screenshot of the MetaboAnalyst "clock" with each of the 12 analysis modules arrayed around a circle.

very graphical and highly interactive (see Figure 9.3 for some examples of MetaboAnalyst's output). Many of the images and plots generated by MetaboAnalyst are colorful, editable, clickable or zoomable. They can also be saved as high-resolution image files for use in publications or posters. Indeed, nearly every table, graph or image generated by MetaboAnalyst can be downloaded and a detailed record, through the R command history of all the analytical steps conducted is always kept. This helps ensure consistency and reproducibility when users are analyzing multiple data sets over multiple sessions spanning multiple weeks. Furthermore, users can take their R command history and use this interactively generated code (typically generated on some pilot data) to run a high throughput data analysis pipeline on a much larger data set *via* MetaboAnalystR. Interestingly, many users of MetaboAnalyst are now using the R command history as a quick and easy way of learning the R language.

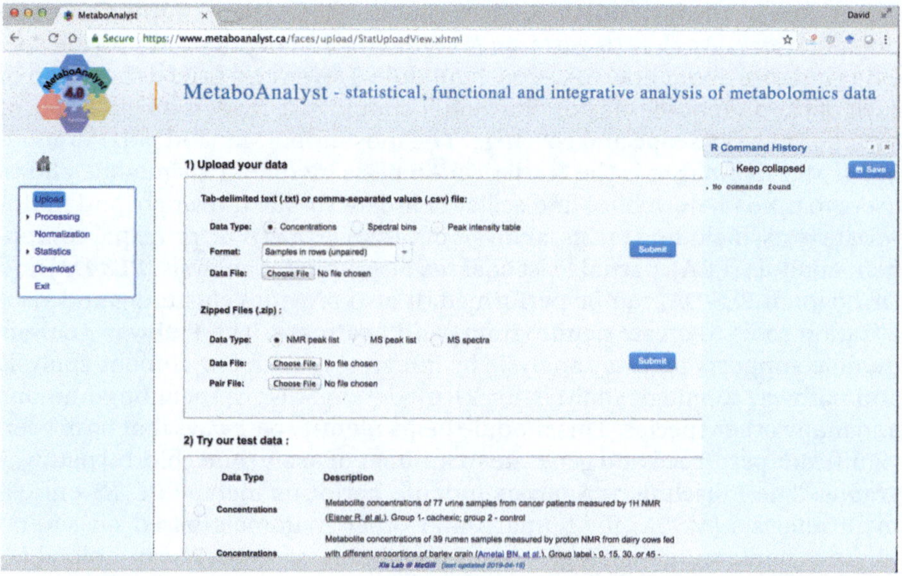


**Figure 9.2** A screenshot of MetaboAnalyst's Statistical Analysis module "start" page with the three main panels displayed (the processing options menu, the main panel and the R Command History panel).

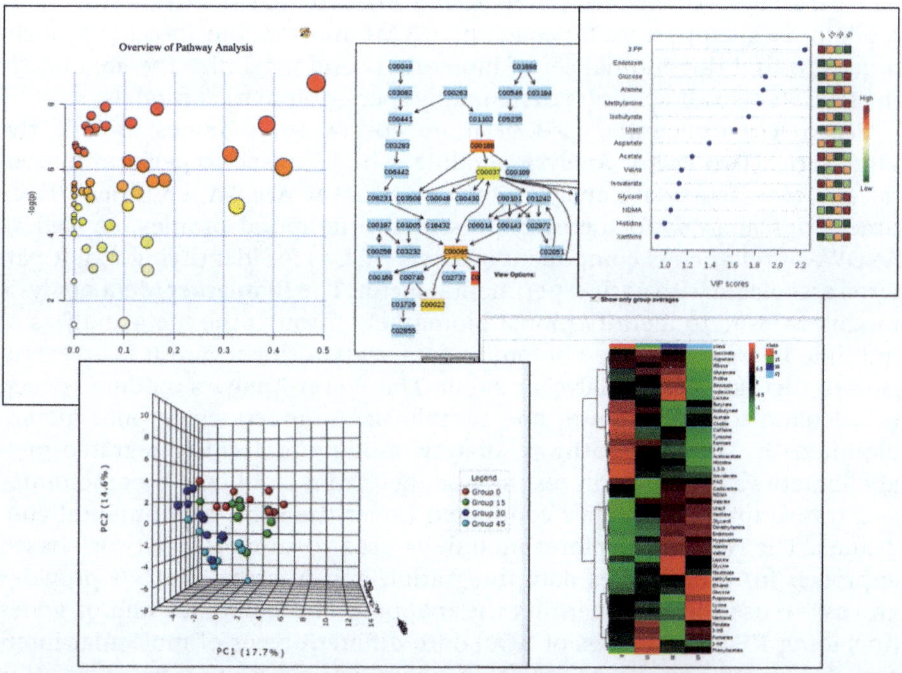


**Figure 9.3** A montage of different graphical displays generated by MetaboAnalyst's various modules including heatmaps, PCA plots and various bar graphs.

As noted in the introduction, there is not enough space in this short section on MetaboAnalyst to describe every module in depth or to provide specific guidance about how to use each module. However, a brief description of each of the 12 modules is possible and it should give readers a better idea of MetaboAnalyst's scope and coverage. The most important and popular module in MetaboAnalyst is the **Statistical Analysis** module. This module allows users to upload, normalize and scale their data so that univariate and multivariate tests, including *t*-tests, analysis of variance (ANOVA), principal component analysis (PCA), partial least squares discriminant analysis (PLS-DA) and Orthogonal PLS-DA, can be performed. It also provides clustering and visualization tools to create dendrograms and heatmaps. The **Pathway Analysis** module supports pathway analysis by integrating both enrichment analysis and pathway topology analysis for 21 model organisms, including humans and many other species. This module helps identify pathways that have been significant perturbed and generates a number of useful and colorful pathway graphs. The **Enrichment Analysis** module performs metabolite set enrichment analysis (MSEA) for humans and other mammals based on several libraries containing ~6300 groups of metabolite sets. Users can upload (1) a list of identified compounds, (2) a list of compounds with concentrations, or (3) a concentration table, to identify those metabolites enriched in certain pathways, diseases or biological processes. The **Biomarker Analysis** module is designed to help users identify one or more biomarkers from metabolomic data *via* analysis of the receiver operating characteristic (ROC) curves.[11] This module uses support vector machine (SVM) and random forest (RF) techniques to find the optimal set of biomarkers and maximize the area under the ROC. It also allows users to manually specify biomarker models.

Other, less frequently used or more specialized modules include the **Time-series/Two Factor Analysis** module, which is used to perform temporal and two-factor data analysis such as two-way ANOVA, empirical Bayes time-series analysis for detecting distinctive temporal profiles, as well as ANOVA-simultaneous component analysis (ASCA) for identifying major patterns associated with each experimental factor. The **Biomarker Meta-analysis** module is used to identify robust biomarkers through the meta-analysis of multiple independent metabolomics data sets. In this regard it is an extension of the biomarker analysis module The **Power Analysis** module is used to calculate statistical power and sample size numbers using pilot metabolomic data. The **Joint Pathway Analysis** module performs integrated metabolic pathway analysis on results obtained from combined metabolomics and transcriptomics studies conducted under the same experimental conditions. The **Network Explorer** module is used to support a network-based approach for multi-omics data integration and interpretation. It provides an easy-to-use tool that permits the mapping of metabolites and/or genes (including KEGG orthologs or KOs) onto different types of molecular interaction networks, as well as molecule–phenotype association networks. It is somewhat similar to Cytoscape[12] in concept and design, but is specifically designed for metabolomics. The MS **Peaks-to-Pathways** module uses the mummichog algorithm to take high-resolution LC-MS spectral peak data and perform metabolic pathway enrichment analysis and visualization. It currently supports 21 model organisms. The Other Utilities module provides tools for batch effect corrections, compound identifier interconversion and lipidomics analysis. The last module is the **Spectral Analysis** module, which provides links to several freely available web-based tools for NMR spectral deconvolution (Bayesil),[13] GC-MS spectral analysis (GC-Autofit) and LC-MS spectral analysis (XCMS-Online).[14]

This last module highlights both a key strength and a key limitation of MetaboAnalyst. As a metabolomics data analysis tool, MetaboAnalyst is compatible with essentially all major instrument platforms for metabolomics (NMR, GC-MS, LC-MS). However, MetaboAnalyst was originally developed for targeted metabolomics where data processing leads to small data files of compound names and concentrations. Partial support for untargeted metabolomics only started appearing in later versions of MetaboAnalyst and the web server still does not support local LC-MS data processing and metabolite annotation of untargeted LC-MS data. This is because it is highly instrument and even method-dependent as well as highly CPU intensive. As a result, MetaboAnalyst outsources this aspect to other web servers (such as XCMS-online) or it encourages users to employ locally installable, freely available spectral preprocessing tools. For LC-MS data preprocessing this includes tools such as MetAlign,[15] OpenMS,[16] MZmine,[17] XCMS,[18] and MS-DIAL.[19] Very recently, a new downloadable R version of MetaboAnalyst (MetaboAnalystR 2.0) has been released and this command-line version now supports comprehensive offline LC-MS data processing (*via* XCMS and CAMERA) along with other MetaboAnalyst functions.[20] So, while the command-line, offline version of MetaboAnalyst (MetaboAnalystR 2.0) does support LC-MS data processing, the MetaboAnalyst webserver remains mostly a downstream data processing tool. This simply reflects the limitations of web servers, the cloud and internet bandwidths for handling large data files and performing interactive data file manipulations. Despite this limitation with regard to data preprocessing (especially for untargeted LC-MS-based metabolomics), MetaboAnalyst still appears to be used equally by both targeted and untargeted metabolomics researchers.

## 9.3 Data Formats and Data Requirements

MetaboAnalyst accepts and processes a number of common data file formats including comma separated value (.csv) files, tab delimited text (.txt) files, compressed (.zip) files, or even simple lists of compound names or MS peaks. A .csv or .txt file is typically used when the whole data set has already been preprocessed to a tabular form such as a compound concentration table, a spectral bin table, or a peak intensity table. The zip file format is typically used to upload multiple MS spectra or multiple peak list files. However, due to bandwidth constraints on the internet, it is generally impractical to upload a large number of raw spectral files (>50 Megabytes) remotely to MetaboAnalyst. Rather, raw spectra should be preprocessed (peak picked and/or aligned) using locally installed software[15–20] to produce the necessary (and smaller) peak list files or peak intensity tables before uploading to MetaboAnalyst.

Uploading an improperly formatted input file is the most common mistake for new users of MetaboAnalyst. As a result a "Data Formats" page has been created (accessible on the left menu in the home page) that provides detailed instructions, screenshots as well as downloadable sample data sets for all the data formats for all the different modules that it supports. It is VERY IMPORTANT that new users read through this before preparing their data files. To further assist users, all MetaboAnalyst modules have data upload pages where example data files typically titled as: "Try Our Test Data" or "Use Example Data" allow users to upload (or download) and explore authentic metabolomics data sets that are specifically suited to that module. The general format for most MetaboAnalyst files is a simple (.txt or .csv) file in which the biosamples are listed in rows and the compounds listed in columns (the default format). The first column should contain unique sample identifiers, the second column should contain phenotype labels and the remaining columns should contain the metabolite levels or concentrations. It is also important to remember that MetaboAnalyst is intended to compare two or more data sets (*i.e.* cases *versus* controls or before and after treatment). Uploading a data set with only one phenotype (or condition or disease) label will lead to relatively meaningless output. On the other hand, it is certainly possible to upload data sets with three, four, or five different phenotype labels. It is also possible to upload data other than metabolite concentrations (in columns three and beyond). For instance, age, weight, height, protein concentrations, gene expression levels or any other numeric data type can be put into these columns. In this regard, MetaboAnalyst is frequently used by other omics researchers to perform gene expression, metagenomic and proteomic data analysis.

## 9.4 General Statistical Analysis with MetaboAnalyst

The most commonly used operation or module in MetaboAnalyst is the Statistical Analysis module located at the top of the MetaboAnalyst "clock" (Figure 9.1). Clicking on this circle will generate a new page, which is the "Data Upload" page (Figure 9.2). MetaboAnalyst uses a navigation tree (on the left side) to guide users through each of the data processing, analysis and interpretation steps. Some links will be enabled or disabled depending on the context. The "Upload" hyperlink is highlighted in blue, representing the current step. On the right-hand side is the R Command History panel, which displays the step-by-step R command workflow of user's analysis in real-time. Each time a user presses a button to perform a function or generate a plot in MetaboAnalyst 4.0, it generates a series of R commands that tell the computer what to do. In most web servers the computer commands are hidden from view, but in MetaboAnalyst, these are made visible to help users better understand the R programming language and to ensure greater reproducibility. In particular, MetaboAnalyst's R Command History can be used to reproduce a user's analysis locally in R *via* the downloadable MetaboAnalystR package.[7]

Users may upload several types of data sets, using the formats described on the MetaboAnalyst pages, including: (1) data concentration lists (from targeted LC-MS or GC-MS studies); (2) MS (from GC or LC studies) peak intensities, or (3) MS (from GC or LC studies) peak lists. Earlier versions and locally installed versions of MetaboAnalyst can support uploads of LC-MS spectra (in netCDF formats) and GC-MS spectra (in netCDF formats), but the current version is not able to support this due to internet bandwidth issues (the files are often gigabytes in size).

Once the data are uploaded, MetaboAnalyst will perform data integrity checks and missing value estimates. MetaboAnalyst checks to see if the sample and variable names are unique and contain no special characters (*i.e.* Latin or Greek letters), that at least three replicates are provided for each group and that all data values except phenotype labels are numeric. Missing values are allowed and can be indicated as a blank or marked as NA (without quotes). By default, missing values are considered by MetaboAnalyst to be caused by signals below the detection limit and are normally replaced by a very small value (half of the minimum positive value found in the data set). However, users may also perform missing value imputation. If peak list data is being used (from untargeted metabolomic data) a data filtering step is normally done to remove low quality data by looking at interquartile ranges or by filtering *via* coefficients of variance. The next step in the statistical analysis process is data normalization, transformation and scaling.

Normalization aims to make each sample (row) comparable to each other. For instance, normalization adjusts metabolite concentration levels in urine samples with different dilution effects due to a patient's water intake. A common route to urine normalization is to divide by the urine's creatinine concentration (normalization by reference feature). MetaboAnalyst offers five options for sample normalization – no normalization, normalization by sum, normalization by median, normalization by a reference sample, normalization by a reference feature, and sample-specific normalization. Transformation is a method to change variables so that they exhibit a more normal or Gaussian distribution as opposed to a skewed or logarithmic distribution. Three different data transformation methods are offered, no transformation, log transformation and cube-root transformation. Scaling aims to make each variable comparable to each other. This procedure is useful when variables are of very different orders of magnitude (some metabolites are at micromolar levels while other metabolites are at millimolar levels). Four scaling methods are offered in MetaboAnalyst – no scaling, auto scaling, Pareto scaling and range scaling.

MetaboAnalyst allows users to interactively "guess and check" which scaling, normalization and transformations are most suitable for their data. Visual displays show the data distributions before and after the parameters

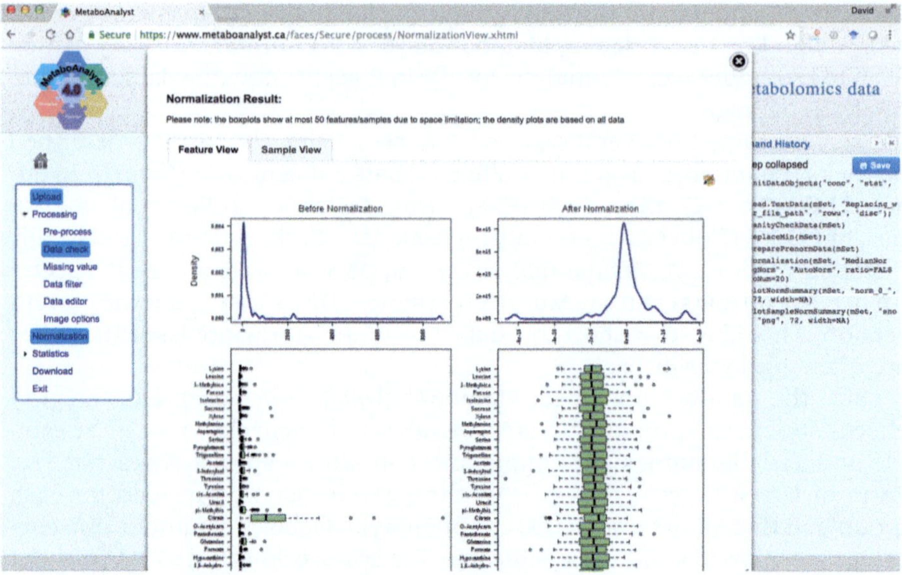


**Figure 9.4** A screenshot of MetaboAnalyst's interactive normalization/scaling graphics viewer.

have been chosen. In particular MetaboAnalyst plots the overall data distribution based on kernel density estimation (before normalization – on the left, and after normalization – on the right) along with box plots showing the distributions of individual variables or metabolite concentrations (before and after normalization) (Figure 9.4). Users should compare the graphical summary on the left and the right (before and after normalization) to guide them towards choosing the methods that work best with their data. Each metabolomic data set is different so there is no hard and fast rule about which combination of parameters works best, however log transformation is usually required for most metabolomic data sets. Typically, a user will have to try two or three combinations before a sufficiently symmetric Gaussian or normal distribution is achieved.

After normalization, scaling and transformation, the data is now ready to be statistically analyzed. The options available are displayed on MetaboAnalyst's "Data Analysis Exploration" page (Figure 9.5). On this page five major analysis categories are presented: (1) Univariate Analysis; (2) Chemometrics Analysis; (3) Feature Identification; (4) Cluster Analysis and (5) Classification and Feature Selection. Listed under each of these five categories are many other subcategories containing individual methods. Under Univariate Analysis users will find six options (Fold Change Analysis, *t*-test, Volcano plots, One-way ANOVA, Correlation Analysis and Pattern Searching). Under Chemometric Analysis users will find four options (PCA, PLS-DA, Sparse PLS-DA and Orthogonal PLS-DA). Under Feature Identification, users can find Significant Analysis of Microarrays (and Metabolites) (SAM) and Empirical Bayesian

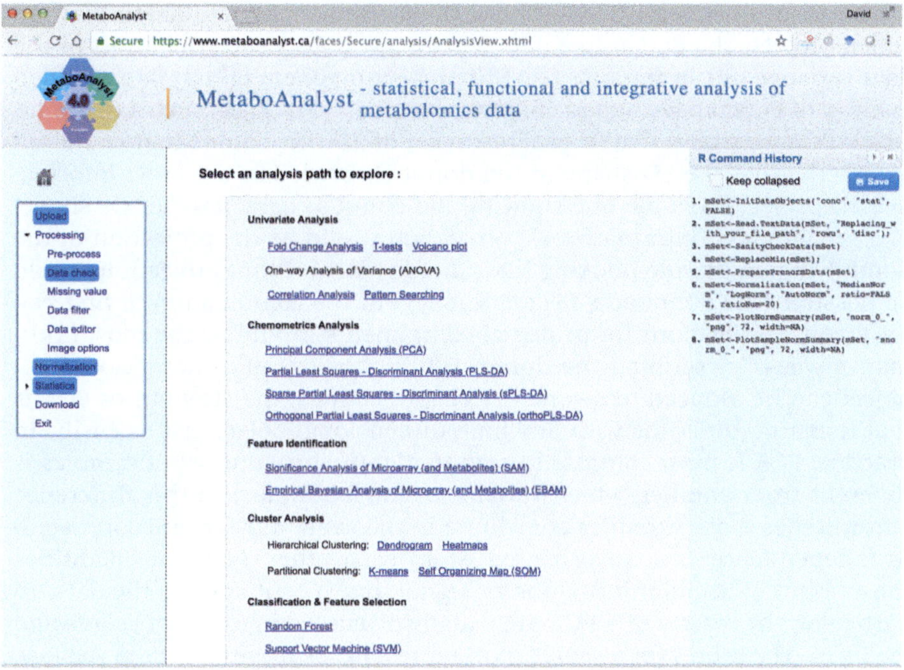


**Figure 9.5** A screenshot of MetaboAnalyst's Data Analysis Exploration page.

Analysis of Microarrays (and Metabolites) (EBAM).[21,22] Under Cluster Analysis, MetaboAnalyst offers Dendograms and Heatmaps as well as K-means and Self-Organizing Feature Maps (SOMs). For the Classification and Feature Selection category MetaboAnalyst offers Random Forest and Support Vector Machine (SVM) options.

Depending on whether the data set consists of two groups or more than two groups, MetaboAnalyst will enable (or disable) various statistical operations. For univariate, two-group analysis MetaboAnalyst will offer *t*-tests, Fold Change Analysis, Volcano plots and Correlation Analysis. For more than two groups MetaboAnalyst will offer ANOVA and pattern recognition or group trend analysis (*via* Pattern Searching with PatternHunter). Univariate methods (such as *t*-tests and ANOVA) are simple to use and the results are usually easy to understand. They are widely used in metabolomics studies for selecting important features from metabolomic data sets. However, univariate approaches are often considered suboptimal as they ignore the correlations that are known to be present among variables (*i.e.* peaks or metabolites). Multivariate methods, which simultaneously take all variables into consideration, are generally considered more suitable for high-dimensional "omics" data analysis. Most of the multivariate statistical options are offered under the Chemometric Analysis option. The two most important multivariate options are PCA and PLS-DA.

PCA is an unsupervised clustering or dimensionality reduction method which projects the data into a new coordinate system such that most of the data variance lies in the first few principal components (PCs). A simplified analogy of PCA can be made as follows: consider a 3D object such as a donut. Using PCA we would like to generate a set of 2D projections that could tell us exactly what the 3D shape of the donut is. Obviously the most informative 2D projection is the one showing the donut's characteristic "O" shape. The second most informative 2D projection would be the projection of the donut lying on its side (looking like a hotdog bun). Other projections could be possible (the donut at a 45° angle, say) but the combination of first two orthogonal projections (or principal components) would be the most informative way of describing the donut. While this example shows how a 3D object can be reduced to a series of 2D projections, the strength of PCA is that it can do the same with any high-dimensional object just as easily. In practice, PCA is most commonly used to identify how one set of samples is different from another, which variables contribute most to this difference and whether those variables contribute in the same way (*i.e.* are correlated) or independently (*i.e.* uncorrelated) from each other. PCA also quantifies the amount of useful information or signal that is contained in the data. In particular, the results of a PCA are usually discussed in terms of scores and loadings. The scores (or a scores plot) represent the original data in the new coordinate system and the loadings (or a loadings plot) represent the weights applied to the original data during the projection process. As a dimensionality reduction technique, PCA is particularly appealing as it allows one to visually detect sample patterns or groupings. An example of a PCA scores plot generated from MetaboAnalyst is shown in Figure 9.6.

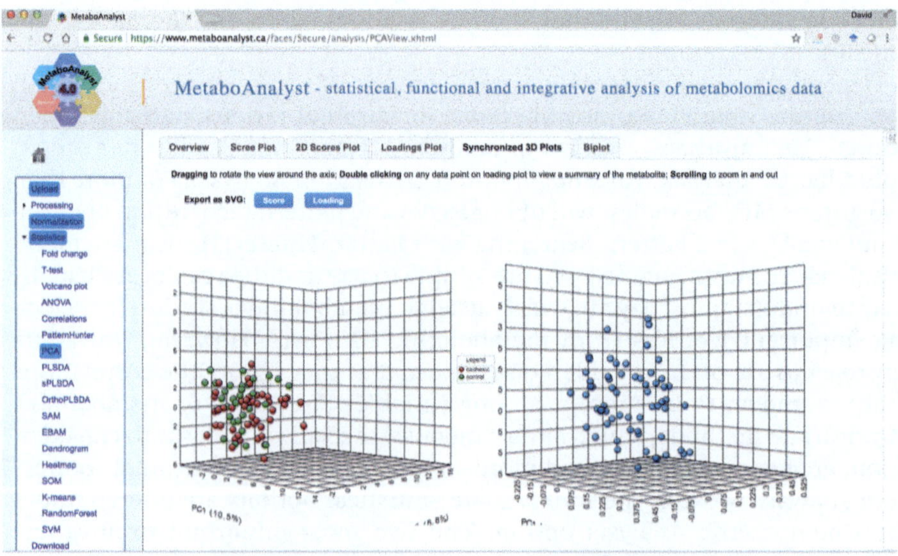


**Figure 9.6** A screenshot of a typical PCA scores plot generated *via* MetaboAnalyst.

Unlike PCA, which is an unsupervised clustering method, PLS-DA is a supervised clustering/classification technique. It uses the group labels to maximize the separation between different groups. PLS-DA methods are excellent for clarifying separations that are only hinted at *via* PCA and for identifying features (metabolites) that contribute to that separation. One important issue associated with PLS-DA is overfitting. MetaboAnalyst provides two approaches to address this issue – cross validation and permutation testing. In cross validation MetaboAnalyst tries to determine the optimal number of components needed to build the PLS-DA model. There are three common performance measures – the sum of squares captured by the model ($R^2$), the cross-validated $R^2$ (also known as $Q^2$), and the prediction accuracy (Accuracy). An $R^2$ value greater than 0.67 indicates a high predictive accuracy, a range of 0.33–0.67 indicates moderate accuracy, and an $R^2$ between 0.19 and 0.33 usually indicates low accuracy or reliability. The same is roughly true for $Q^2$. In permutation testing MetaboAnalyst randomizes the labels in the data set and attempts to determine if it can achieve a better separation *via* PLS-DA (using the ratio of the between-group sum of the squares and the within-group sum of squares, or the B/W-ratio as the distance measure). This process is repeated thousands of times to determine the distribution of separations and to calculate how frequently a random set of labels might produce a better classification. A permutation score of <0.05 is usually considered statistically significant. PLS-DA analyses also produce a list of compounds ranked on the basis of the variable importance in projection (VIP) score (Figure 9.7). These plots are particularly useful for identifying key metabolites that are driving the PLS-DA or PCA separations or which are most perturbed through a given intervention. Often the top ranked VIP

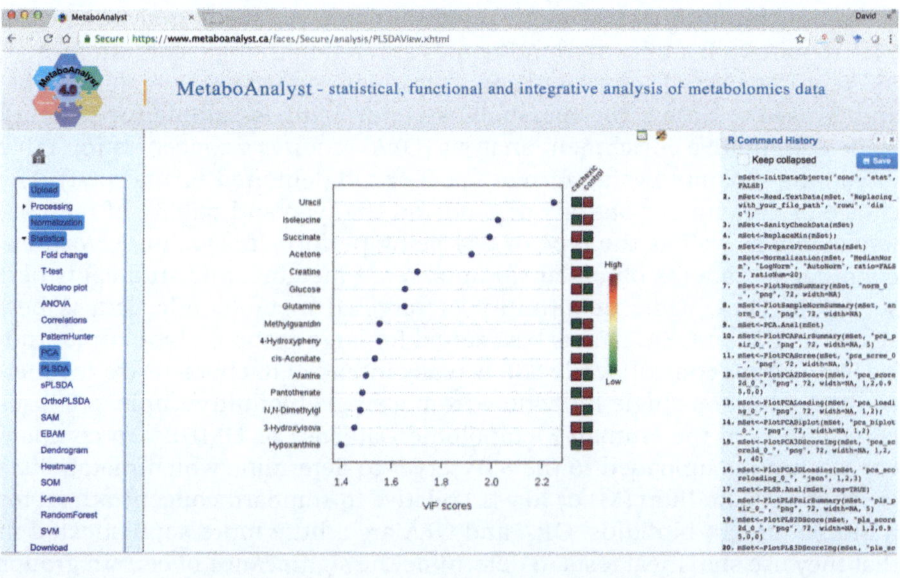


**Figure 9.7** A screenshot of a typical VIP score plot generated *via* MetaboAnalyst.

metabolites are similar (but not identical) to the top-ranked ones in *t*-tests or in PCA loadings plots.

Of course, many other statistical options are available in MetaboAnalyst, as seen in Figure 9.5. Indeed, depending on the questions being asked or analyses being performed other types of techniques or plots (such as heat-maps, SAM or EBAM) might be preferable over the options mentioned here. Those readers who are interested in more details about the various statistical analysis options in MetaboAnalyst are encouraged to look at the MetaboAnalyst tutorials published in Current Protocols in Bioinformatics[9] or to download tutorials that are available on the MetaboAnalyst website.

## 9.5 Enrichment Analysis and Pathway Analysis

A long list of features (or metabolites) that have changed significantly or showed interesting patterns of coordinated change under the different conditions is typically the result of the statistical analyses described above. Such a list is usually not the end of the analysis; rather it is the beginning. From this list of significantly changed metabolites one can start to explore the biological significance of these changes or ask questions about what effects they may have or what is driving these metabolic changes. One of the best routes to link metabolite lists to biological function or biological consequences is through pathway analysis and group-based significance tests. In metabolomics these methods include metabolite set enrichment analysis (MSEA)[4] and metabolite pathway analysis (MetPA).[3] In MetaboAnalyst MSEA and MetPA are primarily oriented towards human metabolome analysis.

The MSEA or Enrichment Analysis module in MetaboAnalyst performs three types of assessment: (1) Over representation analysis or ORA; (2) single sample profiling or SSP and (3) quantitative enrichment analysis or QEA. ORA requires a list of compound names entered in a single-column format, SSP requires a list of compound concentrations entered as two-column table and is used to assess the metabolic state of a single sample (or patient), while quantitative enrichment analysis (QEA) requires a concentration table in comma-separated value format (.csv) or tab delimited format (.txt). The choice of enrichment options depends on the type and quality of metabolomic data as well as the type of test being performed. Qualitative or non-quantitative data (as obtained *via* untargeted metabolomic studies) is best suited for ORA, while quantitative or targeted metabolomic data is best suited for SSP or QEA. QEA is best suited for large group metabolomic studies (case *versus* control) while SSP is really intended to characterize the metabolic health of a single individual from a single biofluid sample. SSP uses clinical data in the Human Metabolome Database or HMDB[23] to compare concentrations uploaded to the SSP server to determine which metabolites are high (H), medium (M) or low (L) relative to standard concentrations for multiple human biofluids. ORA and QEA are a little more sophisticated in that they use statistical tests to determine the significance of certain groups of metabolites by comparing the input data with specially developed metabolite library.

There are two options users can specify for enrichment analysis (for ORA or QEA). The first one is to select a metabolite set library and the second one is to choose a reference metabolome. Both can be used to calculate a background distribution to determine if the matched metabolite set is more enriched for certain metabolites compared to random chance and both are currently limited to human metabolites. If users wish to perform Enrichment Analysis for other species they can prepare and upload their own library or reference metabolome. The current MSEA database includes eight human metabolite libraries covering disease sets in blood (344 diseases), cerebral spinal fluid (166 diseases), and urine (384 diseases), as well as location-based metabolite sets (73 organs, biofluids and tissues), pathway-based metabolite sets (147 metabolic pathways), a single nucleotide polymorphism (SNP) metabolite set (4598 SNPs), a set consisting of drug-related pathways (461 pathways) and a predicted metabolite data set containing metabolites predicted to be changed as a result of dysfunctional enzymes using a genome-scale network model of human metabolism (912 metabolic sets).

The typical result of a QEA or ORA analysis is an interactively viewable/zoomable enrichment network (Figure 9.8) and a table describing the statistics of each node. In the network, each node represents a metabolite set, with its color corresponding to its *p*-value in the enrichment analysis and its size corresponding to its fold enrichment (hits/expected) to the query. Two metabolite sets are connected by an edge (line) if the number of shared metabolites between the two sets is greater than 20%. The nodes can be

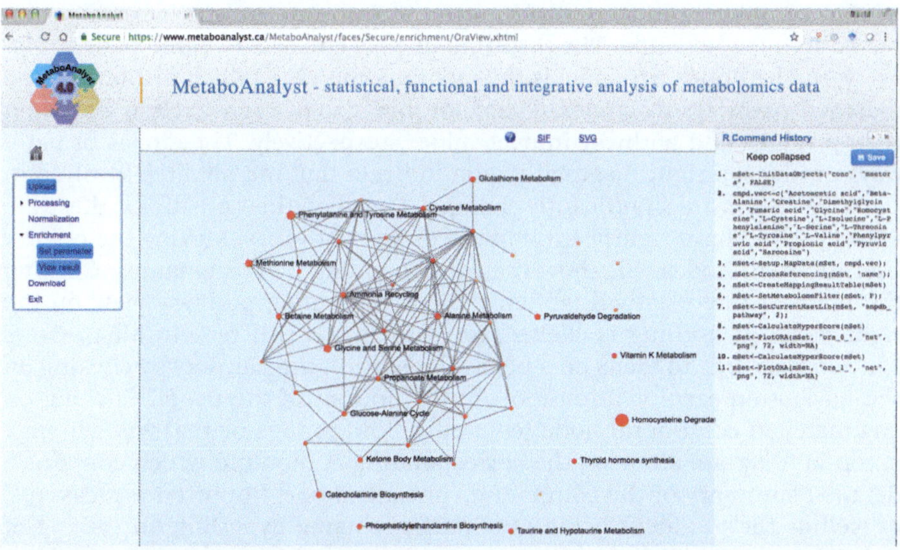


**Figure 9.8** A screenshot of a typical ORA analysis enrichment network generated *via* MetaboAnalyst.

labeled with pathways or diseases or SNPs depending on the library chosen. More detailed statistical information about the enrichment analysis is provided in the table below the network diagram.

Pathway Analysis in MetaboAnalyst is somewhat similar in terms of data input as Enrichment Analysis. Users have two options of either uploading lists of significant metabolite names, as determined *via* untargeted metabolomics, or lists of significant metabolites and their concentrations or relative concentrations, as determined by targeted metabolomics. If users choose the option of a concentration list, they must perform the usual data scaling, transformation and normalization as described in Section 9.4 (above). Three parameters need to be specified for pathway analysis – the pathway library, the algorithm for pathway enrichment analysis, and the algorithm for topological analysis. MetaboAnalyst currently contains KEGG[24] pathways libraries for 21 model organisms and SMPDB[25] pathway libraries for 2 model organisms. Pathway Analysis for quantitative (concentration) requires that one choose between Globaltest[26] and GlobalANCOVA,[27] which are two similar algorithms designed for testing differentially expressed genes or metabolites in functionally-related groups. For name-only lists, Pathway Analysis offers the option of choosing between the Hypergeometric test and Fischer's exact test. Degree centrality and betweenness centrality are two measures to estimate the importance of a compound within a given metabolic network. The former measures the number of connections the node of interest has to other nodes, while the latter measures the number of shortest paths going through the node of interest. The default choices for pathway enrichment analysis and for topological analysis offered by MetaboAnalyst are usually the safest choices, especially for new users.

After completing the required inputs, the Pathway Analysis is usually completed in a few seconds. The resulting display consists of three panels (Figure 9.9). The upper left panel is the metabolome view, which displays all the matched pathways as circles. The color and size of each circle is based on their *p*-values and pathway impact values, respectively. The circles or pathways on the top right diagonal region indicate that metabolites involved in those pathways are significantly changed. Based on their positions, these are more likely to have significant impacts on the pathways. Moving the pointer over different nodes will show the corresponding pathway names. Clicking the node of interests will launch the corresponding pathway view on the right panel (if nothing is clicked, the right panel will remain blue). Users can zoom or drag to focus on a particular section the pathway by clicking on the navigation/manipulation tools at the bottom of the panel. Clicking on any matched compound node (colored yellow, orange or red) will generate a pop-up diagram showing the corresponding compound which contains a detailed summary of the compound concentrations, importance measures, as well as the *p*-value. Clicking the pathway name hyperlink on the top of the right panel will direct the users to the corresponding pathway (either KEGG or SMPDB, depending on the initial choice of the pathway library). The lower table lists the pathway names (clicking a pathway name will update the

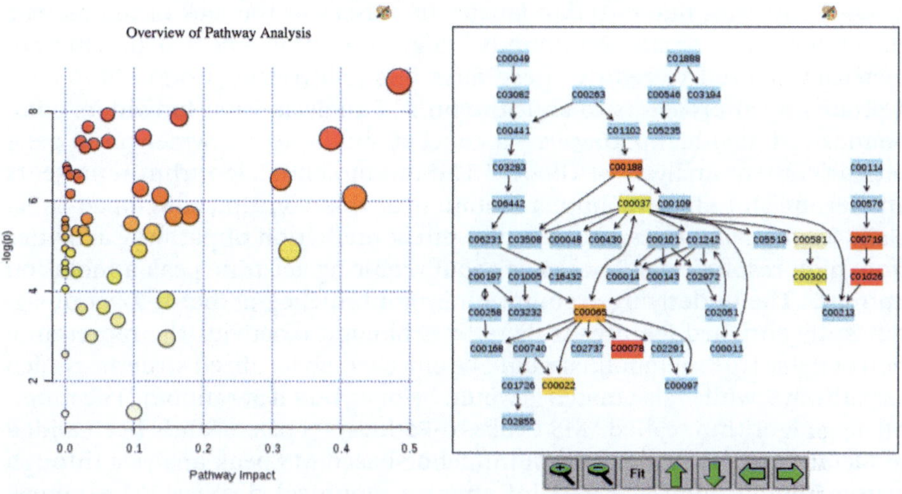


**Figure 9.9** A screenshot of a typical Pathway Analysis output as generated *via* MetaboAnalyst.

pathway diagram on the upper right), the matched metabolites (which will be displayed in a pop-up window as a highlighted list), the statistical values ($p$, log-$p$, false discovery rate or FDR, impact) as well as links to two different pathway databases (SMPDB and KEGG). Clicking on the pathway database links will generate a much more detailed display of the selected pathway.

Both the Pathway Analysis and Enrichment Analysis modules are designed to encourage interactive exploration and assessment of MetaboAnalyst's findings. Users should combine the results of these analyses with their own biological knowledge of the system they are working with and literature research to help guide their interpretations. One limitation of both the Pathway and Enrichment Analysis modules is their focus on the pure "catabolic" and "anabolic" pathways associated with primary metabolites. As many readers may know, metabolites also play key roles in cell signaling, protection, transport and homeostasis. Likewise, the metabolome includes many more compounds that simply the ~1000 or so primary metabolites depicted in KEGG pathways. Efforts are underway to expand these pathways and their metabolite coverage through the SMPDB and further incorporation of new SMPDB elements into MetaboAnalyst.

## 9.6 MS Peaks-to-Pathways and Mummichog

Conventional untargeted LC-MS-based procedures in metabolomics typically include peak identification, peak merging and alignment, and peak annotation. Peak annotations are often performed manually by searching through a variety of spectral or compound databases for *m*/*z* matches, or *m*/*z* and retention time matches. This process can often generate a number of false positives, due to redundancies in masses or the lack of unique *m*/*z* values for many small compounds[28] high-resolution MS instruments are increasingly used to reduce these false hits analytically. Computationally, a promising approach is to shift the unit of analysis from individual compounds to individual pathways – a concept similar to the widely used gene set enrichment analysis or GSEA.[29] The mummichog algorithm represents an elegant and efficient implementation of this concept, developed especially for metabolomics, that enables direct prediction of pathway activities from high-resolution MS peaks, without requiring accurate peak annotation upfront.[8] The underlying assumption in mummichog is that if a list of significantly enriched features truly reflects biological activity, the representation of these true metabolites would be enriched on localized structures such as pathways, while false matches would be distributed at random. The mummichog algorithm (called "MS Peaks-to-Pathways") has recently been added to MetaboAnalyst to support mummichog-based MS peak analysis through a user-friendly interface and informative graphical displays. The knowledgebase for this module consists of five genome-scale metabolic models obtained from the original Python implementation of mummichog that have either been manually curated or downloaded from BioCyc,[30] as well as an expanded library of 21 organisms derived from KEGG metabolic pathways. While compound identification is generally de-emphasized in mummichog, the *post hoc* analysis of the matched compounds is critical for downstream validation and interpretation. To address these needs, we implemented a KEGG style global metabolic network to allow users to visualize the global peak matching patterns and to further allow users to interactively zoom into a particular candidate compound to examine all of its matched isotopic or adduct forms.

To use this module, users must upload a table containing three columns – *m*/*z* features, *p*-values, and statistical scores (*e.g.* t-scores or fold-change values). If these values have not yet been calculated, users can use MetaboAnalyst's Statistical Analysis module (see Section 9.3) to upload their raw *m*/*z* peak tables and perform their statistical analysis of choice, then upload these results into the Peaks-to-Pathways module. Users also need to specify the mass accuracy, the ion mode (positive or negative), and the *p*-value cutoff to delineate between significantly enriched and non-significantly enriched *m*/*z* features. After the required data has been uploaded, users must select an organism (metabolome library) from which to perform the untargeted pathway analysis. This is similar to the Pathway Enrichment module described in Section 9.5.

The output of the Peaks-to-Pathways module consists of a table of results containing ranked pathways that are enriched in the user-uploaded data. The table includes the total number of hits, their raw *p*-values (Fisher's exact test or Hypergeometric value), their EASE score, and the *p*-value modeled on user data using a Gamma distribution. Users can click the "View" link to view the detailed hits for each pathway. A comprehensive table containing the compound matching information for all user-uploaded *m*/*z* features is

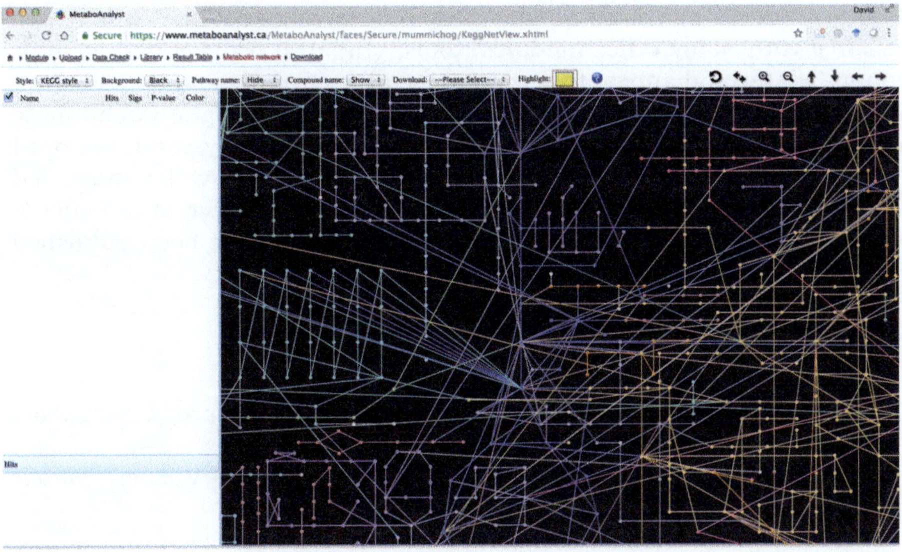


**Figure 9.10** A screenshot of a typical global metabolic network generated *via* mummichog and MetaboAnalyst.

also available for download. All of this information (pathways, compounds, and matched hits) can be interactively explored within the KEGG global metabolic network (Figure 9.10). The page consists of three panels: (1) a top toolbar containing different menus to control various visualizations, (2) a left-hand panel showing the pathway analysis results, and (3) a central viewing panel for interactive visual exploration of the metabolic network. Users can scroll through the global pathway with their mouse to zoom in and out of the network view. Clicking on a pathway name will highlight all of its compounds within the network. Double-clicking a node will show all the matching details for the corresponding compound as shown in the dialog (Figure 9.10).

The MS Peaks-to-Pathways module is one of the newest modules in MetaboAnalyst, however, its popularity is growing rapidly. This is because it offers additional biological insights that are not possible with standard metabolite annotation. Improvements to this module will include increased linkages to other small molecule pathways being generated through the SMPDB project.

## 9.7 Summary

This chapter was intended to provide a brief introduction to the latest version of the MetaboAnalyst web server (version 4.0) along with some of the more frequently used or more important modules for MS-based metabolomic analysis. Specifically, the major focus was on the Statistical Analysis, Enrichment Analysis, Pathway Analysis and MS-Peaks-to-Pathways modules.

As most readers may have already gathered, MetaboAnalyst is a rich web resource with many other modules, tools, graphical widgets and analysis options that would appeal to a wide range of user needs and applications. Space limitations do not allow for a comprehensive description of all these tools or a detailed explanation of the all the options. However, it is hoped that this chapter has provided a sufficiently detailed overview that interested readers will be drawn to explore MetaboAnalyst on their own or to read the many other detailed tutorials on MetaboAnalyst that have been published elsewhere.

## References

1. J. Xia, N. Psychogios, N. Young and D. S. Wishart, *Nucleic Acids Res.*, 2009, **37**(Web Server issue), W652–W660.
2. J. Xia, R. Mandal, I. V. Sinelnikov, D. Broadhurst and D. S. Wishart, *Nucleic Acids Res.*, 2012, **40**(Web Server issue), W127–W133.
3. J. Xia and D. S. Wishart, *Bioinformatics*, 2010, **26**, 2342–2344.
4. J. Xia and D. S. Wishart, *Nucleic Acids Res.*, 2010, **38**(Web Server issue), W71–W77.
5. J. Xia, I. V. Sinelnikov, B. Han and D. S. Wishart, *Nucleic Acids Res.*, 2015, **43**(W1), W251–W257.
6. J. Chong, O. Soufan, C. Li, I. Caraus, S. Li, G. Bourque, D. S. Wishart and J. Xia, *Nucleic Acids Res.*, 2018, **46**(W1), W486–W494.
7. J. Chong and J. Xia, *Bioinformatics*, 2018, **34**, 4313–4314.
8. S. Li, Y. Park, S. Duraisingham, F. H. Strobel, N. Khan, Q. A. Soltow, D. P. Jones and B. Pulendran, *PLoS Comput. Biol.*, 2013, **9**, e1003123.
9. J. Xia and D. S. Wishart, *Curr. Protoc. Bioinf.*, 2016, **55**, 14.10.1–14.10.91.
10. J. Xia and D. S. Wishart, *Nat. Protoc.*, 2011, **6**, 743–760.
11. J. Xia, D. I. Broadhurst, M. Wilson and D. S. Wishart, *Metabolomics*, 2013, **9**, 280–299.
12. P. Shannon, A. Markiel, O. Ozier, N. S. Baliga, J. T. Wang, D. Ramage, N. Amin, B. Schwikowski and T. Ideker, *Genome Res.*, 2003, **13**, 2498–2504.
13. S. Ravanbakhsh, P. Liu, T. C. Bjorndahl, R. Mandal, J. R. Grant, M. Wilson, R. Eisner, I. Sinelnikov, X. Hu, C. Luchinat, R. Greiner and D. S. Wishart, *PLoS One*, 2015, **10**, e0124219.
14. E. M. Forsberg, T. Huan, D. Rinehart, H. P. Benton, B. Warth, B. Hilmers and G. Siuzdak, *Nat. Protoc.*, 2018, **13**, 633–651.
15. A. Lommen and H. J. Kools, *Metabolomics*, 2012, **8**(4), 719–726.
16. H. L. Röst, T. Sachsenberg, S. Aiche, C. Bielow, H. Weisser, F. Aicheler, S. Andreotti, H. C. Ehrlich, P. Gutenbrunner, E. Kenar, X. Liang, S. Nahnsen, L. Nilse, J. Pfeuffer, G. Rosenberger, M. Rurik, U. Schmitt, J. Veit, M. Walzer, D. Wojnar, W. E. Wolski, O. Schilling, J. S. Choudhary, L. Malmström, R. Aebersold, K. Reinert and O. Kohlbacher, *Nat. Methods*, 2016, **13**, 741–748.

17. T. Pluskal, S. Castillo, A. Villar-Briones and M. Oresic, *BMC Bioinf.*, 2010, **11**, 395.
18. C. A. Smith, E. J. Want, G. O'Maille, R. Abagyan and G. Siuzdak, *Anal. Chem.*, 2006, **78**, 779–787.
19. H. Tsugawa, T. Cajka, T. Kind, Y. Ma, B. Higgins, K. Ikeda, M. Kanazawa, J. VanderGheynst, O. Fiehn and M. Arita, *Nat. Methods*, 2015, **12**, 523–526.
20. J. Chong, M. Yamamoto and J. Xia, *Metabolites*, 2019, **9**, E57.
21. B. Efron, R. Tibshirani, J. D. Storey and V. Tusher, *J. Am. Stat. Assoc.*, 2001, **96**, 1151–1160.
22. V. G. Tusher, R. Tibshirani and G. Chu, *Proc. Natl. Acad. Sci. U. S. A.*, 2001, **98**, 5116–5121.
23. D. S. Wishart, D. Tzur, C. Knox, R. Eisner, A. C. Guo, N. Young, D. Cheng, K. Jewell, D. Arndt, S. Sawhney, C. Fung, L. Nikolai, M. Lewis, M. A. Coutouly, I. Forsythe, P. Tang, S. Shrivastava, K. Jeroncic, P. Stothard, G. Amegbey, D. Block, D. D. Hau, J. Wagner, J. Miniaci, M. Clements, M. Gebremedhin, N. Guo, Y. Zhang, G. E. Duggan, G. D. Macinnis, A. M. Weljie, R. Dowlatabadi, F. Bamforth, D. Clive, R. Greiner, L. Li, T. Marrie, B. D. Sykes, H. J. Vogel and L. Querengesser, *Nucleic Acids Res.*, 2007, **35**(Database issue), D521–D526.
24. M. Kanehisa, M. Furumichi, M. Tanabe, Y. Sato and K. Morishima, *Nucleic Acids Res.*, 2017, **45**(D1), D353–D361.
25. A. Frolkis, C. Knox, E. Lim, T. Jewison, V. Law, D. D. Hau, P. Liu, B. Gautam, S. Ly, A. C. Guo, J. Xia, Y. Liang, S. Shrivastava and D. S. Wishart, *Nucleic Acids Res.*, 2010, **38**(Database issue), D480–D487.
26. J. J. Goeman, S. A. van de Geer, F. de Kort and H. C. van Houwelingen, *Bioinformatics*, 2004, **20**, 93–99.
27. M. Hummel, R. Meister and U. Mansmann, *Bioinformatics*, 2008, **24**, 78–85.
28. T. Kind and O. Fiehn, *BMC Bioinf.*, 2007, **8**, 105.
29. A. Subramanian, P. Tamayo, V. K. Mootha, S. Mukherjee, B. L. Ebert, M. A. Gillette, A. Paulovich, S. L. Pomeroy, T. R. Golub, E. S. Lander and J. P. Mesirov, *Proc. Natl. Acad. Sci. U. S. A.*, 2005, **102**, 15545–15550.
30. M. Latendresse, S. Paley and P. D. Karp, *Methods Mol. Biol.*, 2012, **804**, 197–216.

CHAPTER 10

# *Modular metaX Pipeline for Processing Untargeted Metabolomics Data*†

BO WEN*[a,b]

[a]Lester and Sue Smith Breast Center, Baylor College of Medicine, Houston, Texas 77030, USA; [b]Department of Molecular and Human Genetics, Baylor College of Medicine, Houston, Texas 77030, USA
*E-mail: bo.wen@bcm.edu

## 10.1 Introduction

Mass spectrometry coupled with either liquid chromatography (LC-MS) or gas chromatography (GC-MS) has become a popular and powerful technique in metabolomics studies, which aims at comprehensive profiling of all small molecule metabolites (<1500 Da) in biological systems. The global metabolomics technique, also known as untargeted metabolomics, typically generates large datasets with thousands of signals including true biological signals from metabolites as well as noise signals from contaminants and artifacts. Moreover, signal drift and batch effect are frequently encountered in large-scale untargeted metabolomics studies. Computationally intensive processing and analyses are required to handle these issues and generate biologically meaningful results.

†The source code of metaX is available at https://github.com/wenbostar/metaX/.

New Developments in Mass Spectrometry No. 8
Processing Metabolomics and Proteomics Data with Open Software: A Practical Guide
Edited by Robert Winkler

Published by the Royal Society of Chemistry, www.rsc.org

The steps involved in analyzing MS-based metabolomics data usually include peak picking, quality control, data cleaning, pre-processing, univariate and multivariate statistical analysis, and data visualization. Many open-source tools and algorithms have been developed for various aspects of the metabolomics data analysis pipeline.[1–4] Some of them cover limited processing steps while others offer comprehensive pipelines for metabolomics data analysis. An overview of software and methods for metabolomics data analysis can be found in recent reviews.[1–3] However, because metabolomics studies cover a wide range of purposes and experimental methods, the data analysis steps used for one study may differ from those used for another.[5]

metaX, a comprehensive pipeline allowing for the flexible and transparent analysis of untargeted metabolomics data, is one software solution for analysis. Basically, the pipeline aims to allow users to easily perform end-to-end metabolomics data analysis with a flexible combination of different methods, allowing the efficient integration of new modules and customization of the pipeline in multiple ways. At the present time, R is a popular statistical programming environment that provides a convenient environment for statistical analysis of metabolomic and other -omics data. metaX was designed as an R package that automates analysis of untargeted metabolomics data acquired from LC-MS or GC-MS. This package, which is open source and rich in functions, encourages experienced programmers to improve the relevant functions or to build their own pipelines within the R framework. The latest source code of metaX is available at https://github.com/wenbostar/metaX/.

## 10.2 Data Preparation

metaX requires two types of files as input. The first is a sample meta information file in tab-delimited text format which contains the sample names (column name: sample), experiment batch (column name: batch), sample injection order (column name: order), and sample class (column name: class). If quality control (QC) samples are present in the file, the class for QC samples should be set as "NA". Typically, QC samples are comprised of a pool of the study (experiment) samples, are processed the same as the study samples, and are periodically analyzed throughout an analytical run.[6–8] The second file is a peak table file in tab-delimited text format which contains detected metabolite features (or peaks) and their intensity values. The first column in this file should be peak IDs and titled "ID" with the remaining columns containing the peak intensity values for different samples. Alternatively, the peak table files generated by Progenesis QI (exported comma separated value (csv) format file), XCMS, MZmine, and MetaboAnalyst and open file format MS data (such as mzXML, NetCDF) are supported as input. If taking mzXML files as input, metaX will use the R package XCMS[9] to detect peaks, and then use the CAMERA[10] package to perform peak annotation. The functions implemented in metaX for importing data are shown in Table 10.1. Once

**Table 10.1** Data import functions in metaX.

| Function name | Description |
|---|---|
| importDataFromMetaboAnalyst | Import MetaboAnalyst format data |
| importDataFromQI | Import progenesis QI exported data |
| importDataFromXCMS | Import XCMS generated data |
| importDataFromTable | Import metaX compatible tab-delimited format data |
| importMSdata | Import MS/MS data |

imported, metaX stores processing parameters, sample meta information, and peak data in an S4 class "metaXpara" object. An example input dataset is available at https://github.com/wenbostar/metaX/wiki.

## 10.3 Data Pre-processing

The first step in metabolomics data analysis is typically data pre-processing. In metaX, these functions include peak filtering, missing value imputation, data transformation, data scaling, and outlier sample detection (Table 10.2). Missing values, both for biological and technical reasons, are widespread in LC-MS or GC-MS metabolomics data and should be handled depending on their cause to avoid their adverse effects on statistical analysis.[11,12] To reduce the effect of missing values, the peaks can first be filtered based on the percentage of missing values in experimental samples, or QC samples when available. The default in metaX is to remove metabolite features detected in fewer than 50% of QC samples when QC samples are present in the dataset or detected in fewer than 80% of all experimental samples ("80% rule").[11,13] The "modified 80% rule" is also supported in metaX, in which a peak can be excluded from the data when the proportion of non-missing values is fewer than 80% among each biological group.[14] However, the threshold of missing value percentage can be set by users. If QC samples are present, a further peak filtering step can be performed based on the coefficient of variation (CV; also termed relative standard deviation, RSD) of peaks in QC samples. A metabolite feature with a CV over a predetermined value is excluded from the data after data normalization. The default CV threshold of 30% can be modified by users.

After the missing value filtering steps, the remaining missing values can be imputed if complete data are required. This is necessary for some analysis methods, such as clustering, principal component analysis (PCA), and partial least squares-discriminant analysis (PLS-DA). The choice of missing value imputation method on metabolomics data analysis can significantly affect results.[11,15–17] In metaX, four methods are implemented to perform missing value imputation: k-nearest neighbor (KNN), Bayesian principal component analysis replacement (BPCA), svdImpute, and random forest imputation (missForest).[18] The default method in metaX is KNN due to its speed and good performance.[16]

**Table 10.2** Data pre-processing functions in metaX.

| Function name | Description |
|---|---|
| filterPeaks | Filtering peaks based on the percentage of missing values in experiment samples |
| filterQCPeaks | Filtering peaks based on the percentage of missing values in QC samples |
| filterQCPeaksByCV | Filtering peaks based on CV in QC samples |
| missingValueImpute | Missing value imputation |
| preProcess | Data transformation and data scaling |
| autoRemoveOutlier | Outlier samples detection |

Data transformation and data scaling are usually required prior to further analysis. In terms of data transformation, three approaches are available in metaX: log, generalized logarithm (glog), and cube root transformation. Five different scaling approaches are available in metaX: pareto scaling, vast scaling, range scaling, autoscaling, and level scaling. The formulas of these scaling approaches and selection of an appropriate scaling method for a dataset are described in detail elsewhere.[19,20]

Finally, metaX can automatically detect outlier samples based on expansion of the Hotelling's T2 distribution ellipse.[21] A sample within the first and second component of the PCA score plot beyond the expanded ellipse is removed, and then the PCA model is recalculated. By default, three rounds of outlier removal are performed.

## 10.4 Data Quality Assessment

In metabolomics studies, data quality checks are crucial prerequisites to achieve reliable results. Insufficient data quality may result in the failure to generate a hypothesis, or in the worst case, a false or skewed hypothesis.[22] metaX offers a quick and easy data quality check of metabolomics data through visualizations. This can be done using the R function "QCBench" in metaX or the user-friendly web interface at the website http://metax.genomics.cn/ as shown in our previous study.[23] Using metaX, pre- and post-normalization, the data quality is visually assessed in several aspects:

1. The peak number distribution, *i.e.*, the number of features detected per sample over the analysis time (injection order).
2. The missing value distribution for all features in both the QC and experiment samples.
3. A boxplot of peak intensity across samples. The intensity values of all features per sample before and after normalization over the analysis time can be presented side by side.
4. The total peak intensity distribution across samples. *i.e.*, the sum intensity of all features per sample before and after normalization over the analysis time.

5. A correlation heatmap of QC samples (if available) and experiment samples. The correlation plots of QC samples before normalization can be used to check if there is batch effect present in the data. In general, after data normalization the correlations of QC samples should be improved.
6. The metabolite *m*/*z* (or mass) distribution.
7. A plot of *m*/*z versus* retention time.
8. The CV distribution of features, *i.e.*, the CV distribution of all features before and after normalization for each group. In general, the signal quality should be improved after normalization.
9. The PCA score and loading plots of all samples, which give an overview of the dataset and show trends, groupings, and outliers before and after data normalization. The score plot of PCA for the non-normalized data can also provide a simple and easily interpretable visual check of the presence of batch effects. The score plot of PCA for the normalized data with QC samples present gives an initial evaluation of the data quality by examining if all of the QC samples are clustered tightly.[8,22,24] A few example QC plots can be found in our previous study.[23]

A complete example to demonstrate the data quality assessment analysis is shown below. The metabolomics dataset used in this example was generated in a previous study[25] and contains 116 samples in total (26 QC samples, 45 disease samples, and 45 healthy samples). The data were generated in two batches using a G2-XS QTOF mass spectrometer (Waters, Manchester, UK). The raw MS/MS data, available in both the MetaboLights (Study Identifier: MTBLS408) and GigaDB[26] databases, were processed using Progenesis QI (Waters, Nonlinear Dynamics, USA) software as described in the previous publication[25] and over 10 000 features were detected. Both the input files and generated QC report are available at https://github.com/wenbostar/metaX.

```
library(metaX)
# x: a file contains peak intensity data.
# file_type: the file type of the peak intensity data file. 3: Progenesis QI csv format.
QCBench(x ="measurement_pos.csv",file_type =3,sample_list_file ="sample_list_pos.txt")
```

## 10.5 Normalization Evaluation

Various forms of unwanted signal variations, which arise from steps during sample processing and data generation, can be present in MS-based metabolomics data.[27,28] These variations can detrimentally impact significant signal discovery, data interpretation, and statistical power and must be removed in order to focus on biological signals of interest. A number of normalization methods have been developed to remove these unwanted signal variations. Recently, a few studies systematically evaluated different normalization methods on metabolomics data[27–30] and showed that the optimal normalization method may differ based on the dataset. In metaX, two types of normalization methods are provided: (1) Sample-based normalization, such as normalization to total sum, median-based normalization,[31] probabilistic quotient normalization (PQN),[32] variance stabilizing normalization (VSN),[33] quantile-based methods, and trimmed mean of M-values normalization (TMM),[34] is used to correct different sample concentrations. (2) Peak-based normalization is implemented to correct analytical variation within an experiment batch and batch-to-batch variation in large-scale studies.[35] If a study contains QC samples, the widely accepted QC-robust spline batch correction (QC-RSC) can be used.[35] During normalization, the degree of smoothening is controlled by a parameter that sets the proportion of points for smoothening at each point. In metaX, this parameter is automatically determined using leave-one-out cross validation to avoid overfitting. The QC-based support vector regression normalization (QC-SVR)[36] and ComBat[37] methods are also available in metaX. When performing normalization using the ComBat method, the class information for experiment samples can be used as a covariate.

A total of 10 normalization methods are implemented in metaX as shown in Figure 10.1. In order to select an appropriate normalization method for a given dataset, different normalization methods should be evaluated. In metaX, this can be performed using the shiny-based web server interface (http://metax.genomics.cn) introduced in our previous study.[23] It can also be performed using the R function MetBench, which can be used with a few parameters in command line mode. Furthermore, this function can be used in other metabolomics data analysis workflows. An HTML-based evaluation report with all evaluation results included can be generated by this function. This enables users to intuitively select an optimal normalization method.

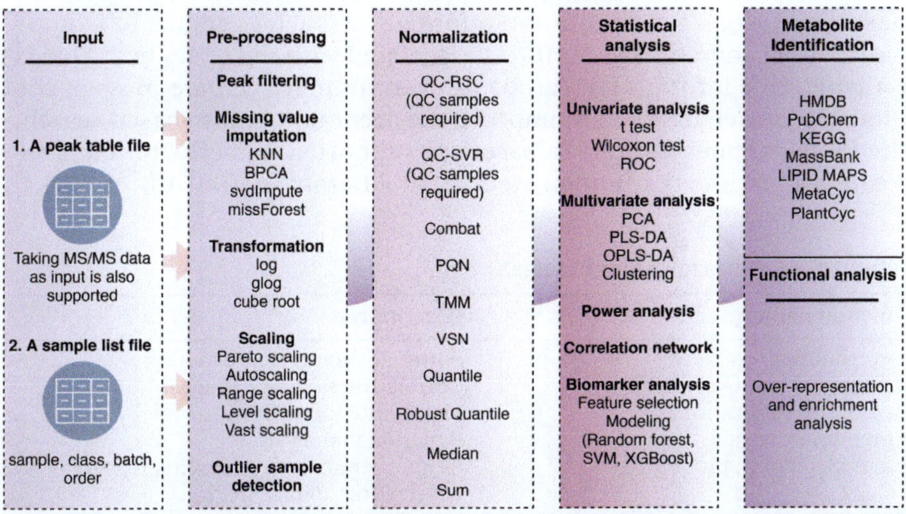


**Figure 10.1** Overview of the metaX framework.

A complete example demonstrating normalization evaluation analysis is shown below. The same dataset as above was used and the generated normalization evaluation report is available at https://github.com/wenbostar/metaX.

```
library(metaX)
# x: a file contains peak intensity data.
# file_type: the file type of the peak intensity data file. 3: Progenesis QI csv format.
MetBench(x = "measurement_pos.csv",file_type =3,sample_list_file = "sample_list_pos.txt")
```

## 10.6 Other Functions in metaX

In addition to the functions described above, metaX provides other functions as shown in Table 10.3 and Figure 10.1, including power and sample size analysis, metabolite correlation network analysis, metabolite identification, functional analysis, and biomarker analysis using machine learning algorithms. Specifically, power and sample size analysis are mainly performed based on the functions implemented in the Bioconductor package SSPA[38] and a figure is generated to show the distribution curve of sample size *versus* the estimated power.

For downstream analysis, accurate identification of metabolites is essential for assessing biological meaning. The metaX package provides functions for metabolite identification as well as functional analysis. Currently, metaX provides a function for fast metabolite identification based on mass search using a Java program implemented in metaX. Specifically, each query *m*/*z* value is searched against a metabolite database while considering user-specified adduct information. Candidate metabolites from the database are retrieved based on the mass difference between a metabolite and the query *m*/*z* value and a pre-specified mass tolerance. If ion annotation information (such as adduct or isotope annotation generated by CAMERA) associated with the query *m*/*z* value is available, this information can be utilized to infer accurate mass of the query ion, which in turn can improve the accuracy of mass-based search. The following metabolite databases are supported in metaX for metabolite identification: the Human Metabolome Database (HMDB),[39] KEGG,[40]

**Table 10.3** Other functions in metaX.

| Function name | Description |
|---|---|
| powerAnalyst | Power and sample size analysis |
| cor.network, plotNetwork | Correlation network analysis |
| metaboliteAnnotation | Metabolite identification |
| FunctionalAnalysis | Functional analysis |
| featureSelection, plotROC | Biomarker analysis including feature selection and modeling |

MassBank,[41] PubChem,[42] LIPID MAPS,[43] MetaCyc,[44] and PlantCyc (www.plantcyc.org). Moreover, metaX can easily be extended to support other databases.

In terms of functional analysis, metaX provides a function for metabolite functional analysis based on ConsensusPathDB, which is a web-based integrated functional analysis tool that can be used to perform over-representation or enrichment analysis of metabolites.[45] metaX directly retrieves data from the ConsensusPathDB online database at each time of analysis using the functions from the R package RCurl. Different types of metabolite identifiers, such as HMDB, KEGG, PubChem Compound, and PubChem Substance, are supported. For biomarker analysis, metaX makes use of the functions from the R package caret to perform the biomarker selection, model creation, and performance evaluation.[46] Currently, two methods, random forest and support vector machine (SVM), are implemented to automatically select a set of best performing features. After the best set of features are selected, a randomForest or XGBoost model can be created and the ROC curve can be plotted.

Network analysis can be a complementary method to univariate and multivariate statistical analysis methods and correlation network analysis has been demonstrated to be a useful method in metabolomics data mining.[47–49] Two types of network analysis methods are implemented in metaX. One is a correlation network analysis without regard for experimental groups information, and the other is a differential correlation network analysis, which aims to identify metabolite correlation differences across conditions. The first method is implemented using the cor function from the stats package to calculate the correlation coefficient, and the second method is implemented using the function comp.2.cc.fdr from the DiffCorr package[50] to calculate the significantly differential correlations. The igraph package[51] is used for network analysis and visualization of the results. In addition, the network can be exported as a file in formats such as gml or pajek, which can be imported into Cytoscape[52] and Gephi[53] for further analysis and visualization. Both of these correlation network analyses aim to describe the correlation patterns among metabolites across samples with a pre-defined correlation coefficient threshold, in which nodes represent metabolites and edges represent the correlation between different metabolites.

## 10.7 Integrated Function metaXpipe

In order to facilitate users to perform end-to-end metabolomics data analysis, a function metaXpipe consisting of different processing steps is provided in metaX. In general, users can use this function to do most of the commonly used analyses. It includes the following steps: (1) peak picking and annotation when MS/MS data is the input; (2) data pre-processing including peak filtering, missing value imputation, and outlier sample removal; (3) data normalization; (4) data transformation and data scaling; (5) commonly used univariate statistical analysis and multivariate statistical analysis such as PCA and PLS-DA analysis; and (6) data quality assessment before and after data normalization. This function generates an HTML-based report which presents the analysis result. Users can easily transition to their downstream analyses of choice using the processed data generated by this function.

Finally, a complete example to show how to run the integrated analysis is shown below. The generated result is available at https://github.com/wenbostar/metaX.

```
library(metaX)
para = importDataFromQI(x = " measurement_pos.csv ", sample_list_file = "sample_list_pos.txt")
metaXpipe(para, cvFilter=0.3, qcsc=1, scale="pareto", t=1, remveOutlier=TRUE, nor.method="combat")
```

## 10.8 Applications

The flexibility of metaX has been demonstrated based on its use by a variety of different metabolomics studies, including studies on lipid profiles of patients with type 2 diabetes mellitus (T2D)[54] and psoriasis,[25] metabolites in yeast strains,[55] and liver metabolites associated with *Toxoplasma gondii* infection.[56] More recently, some of the functions in metaX were also used to analyze proteomics data,[57,58] highlighting the potential use of the tool in analysis of multi-omics studies.

## Acknowledgements

We thank Sara R. Savage for proofreading the manuscript. This work was partly supported by the National Cancer Institute (NCI) CPTAC award U24 CA210954.

## References

1. B. B. Misra and S. Mohapatra, *Electrophoresis*, 2019, **40**, 227–246.
2. B. B. Misra, *Electrophoresis*, 2018, **39**, 909–923.
3. R. Spicer, R. M. Salek, P. Moreno, D. Canueto and C. Steinbeck, *Metabolomics*, 2017, **13**, 106.
4. K. Peters, J. Bradbury, S. Bergmann, M. Capuccini, M. Cascante, P. de Atauri, T. M. D. Ebbels, C. Foguet, R. Glen, A. Gonzalez-Beltran, U. L. Gunther, E. Handakas, T. Hankemeier, K. Haug, S. Herman, P. Holub, M. Izzo, D. Jacob, D. Johnson, F. Jourdan, N. Kale, I. Karaman, B. Khalili, P. Emami Khonsari, K. Kultima, S. Lampa, A. Larsson, C. Ludwig, P. Moreno, S. Neumann, J. A. Novella, C. O'Donovan, J. T. M. Pearce, A. Peluso, M. E. Piras, L. Pireddu, M. A. C. Reed, P. Rocca-Serra, P. Roger,

A. Rosato, R. Rueedi, C. Ruttkies, N. Sadawi, R. M. Salek, S. A. Sansone, V. Selivanov, O. Spjuth, D. Schober, E. A. Thevenot, M. Tomasoni, M. van Rijswijk, M. van Vliet, M. R. Viant, R. J. M. Weber, G. Zanetti and C. Steinbeck, *GigaScience*, 2019, **8**, 1–12.

5. S. Ren, A. A. Hinzman, E. L. Kang, R. D. Szczesniak and L. J. Lu, *Metabolomics*, 2015, **11**, 1492–1513.
6. G. Quintás, Á. Sánchez-Illana, J. D. Piñeiro-Ramos and J. Kuligowski, in *Data Analysis for Omic Sciences: Methods and Applications*, 2018, pp. 137–164.
7. W. B. Dunn, D. Broadhurst, P. Begley, E. Zelena, S. Francis-McIntyre, N. Anderson, M. Brown, J. D. Knowles, A. Halsall, J. N. Haselden, A. W. Nicholls, I. D. Wilson, D. B. Kell, R. Goodacre and Human Serum Metabolome (HUSERMET) Consortium, *Nat. Protoc.*, 2011, **6**, 1060–1083.
8. D. Broadhurst, R. Goodacre, S. N. Reinke, J. Kuligowski, I. D. Wilson, M. R. Lewis and W. B. Dunn, *Metabolomics*, 2018, **14**, 72.
9. C. A. Smith, E. J. Want, G. O'Maille, R. Abagyan and G. Siuzdak, *Anal. Chem.*, 2006, **78**, 779–787.
10. C. Kuhl, R. Tautenhahn, C. Bottcher, T. R. Larson and S. Neumann, *Anal. Chem.*, 2012, **84**, 283–289.
11. R. Wei, J. Wang, M. Su, E. Jia, S. Chen, T. Chen and Y. Ni, *Sci. Rep.*, 2018, **8**, 663.
12. O. Hrydziuszko and M. R. Viant, *Metabolomics*, 2011, **8**, 161–174.
13. S. Bijlsma, I. Bobeldijk, E. R. Verheij, R. Ramaker, S. Kochhar, I. A. Macdonald, B. van Ommen and A. K. Smilde, *Anal. Chem.*, 2006, **78**, 567–574.
14. J. Yang, X. Zhao, X. Lu, X. Lin and G. Xu, *Front. Mol. Biosci.*, 2015, **2**, 4.
15. K. T. Do, S. Wahl, J. Raffler, S. Molnos, M. Laimighofer, J. Adamski, K. Suhre, K. Strauch, A. Peters, C. Gieger, C. Langenberg, I. D. Stewart, F. J. Theis, H. Grallert, G. Kastenmuller and J. Krumsiek, *Metabolomics*, 2018, **14**, 128.
16. R. Di Guida, J. Engel, J. W. Allwood, R. J. Weber, M. R. Jones, U. Sommer, M. R. Viant and W. B. Dunn, *Metabolomics*, 2016, **12**, 93.
17. P. S. Gromski, Y. Xu, H. L. Kotze, E. Correa, D. I. Ellis, E. G. Armitage, M. L. Turner and R. Goodacre, *Metabolites*, 2014, **4**, 433–452.
18. D. J. Stekhoven and P. Buhlmann, *Bioinformatics*, 2012, **28**, 112–118.
19. R. A. van den Berg, H. C. Hoefsloot, J. A. Westerhuis, A. K. Smilde and M. J. van der Werf, *BMC Genomics*, 2006, **7**, 142.
20. P. S. Gromski, Y. Xu, K. A. Hollywood, M. L. Turner and R. Goodacre, *Metabolomics*, 2014, **11**, 684–695.
21. W. M. Edmands, D. K. Barupal and A. Scalbert, *Bioinformatics*, 2015, **31**, 788–790.
22. M. K. R. Engskog, J. Haglöf, T. Arvidsson and C. Pettersson, *Metabolomics*, 2016, **12**, 114.
23. B. Wen, Z. Mei, C. Zeng and S. Liu, *BMC Bioinf.*, 2017, **18**, 183.

24. D. Dudzik, C. Barbas-Bernardos, A. Garcia and C. Barbas, *J. Pharm. Biomed. Anal.*, 2018, **147**, 149–173.
25. C. Zeng, B. Wen, G. Hou, L. Lei, Z. Mei, X. Jia, X. Chen, W. Zhu, J. Li, Y. Kuang, W. Zeng, J. Su, S. Liu, C. Peng and X. Chen, *GigaScience*, 2017, **6**, 1–11.
26. C. Zeng, B. Wen, G. Hou, L. Lei, Z. Mei, X. Jia, X. Chen, W. Zhu, J. Li, Y. Kuang, W. Zeng, J. Su, S. Liu, C. Peng and X. Chen, *GigaScience Database*, 2017, DOI: 10.5524/100341.
27. B. Li, J. Tang, Q. Yang, S. Li, X. Cui, Y. Li, Y. Chen, W. Xue, X. Li and F. Zhu, *Nucleic Acids Res.*, 2017, **45**, W162–W170.
28. A. M. De Livera, D. A. Dias, D. De Souza, T. Rupasinghe, J. Pyke, D. Tull, U. Roessner, M. McConville and T. P. Speed, *Anal. Chem.*, 2012, **84**, 10768–10776.
29. S. Wang, X. Chen, D. Dan, W. Zheng, L. Hu, H. Yang, J. Cheng and M. Gong, *Anal. Chem.*, 2018, **90**, 11124–11130.
30. A. M. De Livera, G. Olshansky, J. A. Simpson and D. J. Creek, *Metabolomics*, 2018, **14**, 54.
31. B. Wen, R. Zhou, Q. Feng, Q. Wang, J. Wang and S. Liu, *Proteomics*, 2014, **14**, 2280–2285.
32. F. Dieterle, A. Ross, G. Schlotterbeck and H. Senn, *Anal. Chem.*, 2006, **78**, 4281–4290.
33. W. Huber, A. von Heydebreck, H. Sultmann, A. Poustka and M. Vingron, *Bioinformatics*, 2002, **18**(Suppl. 1), S96–S104.
34. M. D. Robinson and A. Oshlack, *Genome Biol.*, 2010, **11**, R25.
35. J. A. Kirwan, D. I. Broadhurst, R. L. Davidson and M. R. Viant, *Anal. Bioanal. Chem.*, 2013, **405**, 5147–5157.
36. X. Shen, X. Gong, Y. Cai, Y. Guo, J. Tu, H. Li, T. Zhang, J. Wang, F. Xue and Z.-J. Zhu, *Metabolomics*, 2016, **12**, 89.
37. W. E. Johnson, C. Li and A. Rabinovic, *Biostatistics*, 2007, **8**, 118–127.
38. M. van Iterson, P. A. t Hoen, P. Pedotti, G. J. Hooiveld, J. T. den Dunnen, G. J. van Ommen, J. M. Boer and R. X. Menezes, *BMC Genomics*, 2009, **10**, 439.
39. D. S. Wishart, Y. D. Feunang, A. Marcu, A. C. Guo, K. Liang, R. Vazquez-Fresno, T. Sajed, D. Johnson, C. Li, N. Karu, Z. Sayeeda, E. Lo, N. Assempour, M. Berjanskii, S. Singhal, D. Arndt, Y. Liang, H. Badran, J. Grant, A. Serra-Cayuela, Y. Liu, R. Mandal, V. Neveu, A. Pon, C. Knox, M. Wilson, C. Manach and A. Scalbert, *Nucleic Acids Res.*, 2018, **46**, D608–D617.
40. M. Kanehisa, S. Goto, Y. Sato, M. Furumichi and M. Tanabe, *Nucleic Acids Res.*, 2012, **40**, D109–D114.
41. H. Horai, M. Arita, S. Kanaya, Y. Nihei, T. Ikeda, K. Suwa, Y. Ojima, K. Tanaka, S. Tanaka, K. Aoshima, Y. Oda, Y. Kakazu, M. Kusano, T. Tohge, F. Matsuda, Y. Sawada, M. Y. Hirai, H. Nakanishi, K. Ikeda, N. Akimoto, T. Maoka, H. Takahashi, T. Ara, N. Sakurai, H. Suzuki, D. Shibata, S. Neumann, T. Iida, K. Tanaka, K. Funatsu, F. Matsuura, T.

Soga, R. Taguchi, K. Saito and T. Nishioka, *J. Mass Spectrom.*, 2010, **45**, 703–714.

42. S. Kim, P. A. Thiessen, E. E. Bolton, J. Chen, G. Fu, A. Gindulyte, L. Han, J. He, S. He, B. A. Shoemaker, J. Wang, B. Yu, J. Zhang and S. H. Bryant, *Nucleic Acids Res.*, 2016, **44**, D1202–D1213.
43. M. Sud, E. Fahy, D. Cotter, A. Brown, E. A. Dennis, C. K. Glass, A. H. Merrill Jr, R. C. Murphy, C. R. Raetz, D. W. Russell and S. Subramaniam, *Nucleic Acids Res.*, 2007, **35**, D527–D532.
44. R. Caspi, R. Billington, L. Ferrer, H. Foerster, C. A. Fulcher, I. M. Keseler, A. Kothari, M. Krummenacker, M. Latendresse, L. A. Mueller, Q. Ong, S. Paley, P. Subhraveti, D. S. Weaver and P. D. Karp, *Nucleic Acids Res.*, 2016, **44**, D471–D480.
45. A. Kamburov, U. Stelzl, H. Lehrach and R. Herwig, *Nucleic Acids Res.*, 2013, **41**, D793–D800.
46. M. Kuhn, *J. Stat. Softw.*, 2008, **28**, 1–26.
47. Z. Li, Y. Zhang, T. Hu, S. Likhodii, G. Sun, G. Zhai, Z. Fan, C. Xuan and W. Zhang, *PLoS One*, 2018, **13**, e0207775.
48. A. Rosato, L. Tenori, M. Cascante, P. R. De Atauri Carulla, V. A. P. Martins Dos Santos and E. Saccenti, *Metabolomics*, 2018, **14**, 37.
49. A. Mock, R. Warta, S. Dettling, B. Brors, D. Jager and C. Herold-Mende, *Bioinformatics*, 2018, **34**, 3417–3418.
50. A. Fukushima, *Gene*, 2013, **518**, 209–214.
51. G. Csardi and T. Nepusz, *InterJ. Complex Syst.*, 2006, **1695**, 1–9.
52. P. Shannon, A. Markiel, O. Ozier, N. S. Baliga, J. T. Wang, D. Ramage, N. Amin, B. Schwikowski and T. Ideker, *Genome Res.*, 2003, **13**, 2498–2504.
53. M. Bastian, S. Heymann and M. Jacomy, Third International AAAI Conference on Weblogs and Social Media, 2009.
54. H. Zhong, C. Fang, Y. Fan, Y. Lu, B. Wen, H. Ren, G. Hou, F. Yang, H. Xie, Z. Jie, Y. Peng, Z. Ye, J. Wu, J. Zi, G. Zhao, J. Chen, X. Bao, Y. Hu, Y. Gao, J. Zhang, H. Yang, J. Wang, L. Madsen, K. Kristiansen, C. Ni, J. Li and S. Liu, *GigaScience*, 2017, **6**, 1–12.
55. Y. Shen, Y. Wang, T. Chen, F. Gao, J. Gong, D. Abramczyk, R. Walker, H. Zhao, S. Chen, W. Liu, Y. Luo, C. A. Muller, A. Paul-Dubois-Taine, B. Alver, G. Stracquadanio, L. A. Mitchell, Z. Luo, Y. Fan, B. Zhou, B. Wen, F. Tan, Y. Wang, J. Zi, Z. Xie, B. Li, K. Yang, S. M. Richardson, H. Jiang, C. E. French, C. A. Nieduszynski, R. Koszul, A. L. Marston, Y. Yuan, J. Wang, J. S. Bader, J. Dai, J. D. Boeke, X. Xu, Y. Cai and H. Yang, *Science*, 2017, **355**, eaaf4791.
56. X. Q. Chen, H. M. Elsheikha, R. S. Hu, G. X. Hu, S. L. Guo, C. X. Zhou and X. Q. Zhu, *Front. Cell. Infect. Microbiol.*, 2018, **8**, 189.
57. S. Vasaikar, C. Huang, X. Wang, V. A. Petyuk, S. R. Savage, B. Wen, Y. Dou, Y. Zhang, Z. Shi, O. A. Arshad, M. A. Gritsenko, L. J. Zimmerman, J. E. McDermott, T. R. Clauss, R. J. Moore, R. Zhao, M. E. Monroe, Y. T. Wang,

M. C. Chambers, R. J. C. Slebos, K. S. Lau, Q. Mo, L. Ding, M. Ellis, M. Thiagarajan, C. R. Kinsinger, H. Rodriguez, R. D. Smith, K. D. Rodland, D. C. Liebler, T. Liu, B. Zhang and Clinical Proteomic Tumor Analysis Consortium, *Cell*, 2019, **177**, 1035–1049.e1019.

58. Q. Hu, C. Li, S. Wang, Y. Li, B. Wen, Y. Zhang, K. Liang, J. Yao, Y. Ye, H. Hsiao, T. K. Nguyen, P. K. Park, S. D. Egranov, D. H. Hawke, J. R. Marks, L. Han, M. C. Hung, B. Zhang, C. Lin and L. Yang, *Cell Res.*, 2019, **29**, 286–304.

CHAPTER 11

# *Metabolite Annotation With CEU Mass Mediator*

ALBERTO GIL-DE-LA-FUENTE*[a,b], JOANNA GODZIEN[a,c], ABRAHAM OTERO[a,b] AND CORAL BARBAS[a]

[a]Universidad San Pablo-CEU, CEU Universities, Centre for Metabolomics and Bioanalysis (CEMBIO), Facultad de Farmacia, Campus Monteprı́ncipe, Madrid, 28668, Spain; [b]Universidad San Pablo-CEU, CEU Universities, Department of Information Technology, Escuela Politécnica Superior, Campus Monteprı́ncipe, Madrid, 28668, Spain; [c]Clinical Research Centre, Medical University of Bialystok, Bialystok, 15-001, Poland
*E-mail: alberto.gilf@gmail.com

## 11.1 Introduction

Metabolite identification is one of the most challenging steps in metabolomics and is often the main bottleneck in the entire workflow.[1] A low identification rate hinders biological interpretation of the experiment due to many pieces of the puzzle being missing. A significant misidentification rate could lead to inconsistent analysis of the results and even wrong biological interpretations.[2] Achieving a high percentage of correct metabolite identifications is fundamental for subsequent biological interpretation.

There are several classifications of confidence levels for the identification of metabolites;[3] the most popular is that proposed by the Metabolomics Society. It includes five different confidence levels (see Table 11.1)[4] for the identification of each compound derived from metabolomic experiments.

New Developments in Mass Spectrometry No. 8
Processing Metabolomics and Proteomics Data with Open Software: A Practical Guide
Edited by Robert Winkler

Published by the Royal Society of Chemistry, www.rsc.org

**Table 11.1** Updated confidence levels proposed by the Metabolomics Society (2017).[a]

| Confidence level | Description | Matching requirement |
|---|---|---|
| Level 0 | Unequivocal 3D structure, including full stereochemistry. | Determination of 3D structure following natural product guidelines. |
| Level 1 | Confident 2D structure, using reference standard or full 2D structure elucidation. | At least two orthogonal characteristics, such as MS/MS fragmentation pattern, RT or CCS. |
| Level 2 | Probable structure using literature data and/or fragmentation spectra and/or knowledge over the RT. | At least two orthogonal characteristics matching and evidences for excluding the rest of candidates. |
| Level 3 | Possible structure, isomers or class. | More than one candidate, only one characteristic matched is required for supporting the proposed candidate. |
| Level 4 | Unknown. | Detectable feature in a sample. |

[a]RT: retention time, CCS: collision cross section.

A number of tools exist to annotate the metabolites with different confidence levels.[4] Commonly, the first step in metabolite identification is the assignment of metabolite IDs to the *m*/*z* values obtained by $MS^1$ (Table 11.1 confidence level 3). Then, further analysis using $MS^n$ or exploiting information from other (orthogonal) sources, such as collision cross section (CCS) ion mobility, retention index from GC analysis, retention time (RT) from LC measurements or migration time for CE data, is carried out.[5] This additional information enables researchers to apply their chemistry knowledge to support or refute the identifications, and possibly achieve confidence level 2 (Table 11.1).

Assignment of *m*/*z* values to a unique or a set of putative metabolite candidates is performed by querying accessible databases, which ideally makes this process very accurate and efficient.[6] However, the vast amount of data repositories and the low overlap requires also manual querying and integration of the results from different sources.[7] Data sources can be specific for certain types of compounds; this is the case for the Human Metabolome Database (HMDB), which covers the human metabolome, LipidMaps[8] and LipidBank, which contain only lipids, the Universal Natural Product Database (UNPD) devoted to primary and secondary plant metabolites, or the Milk Composition Database (MCDB) made up by compounds present in milk. Other sources, such as Metlin, KEGG, MassBank or mzCloud, contain all kinds of compounds. That has caused the emergence of a number of software tools offering a common interface to query multiple databases, with some of them providing additional processing features unavailable in the original databases.

## 11.2 Non-maintained Resources

Even though the resources mentioned below are not up-to-date, they are important from a historical perspective because they provided the basis and principles for the development of other tools and databases.

The first tool allowing the simultaneous query of different database resources (HMDB, KEGG, LipidMaps, MetaCyc, RiceCyc or AraCyc) was MZedDB. It was created in 2008 by the Aberystwyth University High Resolution Mass Spectrometry Laboratory for metabolite signal annotation. MZedDB is based on the principle that many ionization products will represent adducts, isotopes, and neutral loss fragments of parent metabolites following a specific set of generalized physical rules.[9] Its main limitation is that it only allows users to perform one-by-one queries, rather than batch queries.

MassTRIX was also created in 2008 and launched again in 2012. It aimed to boost the correct annotation rate in metabolomics by exploiting biological and genomic information coming from putative annotations. Initially, it matched the annotations with KEGG pathway maps and, since the 2012 update, it also incorporates information regarding gene expression.[10,11] MassTRIX could query simultaneously KEGG, HMDB, and LipidMaps databases, making it possible to exploit also MetaCyc. However, this tool has not been updated since 2012.

In 2012, MetaboSearch was born with the purpose of providing a common interface query for Metlin, HMDB, Madison Metabolomics Consortium Database (MMCD), and LipidMaps.[12] However, since its creation it has not been updated, and the access methods used by the original databases have changed. Therefore, MetaboSearch is no longer able to retrieve results from the original sources.

Since the beginning of the omic sciences, the metabolomics community has made an effort to create appropriate software to analyze metabolomic data. However, as the three examples presented above demonstrate, some of them have not been maintained over time, the information provided has become outdated, and – in some cases – they do not work anymore. This can cause setbacks for users, who have integrated these tools into their data analysis workflows, and who are forced to stop using them. A good track record in maintaining and updating a tool is undoubtedly something to evaluate when considering its integration into the metabolomic data analysis workflow. The availability of APIs to access tools in automated ways is also a plus as it eases its integration into data analysis workflows or third-party tools.

## 11.3 CEU Mass Mediator

CEU Mass Mediator (CMM) was created in 2012 with the goal of providing a single interface to query distinct databases. In 2017, a major update was released to assist in metabolite annotation with new functionalities using a knowledge approach, which was again significantly expanded in 2018.[13] Currently, CMM allows the user to simultaneously query KEGG, HMDB, LipidMaps, Metlin, MINE, and an in-house library of oxidized glycerophosphocholines (oxPCs). It scores the annotations based on the probability of ionization and adducts formation, the presence or absence of other expected adducts originating from the same signal, and the elution order of lipids belonging to the same class when working in reversed-phase (RP) mode.[7] CMM is an open source J2EE application (https://github.com/albertogilf/ceuMassMediator) running on TomEE 7.0.2 and MySQL server 5.7.24 that can be accessed through any web browser supporting javaScript (http://ceumass.eps.uspceu.es/) or through its REST API (http://ceumass.eps.uspceu.es/mediator/api/v3). CMM updates the data from the original sources approximately every six months and provides a JSON-based REST API for all its services to facilitate communication with other tools in an automated way (see Figure 11.1).

The main purpose of CMM is advanced metabolite annotation; *i.e.*, assigning putative compounds to the experimental masses obtained by MS. CMM unifies the compounds from different sources based on the IUPAC International Chemical Identifier (InChI) Key, which represents a unique identifier of chemical substances. In this way, CMM avoids returning several putative annotations that correspond to the same metabolite present in several data sources (possibly with different names in each of them). This unification saves a considerable amount of time by avoiding the need to manually unify the results retrieved from different sources, and reduces possible errors arising during manual unification.

CMM's simple and batch search options allow users to find the putative compounds for the *m*/*z* values acquired using any type of accurate mass spectrometer. It also enables filtering the compounds based on the data source and/or the type of metabolite searched. The advanced search is designed specifically for LC/MS (see Figure 11.2).

CMM can distinguish between compared groups and the complete set of acquired signals. All measured signals can provide information to support or refute putative annotations of those signals, which are statistically significant and, by being potential biomarkers, are of greatest interest to annotate. If the user provides the complete signal matrix, CMM will try to extract evidence from it to achieve confidence level 2 in the annotation of significant signals. CMM also exploits information from the Composite Spectrum (CS), the set of all related co-eluting *m*/*z* ions, including isotopes, adducts, and dimers formed by the same compound. The format is used by Agilent Technologies, and CMM takes advantage of this grouping of signals corresponding to the

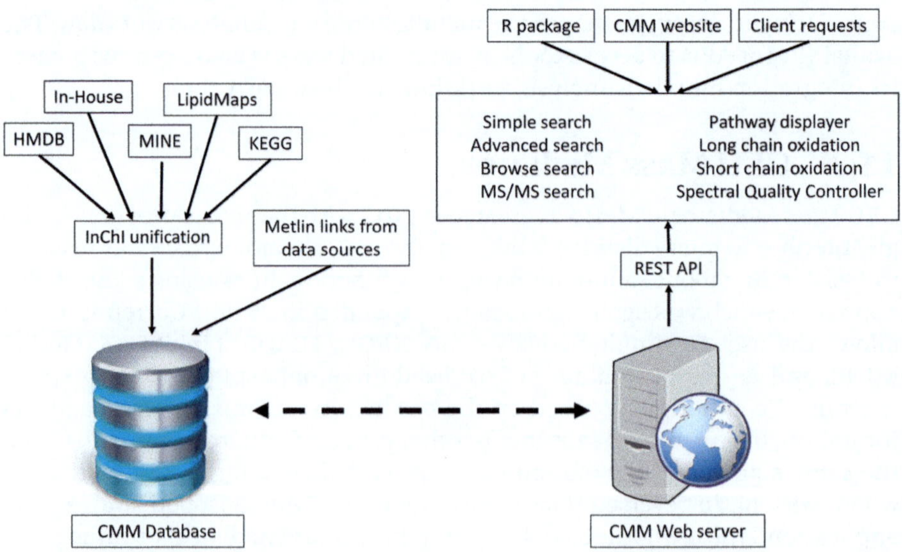


**Figure 11.1** CEU Mass Mediator architecture and list of the services (end-points) available.

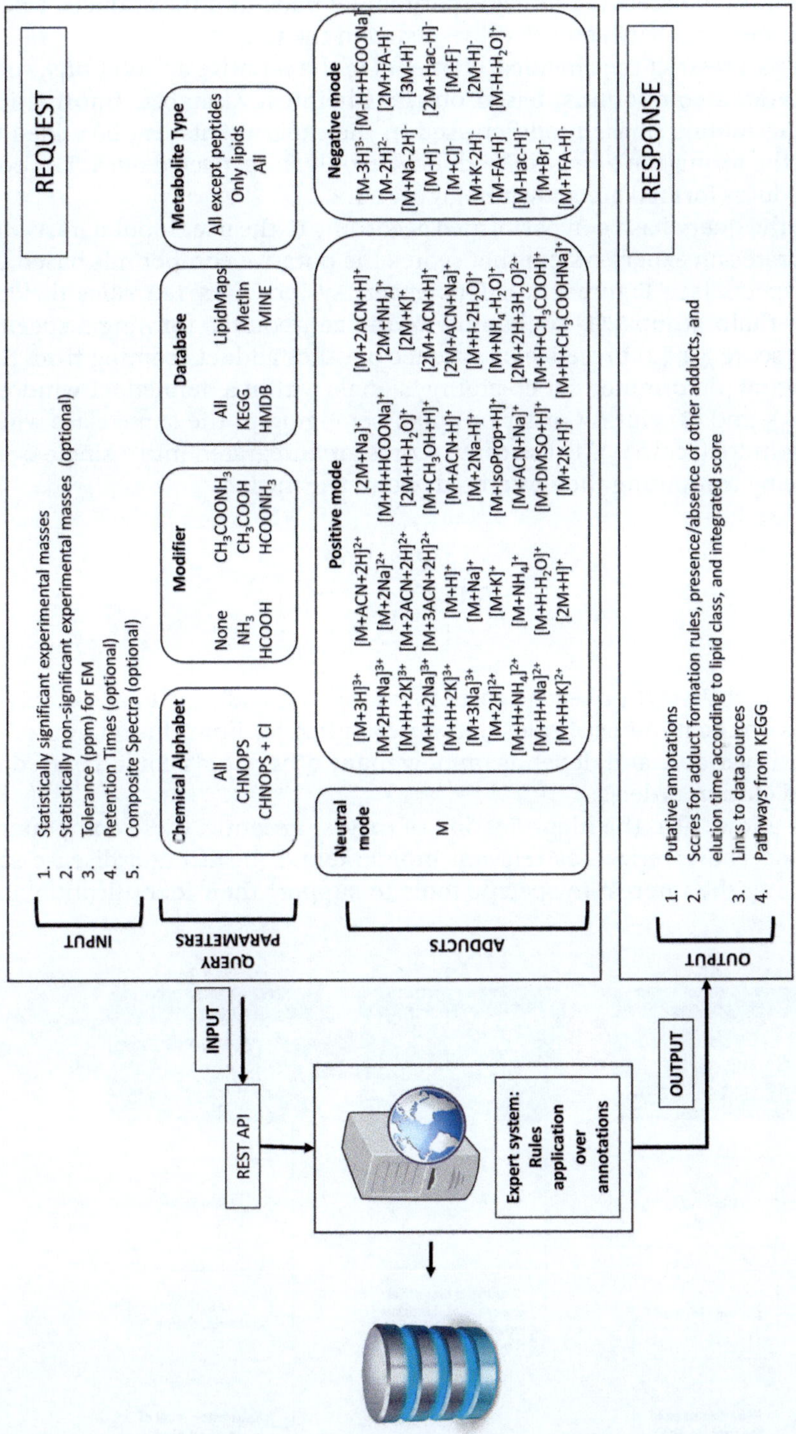


**Figure 11.2** Workflow of CEU Mass Mediator MS[1] batch advanced search.

same feature to detect the target experimental mass and the adducts, calculating differences between $m/z$ values listed in the CS.

Users can restrict the chemical elements of the putative annotations, such as deuterated compounds, based on the Chemical Alphabet. Information about the mobile phase modifier used in the experiments can be added to restrict the formation of possible adducts to only the expected ones. The possible adducts formed are shown in Figure 11.2.

Once the query has been performed according to the user input data, CMM incorporates an expert system that scores the putative compounds based on three aspects (see Figure 11.3). This expert system uses 122 rules divided in three main groups: (1) probability of the compounds forming a specific adduct (score $\chi_1$), (2) presence or absence of other adducts coming from the same signal (determined for co-eluting signals within a defined RT window) (score $\chi_2$), and (3) elution order of lipids belonging to the same class when working in RP (score $\chi_3$). These three scores are integrated into a single overall score by computing their weighted geometric mean:

$$\chi = \exp\left(\frac{\sum_{i=1}^{3} \omega_i \cdot \ln \chi_i}{\sum_{i=1}^{3} \omega_i}\right)$$

where $\omega_i$ is the weight of each score; $\omega_1 = 1$, $\omega_2 = 1$, and $\omega_3 \in [0, 2]$. $\omega_3$ weight depends on the number of rules that were applied for lipid elution time (this number is variable and depends on how many other lipids could be used in the lipid elution order).

CMM allows also the identification of oxPCs. Recently, these compounds have been characterized as relevant biomarkers of health and disease status, driving the interest in specific tools to support their identification and

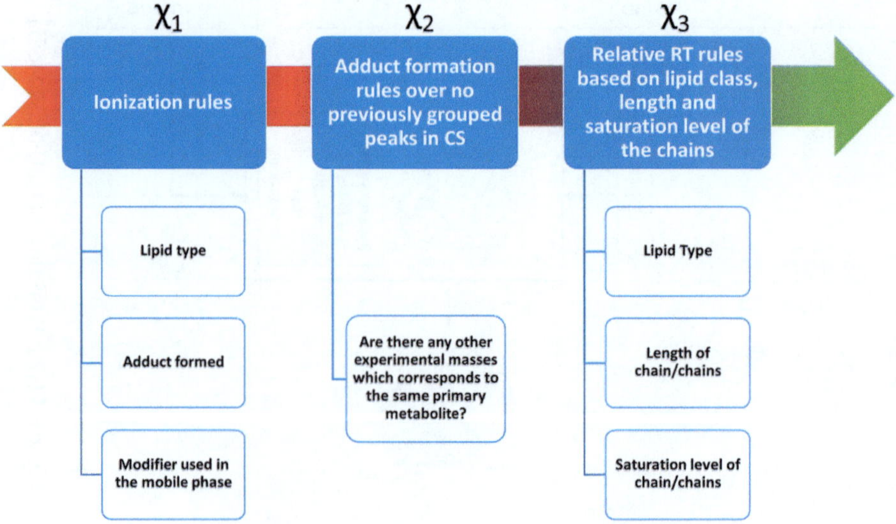


**Figure 11.3** Rules of CEU Mass Mediator expert system.

**Table 11.2** Flowchart of oxidized lipids identification in CEU Mass Mediator.

| | | |
|---|---|---|
| *Long-chain oxidation* | | |
| Step 1 | *m*/*z* of FA1 − Δ oxidation mass = *m*/*z* of native FA | Search against FA library → no hits conclusion: native FA |
| Step 2 | *m*/*z* of FA1 − Δ oxidation mass = *m*/*z* of non-oxidised FA | Search against FA library → no hits conclusion: non-oxidised FA |
| Step 3 | Search of *m*/*z* precursor against databases | |
| Step 4 | Tentative annotation → PC (16:0/20:4 (OH)) | |
| *Short-chain oxidation* | | |
| Step 1 | *m*/*z* of precursor − *m*/*z* of native FA − *m*/*z* of head group = *m*/*z* of oxidised FA | Search against FA library |
| Step 2 | *m*/*z* of oxidised FA − Δ oxidation mass = *m*/*z* of non-oxidised FA | Search against FA library |
| Step 3 | Search of *m*/*z* precursor against databases | |
| Step 4 | Tentative annotation → PC (16:0/5:4 (CHO)) | |

understanding their biological function. Thus, CMM aids in the identification of oxPCs from ESI-LC-MS/MS experimental data. It integrates knowledge about fragmentation of oxPCs as either long- or short-chain oxidized lipids, characterized by different oxidation and fragmentation processes, as well as a different handling of their oxidized derivatives. Based on the fragmentation patterns of oxPCs, the procedure compares the input spectrum introduced by the user with an internal database of oxPCs containing both curated and computationally generated records. The steps for the identification of short-chain and long-chain oxPCs are shown in Table 11.2.

CMM enables sorting the compounds for their subsequent biological interpretation based on the number of compounds from a specific pathway present in the experimental data and the compound's relevance for a given pathway. Relevance is determined by the number of pathways in which a compound is present. Water is an example of a compound with low relevance because, due to its ubiquity in most pathways, its presence does not yield specific biological relevance.

Finally, CMM can be applied also to MS/MS data. It supports MS/MS identification based on spectral similarity measurements between experimental spectra and library spectra of standards and/or predicted spectra contained in HMDB.

Another unique functionality of CMM is its ability to calculate the quality of a MS/MS spectrum for identification purposes. Experimental conditions are key to obtaining a clear spectrum that allows for further identification. A spectrum of inferior quality usually leads to too many unknowns or, even worse, to misidentifications. CMM ranks the quality of a spectrum considering the intensity of the signal in both MS and MS/MS analysis, noise level, number of scans performed to acquire the spectrum, number of samples analyzed (correspondence between different samples provides more confidence to the fragments obtained), presence of more than one compound in the collision cell (arising from chromatographic co-elution), crosstalk, and any spectrum contamination by ions present in the collision cell but originating from previous scans.

Summarizing, CMM offers metabolite annotation by querying unified compounds from distinct data sources, applying expert knowledge to score them, and guide researchers in their annotation process, MS/MS metabolite identification, identification of oxPCs, and for spectral quality control to determine how likely the obtained spectrum is to be identified. Furthermore, all these services can be accessed through a REST API, enabling users to include them in their data analysis and workflows.

## Acknowledgements

The authors are grateful for the support received from the Spanish Ministry of Economy and Competition (CTQ2014-55279-R-, RTI2018-095166-B-I00) and the Community of Madrid (B2017/BMD3751). AGF acknowledges Fundación Universitaria San Pablo CEU for his PhD fellowship. The authors would like to thank Yaoxiang Li and Amrita K. Cheema, from the Department of Oncology at Georgetown University Medical Center, Washington DC, USA, for developing the R client to run CMM services.

## References

1. K. Uppal, D. I. Walker, K. Liu, S. Li, Y. M. Go and D. P. Jones, *Chem. Res. Toxicol.*, 2016, **29**(12), 1956.
2. S. K. Bharti and R. Raja, *Curr. Metabolomics*, 2014, **2**(3), 163.
3. J. Godzien, A. Gil-de-la-Fuente, A. Otero and C. Barbas, in *Data Analysis for Omic Sciences: Methods and Applications*, ed. J. Jaumot, C. Bedia and R. Tauler, Comprehensive Analytical Chemistry, Elsevier, 2018, p. 415.
4. I. Blaženovic, T. Kind, J. Ji and O. Fiehn, *Metabolomics*, 2018, **8**(2), 1989.
5. J. Feng-Xiao, B. Zhou and H. W. Ressom, *TrAC, Trends Anal. Chem.*, 2012, **32**, 1.
6. K. Bingol, L. Bruschweiler-Li, D. Li, B. Zhang, M. Xie and R. Brüschweiler, *Bioanalysis*, 2016, **8**(6), 557.
7. A. Gil-De-La-Fuente, J. Godzien, S. Saugar, R. Garcia-Carmona, H. Badran, D. S. Wishart, C. Barbas and A. Otero, *J. Proteome Res.*, 2019, **18**(2), 797.
8. M. Sud, E. Fahy, D. Cotter, A. Brown, E. A. Dennis, C. K. Glass, A. H. Merrill, R. C. Murphy, C. R. H. Raetz, D. W. Russell and S. Subramaniam, *Nucleic Acids Res.*, 2007, **35**(1), D527.
9. J. Draper, D. P. Enot, D. Parker, M. Beckmann, S. Snowdon, W. Lin and H. Zubair, *BMC Bioinf.*, 2009, **10**, 227.
10. K. Suhre and P. Schmitt-Kopplin, *Nucleic Acids Res.*, 2008, **36**, W481.
11. B. Wagele, M. Witting, P. Schmitt-Kopplin and K. Suhre, *PLoS One*, 2012, **7**(7), e39860.
12. B. Zhou, J. Wang and H. W. Ressom, *PLoS One*, 2012, **7**(6), e40096.
13. A. Gil-de-la-Fuente, J. Godzien, M. Fernández López, F. J. Rupérez, C. Barbas and A. Otero, *J. Pharm. Biomed. Anal.*, 2018, **154**, 138.

CHAPTER 12

# *Metabolite Annotation Using* In Silico *Generated Compounds: MINE and BioTransformer*

ALBERTO GIL-DE-LA-FUENTE*[a,b], JOANNA GODZIEN[a], ABRAHAM OTERO[a,b] AND CORAL BARBAS[a]

[a]Universidad San Pablo-CEU, CEU Universities, Centre for Metabolomics and Bioanalysis (CEMBIO), Facultad de Farmacia, Campus Monteprı́ncipe, Madrid, 28668, Spain; [b]Universidad San Pablo-CEU, CEU Universities, Department of Information Technology, Escuela Politécnica Superior, Campus Monteprı́ncipe, Madrid, 28668, Spain; [c]Clinical Research Centre, Medical University of Bialystok, Bialystok, 15-001, Poland
*E-mail: alberto.gilf@gmail.com

## 12.1 Introduction

Untargeted metabolomics seeks to detect and identify as many metabolites as possible. The most common analysis technique to maximize the number of metabolites detected is chromatography coupled with mass spectrometry (MS), but this method is hampered by the high number of metabolites detected that remain unidentified.[1]

Compound identification requires the matching and assignation of structures to the features detected during the metabolomics experiment. Although the size of metabolomic databases is growing continuously and considerably, many of the signals measured in untargeted metabolomics remain

New Developments in Mass Spectrometry No. 8
Processing Metabolomics and Proteomics Data with Open Software: A Practical Guide
Edited by Robert Winkler

Published by the Royal Society of Chemistry, www.rsc.org

unannotated, in many cases due to the lack of matches with the known compounds present in the databases. This problem causes an untoward situation: there are no compounds in the databases that match the observed features and, therefore, they cannot be identified, and the databases cannot include these experimentally detected compounds because they have not been identified.[2]

There are a large number of software tools using *In silico* fragmentation models to improve the identification rate of features in metabolomic experiments.[3–6] However, these tools do not address the problem of identifying metabolites that are not present in the experimental databases. Their aim is to provide the fragmentation patterns of metabolites previously known.

To boost the metabolome coverage and to overcome the lack of experimentally detected compounds caused by a limited number of commercially available standards, databases have started to include *In silico* generated compounds.[7] These compounds have been predicted using algorithms that apply chemical transformations to previously known compounds, thus increasing metabolome coverage and reducing the number of unknowns.[8] The number of tools to predict new *In silico* compounds is still limited. Here we present the two main alternatives that perform transformations over known compounds to produce new ones that can help researchers identifying compounds that have not been detected experimentally before.

## 12.2 Non-maintained Resources

MyCompoundId[9] was one of the first tools providing putative metabolite annotations using *In silico* compounds. It encompasses 8021 known compounds from HMDB and their predicted metabolic products (375 809 for one metabolic reaction and 10 583 901 for two metabolic reactions). MyCompoundId was a major contribution when it was released, and it allowed the identification of at least twice as many metabolites than when restricting the search to the experimentally known compounds.[9,10] However, this resource has not been updated since 2013, while new software tools with a similar purpose have appeared, and HMDB has grown considerably.

## 12.3 Metabolic *In Silico* Network Expansion Databases

Metabolic *In silico* Network Expansion Databases (MINE) represent an enzymatic expansion of metabolites from the Kyoto Encyclopedia of Genes and Genomes (KEGG), EcoCyc, Yeast Metabolome Database version 1.0 (YMDB), and Chemical Damage SEED. They contain compounds generated by the Biochemical Network Integrated Computational Explorer (BNICE) algorithm and from hand-curated reaction rules generalized according to the Enzyme Commission classification system.[11] To ensure canonical valences and proper placement of charge, the products generated by the BNICE framework have been processed with ChemAxon's Standardizer & Structure Checker (JChem 6.0.4, 2013).

*In silico* generated compounds are ranked according to the Natural Product Likeness scores[12] calculated by adding the Natural Product-likeness scores of every structural fragment of the molecule. The individual score is calculated by:

$$\text{Fragment}_i = \log\left[\frac{\text{NP}_i}{\text{SM}_i} \times \frac{\text{SM}_t}{\text{NP}_t}\right]$$

where $\text{NP}_i$ is the total number of molecules in the natural products dataset, in which the $\text{Fragment}_i$ occurs; $\text{SM}_i$ is the total number of molecules present in the synthetic molecules dataset, in which the $\text{Fragment}_i$ occurs; $\text{SM}_t$ is the total number of molecules in the synthetic molecules dataset; and $\text{NP}_t$ is the total number of molecules in the natural product dataset12.

MINE contain over 571 000 compounds, of which 93% are not present in the PubChem database, which makes them the largest repository of metabolites. Although a high percentage of these compounds have not been experimentally detected yet, they have on average higher structural similarity to natural products than compounds from KEGG or PubChem databases. Compound and reaction lists are stored in a Mongo Database (v2.6.2). Reactions contain arrays of reactants and products as tuples of the stoichiometric coefficient, the compound ID, and a list of the operators that predicted the reaction. MINE avoids generalized structures because they cannot be assigned to the experimentally accurate mass detected by MS and it unifies the duplicated compounds using the Standard InChIKey.[13] MINE can be downloaded as SDF files and is freely accessible through the web-page (http://minedatabase.mcs.anl.gov/) or the APIs available in Perl, JavaScript, and Python (https://github.com/JamesJeffryes/MINE-API) to ease automated access (see Figure 12.1).

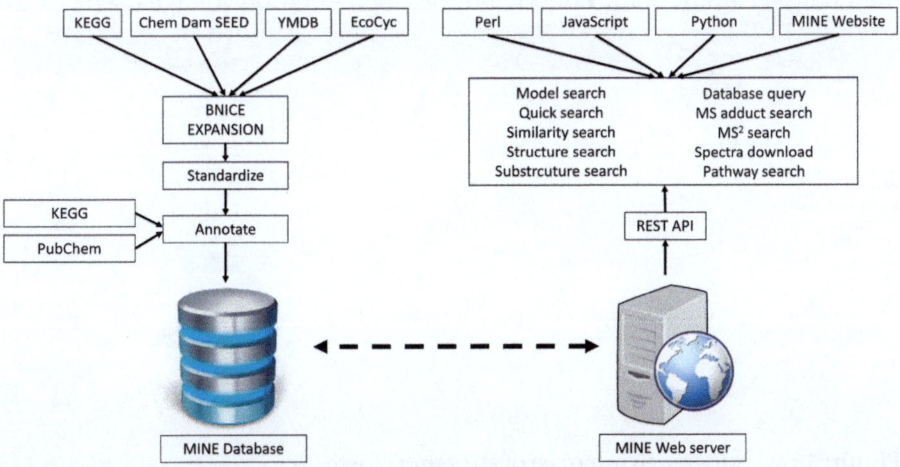


**Figure 12.1** MINE architecture schema and list of the services (end-points) available.

MINE's web interface has been designed with the aim of: (1) investigation of potential enzymatic transformations, (2) annotation of accurate masses and (3) chemical structure search. In addition to expanding metabolome coverage, the MINE framework also offers a pipeline for illuminating the synthesis and degradation of poorly annotated secondary metabolites by focusing the search on a region of interest in the metabolic network.[11] MINE's API provides three main search services.

### 12.3.1 Structure Search

The structure search can be used to look for compounds with similar functional groups that share common substructure, or to retrieve exact matches to the input structure introduced. The input structure can use any format recognized by OpenBabel (*e.g.*, MOL file, SMILES, InChI) or it can be drawn *via* the web interface using the ChemAxon plugin for a more visual representation of molecules (see Figure 12.2).

### 12.3.2 MS Adduct Search

The MS adduct search allows batch query based on *m*/*z* values and a preselected set of adducts, and can be restricted by the Kovats Retention Index, by logP and/or to include/exclude halogenated compounds (see Figure 12.3). Moreover, an organism can be chosen to select the Likelihood Scoring model in KEGG.

### 12.3.3 MS/MS Search

The MS/MS search supports information retrieved *via* MS search and couples it to data on product ions. The user can choose the type of spectral scoring method (Jaccard or dot product) and the fragmentation voltage used (10, 20

**Figure 12.2** MINE web interface of structure search.

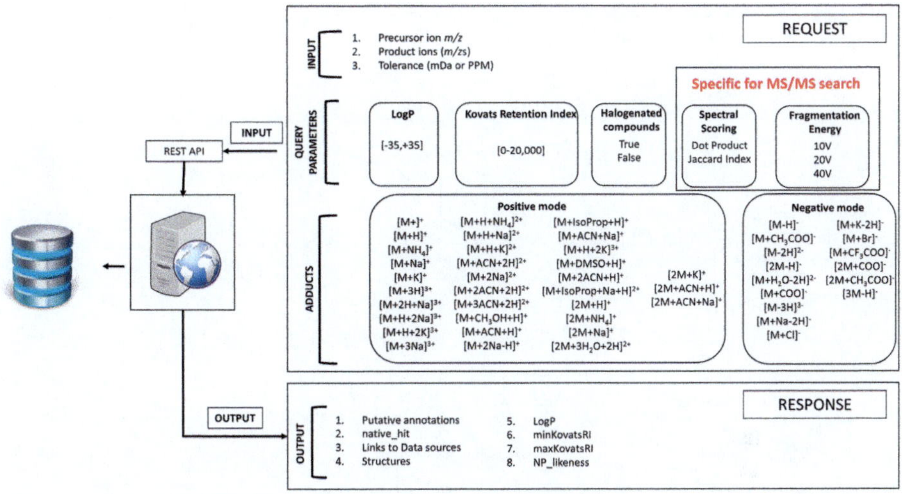


**Figure 12.3** Workflow of MINE MS$^1$ and MS/MS search.

or 40 V; see Figure 12.3). The user can also filter the results by the Kovats RI, the logP and/or include/exclude the halogenated compounds from the results list.

### 12.3.4 Compound Page

The compound page of MINE displays all the data available for the specific compound: formula, name, native charge, different identifiers (InChIKey, Smiles, data sources Ids), NP likeness, logP, Kovats RI, the enzymatic reactions where it is present, the pathways where it has been detected, and all the predicted reactions that produce or consume this compound (see Figure 12.4).

MINE comprise a considerable number of compounds not present in other sources and, therefore, they conveniently reduce the number of unknowns in metabolomic studies while increasing metabolome coverage. It should nevertheless be kept in mind that annotations corresponding to *In silico* compounds are often impossible to confirm, as there are no standards available to ensure confidence level 1 or 0 (see Table 11.1). MINE services can be accessed through their APIs, which enables developers to use them in an automated way.

## 12.4 BioTransformer

Here we present an alternative tool that predicts metabolites based on the application of machine learning models, biotransformations and precedence rules to generate new compounds from a given structure.

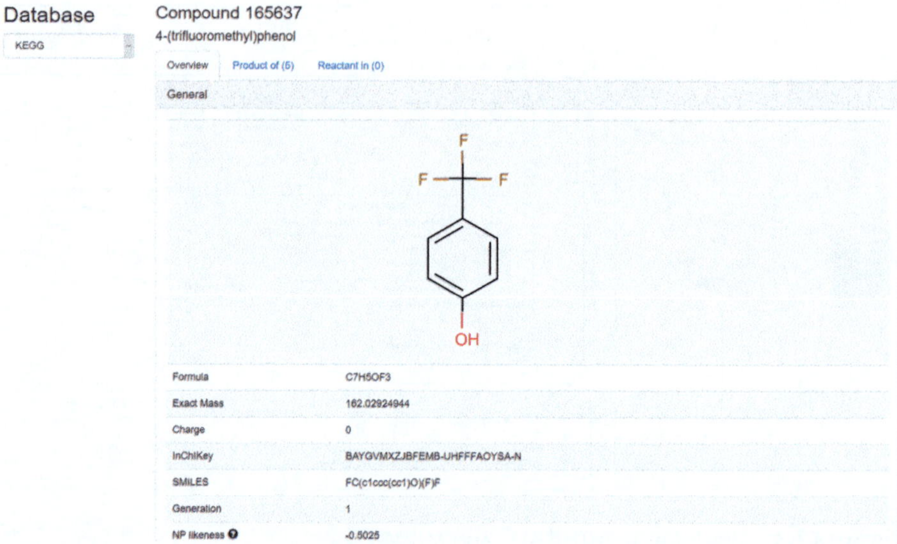


**Figure 12.4** MINE compound page for 4-(trifluoromethyl)phenol.

BioTransformer is a software tool comprising a Metabolism Prediction Tool (BMPT) and a Metabolite Identification Tool (BMIT). It predicts small molecule metabolism in mammals, their gut microbiota, as well as the soil/aquatic microbiota.[14] Moreover, it assists scientists in the identification of metabolites based on the metabolism prediction and using MS data, monoisotopic mass or chemical formula.

### 12.4.1 The BioTransformer Metabolism Prediction Tool (BMPT)

BioTransformer uses both a knowledge-based (KB) approach and machine learning (ML) algorithms to predict small molecules metabolism. The knowledge-based system consists of three major components:

1. A biotransformation database (MetXBioDB) containing annotations of experimentally confirmed metabolic reactions.
2. A reaction knowledgebase containing generic biotransformation rules, preference rules, and other constraints for metabolism prediction.
3. A reasoning engine implementing generic and transformer-specific algorithms for metabolite prediction and selection.

The machine learning system uses a set of random forest and ensemble prediction models for the prediction of CYP450 substrate selectivity and for the Phase II filtering of molecules. These two options allow for a comprehensive coverage of metabolism in the human superorganism. The *allHuman* option covers biotransformations occurring in human tissues, gut microbiota; and

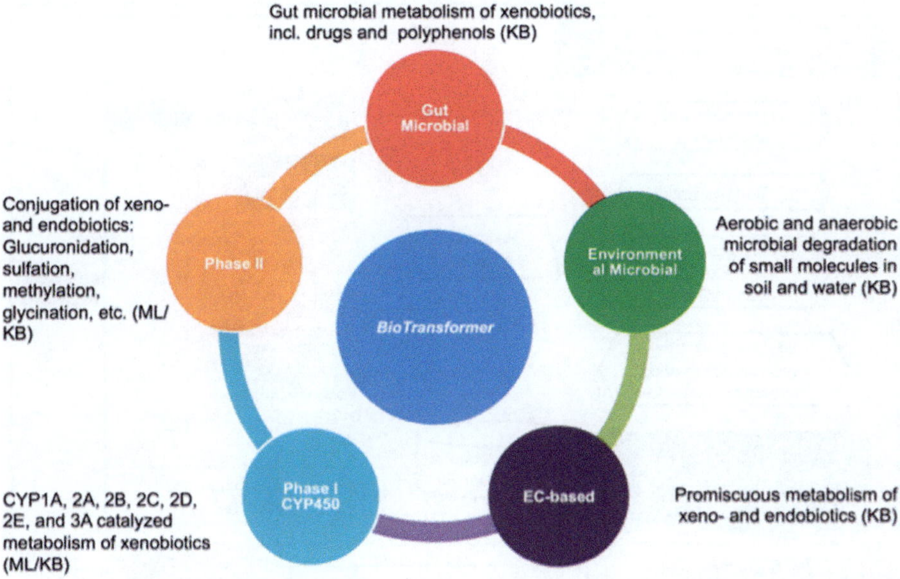


**Figure 12.5** Biotransformer's five metabolism prediction modules. Reproduced from ref. 14, https://doi.org/10.1186/s13321-018-0324-5, under the terms of a CC BY 4.0 license, https://creativecommons.org/licenses/by/4.0/.

the *superbio* option follows a pre-defined sequence of reaction types: starting from promiscuous metabolic reactions, and followed by CYP450-catalyzed reactions, gut microbial degradation, and Phase II reactions.

The BMPT contains five independent modules called "transformers" that apply different biotransformations to metabolites (see Figure 12.5):

1. EC-based: promiscuous metabolism of xeno- and endobiotics; Knowledge-based (KB) system.
2. CYP450: CYP1A, 2A, 2B, 2C, 2D, 2E and 3A catalyzed metabolism of xenobiotics; Knowledge-based (KB)/Machine Learning-based (ML) system.
3. Phase II: conjugation of xeno- and endobiotics: glucuronidation, sulfation, methylation, glycination, *etc.*; KB/ML system.
4. Human Gut Microbial: gut microbial metabolism of xenobiotics, including drugs and polyphenols; KB system.
5. Environmental Microbial: aerobic and anaerobic microbial degradation of small molecules in soil and water; KB system.

The BMPT accepts molecule structures in SMILES, InChI, MOL or SDF format. The structures should correspond to an organic molecule (excluding mixtures and salts). Then, the BMPT standardizes the input structure removing charges from functional groups, checking and validating bond types and

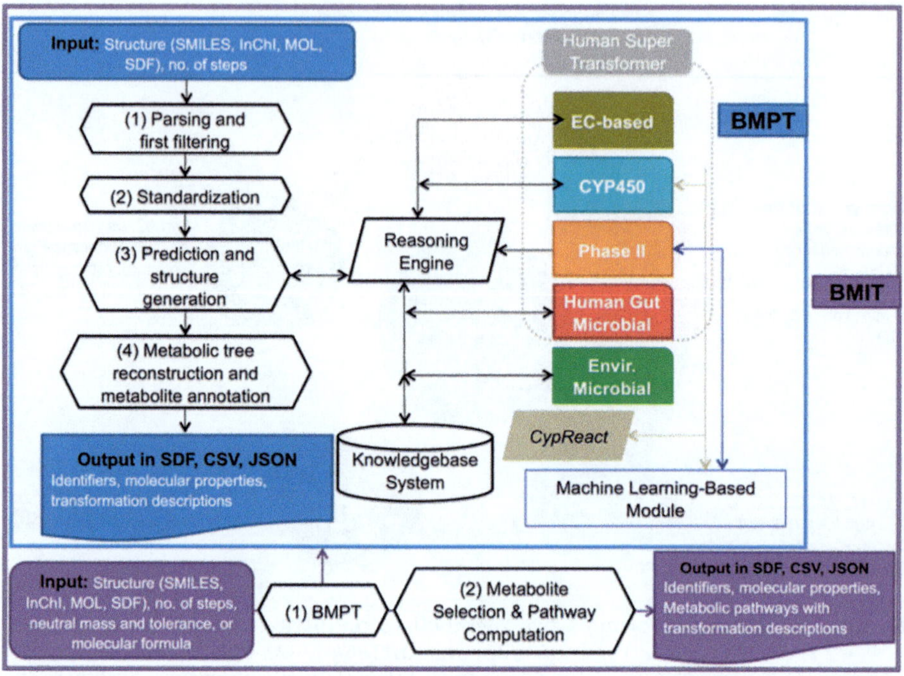


**Figure 12.6** Biotransformer's workflow for the metabolism prediction tool (BMPT) and the metabolism identification tool (BMIT). Reproduced from ref. 14, https://doi.org/10.1186/s13321-018-0324-5, under the terms of a CC BY 4.0 license, https://creativecommons.org/licenses/by/4.0/.

adding explicitly hydrogen atoms. The standardized structure is processed by the reasoning engine according to the biotransformer chosen (EC based, CYP450, Phase II, Human gut microbial, Environmental microbial, *AllHuman* or *Superbio*) and the structures and biotransformations are annotated. Biotransformer provides structural information, physiochemical properties and an explanation about its origin for each predicted metabolite. The output result can be downloaded as JSON, SDF or CSV file (see Figure 12.6).

### 12.4.2 The BioTransformer Metabolism Identification Tool (BMIT)

The BMIT is built from the BMPT. Given the input structure and a set of molecular masses (optionally the user can specify the mass tolerance) or chemical formulas, the BMIT identifies the potential metabolites for each mass or chemical formula. The BMIT links the starting structure with the metabolites returned based on the metabolic tree obtained upon metabolism prediction.[14] Moreover, it returns a biosynthetic pathway leading to each identified metabolite, starting from the query molecule. The output result can be downloaded as JSON, SDF or CSV file (see Figure 12.6).

BioTransformer was implemented in Java and it is freely accessible. Both BMPT and BMIT can be used by the user through a web interface (www.biotransformer.ca), using the available java library (a jar file -https://bitbucket.org/djoumbou/biotransformerjar/-) or through their RESTful API (www.biotransformer.ca).

## Acknowledgements

The authors are grateful for the support received from the Spanish Ministry of Economy and Competition (CTQ2014-55279-R-, RTI2018-095166-B-I00) and the Community of Madrid (B2017/BMD3751). AGF acknowledges Fundación Universitaria San Pablo CEU for his PhD fellowship. The authors would like to thank James G. Jeffryes for developing MINE, Yannick Djombou-Feunang for the development of Biotransformer and the developers of all the metabolomic software tools.

## References

1. A. Scalbert, L. Brennan, O. Fiehn, T. Hankemeier, B. S. Kristal, B. van Ommen, E. Pujos-Guillot, E. Verheij, D. S. Wishart and S. Wopereis, *Metabolomics*, 2009, **5**, 435.
2. L. C. Menikarachchi, D. W. Hill, M. A. Hamdalla, I. I. Mandoiu and D. F. Grant, *J. Chem. Inf. Model.*, 2013, **53**(9), 2483.
3. L. Ridder, J. J. J. van der Hooft, S. Verhoeven, R. C. H. de Vos, R. van Schaik and J. Vervoort, *Rapid Commun. Mass Spectrom.*, 2012, **26**(20), 2461.
4. F. Allen, A. Pon, M. Wilson, R. Greiner and D. S. Wishart, *Nucleic Acids Res.*, 2013, **42**, W94.
5. K. Dührkop, H. Shen, M. Meusel, J. Rousu and S. Böcker, *Proc. Natl. Acad. Sci. U. S. A.*, 2015, **112**(41), 12580.
6. C. Ruttkies, E. L. Schymanski, S. Wolf, J. Hollender and S. Neumann, *J. Cheminf.*, 2016, **8**, 3.
7. D. S. Wishart, Y. Djombou-Feunang, A. Marcu, A. C. Guo, K. Liang, R. Vázquez Fresno, T. Sajed, D. Johnson, C. Li, N. Karu, Z. Sayeeda, E. Lo, N. Assempour, M. Berjanskii, S. Singhal, D. Arndt, Y. Lian, H. Badran, J. Grant, A. Serra-Cayuela, Y. Liu, R. Mandal, V. Neveu, A. Pon, C. Knox, M. Wilson, C. Manach and A. Scalbert, *Nucleic Acids Res.*, 2018, **46**(D1), D608.
8. J. Godzien, A. Gil-de-la-Fuente, A. Otero and C. Barbas, in *Data Analysis for Omic Sciences: Methods and Applications*, ed. J. Jaumot, C. Bedia and R. Tauler, Comprehensive Analytical Chemistry, Elsevier, 2018, p. 415.
9. L. Li, R. Li, J. Zhou, A. Zuniga, A. E. Stanislaus, Y. Wu, T. Huan, J. Zheng, Y. Shi, D. S. Wishart and G. Lin, *Anal. Chem.*, 2013, **85**(6), 3401.
10. T. Huan, C. Tang, R. Li, Y. Shi, G. Lin and L. Li, *Anal. Chem.*, 2015, **87**(20), 10619.

11. J. G. Jeffryes, R. L. Colastani, M. Elbadawi-Sidhu, T. Kind, T. D. Niehaus, L. J. Broadbelt, A. D. Hanson, O. Fiehn, K. E. J. Tyo and C. S. Henry, *J. Cheminf.*, 2015, **7**, 44.
12. K. V. Jayaseelan, P. Moreno, A. Truszkowski, P. Ertl and C. Steinbeck, *BMC Bioinf.*, 2012, **13**, 106.
13. S. Heller, A. McNaught, S. Stein, D. Tchekhovskoi and I. Pletnev, *J. Cheminf.*, 2013, **5**, 7.
14. Y. Djoumbou-Feunang, J. Fiamoncini, A. Gil-de-la-Fuente, R. Greiner, C. Manach and D. S. Wishart, *J. Cheminf.*, 2019, **11**, 2.

CHAPTER 13

# Trans-Proteomic Pipeline for the Identification, Validation, and Quantification of Proteins†

ERIC W. DEUTSCH*, LUIS MENDOZA, DAVID D. SHTEYNBERG, ZHI SUN, MICHAEL H. HOOPMANN AND ROBERT L. MORITZ

Institute for Systems Biology, 401 Terry Ave N, Seattle, WA, 98109, USA
*E-mail: edeutsch@systemsbiology.org

## 13.1 Introduction

The Trans-Proteomic Pipeline (TPP) is a suite of software tools that enables start-to-finish analysis of mass spectrometry (MS) proteomics datasets, originally and continually developed at the Institute for Systems Biology in Seattle, USA. While several of the component tools such as PepideProphet[1] and ProteinProphet[2] predated the TPP as an entity itself, the TPP was first presented in 2005 by Keller *et al.*[3] as a uniform analysis platform comprising a set of tools that were all interoperable on account of a shared set of open file formats. Since then the TPP has grown to encompass many additional tools, all still based on the same set of open file formats. The expansion of the TPP has been summarized by reviews in 2010[4] and 2015.[5]

The TPP has always been free and open source, released under the GNU Lesser General Public License (LGPL) (https://en.wikipedia.org/wiki/

†All tutorials can be accessed *via* http://tppms.org/tutorials/.

New Developments in Mass Spectrometry No. 8
Processing Metabolomics and Proteomics Data with Open Software: A Practical Guide
Edited by Robert Winkler

Published by the Royal Society of Chemistry, www.rsc.org

GNU_Lesser_General_Public_License), which is more permissive than the protective GPL. The complete source code of the TPP is available for inspection and reuse at its SourceForge site https://sourceforge.net/projects/sashimi/. Packaged distributions of specific TPP releases for the Linux and Microsoft Windows operating systems are available at the same site. The TPP may be compiled under Apple's OS X, but does need streamlining to improve ease of use. The TPP has also been freely deployed on cloud computing platforms such as Amazon Web services (AWS), the Microsoft Azure cloud computing platform, and the Galaxy-P platform (http://galaxyp.org/). The ability to deploy the TPP across a variety of platforms is a major strength.

It can be a daunting task to become familiar with the many components and capabilities of the TPP. However, there are several resources that are available to assist users in learning to use it effectively. The primary TPP Web site (http://www.tppms.org/) is the overall portal to the many available resources, including a wiki-based documentation system, a set of on-line tutorials, and links to available classroom courses taught all over the world to give students understanding of the tools and experience in using them *via* a series of lectures and hands-on tutorials on each tool.

In this chapter we provide a brief overview of the tools that make up the TPP and highlight the publicly available tutorials that guide a user through the use of each tool or general mode of using of the TPP. Some of the tutorials are maintained as a wiki page at the TPP wiki site. Others are written as a Word document that can be easily printed and followed in a classroom setting or at one's own pace. Some provide questions with answers at the end to lead the user toward a deeper understanding of a tool and how to interpret its output. All tutorials are available under the Creative Commons CC0 license, which permits unhindered access and reuse. Table 13.1 provides a summary listing of tutorials and guides mentioned in this chapter. An online up-to-date version of that table is available at http://tppms.org/tutorials/, including clickable hyperlinks to each tutorial. Hyperlinks to individual tutorials are not provided here as these may change over time, but the latest hyperlinks will always be available at the above URL.

## 13.2 Using the Tools

The TPP comprises an extensive collection of tools for all aspects of analysis of MS/MS datasets, as summarized in Figure 13.1. Native vendor instrument formats are first converted to the Human Proteome Organization (HUPO) Proteomics Standards Initiative (PSI) mzML format,[6] which is used by all downstream tools that need access to the mass spectra. Peptide identification is performed with one or more search engines, either ones bundled with the TPP or available externally. Downstream validation, quantitation, and protein assignment, plus other functionality, are all made interoperable by the TPP open formats pepXML and protXML. All of these tools are maintained together and available as a single installation in order to maximize ease of use.

The first step to using the TPP usually begins with installation on an accessible computer, although in some cases this may already be done centrally for the user. Installation on a computer running Microsoft Windows is generally

**Table 13.1** Summary of current TPP tutorials and guides mentioned in this chapter.

| # | Tag | URL |
|---|---|---|
| 01 | TPPGuides | http://tppms.org/tutorials/ |
| 02 | WinInstall | Installation of TPP 5.2.0 on Windows systems |
| 03 | LinInstall | Collection of installation recipes for various Linux flavours and TPP versions |
| 04 | LinDocker | Run TPP CLI and GUI on Linux *via* Docker instead of performing a local installation |
| 05 | WinDocker | Run TPP CLI and GUI on Windows *via* Docker container |
| 06 | AWS | Launch an Amazon Machine Image running TPP and take a tour |
| 07 | tpp2mzId | Convert existing PepXML and ProtXML output to mzIdentML |
| 08 | ConvAndSearch | Starting with Thermo raw files, convert to mzML format and search with Comet and X!Tandem, followed by examination of results with Pep3D and the PepXMLViewer |
| 09 | SpectraSTSearch | Use SpectraST to search a SILAC dataset using spectral libraries as a reference |
| 10 | KojakSearch | Use Kojak to search and validate a cross-linking experiment |
| 11 | PeptideProphet | Use PeptideProphet to validate Comet search results |
| 12 | iProphet | Use iProphet to validate peptide sequences from PeptideProphet results |
| 13 | ProteinProphet | Use ProteinProphet to generate a validated protein list from iProphet results |
| 14 | PTMProphet | Use PTMProphet to validate and localize PTM sites |
| 15 | reSpect | Use ReSpect to identify multiple peptide ions in spectra |
| 16 | DISCO | Use DISCO to extract de-multiplexed MS2 spectra and search with Comet |
| 17 | SpecLibCreate | Use SpectraST in build mode to create a spectral library from validated search results |
| 18 | SpecLibAlter | Use SpectraST to alter a downloaded library and perform a search with it |
| 19 | IsotopicQuant | Use ASAPRatio and XPRESS to analyse a SILAC dataset |
| 20 | Libra | Use Libra to analyse an iTRAQ-8 dataset |
| 21 | StPeter | Use StPeter to perform label-free quant in a sample dataset |

a very easy wizard-based workflow that takes less than a minute. However, possible complications and considerations are described in the TPP 5.2 Windows Installation Guide (Table 13.1, entry 02 "WinInstall"). Installation on Linux is a bit more involved, owing to the variation of different distributions of Linux. A series of step-by-step tutorials for different versions of TPP and different versions of Linux – primarily Ubuntu Linux – is available *via* Table 13.1 entry 03 "LinInstall". This includes installing prerequisites, downloading and compiling the source code, installing the results, and testing it.

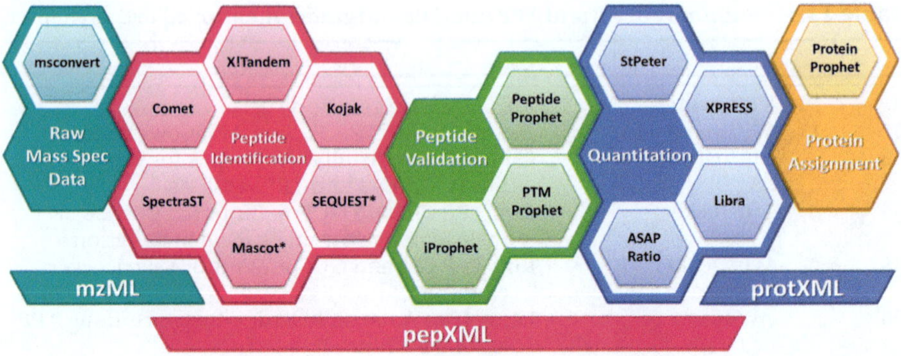


**Figure 13.1** Overview of the software tools and data formats of the TPP. Tools are grouped into broad categories for which there are often alternatives, and underpinned by a set of XML open formats. Most of the depicted tools come bundled with the TPP, but a few, such as SEQUEST and Mascot, are external tools whose output can be used by the TPP.

The native environment for the TPP is the command-line interface (CLI) on Linux, which makes the TPP easily scriptable, compatible with workflow management systems, and deployable on cloud computing platforms. However, for many users a graphical user interface (GUI) greatly enhances the ease of use of software, and thus the TPP provides a GUI that enables selection of parameters, launching of tools, and exploration of the output using an HTML + JavaScript interface code-named Petunia. One benefit of the Petunia interface is that it displays (and keeps as a log) the equivalent CLI commands to launch analyses with the selected options, fostering a transition to the CLI if desired. Although the Web-based HTML + JavaScript interface is not as refined as a native Microsoft Windows application interface would be, its Web-based nature means that the GUI can be used both on local computers as well as remote computers at a collaborator's institution, on a cloud computing instance, or a virtualization container (*e.g.* Docker). There are no dedicated tutorials for the CLI and GUI, but most individual tool tutorials guide the user through using the tool *via* the GUI, and this is covered extensively in the regular in-person tutorial courses held by the developers of the TPP.

Virtualization container systems such as Docker (https://www.docker.com/) have become a popular mechanism for solving the problem that many software packages are difficult to install or have complex dependencies. *Via* this mechanism a tool provider can create a container with the software tool and all dependencies preinstalled and preconfigured such that the tool is guaranteed to run on any system with Docker installed. A Linux-based container can even run on a Microsoft Windows machine with Docker installed. TPP containers are available on Docker Hub at https://hub.docker.com/u/spctools, and tutorials that demonstrate how to run TPP *via* Docker on a Linux host is available *via* Table 13.1 entry 04 "LinDocker", and on a Windows host *via* Table 13.1 entry 05 "WinDocker". These tutorials provide a brief introduction to installing Docker and then the very few steps are needed to pull down the TPP container and launch a CLI example as well as launch the GUI and take it for a brief tour.

As previously mentioned, the TPP is easily deployed on cloud computing platforms, providing a potentially huge capacity for processing large volumes of data in a short period of time. The latest Linux installation guide (Table 13.1 entry 03 "LinInstall") was tested on a fresh instance running on the Microsoft Azure platform. A user may install and customize TPP on such an instance quite rapidly and then stop and restart the instance as needed, paying Azure only for the hours during which the instance is running. This permits TPP to be used on a fast, powerful multi-core instance, yet only accruing charges as it is needed. We have also built Amazon Web Services (AWS) Amazon Machine Images (AMIs), which are snapshots of a fully configured computer running TPP that may be deployed on the many different kinds of AWS instances. See Table 13.1 entry 06 "AWS" for a tutorial on using TPP on AWS.

One of the foundations of the TPP is the interoperability of all the tools *via* open formats. The TPP provides conversion of native vendor formats to the PSI mzML format[6] *via* the bundled ProteoWizard[7] msconvert tool. Most search engines are able to read mzML to produce their search results. The TPP also supports mzXML[8] and other formats (see Deutsch 2012[9] for a review of formats in MS proteomics), but mzML is highly recommended. The output of search engines is converted to the TPP PepXML[3] format if not natively written, and most downstream tools read, refine, and write their results using the PepXML format. Protein inference results are written in the ProtXML[3] format, and this format is also read and written by various tools that operate at the protein level. Once processing is complete, the output may be exported to the official PSI standard format mzIdentML[10,11] *via* the tpp2mzId tool for deposition to data repositories or interoperation with other non-TPP tools that do not support PepXML and ProtXML. See Table 13.1 entry 07 "tpp2mzid" for a tutorial on converting TPP-based PepXML and ProtXML processing results to mzIdentML.

There exists a large variety of search engines available for processing of MS/MS spectra into putative identifications and scores. Although not strictly TPP tools, two general purpose open-source sequence search engines are bundled with the TPP: Comet,[12] which is highly similar to SEQUEST,[13] and X!Tandem.[14] These tools try to identify MS/MS spectra based on an input protein sequence database in the form of a FASTA or PEFF[15] file. The Kojak[16] search engine is similar, but is designed specifically for identifying cross-linked peptide spectra; it writes its output in PepXML using special extensions for cross-linked peptides,[17] although these can be processed as usual with downstream TPP tools. The SpectraST[18] spectral library searching tool also attempts to identify MS/MS spectra, but instead of using a protein sequence database, it uses as input a spectral library, which is a collection of previously identified consensus spectra compiled from previous datasets. A tutorial on converting raw vendor files to mzML, sequence searching the mzMLs with Comet and X!Tandem, and examining the results is provided *via* Table 13.1 entry 08 “ConvAndSearch”. A tutorial on spectral library searching with SpectraST is provided *via* Table 13.1 entry 09 “SpectraSTSearch”. A tutorial on using Kojak on a cross-linking analysis is provided *via* Table 13.1 entry 10 “KojakSearch”. Beyond these mentioned search engines, the TPP supports a variety of third party search engines, as summarized in Figure 13.2.

In general, search engines will provide the best matching answer to each spectrum along with a score or set of scores. Yet only a fraction of those answers are correct and the rest are incorrect matches, often because the correct answer was not in the search space. Most search engines provide scores for each answer, but determining how those scores translate into likelihood of being correct is often difficult. The TPP provides a series of validation tools that apply Bayesian expectation maximization modelling

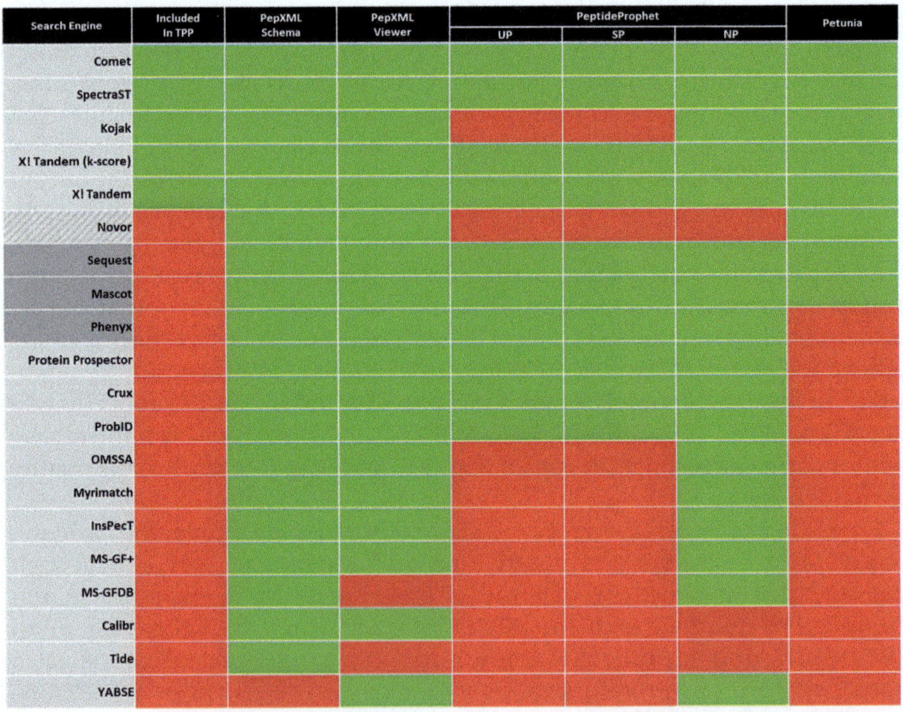


**Figure 13.2** Summary of support of many search engines by TPP components. Different components of the TPP are listed across the top. Search engines are listed as rows. Green indicates support by that component of the search engine. Red indicates lack of direct support for a given search engine by each TPP component. Under PeptideProphet, UP means unsupervized parametric model, SP means supervized parametric, and NP means non-parametric model is supported.

algorithms to assign robust probabilities of being correct to each peptide-spectrum match (PSM), peptide sequence, and inferred protein entry. The PeptideProphet[1] tool models one or more search engine scores as well some additional attributes of each match to assign probabilities to each PSM. The tutorial in Table 13.1 entry 11 "PeptideProphet" leads the reader through executing PeptideProphet on a sample yeast dataset, examining the models that PeptideProphet produced, and looking through the list of PSMs for notable entries. Whereas PeptideProphet computes a probability for each PSM based on the attributes of the match, iProphet[19] refines the probabilities of each PSM based on other corroborating evidence (*e.g.*, other PSMs with the same sequence, but different charge state or different mass modification.) If a dataset is searched with multiple search engines, these results can be combined into a single result with iProphet. A tutorial for iProphet is available in Table 13.1 entry 12 "iProphet." The ProteinProphet[2] tool then performs the peptide to protein inference step, using the PSM probabilities computed by PeptideProphet or iProphet to compute protein-level probabilities. A tutorial for running ProteinProphet and examining the output it in Table 13.1 entry 13 "ProteinProphet."

When most search engines provide a PSM result that includes one or more residues with a mass modification, there is usually no indication about the confidence with which a modification is localized to the proposed site when there are multiple possible sites. The PTMProphet[20] tool takes as input a PepXML file with search results including mass modifications and computes robust localization probabilities for each modification and each site, writing out the resulting probabilities to a new PepXML file. This enables the user to focus only on the PTMs that are localized to a specific residue with high confidence. A tutorial that guides the user through running PTMProphet and examining the results on a synthetic peptide ground-truth dataset is available *via* Table 13.1 entry 14 "PTMProphet".

In standard shotgun proteomics data analysis, it is generally assumed that each MS2 spectrum contains the fragments of a single peptide ion. However, this is rarely completely true, and with many datasets derived from complex samples, there are many spectra that contain more than one identifiable peptide ion. We have developed the reSpect[21] tool that allows iterative rounds of searching of spectra, enabling the identification of multiple peptide ions per spectrum. The reSpect software works by attenuating out the spectrum peaks identified in the initial (or previous) round of identification, which is then followed by another round of database searching without the peaks associated with the initially (or previously) identified peptide. The results from multiple rounds of running reSpect and searching can then be combined into a single result with iProphet for streamlined downstream analysis. A tutorial demonstrating the identification of multiple peptide ions in a dataset containing some chimeric spectra is available *via* Table 13.1 entry 15 "reSpect." Some DIA workflows such as SWATH-MS intentionally create highly multiplexed fragment ion spectra (with dozens of precursor ions fragmented at once) for which traditional search engines are not suitable. However, our DISCO tool

(Shteynberg. *et al.* in preparation) is able to extract demultiplexed spectra from the original scan based on the co-elution profiles of precursors and fragment ions, and these demultiplexed spectra can be searched with traditional search engines, possibly in conjunction with reSpect to further detangle and potentially identify precisely co-eluting precursor ions. The DISCO tutorial (Table 13.1 entry 16 "DISCO") leads the reader through the running DISCO on a small slice of a DIA run, followed by Comet searching and validation with PeptideProphet, iProphet, and ProteinProphet.

Spectral libraries have become important mechanisms for processing both DDA datasets *via* spectral library searching as well as for DIA datasets *via* library-based DIA analysis such as with OpenSWATH.[22] SpectraST is an advanced tool for both spectral library search[18] as well as spectral library creation.[23] In additional to the tutorial described above for spectral library search, we also have developed a tutorial that guides the reader through creating a spectral library from a previous search result (Table 13.1 entry 17 "SpecLibCreate"), as well as manipulating a library downloaded from NIST into a SILAC-suitable library (Table 13.1 entry 18 "SpecLibAlter.")

Following the identification and validation of peptides and proteins in a dataset, the crucial next step is to quantify the abundances of those peptides and proteins, typically in a relative manner between related samples. The three main quantitation strategies typically used are isotopic labelling, isobaric labelling, and label free. The TPP includes two tools for isotopically labelled data, ASAPRatio[24] and XPRESS. In general, ASAPRatio is the more advanced tool, but there are a few cases where the simpler XPRESS tool is a better fit. The isotopic labelling analysis tutorial (Table 13.1 entry 19 "IsotopicQuant") guides the reader through the analysis of a simple yeast dataset with both ASAPRatio and XPRESS to obtain an overview of the features of both. For isobaric labelling strategies (such as iTRAQ[25,26] and TMT[27]), the TPP includes the Libra tool,[28] which measures the reporter ions in each PSM fragment ion spectrum and combines the information up to the peptide and protein level. The Libra tutorial guides the reader through an analysis of a sample iTRAQ reference dataset (Table 13.1 entry 20 "Libra.") TPP includes the StPeter[29] label-free quantitation tool, which uses the SIn method based on summing up the annotated fragments in MS2 spectra to yield robust PSM-level, peptide-level, and protein-level quantitation. The StPeter tutorial (Table 13.1 entry 21 "StPeter") guides the user through the analysis of the reference dataset and examination of the result.

With all open source tools, the expectation is that others can co-develop program enhancements beyond the offering currently released into the public domain and contribute to the pipeline for the benefit of all users. As in the case of the TPP, a number of tools and implementations have been developed by others to take advantage of the functionality of the integrated and interoperability of the TPP. As an example of external modules interfacing with the TPP, a recent Windows product called WinProphet (see http://ms.iis.sinica.edu.tw/Comics/Software_WinProphet.html) has been released that allows a scripting interface for automated sample processing. WinProphet can create and automatically execute TPP analyses supporting various functionalities including database search for protein and peptide identification, spectral library construction and search, DIA (Data-Independent Acquisition) data analysis, isobaric labelling and label-free quantitation all supported with a graphical user interface.

## 13.3 Conclusion

The toolset of the TPP is continually being expanded and refined with continued funding from the US National Institutes of Health, National Institute for General Medical Sciences. The TPP forms the foundation for dataset reanalysis for PeptideAtlas,[30,31] which aims to provide current snapshots of the observed proteome for many species based on a uniform processing of publicly available data. The TPP provides the ability for PeptideAtlas to compute and control the FDR across all datasets to provide a high-quality compendium for each species. The TPP can be run on many different platforms, from a modest laptop computer running Microsoft Windows to large-scale Linux computing environments with thousands of cores *via* cloud-computing machine images or Docker containers and each TPP instance running on a Windows or Linux platform can be controlled remotely *via* any web browser.

We have reviewed here the many current tools and described how to access tutorials that help users become familiar with these tools. Many of these tutorials are used during the TPP courses that are taught throughout the year, and the tutorials are periodically refreshed as the capability of the tools expand or the interfaces are streamlined to make them easier to use. All up-to-date tutorials may be accessed at http://tppms.org/tutorials/ and questions about or suggestions for these tutorials may be posted to our discussion group at spctools-discuss@googlegroups.com. One can search previously asked questions and answers and learn how to join the group at https://groups.google.com/group/spctools-discuss.

## Acknowledgements

This work was funded in part by the National Institutes of Health grants from the National Institute for General Medical Sciences R01GM087221, R24GM127667, 2P41GM103533, and the National Institute On Aging under Award Number U19AG023122.

## References

1. A. Keller, A. I. Nesvizhskii, E. Kolker and R. Aebersold, Empirical Statistical Model to Estimate the Accuracy of Peptide Identifications Made by MS/MS and Database Search, *Anal. Chem.*, 2002, **74**(20), 5383–5392.
2. A. I. Nesvizhskii, A. Keller, E. Kolker and R. Aebersold, A Statistical Model for Identifying Proteins by Tandem Mass Spectrometry, *Anal. Chem.*, 2003, **75**(17), 4646–4658.

3. A. Keller, J. Eng, N. Zhang, X. Li and R. Aebersold, A Uniform Proteomics MS/MS Analysis Platform Utilizing Open XML File Formats, *Mol. Syst. Biol.*, 2005, **1**, 2005.0017.
4. E. W. Deutsch, L. Mendoza, D. Shteynberg, T. Farrah, H. Lam, N. Tasman, Z. Sun, E. Nilsson, B. Pratt and B. Prazen, *et al.*, A Guided Tour of the Trans-Proteomic Pipeline, *Proteomics*, 2010, **10**(6), 1150–1159.
5. E. W. Deutsch, L. Mendoza, D. Shteynberg, J. Slagel, Z. Sun and R. L. Moritz, Trans-Proteomic Pipeline, a Standardized Data Processing Pipeline for Large-Scale Reproducible Proteomics Informatics, *Proteomics: Clin. Appl.*, 2015, **9**(7–8), 745–754.
6. L. Martens, M. Chambers, M. Sturm, D. Kessner, F. Levander, J. Shofstahl, W. H. Tang, A. Römpp, S. Neumann and A. D. Pizarro, *et al.*, MzML–a Community Standard for Mass Spectrometry Data, *Mol. Cell. Proteomics*, 2011, **10**(1), R110.000133.
7. M. C. Chambers, B. Maclean, R. Burke, D. Amodei, D. L. Ruderman, S. Neumann, L. Gatto, B. Fischer, B. Pratt and J. Egertson, *et al.*, A Cross-Platform Toolkit for Mass Spectrometry and Proteomics, *Nat. Biotechnol.*, 2012, **30**(10), 918–920.
8. P. G. A. Pedrioli, J. K. Eng, R. Hubley, M. Vogelzang, E. W. Deutsch, B. Raught, B. Pratt, E. Nilsson, R. H. Angeletti and R. Apweiler, *et al.*, A Common Open Representation of Mass Spectrometry Data and Its Application to Proteomics Research, *Nat. Biotechnol.*, 2004, **22**(11), 1459–1466.
9. E. W. Deutsch, File Formats Commonly Used in Mass Spectrometry Proteomics, *Mol. Cell. Proteomics*, 2012, **11**(12), 1612–1621.
10. A. R. Jones, M. Eisenacher, G. Mayer, O. Kohlbacher, J. Siepen, S. J. Hubbard, J. N. Selley, B. C. Searle, J. Shofstahl and S. L. Seymour, *et al.*, The MzIdentML Data Standard for Mass Spectrometry-Based Proteomics Results, *Mol. Cell. Proteomics*, 2012, **11**(7), M111.014381.
11. J. A. Vizcaíno, G. Mayer, S. Perkins, H. Barsnes, M. Vaudel, Y. Perez-Riverol, T. Ternent, J. Uszkoreit, M. Eisenacher and L. Fischer, *et al.*, The MzIdentML Data Standard Version 1.2, Supporting Advances in Proteome Informatics, *Mol. Cell. Proteomics*, 2017, **16**(7), 1275–1285.
12. J. K. Eng, T. A. Jahan and M. R. Hoopmann, Comet: An Open-Source MS/MS Sequence Database Search Tool, *Proteomics*, 2013, **13**(1), 22–24.
13. J. K. Eng, A. L. McCormack and J. R. Yates, An Approach to Correlate Tandem Mass Spectral Data of Peptides with Amino Acid Sequences in a Protein Database, *J. Am. Soc. Mass Spectrom.*, 1994, **5**(11), 976–989.
14. R. Craig and R. C. Beavis, TANDEM: Matching Proteins with Tandem Mass Spectra, *Bioinformatics*, 2004, **20**(9), 1466–1467.
15. P.-A. Binz, J. Shofstahl, J. A. Vizcaino, H. Barsnes, R. J. Chalkley, G. Menschaert, E. Alpi, K. R. Clauser, J. K. Eng and L. Lane, *et al.*, Proteomics Standards Initiative Extended FASTA Format (PEFF), *J. Proteome Res.*, 2019, **18**(6), 2686–2692.
16. M. R. Hoopmann, A. Zelter, R. S. Johnson, M. Riffle, M. J. MacCoss, T. N. Davis and R. L. Moritz, Kojak: Efficient Analysis of Chemically Cross-Linked Protein Complexes, *J. Proteome Res.*, 2015, **14**(5), 2190–2198.

17. M. R. Hoopmann, L. Mendoza, E. W. Deutsch, D. Shteynberg and R. L. Moritz, An Open Data Format for Visualization and Analysis of Cross-Linked Mass Spectrometry Results, *J. Am. Soc. Mass Spectrom.*, 2016, **27**(11), 1728–1734.
18. H. Lam, E. W. Deutsch, J. S. Eddes, J. K. Eng, N. King, S. E. Stein and R. Aebersold, Development and Validation of a Spectral Library Searching Method for Peptide Identification from MS/MS, *Proteomics*, 2007, **7**(5), 655–667.
19. D. Shteynberg, E. W. Deutsch, H. Lam, J. K. Eng, Z. Sun, N. Tasman, L. Mendoza, R. L. Moritz, R. Aebersold and A. I. Nesvizhskii, IProphet: Multi-Level Integrative Analysis of Shotgun Proteomic Data Improves Peptide and Protein Identification Rates and Error Estimates, *Mol. Cell. Proteomics*, 2011, **10**(12), M111.007690.
20. D. D. Shteynberg, E. W. Deutsch, D. S. Campbell, M. R. Hoopmann, U. Kusebauch, D. Lee, L. Mendoza, M. K. Midha, Z. Sun and A. Whetton, *et al.*, PTMProphet: Fast and Accurate Mass Modification Localization for the Trans-Proteomic Pipeline, *J. Proteome Res.*, 2019, **18**(12), 4262–4272.
21. D. Shteynberg, L. Mendoza, M. R. Hoopmann, Z. Sun, F. Schmidt, E. W. Deutsch and R. L. Moritz, ReSpect: Software for Identification of High and Low Abundance Ion Species in Chimeric Tandem Mass Spectra, *J. Am. Soc. Mass Spectrom.*, 2015, **26**(11), 1837–1847.
22. H. L. Röst, R. Aebersold and O. T. Schubert, Automated SWATH Data Analysis Using Targeted Extraction of Ion Chromatograms, *Methods Mol. Biol.*, 2017, **1550**, 289–307.
23. H. Lam, E. W. Deutsch, J. S. Eddes, J. K. Eng, S. E. Stein and R. Aebersold, Building Consensus Spectral Libraries for Peptide Identification in Proteomics. Nat, *Methods*, 2008, **5**(10), 873–875.
24. X.-J. Li, H. Zhang, J. A. Ranish and R. Aebersold, Automated Statistical Analysis of Protein Abundance Ratios from Data Generated by Stable-Isotope Dilution and Tandem Mass Spectrometry, *Anal. Chem.*, 2003, **75**(23), 6648–6657.
25. P. L. Ross, Y. N. Huang, J. N. Marchese, B. Williamson, K. Parker, S. Hattan, N. Khainovski, S. Pillai, S. Dey and S. Daniels, *et al.*, Multiplexed Protein Quantitation in Saccharomyces Cerevisiae Using Amine-Reactive Isobaric Tagging Reagents, *Mol. Cell. Proteomics*, 2004, **3**(12), 1154–1169.
26. L. Choe, M. D'Ascenzo, N. R. Relkin, D. Pappin, P. Ross, B. Williamson, S. Guertin, P. Pribil and K. H. Lee, 8-Plex Quantitation of Changes in Cerebrospinal Fluid Protein Expression in Subjects Undergoing Intravenous Immunoglobulin Treatment for Alzheimer's Disease, *Proteomics*, 2007, **7**(20), 3651–3660.
27. A. Thompson, J. Schäfer, K. Kuhn, S. Kienle, J. Schwarz, G. Schmidt, T. Neumann, R. Johnstone, A. K. A. Mohammed and C. Hamon, Tandem Mass Tags: A Novel Quantification Strategy for Comparative Analysis of Complex Protein Mixtures by MS/MS, *Anal. Chem.*, 2003, **75**(8), 1895–1904.

28. P. G. A. Pedrioli, B. Raught, X.-D. Zhang, R. Rogers, J. Aitchison, M. Matunis and R. Aebersold, Automated Identification of SUMOylation Sites Using Mass Spectrometry and SUMmOn Pattern Recognition Software. Nat, *Methods*, 2006, **3**(7), 533–539.
29. M. R. Hoopmann, J. M. Winget, L. Mendoza and R. L. Moritz, StPeter: Seamless Label-Free Quantification with the Trans-Proteomic Pipeline, *J. Proteome Res.*, 2018, **17**(3), 1314–1320.
30. F. Desiere, E. W. Deutsch, N. L. King, A. I. Nesvizhskii, P. Mallick, J. Eng, S. Chen, J. Eddes, S. N. Loevenich and R. Aebersold, The PeptideAtlas Project, *Nucleic Acids Res.*, 2006, **34**(Database issue), D655–D658.
31. E. W. Deutsch, Z. Sun, D. Campbell, U. Kusebauch, C. S. Chu, L. Mendoza, D. Shteynberg, G. S. Omenn and R. L. Moritz, State of the Human Proteome in 2014/2015 As Viewed through PeptideAtlas: Enhancing Accuracy and Coverage through the AtlasProphet, *J. Proteome Res.*, 2015, **14**(9), 3461–3473.

CHAPTER 14

# Quantitative Proteomics Data Analysis with PANDA, LFAQ and PANDA-view

KAIKUN XU AND CHENG CHANG*

State Key Laboratory of Proteomics, Beijing Proteome Research Center, Beijing Institute of Lifeomics, National Center for Protein Sciences (Beijing), Life Science Park Road No. 38, 102206, Beijing, China
*E-mail: changchengbio@gmail.com

## 14.1 Introduction

Nowadays, quantitative proteomics which focuses on accurate determination of protein abundances and their changes on proteome-wide scale has become an important aspect in proteomics research.[1] In general, there are two main kinds of strategies in quantitative proteomics, *i.e.*, relative and absolute protein quantifications. Relative protein quantification is exploring the abundance changes of the same protein in different samples. While, absolute protein quantification aims to determine the absolute amount of every protein in a single sample. A number of quantitative algorithms and tools have been proposed and developed, which provide a solid support for subsequent research including biomarker discovery and selection of differentially expressed proteins. However, as the techniques of mass spectrometry (MS) and the experimental strategies of quantitative proteomics are improving rapidly, accurate quantification algorithms and flexible tools are still required.[2]

New Developments in Mass Spectrometry No. 8
Processing Metabolomics and Proteomics Data with Open Software: A Practical Guide
Edited by Robert Winkler

Published by the Royal Society of Chemistry, www.rsc.org

Here, we provide some free and easy-to-use software tools for both relative and absolute quantification, as well as the post-processing of quantification results. For relative protein quantification, we present a novel and comprehensive software package, named PANDA[3] which is developed based on our solid foundations in quantitative proteomics over a number of years. The advantage algorithms of our previous works, *i.e.* LFQuant[4] and SILVER[5] have been implemented into PANDA. The core of PANDA was written in standard C++ on the platform of Microsoft Visual Studio ultimate 2017 in the Windows System. The interfaces of PANDA were implemented in C# on the same platform. For absolute protein quantification, we propose a novel algorithm named LFAQ[6] for label-free absolute protein quantification, which can correct the biased MS intensities using the predicted peptide quantitative factors for all identified peptides. LFAQ provides user-friendly interfaces for parameter setting, quantification analysis and result visualization. The core of LFAQ was written in standard C++ on the platform of Microsoft Visual Studio ultimate 2013 in the Windows System. The interface of LFAQ was implemented in C# on the same platform. Moreover, for post-processing of the quantification results, PANDA-view[7] is developed for statistical analysis and data visualization as an affiliated tool of PANDA. PANDA-view can directly read and perform a multi-level representation of PANDA's quantification results. In addition, PANDA-view is compatible with other -omics tools by taking their results in tab-delimited text (*.txt) and CSV (*.csv) file formats as input. The core of PANDA-view was written in Qt C++ on the platform of Microsoft Visual Studio ultimate 2013 under the Windows System. The interface of PANDA-view was implemented in C# on the same platform. Parts of the functions in PANDA-view are developed based on the R statistical environment.

In this chapter, we provide detailed descriptions of the user manuals for PANDA, LFAQ and PANDA-view to facilitate their use for proteomics researchers.

## 14.2 Materials

### 14.2.1 Hardware Requirements

A personal computer with at least 2 GB of RAM is required. It is recommended that the program is run with a Pentium III/800 MHz or higher-performance dual-threaded CPU.

### 14.2.2 Software Requirements

PANDA, LFAQ and PANDA-view can only run on 64 bit versions of the Windows operating system (OS), *i.e.*, Windows 7 SP1, Windows Server 2008 R2 SP1, Windows Server 2008 SP2, Windows 8 and Windows 10. In addition, .NET Framework 4.5 or higher is required. Microsoft Visual C++ Redistributable for Visual Studio 2017 (x64) needs to be installed for PANDA and Microsoft Visual C++ Redistributable for Visual Studio 2013 (x64) needs to be installed for LFAQ and PANDA-view. Moreover, in PANDA, both 32 bit and 64 bit versions of Thermo Fisher Scientific MSFileReader software tools are required to be installed to access Thermo raw files. As PANDA, LFAQ and PANDA-view are Windows-based tools, users using Mac or Linux computers need to use Windows emulation software such as Boot Camp. Besides the above requirements, LFAQ and PANDA-view also demands the installation of R for Windows (version 3.1.0 or newer).

### 14.2.3 Software Installation

PANDA is freely available at https://sourceforge.net/projects/panda-tools/. No installation is required. Just un-compress the downloaded 7z package into a specified file folder and double-click the "PANDA.exe" file to start the software. Once parameters in the graphical user interface (GUI) are set or a configuration file containing all the necessary parameters is loaded, PANDA is ready to perform quantification analysis. LFAQ is freely available at https://LFAQ.github.io/LFAQ/ and PANDA-view is freely available at https://sourceforge.net/projects/panda-view/. Just like PANDA, users can un-compress the compressed tool package and start LFAQ or PANDA-view by clicking the corresponding executable file in the folder. The GUIs of PANDA, LFAQ and PANDA-view are shown in Figure 14.1.

However, before starting LFAQ and PANDA-view, the R for Windows (version 3.1.0 or newer) should be installed. After that, users should add the path of "RScript.exe" into the system environment variable and install some necessary R packages, because LFAQ and PANDA-view implement some R-based methods by calling "Rscript.exe" to execute the R codes. When there are several versions of R installed in a user's computer, LFAQ and PANDA-view will call the "Rscript.exe" whose path is added into the system environment variable. The method for setting a system environment variable can be found at http://www.computerhope.com/issues/ch000549.htm. By default, "RScript.exe" is in a folder such as "c:\Program Files\R\R-3.3.3\bin\". Then, this folder

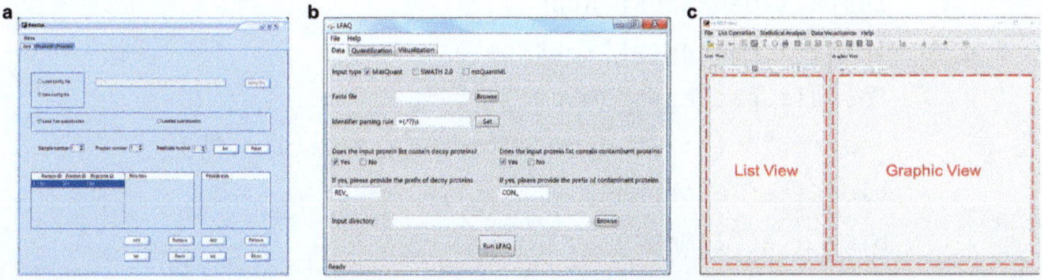


**Figure 14.1** GUIs of PANDA (a), LFAQ (b) and PANDA-view (c).

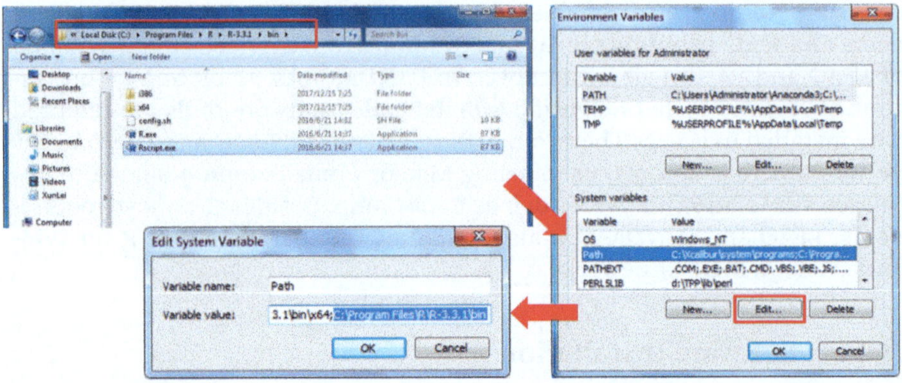


**Figure 14.2** Illustration of adding the "RScript.exe" path into system environment variable.

path should be added into the system environment variable. In addition, the path "c:\Program Files\R\R-3.3.3\bin\x64\" should also be added for the 64 bit OS. See Figure 14.2 for details.

To use LFAQ, users should install the R package "ggplot2" using the following command.

```
install.packages("ggplot2")
```

For PANDA-view, the following R packages should be installed using the commands below.

```
install.packages("rgl")
install.packages("scatterplot3d")
install.packages("RColorBrewer")
install.packages("gplots")
install.packages("survival")
install.packages("coin")
install.packages("Rcpp")
install.packages("lattice")
install.packages("mice")
source("http://bioconductor.org/biocLite.R")
biocLite("Biobase")
biocLite("limma")
biocLite("impute")
biocLite("R.methodsS3")
biocLite("matrixStats")
biocLite("samr")
```

## 14.3 Procedures

In this section we describe the complete set of parameters and configuration options in PANDA, LFAQ and PANDA-view. For routine analysis, only a small part of the parameters should be modified, the remaining parameters are unchanged from the default.

### 14.3.1 Relative Protein Quantification Using PANDA

1. Start PANDA by double-clicking the PANDA.exe file.

#### *14.3.1.1 Set Parameters in the Data Interface*

2. Set experimental design parameters. Users can set experimental design parameters in the GUI by choosing "New config file". When the quantitative method used in the experiment is "Label free quantification", three parameters "Sample number", "Fraction number" and "Replicate number" need to be set (Figure 14.3a). When the quantitative method is "Labeled quantification", "Sample number" parameter is replaced by "Layer number" (Figure 14.3b). Users can click the Reset button to reset the experiment design. The detailed annotations of these parameters are shown in Table 14.1.
3. Upload MS files by clicking the "Add" and "Remove" buttons. PANDA supports Thermo raw (MSFileReader is required at first), mzXML and mzML.

**CAUTION:** It is important to make sure that the file number of the MS data is equal to the number set in the experiment design, *i.e.* "Sample number" × "Fraction number" × "Replicate number" for "Label free quantification" or

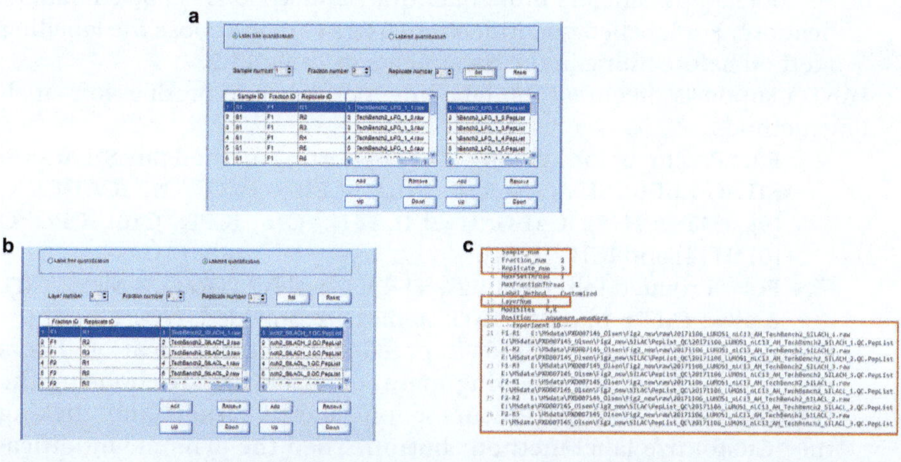

**Figure 14.3** Setting the experimental design parameters in PANDA. (a)–(b) Setting the experimental design parameters in GUI for label free (a) and labelled (b) quantifications, respectively. (c) Setting the experimental design parameters by writing a configuration file directly.

**Table 14.1** Annotations of the parameters about experimental design in PANDA.

| Parameter | Annotation |
|---|---|
| Sample number | Sample (biological replicate) number |
| Fraction number | Fraction number in each sample |
| Replicate number | Technical replicate number in each fraction |
| Layer number | The labelling layer number for a specific labelling method. Default layer number is 2, representing heavy and light labels. |

"Fraction number" × "Replicate number" for "Labeled quantification". Furthermore, the MS files must be in the same order as the experimental design parameters on the left; the order can be adjusted using the two buttons "Up" and "Down" (Figure 14.3b).

4. Upload peptide identification files by clicking the "Add" and "Remove" buttons. The file number and order also need to be the same as the left parameter (Figure 14.3b). PANDA supports mzIdentML[8] and the quality control results of PeptideProphet[9] and PepDistiller.[10] PeptideProphet can be freely obtained from Trans-Proteomic Pipeline. PepDistiller can be freely downloaded from https://sourceforge.net/projects/pepdistiller/.

▲ **Critical:** It is recommended that parameters are set by writing a configuration file (*.config) and loading it into PANDA *via* the "Load config file" button when there are a large number of MS files (Figure 14.3c).

### 14.3.1.2 Set Parameters in the Parameter Interface

5. Set peptide quantification parameters. Users need to fill out the parameters listed in Table 14.2. However, most parameters will already have been filled in by default (Figure 14.4a).
6. Set labelled parameters if the quantitative method is "Labeled quantification". For labelled quantification, users should choose the labelling method before filling in the parameters in Table 14.2.

PANDA supports precursor ion labelling methods and product ion labelling methods.

- For "Precursor ion labeling", the supporting methods are SILAC-K6, SILAC-K6R6, SILAC-K6R10, SILAC-K8R10, $^{18}O$, $^{15}N$, ICAT-C:$^{13}C$ (9), ICAT-D:$^{2}H$ (8), ICAT-G:$^{2}H$ (8), ICAT-H:$^{13}C$ (6), ICPL:$^{13}C$ (6), ICPL:$^{13}C$ (6) $^{2}H$ (4) and ICPL:$^{2}H$ (4).
- For "Product ion labeling", iTRAQ 4-plex, iTRAQ 8-plex, TMT 2-plex, TMT 6-plex and TMT 10-plex are provided (Figure 14.4b).

7. Customize labelling methods in precursor ion quantification. Users can define their own labelling methods in PANDA by choosing the "Customized" item in "Precursor ion quantification" and clicking the "Customize label method" button. Then the dynamic modifications read from the peptide identification files will be shown in the

**Table 14.2** Annotations of the parameters about peptide quantification in PANDA.

| Parameter | Annotation |
|---|---|
| XIC peak tolerance min | The peptide *m*/*z* tolerance in ppm used during XIC construction, −10 is recommended. |
| XIC peak tolerance max | The peptide *m*/*z* tolerance in ppm used during XIC construction, 10 is recommended. |
| SN cut-off | The isotope signal-to-noise ratio cut-off during XIC construction, 2 is recommended. |
| Goodness of fit cut-off | The goodness of the match between the observed and theoretical isotopic distribution, 0.6 is recommended. |
| XIC RT truncation min | The peptide retention time minimum (in minutes) during XIC construction, −5 is recommended. |
| XIC RT truncation max | The peptide retention time maximum (in minutes) during XIC construction, 5 is recommended. |
| XIC RT truncation gap | The peptide retention time gap (in minutes) during XIC construction, 0.2 is recommended. |
| Working path | The file folder path to store the results of PANDA |
| Peptide FDR | The FDR at peptide level, 0.01 is recommended. |
| Do cross quantification for label-free | A bool value. Check if we need cross search quantification for label-free. |

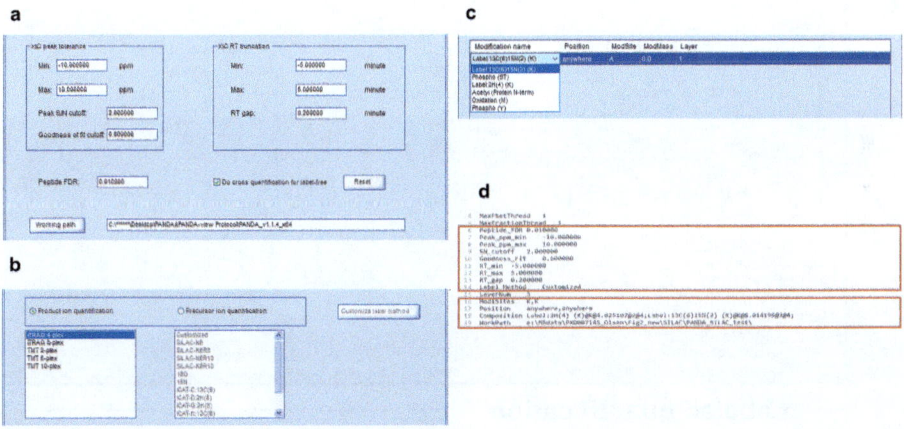

**Figure 14.4** Setting peptide quantitation parameters in PANDA. (a) Setting general parameters. (b) Choosing labelling method. (c) Setting user-defined labelling method. (d) Setting labelling method in a configuration file.

dialog for users to choose (Figure 14.4c). For a user-defined labelling method, the parameters, including its position, modification site, modification mass and labelling layer are required to be filled in manually.

▲ **Critical**: Peptide quantification parameters written in the configuration file are shown in the box in Figure 14.4d.

### 14.3.1.3 Set Parameters in the Progress Interface

8. In the Progress interface, users can set the threads during quantitative process. As shown in Table 14.3, for label-free quantification users can set the thread number for each fraction set, each fraction and cross search quantification (Figure 14.5a). For labelled quantification, only the thread in each fraction set and fraction can be set (Figure 14.5b).

**Table 14.3** Annotations of the parameters in the Progress interface of PANDA.

| Parameter | Annotation |
|---|---|
| Max fraction set thread | Max thread number in each fraction set |
| Max fraction thread | Max thread number in each fraction |
| Max cross-search thread | Max thread number in cross search quantification for label-free |
| Cached file number | File number read in the cache (label-free quantification only) |

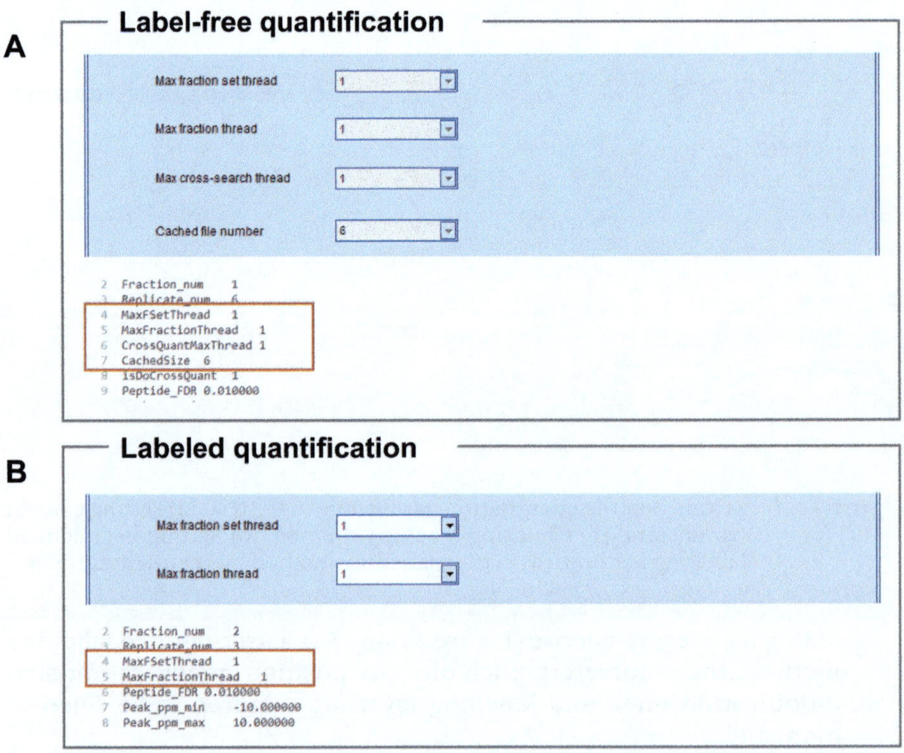


**Figure 14.5** Setting thread parameters in PANDA. (a)–(b) Setting the experimental design parameters in GUI or in a configuration file for label free (a) and labelled (b) quantifications, respectively.

**CAUTION:**

- The max thread to choose in the GUI is the max thread in the computer by default.
- The max cached file number is the MS file number loaded in PANDA. Users should choose a proper value in case the program crashes due to memory overflow.
- In PANDA, fraction set is defined as a set of MS data containing all the samples and replicates in each fraction. For example, if the sample number is 5, fraction number is 3, replicate number is 2. Then we have three fraction sets. Each contains 10 (5 sample × 2 replicate) MS files.

9. Run PANDA. After all the parameters in the Data, Parameter and Progress interfaces are set, users can click the Run button to run PANDA directly and click the Stop button to force PANDA to stop. Once the Run button is clicked, all the parameters set in GUI will be written as a configuration file named using the computer local time (*e.g.* PANDA_2019_04_26_22_45_36.config). The configuration file is saved in the same folder as the PANDA program. Meanwhile, the basic information during the quantification will be shown in the Quantification progress text box (Figure 14.6a). When all the calculations are done, an alert dialog will be shown (Figure 14.6b).

#### *14.3.1.4 Anticipated Results*

Once the calculation is complete, PANDA generates different levels of quantification result files in the result directory (Table 14.4), *i.e.* the peptide-sequence-match (PSM) level (PeptideIons_FX.txt), peptide level (Peptides.txt) and protein level (Proteins.txt).

### 14.3.2 Absolute Protein Quantification Using LFAQ

Users can double-click "LFAQ.exe" and set parameters to run LFAQ (Figure 14.1b). A parameter file including the input data file path and quantification parameters is required before starting LFAQ.

#### *14.3.2.1 Set Parameters and Generate a New Parameter File in GUI*

In GUI, the LFAQ parameters are separated into two groups: the parameters about input data (Figures 14.7 and 14.8) and the parameters about quantification (Figures 14.9 and 14.10).

##### 14.3.2.1.1 Set Parameters About the Input Data

1. Choose the input of LFAQ. LFAQ quantifies proteins based on the peptide quantification results from other software tools. Currently, LFAQ

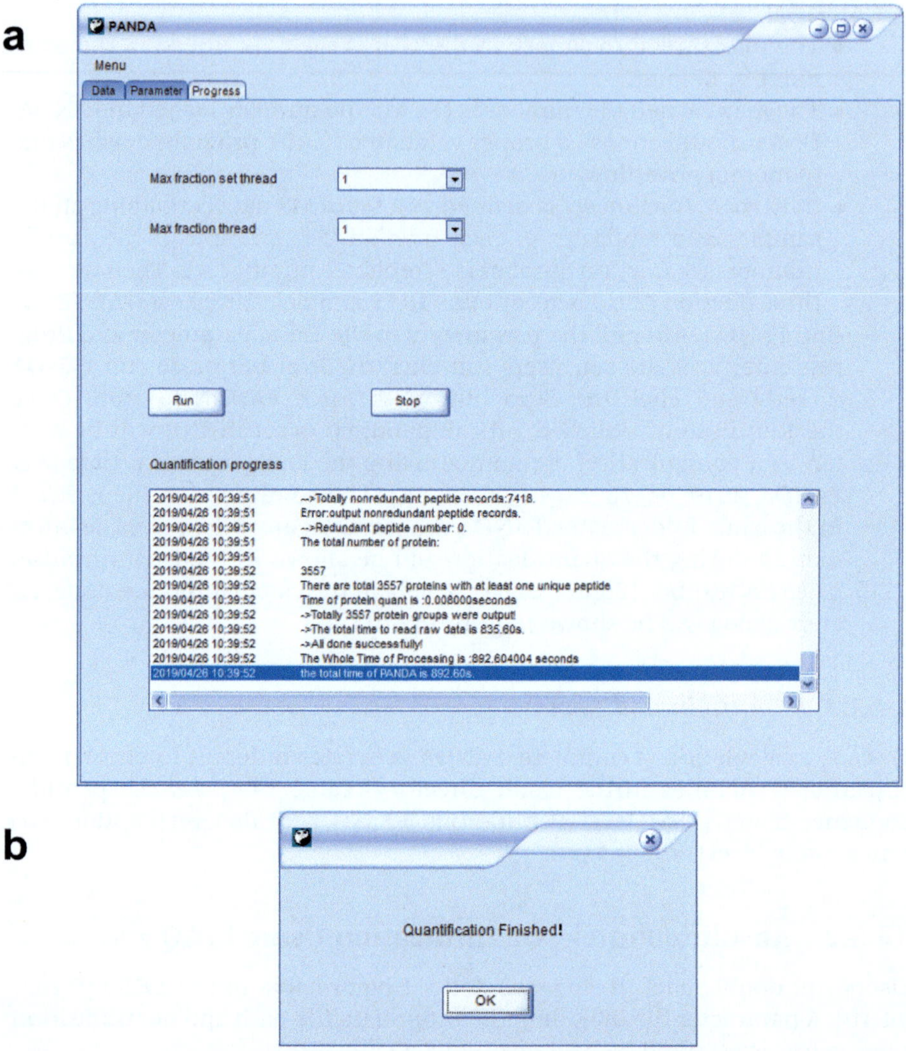


**Figure 14.6** The progress interface in PANDA. (a) The quantification progress will be shown in the GUI when PANDA is running. (b) When PANDA finishes all the calculations, a dialog will pop up.

**Table 14.4** Annotations of PANDA quantification result files.[a]

| File name | Annotations |
|---|---|
| Proteins.txt | The protein quantification information. |
| Petides.txt | The non-redundant peptide quantification information. |
| PeptideIons_FX.txt | The peptide ion quantification information in the X fraction. |
| PeptideIons_FX_XIC.xml | The XIC data information for every peptide ion quantified in the X fraction. The XML file is used for XIC visualization. |
| Config file | The configuration file containing all the parameters set in GUI. |

[a]Note that the X represents the fraction index of parameter setting in the *Data* interface.

is compatible with the results of MaxQuant,[11] SWATH 2.0 or the public standard quantification data format mzQuantML[12] proposed by HUPO/PSI (Figure 14.7a). For MaxQuant results, the following files are required: (1) proteinGroups.txt, (2) peptides.txt, (3) experimentalDesignTemplate.txt. For SWATH 2.0 results, the second sheet "Area – peptides" of the result excel file should be saved as a separate CSV file at first. For mzQuantML format, just one file suffixed with "mzq" is required.

1) Load the protein sequence file (in fasta format) used in identification. Click the "Browse" button to select the fasta file. Note: the rows of protein headers must begin with the symbol ">" (Figure 14.7b).
2) Provide a regular expression to properly extract the protein identifiers from the protein sequence file. Click the "Set" button (Figure 14.7c) to open the "Set Regular Expression" dialog (Figure 14.8). Note that the "Input directory" should be set in advance when the input type is MaxQuant.

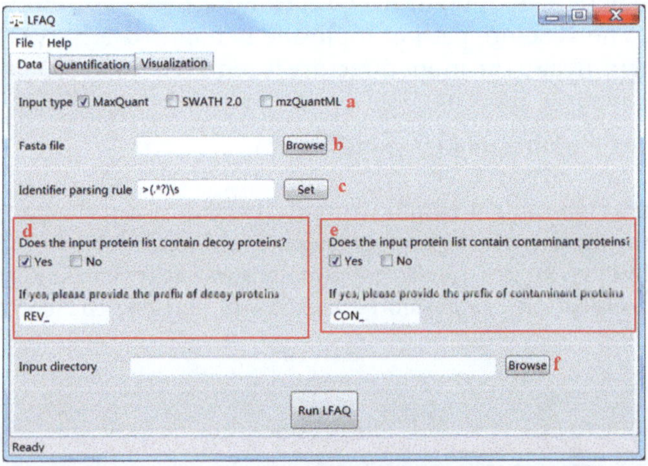


**Figure 14.7** Parameter setting interface for input data in LFAQ.

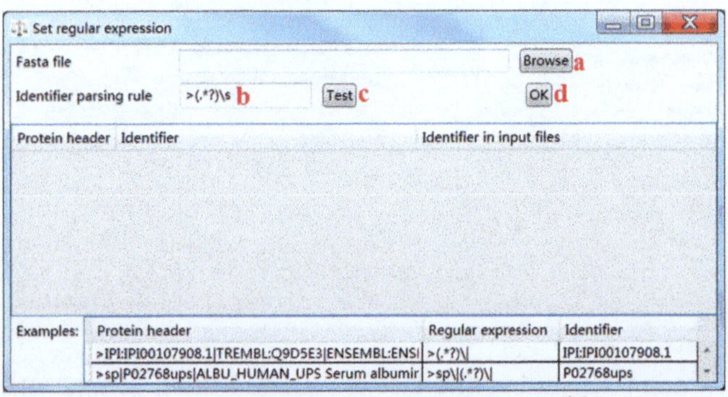


**Figure 14.8** Parameter setting interface for the regular expression in LFAQ.

2. In the "Set Regular Expression" dialog (Figure 14.8), the user can specify the regular expression used to extract the protein identifiers from the protein sequence.
   1) Click the "Browse" button to choose the fasta file (Figure 14.8a).
   2) Fill in the regular expression (Figure 14.8b). There are some examples below for reference.
   3) Click the "Test" button to check the regular expression (Figure 14.8c). Note that the identifiers extracted from the fasta file by the regular expression should be the same as the protein IDs in the input files.
   4) Click the "OK" button if the regular expression is confirmed (Figure 14.8d).
3. Choose if the input protein list contains decoy proteins or not. If yes, provide the prefix of decoy proteins, otherwise leave it blank (Figure 14.7d).
4. Choose if the input protein list contains contaminant proteins or not. If yes, provide the prefix of contaminant proteins, otherwise leave it blank (Figure 14.7e).
5. Choose the location (directory) where the input files were located or the absolute path of input file by clicking the "Browse" button (Figure 14.7f). The absolute path of input file is required for SWATH 2.0 and mzQuantML and the input directory is required for MaxQuant.

#### 14.3.2.1.2 Set Parameters for Quantification

6. Choose a regression method. To predict peptide quantification efficiency, we have implemented two regression methods in LFAQ: BART (Bayesian Addictive Regression Trees) and stepwise regression (Figure 14.9a). When one of the regression methods is chosen, the corresponding

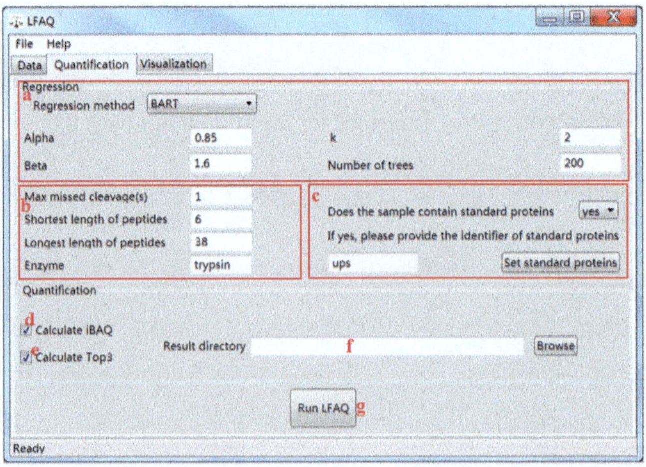


**Figure 14.9** Parameter setting interface for quantification using BART regression method in LFAQ.

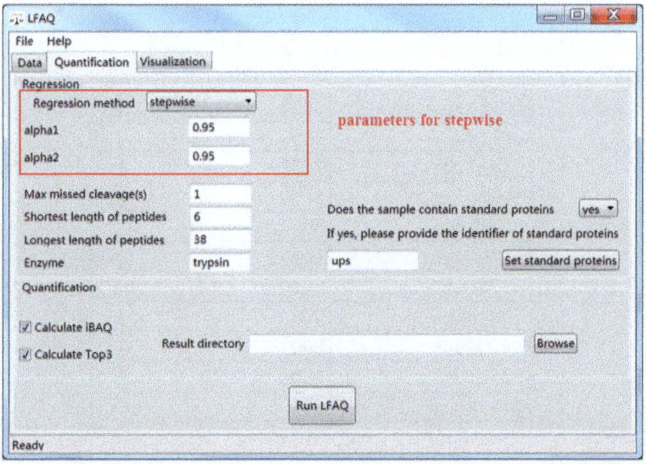


**Figure 14.10** Parameter setting interface for quantification using stepwise regression method in LFAQ.

parameters will appear (Figure 14.10). Based on our experience, we suggest users choose the BART as the preferred selection.

7. Choose an enzyme (Figure 14.9b). Until now, LFAQ only supports the "trypsin" enzyme.
8. Choose if the sample contains standard proteins or not (Figure 14.9c). If yes, provide the identifiers which can be used to distinguish standard proteins from other proteins. For example, if UPS2 proteins were used as standard proteins, just fill in "ups" for the textbox because all UPS2 protein names contain "ups" (case-insensitive here). Then click the "Set standard proteins" button to open a dialog to specify the injection amounts of standard proteins.
9. Calculate iBAQ value for every protein (optional) (Figure 14.9d).
10. Calculate Top3 value for every protein (optional) (Figure 14.9e).
11. Choose the directory where the quantification results will be saved by clicking the "Browse" button (Figure 14.9f).
12. In the dialog of "Set standard proteins", the user can edit the table of standard proteins (Figure 14.11). The first column represents protein IDs, and the second column represents the injection amounts of the proteins. Fill in the amounts of standard proteins, then click the "Save" button to save the information to the "StandardProteins.txt" file. Note that the corresponding standard protein IDs in the input file should contain the standard protein IDs in the "StandardProteins.txt" file.

#### *14.3.2.2 Load an Existing Parameter File*

13. If an existing parameter file (*.params) is available, a convenient way to set parameters is loading this parameter file directly (Figure 14.12).

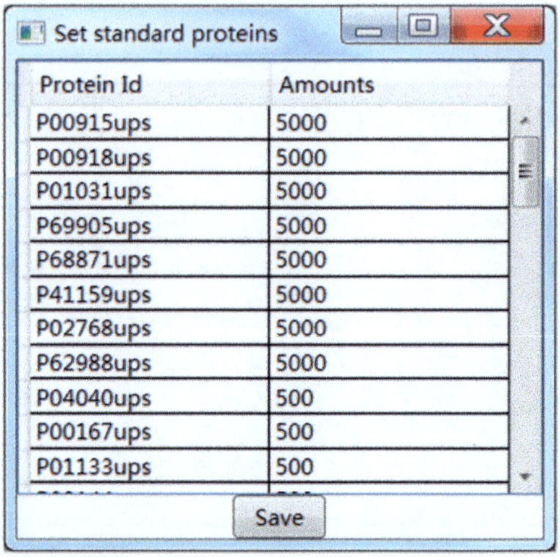


**Figure 14.11** Standard protein setting dialog in LFAQ.

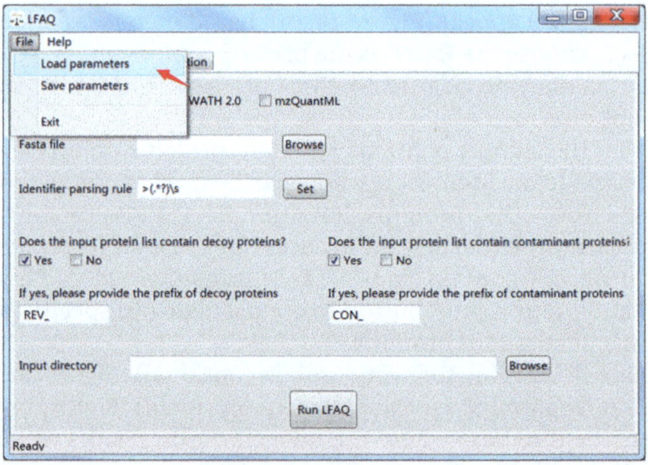


**Figure 14.12** Interface of manually loading a parameter file in LFAQ.

## *14.3.2.3 Run LFAQ*

14. Make sure that all the parameters are appropriate, and click the "Run LFAQ" button to start LFAQ (Figure 14.9g). A parameter file named as the local time (*e.g.* parameters20170713-09-37-07.params) containing all the parameters set in the GUI is then generated in the result file folder. When LFAQ is running, the status bar at the bottom of the window will show "LFAQ is running!" (Figure 14.13). If the background program is completed successfully, a dialog showing "LFAQ quantification finished successfully!" will appear (Figure 14.14) and the status bar will show "Finished!".

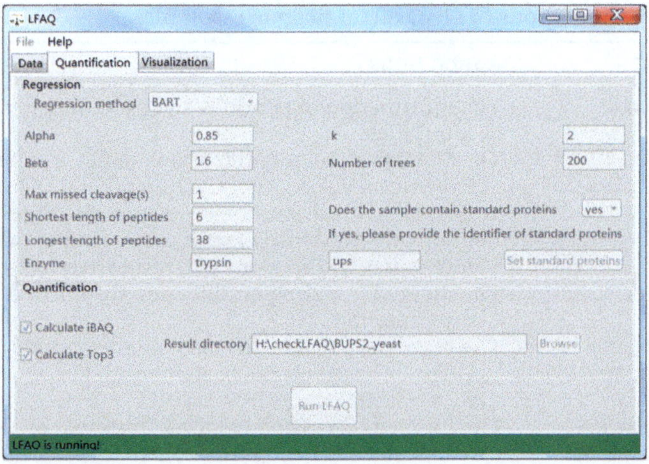


**Figure 14.13** The interface when LFAQ is running.

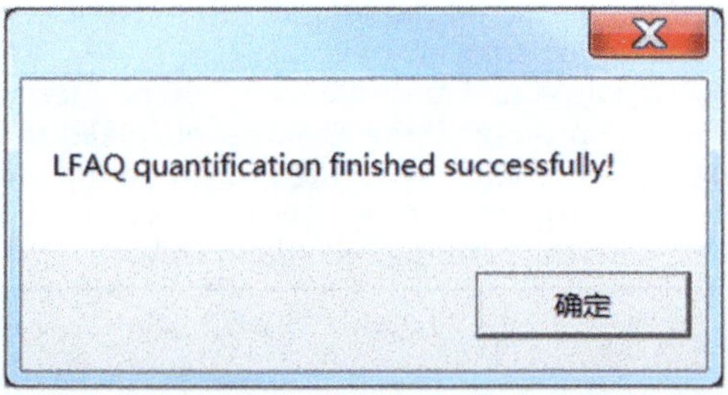


**Figure 14.14** The pop-up dialog when LFAQ finished successfully.

### *14.3.2.4 Anticipated Results*

Once the calculation is done, LFAQ generates several files in the result directory (Table 14.5).

▲ **Critical:** Note that the *ExperimentName* represents the name of the corresponding experiment.

**14.3.2.4.1 Annotation of ProteinResultsExperimentName.txt.** This file contains the quantification results of proteins from the experiment *ExperimentName*. A detailed description of every column in this file is given in Table 14.6.

**14.3.2.4.2 Annotation of ProteinMergedResults.txt.** This file contains the merged protein quantification results of LFAQ from different experiments. Note that for now, LFAQ only supports the SWATH 2.0 result containing a single experiment with one or more replicates (Table 14.7).

**14.3.2.4.3 Annotation of LFAQResultsForStandardProteinsExperimentName.txt.** This file contains the actual abundances and the predicted

**Table 14.5** Annotations of LFAQ quantification result files.

| File name | Annotations |
|---|---|
| ProteinsInfo.txt | The information of peptides and proteins extracted from input files. |
| ProteinResults *ExperimentName*.txt | The LFAQ protein quantification results of experiment *ExperimentName*. Note: CVOriginal and CVAfter Corrected represent the CV of peptide original intensities and the Q-factor corrected peptide intensities within a protein, respectively. |
| ProteinMergedResults.txt | The merged protein quantification results of different experiments. |
| RegressionResults *ExperimentName*.txt | The regression results to calculate the peptide Q-factor based on the protein input list of experiment *ExperimentName*. |
| MergedRegression Results.txt | The merged regression results of different experiments. |
| SelectedFeatures *ExperimentName*.txt | The selected features by regression based on the protein input list of experiment *ExperimentName*. |
| PeptideFeatures *ExperimentName*.txt | All the 587 features of the peptides used in LFAQ based on the protein input list of experiment *ExperimentName*. |
| LFAQResultsFor StandardProteins *ExperimentName*.txt | The actual and calculated abundances of standard proteins if the amounts of spiked-in standard proteins is provided based on the protein input list of experiment *ExperimentName*. |
| log.txt | All the running records in LFAQ. |

abundances of standard proteins if the amounts of spiked-in standard proteins are provided based on the protein input list of experiment *ExperimentName* (Table 14.8).

## 14.3.3 Post-processes of the Quantification Results Using PANDA-view

### *14.3.3.1 Missing Value Imputation and Normalization*

1. Start PANDA-view by double-clicking the PANDA-view.exe file.

▲ **Critical:** It is recommended you check the R version and the required R packages by clicking "R requirement" item in the "help" menu.

- PANDA-view will automatically check if the path of "Rscript.exe" is added into the system environment variable and if all the necessary R packages are installed correctly. Then, the R version and all the necessary R packages used in PANDA-view will be shown in a pop-up dialog (Figure 14.15a)
- If the system environment variable does not contain the path of "Rscript.exe", an error will pop up and the "R requirement" dialog will be empty (Figure 14.15b).

**Table 14.6** Descriptions of the headers in ProteinResultsExperimentName.txt.

| Column name | Description |
|---|---|
| Protein IDs | All the protein IDs in a protein group. |
| Majority protein IDs | Representative protein IDs of this protein group. |
| NumberOfUnique Peptides | The number of unique peptides in the protein group. |
| LFAQ | The LFAQ value of the protein group, in which the Qfactor was used to correct the MS intensities at peptide level. |
| iBAQ | The iBAQ value is provided when the user chooses to calculate iBAQ. If MaxQuant is chosen as input, LFAQ directly outputs the iBAQ value calculated by MaxQuant, otherwise, the iBAQ value is calculated by LFAQ. |
| Top3 | The Top3 value is provided when the user chooses to calculate Top3. |
| CVOriginal | The CV of native intensities of peptides in this protein group. |
| CVAfterCorrected | The CV of Q-factor corrected peptide intensities within this protein group. |
| NumberOf TheoreticEnzyme | The number of theoretically digested peptides. |
| PredictedMol(LFAQ) | The predicted amount of the protein group is calculated using LFAQ, when the user provides the amount of standard proteins. |
| PredictedMol(iBAQ) | The predicted amount of the protein group is calculated using iBAQ if checkbox "Calculate iBAQ" is chosen, and the user provides the amount of standard proteins. |
| PredictedMol(Top3) | The predicted amount of the protein group is calculated using Top3 if checkbox "Calculate Top3" is chosen, and the user provides the amount of standard proteins. |
| Coverage | The represent protein sequence coverage. |
| PeptideSequences | The sequences of the quantified peptides in this protein group. |
| PeptideOriginal Intensities | The native intensities of peptides in this protein group. |
| PeptideQfactors | The peptide Q-factor used to correct the original intensities at peptide level. |
| PeptideCorrected Intensities | The corrected intensities of peptides in this protein group. |
| PeptideMWs | The molecule weight of peptides in this protein group. |

2. Upload data to PANDA-view. Users can upload data by clicking the "Open file" item in File or click the "Open file" button in the Menu. Both txt and csv files are supported. Users can choose the data columns from the candidate columns for subsequent analysis. Click OK to display the selected data (Figure 14.16).

**CAUTION:** PANDA-view can read the results of PANDA directly as "Protein QuantResults", "Peptide QuantResults" or "PeptideIons QuantResults" to perform a multi-level representation of the PANDA's results from protein to peptide, then to peptide ion, then to its extracted ion chromatography (XIC) view (Figure 14.17). Here, peptide ion indicates a specific peptide with certain modifications and charges detected in MS.

**Table 14.7** Descriptions of the headers in ProteinMergedResults.txt.

| Name | Description |
|---|---|
| Protein IDs | All the protein IDs in a protein group. |
| Majority protein IDs | Representative protein IDs of this protein group. |
| Experiments | The names of experiments where the representative proteins are identified. |
| LFAQ | The LFAQ values separated by semicolon of the protein group of all experiments. |
| iBAQ | The iBAQ values separated by semicolon of this protein group of all experiments provided when the user chooses to calculate iBAQ. If MaxQuant is chosen as input, LFAQ directly outputs the iBAQ value calculated by MaxQuant, otherwise, the iBAQ value is calculated by LFAQ. |
| Top3 | The Top3 values separated by semicolon of this protein group of all experiments provided when the user chooses to calculate Top3. |
| PredictedMol(LFAQ) | The predicted amounts separated by semicolon of the protein group of all experiments provided based on LFAQ and the amounts of standard proteins. |
| PredictedMol(iBAQ) | The predicted amounts separated by semicolon of the protein group of all experiments provided based on iBAQ and the amounts of standard proteins. |
| PredictedMol(Top3) | The predicted amounts separated by semicolon of the protein group of all experiments provided based on Top3 and the amounts of standard proteins. |
| LFAQCV | The protein CV of LFAQ values of all the experiments. |
| iBAQCV | The protein CV of iBAQ values of all the experiments. |
| Top3CV | The protein CV of Top3 values of all the experiments. |
| PeptidesIntensities | The corrected peptide intensities from all the experiments. They are separated by comma in one experiment and by semicolon in different experiments. |

3. Search, sort and filter. Users can search key words, sort the table in ascending or descending order or select the values between the user-defined limits of numeric columns through clicking the corresponding item in "List Operation". Intensity information of entries (proteins or peptides) will not change after these steps, for which the parameters are shown in Figure 14.18a.
4. Logarithm, normalization and missing value imputation. The parameters of these steps are shown in Figure 14.18b. These steps will change the intensity information of entries in order to meet the prerequisites for subsequent statistical analysis. These functions can be used through corresponding items in "List Operation" ("Imputation" in "Statistical Analysis").

▲ **Critical:** The choice of parameters needs to conform to mathematical logic. For example, the base of logarithms cannot be set to any negative values. Another example, in the normalization step, ensure that the denominator in the calculation formula is not zero.

**CAUTION:** PANDA-view supports seven normalization methods, *i.e.* z-score normalization, median normalization, maximum normalization, global normalization, IQR normalization, quantile normalization[13,14] and VSN,[15] for numeric columns. Detailed descriptions of these normalization methods are shown in Table 14.9.

**Table 14.8** Descriptions of the headers in LFAQResultsForStandardProteinsExperimentName.txt.

| Name | Description |
|---|---|
| Protein ID | The standard protein ID. |
| NumberOfUnique Peptides | The number of unique peptides of this protein. |
| SpikedInMols | The spiked-in amount of this standard protein. |
| LFAQ | The LFAQ value of this standard protein. |
| iBAQ | The iBAQ value of this standard protein is provided if the user chooses to calculate iBAQ. |
| Top3 | The Top3 value of this standard protein is provided if the user chooses to calculate Top3. |
| PredictedMol(LFAQ) | The predicted amount of this standard protein is calculated using LFAQ |
| PredictedMol(iBAQ) | The predicted amount of this standard protein is calculated using iBAQ if checkbox "Calculate iBAQ" is chosen, and the user provides the amount of standard proteins. |
| PredictedMol(Top3) | The predicted amount of this standard protein is calculated using Top3 if checkbox "Calculate Top3" is chosen, and the user provides the amount of standard proteins. |
| Coverage | The represent protein sequence coverage of the used unique peptides. |
| ProteinSequenceLength | The length of represent protein sequence. |

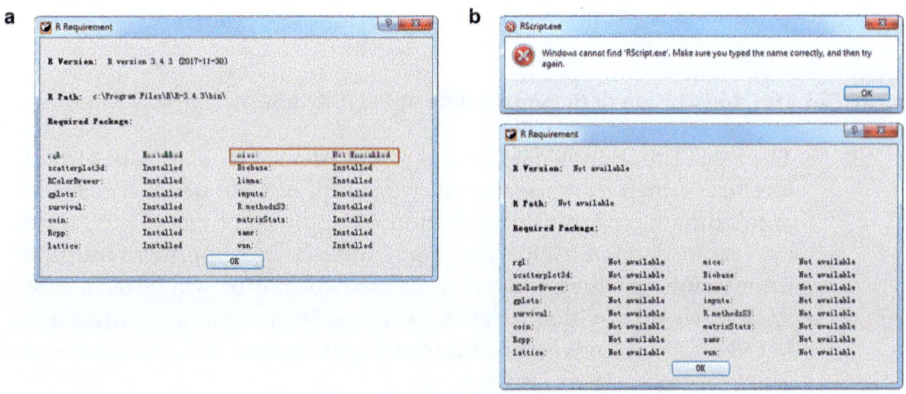


**Figure 14.15** The initial check in PANDA-view. (a) Initial check of R packages installation. (b) Initial check of environment variable and R installation.

### *14.3.3.2 Statistical Analysis of Quantification Result*

5. Statistical analysis. There are seven kinds of statistical tests implemented in PANDA-view (Figure 14.19), for which the methods are all based on R. PANDA-view will add the *p*-value information (such as anova.*p*, krus.*p*) corresponding to the statistical test at the end of the input form.

▲ **Critical:** It is essential to understand the assumptions and application of statistical methods.

- The Fisher exact test is used to test the independence between samples with discretization of each sample's information (such as spectral counts data).

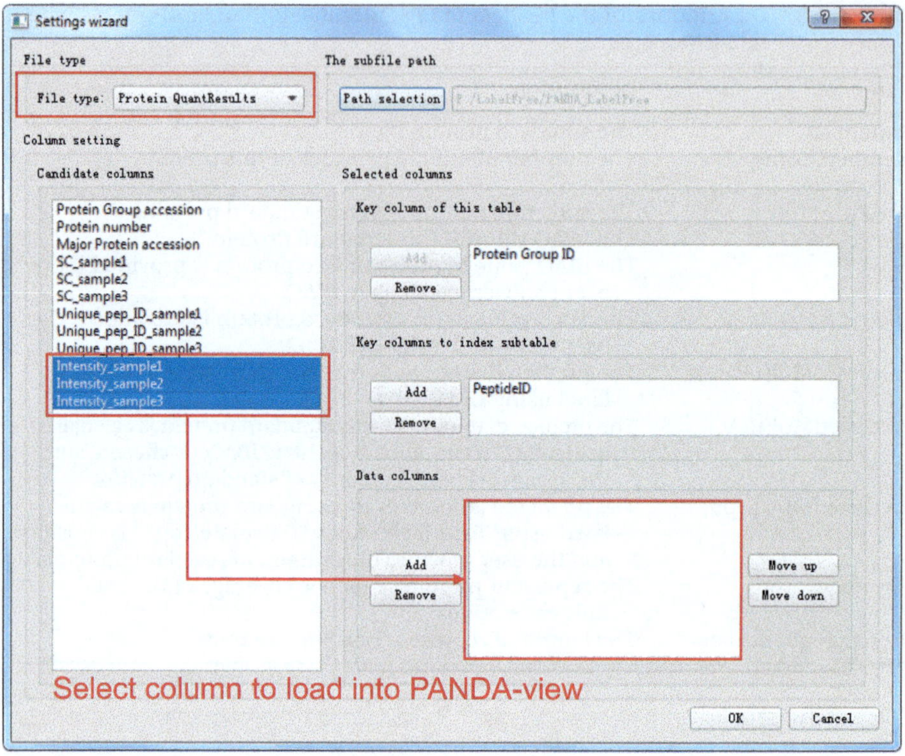


**Figure 14.16** Illustration of uploading data in PANDA-view.

- The *t* test function is used to judge whether there is a statistical difference of each protein/gene in two samples. Data needs to have an approximately normal distribution to apply the *t* test. Thus, users are recommended to take the logarithm of data or perform multiple imputation, in which a logarithmic transformation will be done first.
- PANDA-view provides ANOVA function to analyze the difference between group means of different experimental conditions in single factor experiment design.
- PANDA-view provides a Rank-sum test function to judge whether samples have originated from the same distribution. There is no requirement for data distribution status in this function.
- PANDA-view provides a Permutation test function to judge whether the means of two groups are equal or which is larger. This function is recommended to be used in cases of insufficient sample size or unknown distribution.
- Significance analysis of microarrays (SAM) is provided to determine whether the abundance changes of proteins/genes are statistically significant. This analysis uses non-parametric statistics, so the data may not be required to take logarithmic conversion.
- Multiple comparisons, multiplicity or multiple testing problems frequently occur in biological data analysis. In PANDA-view, several

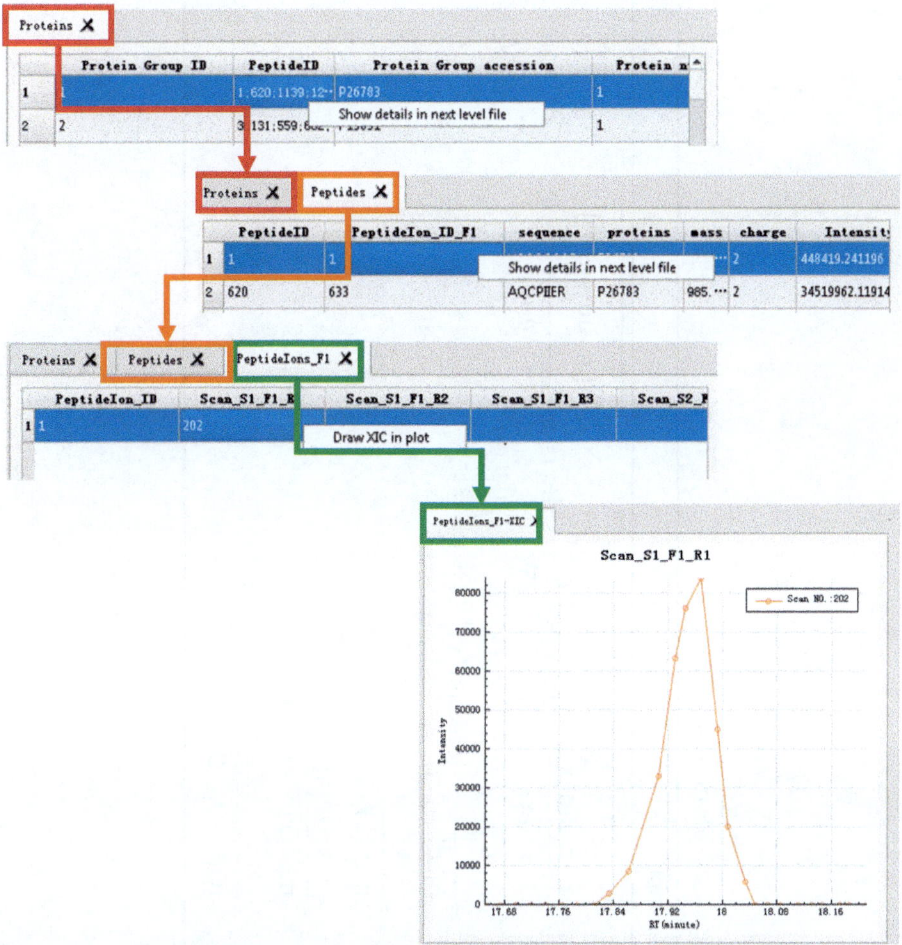


**Figure 14.17** Multi-level representation of quantitative data: from protein list, to peptide list, to peptide ion list, to XIC view.

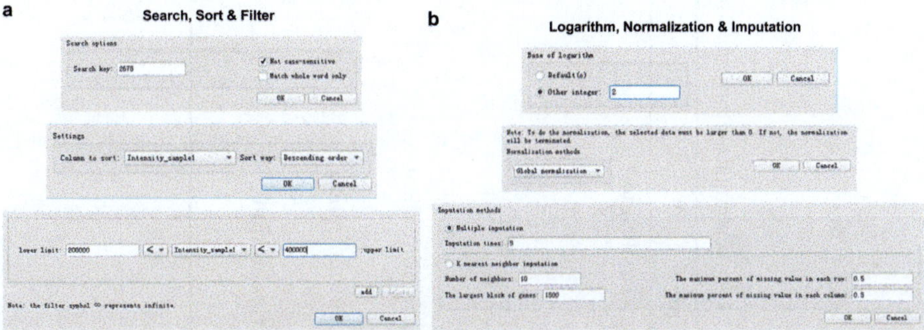


**Figure 14.18** Post-processes of quantification results in PANDA-view, including (a) search, sort and filter; (b) logarithm, normalization and missing value imputation.

**Table 14.9** Calculation formula for normalization methods in PANDA-view.

| Parameter | Annotation |
|---|---|
| *z*-score normalization | $x_{\text{norm}} = \frac{x - \text{mean}(x)}{\text{std}(x)}, \text{mean}(x) = \frac{1}{n}\sum_{i=1}^{n} x_i, \text{std}(x) = \sqrt{\frac{\sum_{i=1}^{n}(x_i - \text{mean}(x))^2}{n-1}}$ |
| Median normalization | $x_{\text{norm}} = \frac{x}{\text{median}(x)}, \text{median}(x) = \frac{x_{\left\lfloor \frac{n}{2} \right\rfloor} + x_{\left\lfloor \frac{n}{2} + \frac{1}{2} \right\rfloor}}{2}$ |
| Maximum normalization | $x_{\text{norm}} = \frac{x}{\max(x)}$ |
| Global normalization | $x_{\text{norm}} = \frac{x - \min(x)}{\max(x) - \min(x)}$ |
| IQR normalization | $x_{\text{norm}} = \frac{x}{\text{IQR}} = \frac{x}{Q_3 - Q_1}$ |

prevalent methods to adjust the *p*-value in multiple hypothesis test are implemented, such as the Bonferroni method,[16] the Benjamini–Hochberg method[17] and the Benjamini–Yekutieli method.[18]

6. Save the table as a file. Users can save the current table in the GUI in the format of txt or csv by clicking the "save current table" in File or clicking the "Save" button in the GUI.

### *14.3.3.3 Visualization of Quantification Results*

7. PANDA-view provides an implementation of nine visualization methods commonly used for proteomic analysis (Figure 14.20).

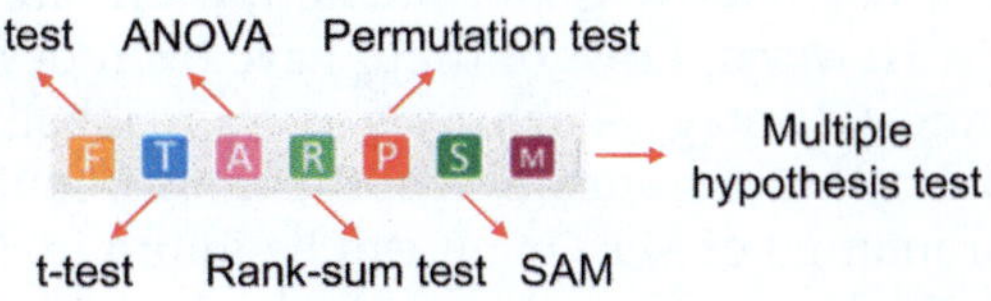


**Figure 14.19** Shortcuts of the statistical analysis functions in PANDA-view.

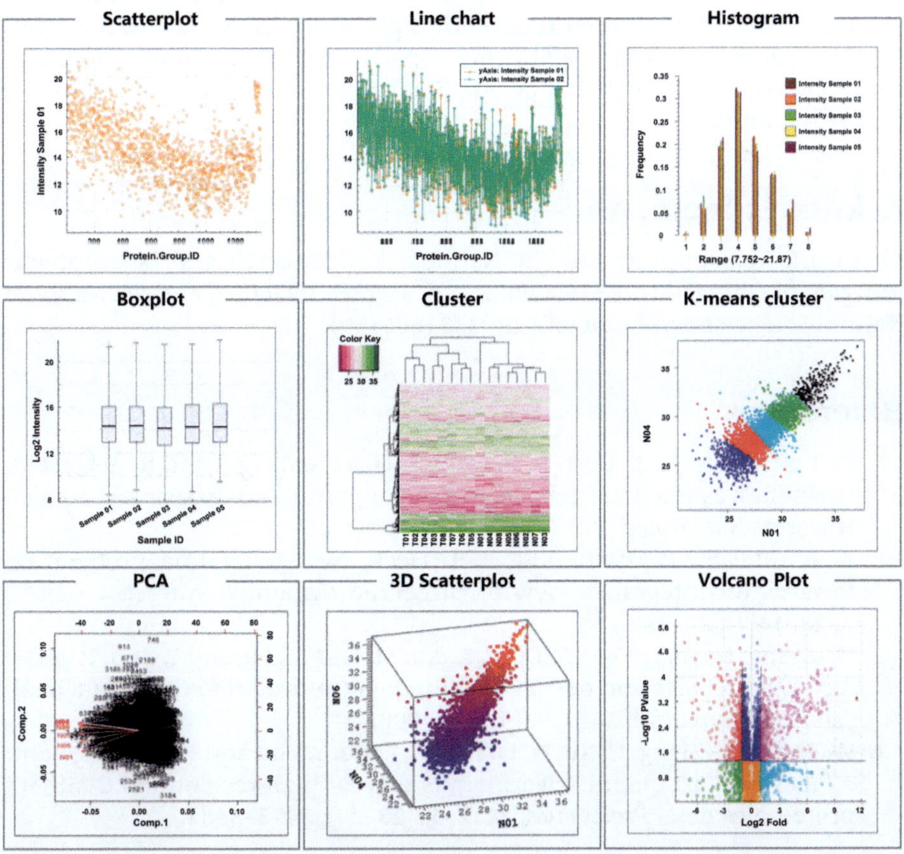


**Figure 14.20** Screenshots of the data visualization functions in PANDA-view.

8. Users can save the current image in Graphic View of GUI by clicking "Export current image" in File or clicking the same button in Menu. In Export settings dialog, users can set the image size and the resolution with default resolution, standard resolution, high resolution and ultra-resolution. Images can be saved in PNG, JPG, BMP or PDF format.

## 14.4 Discussions and Conclusions

In summary, we provide detailed descriptions of three software tools for quantitative proteomics, *i.e.*, PANDA for relative quantification, LFAQ for absolute quantification and PANDA-view for post-processes of quantification results. For relative quantification, there are already a number of existing tools. However, most of them have been developed only for a specific quantitative strategy, such as a certain labelling methods. Among them, MaxQuant[11] is currently one of the most popular software tools. The detailed user manual of MaxQuant can be found in.[19,20] However, it cannot handle the $^{15}N$ labelling data due to the limitation of its algorithm. Moreover, MaxQuant is not compatible with other quantitative tools. In contrast, PANDA is designed to contain comprehensive quantification methods and can be combined with other tools with high flexibility. Shown in the original paper about PANDA,[3] it performs with comparable accuracy and precision to MaxQuant, which can be considered as a complementary solution for the community.

## Acknowledgements

This work was supported by the National Key Research and Development Program of China [2017YFA0505002 and 2017YFC0906602] and the National Natural Science Foundation of China [21605159].

## References

1. O. T. Schubert, H. L. Rost, B. C. Collins, G. Rosenberger and R. Aebersold, Quantitative proteomics: challenges and opportunities in basic and applied research, *Nat. Protoc.*, 2017, **12**(7), 1289–1294.
2. A. F. Altelaar, J. Munoz and A. J. Heck, Next-generation proteomics: towards an integrative view of proteome dynamics, *Nat. Rev. Genet.*, 2013, **14**(1), 35–48.
3. C. Chang, M. Li, C. Guo, Y. Ding, K. Xu, M. Han, F. He and Y. Zhu, PANDA: A comprehensive and flexible tool for quantitative proteomics data analysis, *Bioinformatics*, 2019, **35**(5), 898–900.
4. W. Zhang, J. Zhang, C. Xu, N. Li, H. Liu, J. Ma, Y. Zhu and H. Xie, LFQuant: A label-free fast quantitative analysis tool for high-resolution LC-MS/MS proteomics data, *Proteomics*, 2012, **12**(23–24), 3475–3484.

5. C. Chang, J. Zhang, M. Han, J. Ma, W. Zhang, S. Wu, K. Liu, H. Xie, F. He and Y. Zhu, SILVER: an efficient tool for stable isotope labeling LC-MS data quantitative analysis with quality control methods, *Bioinformatics*, 2014, **30**(4), 586–587.
6. C. Chang, Z. Gao, W. Ying, Y. Fu, Y. Zhao, S. Wu, M. Li, G. Wang, X. Qian, Y. Zhu and F. He, LFAQ: Toward Unbiased Label-Free Absolute Protein Quantification by Predicting Peptide Quantitative Factors, *Anal. Chem.*, 2019, **91**(2), 1335–1343.
7. C. Chang, K. Xu, C. Guo, J. Wang, Q. Yan, J. Zhang, F. He and Y. Zhu, PANDA-view: an easy-to-use tool for statistical analysis and visualization of quantitative proteomics data, *Bioinformatics*, 2018, **34**(20), 3594–3596.
8. A. R. Jones, M. Eisenacher, G. Mayer, O. Kohlbacher, J. Siepen, S. J. Hubbard, J. N. Selley, B. C. Searle, J. Shofstahl, S. L. Seymour, R. Julian, P. A. Binz, E. W. Deutsch, H. Hermjakob, F. Reisinger, J. Griss, J. A. Vizcaino, M. Chambers, A. Pizarro and D. Creasy, The mzIdentML data standard for mass spectrometry-based proteomics results, *Mol. Cell. Proteomics*, 2012, **11**(7), M111.014381.
9. A. Keller, A. I. Nesvizhskii, E. Kolker and R. Aebersold, Empirical statistical model to estimate the accuracy of peptide identifications made by MS/MS and database search, *Anal. Chem.*, 2002, **74**(20), 5383–5392.
10. N. Li, S. Wu, C. Zhang, C. Chang, J. Zhang, J. Ma, L. Li, X. Qian, P. Xu, Y. Zhu and F. He, PepDistiller: A quality control tool to improve the sensitivity and accuracy of peptide identifications in shotgun proteomics, *Proteomics*, 2012, **12**(11), 1720–1725.
11. J. Cox and M. Mann, MaxQuant enables high peptide identification rates, individualized p.p.b.-range mass accuracies and proteome-wide protein quantification, *Nat. Biotechnol.*, 2008, **26**(12), 1367–1372.
12. M. Walzer, D. Qi, G. Mayer, J. Uszkoreit, M. Eisenacher, T. Sachsenberg, F. F. Gonzalez-Galarza, J. Fan, C. Bessant, E. W. Deutsch, F. Reisinger, J. A. Vizcaino, J. A. Medina-Aunon, J. P. Albar, O. Kohlbacher and A. R. Jones, The mzQuantML data standard for mass spectrometry-based quantitative studies in proteomics, *Mol. Cell. Proteomics*, 2013, **12**(8), 2332–2340.
13. T. Valikangas, T. Suomi and L. L. Elo, A systematic evaluation of normalization methods in quantitative label-free proteomics, *Briefings Bioinf.*, 2018, **19**(1), 1–11.
14. M. E. Ritchie, B. Phipson, D. Wu, Y. Hu, C. W. Law, W. Shi and G. K. Smyth, limma powers differential expression analyses for RNA-sequencing and microarray studies, *Nucleic Acids Res.*, 2015, **43**(7), e47.
15. W. Huber, A. von Heydebreck, H. Sultmann, A. Poustka and M. Vingron, Variance stabilization applied to microarray data calibration and to the quantification of differential expression, *Bioinformatics*, 2002, **18**(Suppl. 1), S96–S104.

16. O. J. Dunn, Multiple Comparisons among Means, *J. Am. Stat. Assoc.*, 1961, **56**(293), 52–64.
17. Y. Benjamini and Y. Hochberg, Controlling the False Discovery Rate: A Practical and Powerful Approach to Multiple Testing, *J. R. Stat. Soc. Ser. B. Methodol.*, 1995, **57**(1), 289–300.
18. Y. Benjamini and D. Yekutieli, The control of the false discovery rate in multiple testing under dependency, *Ann. Stat.*, 2001, **29**(4), 1165–1188.
19. S. Tyanova, T. Temu and J. Cox, The MaxQuant computational platform for mass spectrometry-based shotgun proteomics, *Nat. Protoc.*, 2016, **11**(12), 2301–2319.
20. J. Cox, I. Matic, M. Hilger, N. Nagaraj, M. Selbach, J. V. Olsen and M. Mann, A practical guide to the MaxQuant computational platform for SILAC-based quantitative proteomics, *Nat. Protoc.*, 2009, **4**(5), 698–705.

CHAPTER 15

# *Proteomic Workflows with R/ R Markdown*

MAGNUS PALMBLAD*[a] AND CESARÉ OVANDO-VÁZQUEZ*[b]

[a]Leiden University Medical Center, Center for Proteomics and Metabolomics, Postzone P1-Q, P.O. Box 9600, Leiden, 2300RC, The Netherlands; [b]CONACYT Centro Nacional de Supercómputo, Instituto Potosino de Investigación Científica y Tecnológica A. C., Camino a la presade San José, San Luis Potosí, 78216, Mexico
*E-mail: n.m.palmblad@lumc.nl, cesare.ovando@ipicyt.edu.mx

## 15.1 Introduction

### 15.1.1 R

R is a high-level object-oriented language for statistical computing and graphics. R is primarily written in C, Fortran and itself. It can be installed and runs on a variety of UNIX platforms, macOS, and Windows (https://www.r-project.org/). It is used to develop statistical, data mining, data analysis, and bioinformatics software. It implements linear and non-linear models, statistical tests, time-series analysis, statistical classification, and clustering, among other statistical methodologies.

R works with packages, collection of functions and datasets, to increase its power and applicability. It has an official repository, Comprehensive R Archive Network (CRAN), with thousands of publicly available packages. R can install packages *via* the *install.package()* function from (CRAN, https://cran.r-project.org/), *e.g.*

New Developments in Mass Spectrometry No. 8
Processing Metabolomics and Proteomics Data with Open Software: A Practical Guide
Edited by Robert Winkler

Published by the Royal Society of Chemistry, www.rsc.org

```
install.packages("your_package") # install your package of interest
```

Also, R can install packages from specialized repositories for high-throughput omics data, such as Bioconductor (https://www.bioconductor.org/), *e.g.*

```
if (!requireNamespace("BiocManager", quietly = TRUE)) # install BiocManager
 install.packages("BiocManager")
BiocManager::install("your_package") # install your package of interest using BiocManager
```

The R community has developed many packages (hundreds) to process and analyze Mass Spectrometry data (-MS). In Bioconductor we can find hundreds of -MS R packages and even workflows on -MS, just by searching for the term *mass spectrometry*.

### 15.1.2 Markdown and R Markdown

Markdown is a *lightweight* markup language to format documents from plain text documents and uses a simple syntaxis (http://rmarkdown.rstudio.com). Markdown was created by Jhon Gruber in 2004 (https://daringfireball.net/projects/markdown/). You can edit a markdown plain text document in any text editor application.

R Markdown is an R package embedded in Rstudio. It facilitates the creation of formatted documents combining text, code (R, Python, Rcpp and SQL), and the code execution output. R Markdown helps to provide a history with your data. It converts your analysis into elegant documents or presentations; these documents are easily reproducible. R Markdown helps to produce narrative text and code combined into high-quality outputs. R Markdown can combine multiple languages like R, Python, C++, and SQL. It can generate outputs like HTML, PDF, MS Word, books, scientific articles and presentation slides (http://rmarkdown.rstudio.com).

To generate an Rmd document in Rstudio (https://www.rstudio.com/) just click in *File -> New File -> R Markdown*. An example is opened and ready to be compiled by the knitr R package (https://www.rdocumentation.org/packages/knitr/versions/0.1). Clicking the Knit icon will generate the formatted document (HTML, PDF or MS Word).

## 15.2 Materials, Results, and Discussion

As demonstrated elsewhere in this chapter, several R packages are now available for reading and analyzing LC-MS(-MS) data in the community standard mzML or legacy mzXML (http://www.psidev.info/mzML) formats without having to resort to a generic XML parser, the utilization of which is considerably more cumbersome. Once mass spectra and chromatograms are available as common and easily manipulated data structures in R, such as lists or matrices, the R and RStudio combination provides an excellent environment for quickly prototyping algorithms for mass spectrometry data analyses while having a large battery of statistical and visualization tools at one's disposal. When some part of the code needs to be sped up, refactoring into C/C++[1,2] can be done piecemeal using the Rcpp[3] package and embedding the C or C++ within R.

Developing interactive web applications for processing or exploring mass spectrometry data is made possible by the Shiny R package (https://shiny.rstudio.com). Though mzML is the community standard for mass spectrometry data, mzML files are typically several times larger than the original raw files. In most instances, the time it takes to convert a raw file to mzML is shorter than the extra time it takes to transfer the mzML rather than the raw file. Unless one has a very good reason to allow the upload of mzML, it makes more sense to build the web app to take raw files as input and convert these to mzML on the server-side. This is also the approach taken by most interactive web apps for analyzing mass spectrometry data. The conversion can be done *e.g.* using ProteoWizard[4] by calling *msconvert()* using the R *system()* call, *e.g.*

```
raw_files <-list.files(pattern ='raw$') # set your raw files

for(x in raw_files) {

 system(paste('msconvert', x,

       '--32 --zlib --filter \'peakPicking true 1-\'')) # convert each raw file

}
```

will convert all Thermo raw files in a directory to 32 bit, compressed and centroided mzML files (the default output format of msconvert). From here on, the data can be accessed using functions in the mzR[5] or MSnbase[6] R packages and other external tools called *via* subsequent system() invocations.

R Markdowns and Jupyter notebooks (https://jupyter.org/) support many languages, including R, and can be used to create dynamic, executable, polyglot documents combining code, documentation, and output is a single report. These can be formatted as HTML or PDF using the knitr[7] R package without the author having to know any HTML or LaTeX. Such Markdowns are an excellent way to document a data analysis workflow and can be included as supplemental information to a publication. In the 2016 ABRF iPRG[8] study on inferring proteoforms from bottom-up proteomics data,[9] participants were allowed and encouraged to submit an R Markdown or Jupyter notebook describing their solution.[10] As guides, example but simplistic solutions to the problem were provided in R and Python. New and emerging types of journals, notably Nextjournal (nextjournal.com), provide platforms for writing, coding, analyzing and disseminating scientific results as interactive Markdowns or notebooks. The Markdown or notebook then becomes the publication, versioned and citable with a DOI. For its inclusion in these and many other ongoing efforts, R remains an attractive language for mass spectrometry data analysis, in particular for rapid prototyping of solutions to computationally inexpensive or modestly expensive problems, visualization and disseminating data analysis methods and results in publications.

### 15.2.1 Example of a Simple Workflow

#### *15.2.1.1 Load R Packages and Specify the Search Parameters and Sequence Database*

First, we load R packages, specify the parameters and sequence database:

```
library(vioplot)
COMET_PARAMETERS <- 'comet.params'                                         # comet parameters
FASTA_FILE <- 'ipi.HUMAN.v3.69_plus_contaminants_RAND01.fasta'     # sequence database
LIBRA_CONDITIONS <- 'condition.xml'                                         # Libra for iTRAQ
```

#### *15.2.1.2 Convert Raw Data to mzML*

Then, we convert raw file(s) to mzML:

```
raw_files <- list.files(pattern = '\\.raw$')                               # list all .raw files
for(x in raw_files) system(paste('msconvert', x, '-v --mzML'))   # convert them to mzML
```

#### *15.2.1.3 Perform Database Search*

Next we run comet with the defined parameters on each mzML file separately:

```
mzML_files <- list.files(pattern = '\\.mzML$')
for(x in mzML_files) system(paste('comet','-P', COMET_PARAMETERS, ' -N', sub('\\.mzML$',
                                        '', x),'-D',FASTA_FILE,'',x, sep=''))
```

### 15.2.1.4 Validate Search Results With PeptideProphet[11]

We then validate the comet peptide-spectrum matches using PeptideProphet:

```
pepXML_files <-list.files(pattern = '\\.pep\\.xml$')            # List PepXML files
for(x in pepXML_files) system(paste('xinteract -N', sub('\\.pep\\.xml$',
                    '.interact.pep.xml', x), ' -p0.05 -l7 -PPM -O -D',
          FASTA_FILE,' -L', LIBRA_CONDITIONS, ' ', x,
          sep = ''))
```

### 15.2.1.5 Protein Inference using ProteinProphet[12]

Next, we infer proteins and aggregate peptide-level quantitation to the proteins (Figure 15.1):

```
interact_files <- list.files(pattern='interact\\.pep\\.xml$')
for(x in interact_files) system(paste('ProteinProphet ', x, ' ', x, '.prot.xml LIBRA',
          LIBRA_CONDITIONS, sep = ''))
```

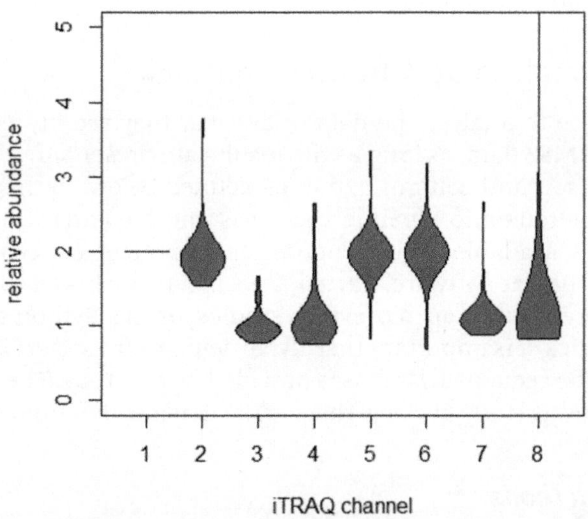


**Figure 15.1** Violin plot showing the relative abundances for a synthetic control, 100 random numbers generated from a normal distribution with mean equal to 2 and standard deviation of $1 \times 10^{-5}$, and 7 of 8 iTRAQ channels.

#### 15.2.1.6 Collect iTRAQ-quantitation from the Quantitation.tsv File and Make Violin Plot

We collect the results in an R data frame, with the proteins as rows and iTRAQ channels as columns:

```
Q <-read.delim('quantitation.tsv')                                    # read Libra results
protratios <- c("protratio0", "protratio1", "protratio2", "protratio3",
       "protratio4", "protratio5", "protratio6", "protratio7")
PR <- Q[!is.na(Q$protratio0), protratios]
rownames(PR) <- Q[!is.na(Q$protratio0), 1]
PR <- PR[PR$protratio0>0, ]                                           # all protein ratios
PR <- PR*2
vioplot(rnorm(100, 2, 1E-5), PR[,2], PR[,3], PR[,4], PR[,5], PR[,6], PR[,7], PR[,8],
   drawRect = FALSE, ylim = c(0, 5))
title(xlab = "iTRAQ channel", ylab = "relative abundance")
```

#### 15.2.1.7 Summary of the Simple Workflow

This R Markdown example shows how to process an 8-channel iTRAQ dataset using the Trans-Proteomic Pipeline. The purpose of the example was to illustrate how an entire analysis and its output can be integrated in one R Markdown document, not necessarily to show the best way to analyze this or any other 8-channel iTRAQ dataset.

### 15.2.2 Example of an Advanced Workflow

Readers are free to analyze the data in any way they see fit, using either or both MS1 and MS2 data, as long as the results are presented in a data matrix of the format, row and column names as defined below. In this example, we use the Trans-Proteomic Pipeline[13] and assume a standard installation of this pipeline is available. This is not an endorsement or recommendation to use this particular software. Participants are also allowed to submit their results in a spreadsheet, as in previous studies, or as a Python script.

In iPRG studies, it is important that all participants must start from the same data and use the sequence databases provided by the iPRG. The raw data files are included in the study package that can be downloaded from iprg2016.org.

#### 15.2.2.1 Methods

**Load Package 'XML' and Specify the Input.** Here we load the required R package(s), set the working directory and specify which sequence database and search engine to use.

```
require(XML)
setwd('C:/Inetpub/wwwroot/ISB/data/iPRG2016')      # where you want to download the data
SEARCH_ENGINE <- 'comet'                           # your locally installed search engine
SEARCH_ENGINE_PARAMETERS <- 'comet_5ppm.params'    # your search parameters
FASTA_FILE <- 'iPRG2016.fasta'                     # the (downloaded) sequence database
```

The particular parameter file used above can be downloaded here. This should not be seen as an endorsement of a particular search engine or settings, in this study or otherwise.

**Convert Raw Data to mzML.** Although study participants are allowed to use any software they wish to from this point forward, we suggest that if converting the raw files to mzML, this is done by calling the *msconvert()* function. The parameters can be modified depending on the search engine and whether MS1 data is used.

```
raw_files <- list.files(pattern = 'raw$')
for(x in raw_files) system(paste('msconvert', x, '--32 --zlib --filter
                              \"peakPicking true 2-\", --filter \"msLevel 2\"'))
```

**Perform Database Search.** Here we simply call the defined search engine with the defined parameters on each data file separately.

```
mzML_files <- list.files(pattern = 'mzML$')
for(x in mzML_files) system(paste(SEARCH_ENGINE, ' -P', SEARCH_ENGINE_PARAMETERS, ' -N'
                              , x,' -D', FASTA_FILE, ' ', x, sep =''))
```

**Validate search results with PeptideProphet and ProteinProphet.** Here we run PeptideProphet[11] and ProteinProphet[12] with one command, using the -Op flag. Again, we do this on each file separately.

```
pepXML_files <- list.files(pattern ='[^to].pep.xml$')   # list original pepXML files
for(x in pepXML_files) system(paste('xinteract -N', x
                              , '.interact.pep.xml -p0.05 -l7 -PPM -Op ', x
                              , sep =''))
```

**15.2.2.1.5 Collect all ProteinProphet results.** Next, we read in all the results from the command line calls into R using the XML package. Another possibility would be to use the mzR parser from the Bioconductor package.

```
ProteinProphet_files <- list.files(pattern='interact.prot.xml$')
protein <- c()
for(x in ProteinProphet_files)
{
  doc <- xmlToList(x)
 for (i inseq(3,length(doc)-1)) protein <-
   c(protein,doc[i]$protein_group$protein$.attrs['protein_name'])
}
protein <- unique(protein)              # protein is now a vector of all proteoforms
probability <- list()
for(x in ProteinProphet_files)
{
  doc <- xmlToList(x)
 for (p in protein) probability[[x]][[p]] <-0
 for (i inseq(3,length(doc)-1))
     probability[[x]][[doc[i]$protein_group$protein$.attrs['protein_name']]] <-
   doc[i]$protein_group$protein$.attrs['probability']
}
```

## *15.2.2.2 Results*

**Export results for study submission.** We place the results in an R probability matrix with the sequences (proteoforms) as rows and samples as columns. The probabilities are the local false discovery rates calculated by ProteinProphet, applying parsimony, also known as "Occam's razor".

```
DF <- as.data.frame(sapply(probability, rbind), rownames <-protein)
DM <- as.matrix(DF[,1:12])
class(DM)='numeric'
colnames(DM) <-unlist(strsplit(colnames(DF),".mzML"))[seq(1,23,2)]
```

Finally, we extract and export the PrEST results to a tab-delimited file, ready for submission on iprg2016.org:

```
PrESTs<-DM[grep("HPRR", rownames(DM)),]

PrESTs<-PrESTs[order(rownames(PrESTs)),]

write.table(data.frame("PEP"=rownames(PrESTs),PrESTs)

           , file='my_iPRG2016_results.txt', row.names=FALSE, sep='\t')
```

#### 15.2.2.3 Summary of the Advanced Workflow

This example solution shows how the iPRG 2016 challenge can be addressed and demonstrates how an entire solution can be integrated into an R Markdown document. The example is not meant to convey the *best* solution and should only be used as an example or a template for study submissions.

## 15.3 Conclusion

In this chapter, we introduced R/R Markdown and their capabilities to produce formatted documents (HMTL, PDF, MS Word, *etc.*) combining text, code, and the code execution output. We demonstrated the R/R Markdown capabilities developing simple and advanced workflows to address two particular challenges, process an 8-channel iTRAQ dataset and address the iPRG 2016 challenge. In conclusion, R/R Markdown are powerful tools to simplify the -MS data processing and reporting results by producing high-quality reports in several formats ready to be shared.

## Acknowledgements

COV acknowledges his CONACYT Cátedra project 809.

## References

1. B. Stroustrup, *The C++ Programming Language*, Addison-Wesley, 1986.
2. D. M. Ritchie, *ACM SIGPLAN Not.*, 1993, **28**(3), 201.
3. D. Eddelbuettel and R. Francois, *J. Stat. Softw.*, 2011, **40**(8), 1–18.
4. D. Kessner, M. Chambers, R. Burke, D. Augus and P. Mallick, *Bioinformatics*, 2008, **24**(21), 2534.
5. M. C. Chambers, *et al.*, *Nat. Biotechnol.*, 2012, **30**, 918.
6. L. Gatto and K. Lilley, *Bioinformatics*, 2012, **28**(2), 288.
7. Y. Xie, *Chapman and Hall/CRC*, 2nd. ed. 2015.
8. https://abrf.org/research-group/proteome-informatics-research-group-iprg.
9. J. Y. Lee, *et al.*, *J. Biomol. Tech.*, 2018, **29**(2), 39.

10. T. Matthew, *et al.*, *J. Proteome Res.*, 2018, **17**(5), 187.
11. A. Keller, A. I. Nesvizhskii, E. Koker and R. Aebersold, *Anal. Chem.*, 2002, **74**(20), 5383.
12. A. Keller, A. I. Nesvizhskii, E. Koker and R. Aebersold, *Anal. Chem.*, 2003, **75**(17), 4646.
13. http://tools.proteomecenter.org/software.php.

CHAPTER 16

# *Python in Proteomics*

HANNES L. RÖST*

Donnelly Centre, University of Toronto, Toronto, Canada
*E-mail: hannes.rost@utoronto.ca

Python is a versatile scripting language that is widely used in industry and academia. In bioinformatics, there are multiple packages supporting data analysis with Python that range from biological sequence analysis with Biopython[1] to structural modeling and visualization with packages like PyMOL and PyRosetta,[2] to numerical computation and advanced plotting with NumPy/SciPy.[3] In the proteomics community, Python began to be widely used around 2012 when several mature Python packages were published including pymzML,[4] Pyteomics[5] and pyOpenMS.[6] This has led to an ever-increasing interest in the Python programming language in the proteomics and mass spectrometry community. The number of publications referencing or using Python has risen eight fold since 2012 (compared with the same time period before 2012), with multiple open-source Python packages now supporting mass spectrometric data analysis and processing.[4,5,7–14] Computing and data analysis in mass spectrometry is very diverse and in many cases must be tailored to a specific experiment. Often, multiple analysis steps have to be performed (identification, quantification, post-translational modification analysis, filtering, FDR analysis *etc.*) in an analysis pipeline, which requires high flexibility in the analysis. This is where Python truly shines, due to its flexibility, visualization capabilities and the ability to extend computation with a large number of powerful libraries. Python can be used to quickly prototype software, combine existing libraries into powerful analysis workflows while avoiding the trap of re-inventing the wheel for a new project.

New Developments in Mass Spectrometry No. 8
Processing Metabolomics and Proteomics Data with Open Software: A Practical Guide
Edited by Robert Winkler

Published by the Royal Society of Chemistry, www.rsc.org

Here, we will describe data analysis with Python using the pyOpenMS package.[6] An extended documentation and tutorial can also be found online at https://pyopenms.readthedocs.io. To allow the reader to follow all steps in the tutorial, we will also describe the installation process of the software. Our installation is based on Anaconda, an open-source Python distribution that includes the Spyder integrated development environment (IDE) that allows development with pyOpenMS in a graphical environment.

## 16.1 Installation

In order to install pyOpenMS, we recommend to use Anaconda and Spyder which provide a fully integrated environment for developing, running and debugging Python code. You can download the latest Python version (currently 3.7) bundled with Anaconda from https://www.anaconda.com/distribution/. After installation, you will have multiple new applications available on your computer. To set up pyOpenMS, use the newly installed Anaconda environment to install pyOpenMS (on Windows: go to the Start Menu and click on the "Anaconda Powershell Prompt" application) and type:

```
pip install pyopenms
```

which will result in a message "Successfully installed pyopenms" including a version number. You should now be set up to use pyOpenMS and you can start the Spyder application (on Windows: go to the Start Menu and click on the "Spyder" application) which will open up and provide you with multiple windows (see Figure 16.1).

In addition, we recommend that you install the OpenMS tool suite from https://www.openms.de/download/openms-binaries/ which will install the visualization tool TOPPView as well as over 180 TOPP tools that are executables distributed with OpenMS (see Chapter 6 for more information on OpenMS and TOPPView).

## 16.2 Getting Started

After installation, you can use the full extent of the OpenMS library. There are multiple ways to get information about the available functions and methods. We can inspect individual pyOpenMS objects through the `help` function:

```
from pyopenms import *
help(MSExperiment)
```

which will display an extensive help text and list all available functions, indicating that `MSExperiment` exposes methods such as `getNrSpectra()` and

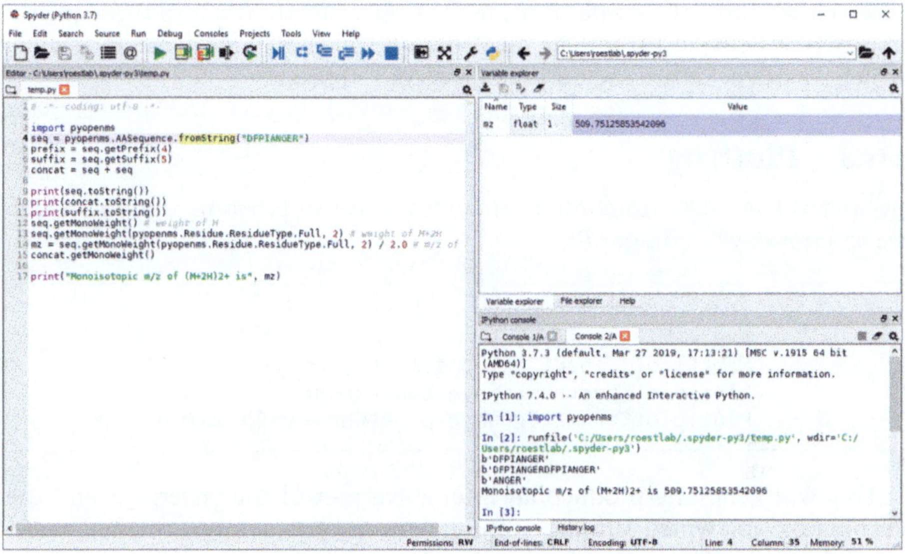

**Figure 16.1** *Spyder graphical interface for Python.* The Spyder software is an open-source integrated development environment (IDE) for Python which allows users to develop, run and debug Python scripts. Here, a screenshot of the software is shown after executing the `import pyopenms` command in the console (lower right). In the left window, a script that is currently active is shown and its result is shown on the command window (lower right). On the top, the variable explorer is active which shows the currently available variables, here the `mz` variable is displayed that has the value 509.75 and is computed on line 14 of the script.

`getSpectrum(id)` where the argument `id` indicates the spectrum identifer. It also points to the base class `ExperimentalSettings` which can be investigated in the same manner for additional information.

pyOpenMS supports a variety of different file formats through the implementations in OpenMS. In order to demonstrate the capabilities of pyOpenMS to read different mass spectrometric data files, we will download two files that have been prepared for this chapter and are available from Zenodo:†

```
from urllib.request import urlretrieve
from pyopenms import *
url = "https://zenodo.org/record/2653155/files/"
urlretrieve (url + "example.mzML", "example.mzML")
urlretrieve (url + "search.fasta", "search.fasta")
exp = MSExperiment()
MzMLFile().load("example.mzML", exp)
exp.getNrSpectra()
exp.getNrChromatograms()
```

†https://zenodo.org/record/2653155

which will load the content of the "example.mzML" file into the `exp` variable of type `MSExperiment`. The file contains 3 spectra and 5 chromatograms, which we can see from the output of the code above.

## 16.3 Plotting

pyOpenMS has basic functionality to plot spectra and chromatograms, which we can now try out on our file:

```
from pyopenms import *
exp = MSExperiment()
MzMLFile().load("example.mzML", exp)
Plot.plotSpectrum(exp.getSpectrum(0))
Plot.plotChromatogram(exp.getChromatogram(0))
```

This will generally produce an interactive plot of the spectrum and the chromatogram, which allows zooming, panning and detailed manual analysis of the data. When using Spyder, by default an inline plot will be generated that is static and cannot be manipulated.

## 16.4 Chemistry

### 16.4.1 Elements

OpenMS has representations for various chemical concepts including molecular formulas, isotopes, amino acid sequences and modifications. First, we look at how elements are stored in OpenMS:

```
from pyopenms import *

edb = ElementDB()
sulfur = edb.getElement("S")
print(sulfur.getName())
isotopes = sulfur.getIsotopeDistribution()
for iso in isotopes.getContainer():
    print (iso.getMZ(), ":", iso.getIntensity())
```

As we can see, OpenMS knows common elements like sulfur as well as their isotopic distribution. These values are stored in `Elements.xml` in the OpenMS share folder and can, in principle, be modified. The above code outputs the isotopes of sulfur and their abundance.

### 16.4.2 Molecular Formula

Elements can be combined to molecular formulas (`EmpiricalFormula`) which can be used to describe small molecules or peptides. The class supports a large number of operations like addition and subtraction. A simple example is given in the next few lines of code.

```
from pyopenms import *

methanol = EmpiricalFormula("CH3OH")
water = EmpiricalFormula("H2O")
ethanol = EmpiricalFormula("CH2" + methanol.toString())
wm = water + methanol
print(wm.toString())
print(wm.getElementalComposition())
```

Note how in lines 5 and 6 we were able to make new molecules by adding existing molecules (either by adding two `EmpiricalFormula` objects or by adding simple strings).

### 16.4.3 Isotopic Distributions

OpenMS can also generate theoretical isotopic distributions from analytes represented as `EmpiricalFormula`. Currently there are two algorithms implemented, CoarseIsotopePatternGenerator which produces unit mass isotope patterns and FineIsotopePatternGenerator which is based on the IsoSpec algorithm:[15]

Note how the result calculated with the `FineIsotopePatternGenerator` contains the hyperfine isotope structure with heavy isotopes of carbon, hydrogen and oxygen clearly distinguished while the coarse (unit resolution) isotopic distribution contains summed probabilities for each isotopic peak without the hyperfine resolution. Also note how the differences between the hyperfine peaks can reach more than 115 ppm (52.041 *versus* 52.047). Note that the FineIsotopePatternGenerator will generate peaks until the total probability not covered by the current result reaches $1 \times 10^{-5}$.

```
    from pyopenms import *

    wm = EmpiricalFormula("CH3OH") +
EmpiricalFormula("H2O")

    print("Coarse Isotope Distribution:")
    gen = CoarseIsotopePatternGenerator(5)
    isotopes = wm.getIsotopeDistribution(gen)
    for iso in isotopes.getContainer():
        print (iso.getMZ(), ":", iso.getIntensity())

    print("Fine Isotope Distribution:")
    gen = FineIsotopePatternGenerator(1e-5)
    isotopes = wm.getIsotopeDistribution(gen)
    for iso in isotopes.getContainer():
        print (iso.getMZ(), ":", iso.getIntensity())
```

### 16.4.4 Amino Acids

An amino acid residue is represented in OpenMS by the class `Residue`. It provides a container for the amino acids as well as some functionality. The class is able to provide information such as the isotope distribution of the residue, the average and monoisotopic weight. The residues can be identified by their full name, their three letter abbreviation or the single letter abbreviation. The residue can also be modified, which is implemented in the `Modification` class.

An amino acid residue modification is represented in OpenMS by the class `ResidueModification`. The known modifications are stored in the `ModificationsDB` object, which is capable of retrieving specific modifications. It contains UniMod as well as PSI modifications.

## 16.5 Peptides and Proteins

### 16.5.1 Amino Acid Sequences

The `AASequence` class handles amino acid sequences in OpenMS. A string of amino acid residues can be turned into an instance of `AASequence` to provide some commonly used operations and data. The implementation supports mathematical operations like addition or subtraction. Also, average and mono isotopic weight and isotope distributions are accessible.

Weights, formulas and isotope distribution can be calculated depending on the charge state (additional proton count in case of positive ions) and ion type. Therefore, the class allows for a flexible handling of amino acid strings.

A very simple example of handling amino acid sequence with AASequence is given in the next few lines, which also calculates the weight of the `(M)` and $(M+2H)^{2+}$ ions.

```
from pyopenms import *
seq = AASequence.fromString("DFPIANGER")
concat = seq + seq

# weight of M
seq.getMonoWeight()
# weight of M+2H
seq.getMonoWeight(Residue.ResidueType.Full, 2)
mz = seq.getMonoWeight(Residue.ResidueType.Full, 2) /
2.0
concat.getMonoWeight()

print("Monoisotopic m/z of (M+2H)2+ is", mz)
```

### 16.5.2 Molecular Formula

Note how we can easily calculate the charged weight of a $(M+2H)^{2+}$ ion and compute *m*/*z* – simply dividing by the charge. We can now combine our knowledge of `AASequence` with what we learned above about

`EmpiricalFormula` to get accurate mass and isotope distributions from the amino acid sequence:

```
    from pyopenms import *
    seq = AASequence.fromString("DFPIANGER")
    seq_formula = seq.getFormula()
    print("Peptide", seq, "has molecular formula",
seq_formula)
    print("="*35)

    gen = CoarseIsotopePatternGenerator(6)
    isotopes = seq_formula.getIsotopeDistribution(gen)
    for iso in isotopes.getContainer():
      print ("Isotope", iso.getMZ(), ":",
iso.getIntensity())

    suffix = seq.getSuffix(3) # y3 ion "GER"
    print("="*35)
    print("y3 ion :", suffix)
    y3_formula =
suffix.getFormula(Residue.ResidueType.YIon, 2) # y3++ ion
    suffix.getMonoWeight(Residue.ResidueType.YIon, 2) /
2.0 # CORRECT
    suffix.getMonoWeight(Residue.ResidueType.XIon, 2) /
2.0 # CORRECT
    suffix.getMonoWeight(Residue.ResidueType.BIon, 2) /
2.0 # INCORRECT

    print("y3 mz :",
suffix.getMonoWeight(Residue.ResidueType.YIon, 2) / 2.0 )
    print(y3_formula)
    print(seq_formula)
```

Note that we need to remember that we are dealing with an ion of the *x*/*y*/*z* series since we used a suffix of the original peptide and using any other ion type will produce a different mass-to-charge ratio (and while "GER" would also be a valid "*x*3" ion, note that it *cannot* be a valid ion from the *a*/*b*/*c* series and therefore the mass cannot refer to the same input peptide "DFPIANGER" since its "*b*3" ion would be "DFP" and not "GER").

### 16.5.3 Modified Sequences

The `AASequence` class can also handle modifications, modifications are specified using a unique string identifier present in the `ModificationsDB` in round brackets after the modified amino acid or by providing the mass of the residue in square brackets. For example `AASequence.fromString(".DFPIAM(Oxidation)GER.")` creates an instance of the peptide "DFPIAMGER" with an oxidized methionine. There are multiple ways to specify modifications, and `AASequence.fromString("DFPIAM(UniMod:35)GER")`, `AASequence.fromString("DFPIAM[+16]GER")` and `AASequence.fromString("DFPIAM[147]GER")` are all equivalent.

```
    from pyopenms import *
    seq =
AASequence.fromString("PEPTIDESEKUEM(Oxidation)CER")
    print(seq.toUnmodifiedString())
    print(seq.toString())
    print(seq.toUniModString())
    print(seq.toBracketString())
    print(seq.toBracketString(False))

    print(AASequence.fromString("DFPIAM(UniMod:35)GER"))
    print(AASequence.fromString("DFPIAM[+16]GER"))
    print(AASequence.fromString("DFPIAM[+15.99]GER"))
    print(AASequence.fromString("DFPIAM[147]GER"))
    print(AASequence.fromString("DFPIAM[147.035405]GER"))
```

Note there is a subtle difference between `AASequence.fromString(".DFPIAM[+16]GER.")` and `AASequence.fromString(".DFPIAM[+15.9949]GER.")` – while the former will try to find the first modification matching to a mass difference of 16 +/− 0.5, the latter will try to find the closest matching modification to the exact mass. The exact mass approach usually gives the intended results while the first approach may or may not.

### 16.5.4 Proteins

Protein sequences can be accessed through the `FASTAEntry` object and can be read and stored on disk using a `FASTAFile`:

```
    from pyopenms import *
    bsa = FASTAEntry()
    bsa.sequence =
"MKWVTFISLLLLFSSAYSRGVFRRDTHKSEIAHRFKDLGE"
    bsa.description = "BSA Bovine Albumin (partial sequence)"
    bsa.identifier = "BSA"
    alb = FASTAEntry()
    alb.sequence =
"MKWVTFISLLFLFSSAYSRGVFRRDAHKSEVAHRFKDLGE"
    alb.description = "ALB Human Albumin (partial sequence)"
    alb.identifier = "ALB"

    entries = [bsa, alb]

    f = FASTAFile()
    f.store("example.fasta", entries)
```

### 16.5.5 TheoreticalSpectrumGenerator

This class implements a simple generator which generates tandem MS spectra from a given peptide charge combination. There are various options which influence the occurring ions and their intensities.

```
from pyopenms import *

tsg = TheoreticalSpectrumGenerator()
spec1 = MSSpectrum()
spec2 = MSSpectrum()
peptide = AASequence.fromString("DFPIANGER")
# standard behavior is adding b-and y-ions of charge 1
p = Param()
p.setValue("add_b_ions", "false")
tsg.setParameters(p)
tsg.getSpectrum(spec1, peptide, 1, 1)
p.setValue("add_b_ions", "true")
p.setValue("add_a_ions", "true")
p.setValue("add_losses", "true")
p.setValue("add_metainfo", "true")
tsg.setParameters(p)
tsg.getSpectrum(spec2, peptide, 1, 2)
print("Spectrum 1 has", spec1.size(), "peaks.")
print("Spectrum 2 has", spec2.size(), "peaks.")

# Iterate over annotated ions and their masses
for ion, peak in zip(spec2.getStringDataArrays()[0],
spec2):
    print(ion, peak.getMZ())
```

The example shows how to put peaks of a certain type, *y*-ions in this case, into a spectrum. Spectrum 2 is filled with a complete spectrum of all peaks (*a*-, *b*-, *y*-ions and losses). The `TheoreticalSpectrumGenerator` has many parameters which have a detailed description located in the class documentation. For the first spectrum, no *b* ions are added. Note how the `add_metainfo` parameter in the second example populates the `StringDataArray` of the output spectrum, allowing us to iterate over annotated ions and their masses.

## 16.6 Digestion

### 16.6.1 Proteolytic Digestion with Trypsin

OpenMS has classes for proteolytic digestion which can be used as follows:

```
    from pyopenms import *
    from urllib.request import urlretrieve
    urlretrieve
("http://www.uniprot.org/uniprot/P02769.fasta",
"bsa.fasta")

    dig = ProteaseDigestion()
    dig.getEnzymeName() # Trypsin
    bsa = "".join([l.strip() for l in
open("bsa.fasta").readlines()[1:]])
    bsa = AASequence.fromString(bsa)
    result = []
    dig.digest(bsa, result)
    print(result[4].toString())
    len(result) # 82 peptides
```

## 16.6.2 Proteolytic Digestion with Lys-C

We can of course also use different enzymes, these are defined `Enzyme.xml` file and can be accessed using the `EnzymesDB`.

```
    names = []
    ProteaseDB().getAllNames(names)
    len(names) # at least 25 by default
    e = ProteaseDB().getEnzyme('Lys-C')
    e.getRegExDescription()
    e.getRegEx()
```

Now that we have learned about the other enzymes available, we can use it to cut the protein of interest:

```
    from pyopenms import *
    from urllib.request import urlretrieve
    urlretrieve
("http://www.uniprot.org/uniprot/P02769.fasta",
"bsa.fasta")

    dig = ProteaseDigestion()
    dig.setEnzyme('Lys-C')
    bsa = "".join([l.strip() for l in
open("bsa.fasta").readlines()[1:]])
    bsa = AASequence.fromString(bsa)
    result = []
    dig.digest(bsa, result)
    print(result[4].toString())
    len(result) # 57 peptides
```

We now get different digested peptides (57 *versus* 82) and the fourth peptide is now `GLVLIAFSQYLQQCPFDEHVK` instead of `DTHK` as with Trypsin (see above).

## 16.7 Simple Data Manipulation

Here we will look at a few simple data manipulation techniques on spectral data, such as filtering.

### 16.7.1 Filtering Spectra

We will filter the "example.mzML" file by only retaining spectra that match a certain identifier:

```
from pyopenms import *
inp = MSExperiment()
MzMLFile().load("example.mzML", inp)

e = MSExperiment()
for s in inp:
  if s.getNativeID().startswith("scan="):
    e.addSpectrum(s)

MzMLFile().store("test_filtered.mzML", e)
```

### 16.7.2 Filtering by MS Level

Similarly, we can filter the `example.mzML` file by MS level, retaining only spectra that are not MS1 spectra (*e.g.* MS2, MS3 or MSn spectra):

```
from pyopenms import *
inp = MSExper iment()
MzMLFile().load("example.mzML", inp)

e = MSExperiment()
for s in inp:
  if s.getMSLevel() > 1:
    e.addSpectrum(s)

MzMLFile().store("test_filtered.mzML", e)
```

Note that we can easily replace line 7 with more complicated criteria, such as filtering by MS level and scan identifier at the same time:

```
    if s.getMSLevel() > 1 and
  s.getNativeID().startswith("scan="):
```

### 16.7.3 Filtering by Scan Number

Or we could use an external list of scan numbers to filter by scan numbers, thus only retaining MS scans in which we are interested in:

```
from pyopenms import *
inp = MSExperiment()
MzMLFile().load("example.mzML", inp)
scan_nrs = [0, 2, 5, 7]

e = MSExperiment()
for k, s in enumerate(inp):
  if k in scan_nrs and s.getMSLevel() == 1:
    e.addSpectrum(s)

MzMLFile().store("test_filtered.mzML", e)
```

### 16.7.4 Filtering Spectra and Peaks

We can now move on to more advanced filtering, suppose we are interested in only a part of all fragment ion spectra, such as a specific *m*/*z* window. We can easily filter our data accordingly:

```
from pyopenms import *
inp = MSExperiment()
MzMLFile().load("example.mzML", inp)

mz_start = 6.0
mz_end = 12.0
e = MSExperiment()
for s in inp:
  if s.getMSLevel() > 1:
    filtered_mz = []
    filtered_int = []
    for mz, i in zip(*s.get_peaks()):
      if mz > mz_start and mz < mz_end:
        filtered_mz.append(mz)
        filtered_int.append(i)
    s.set_peaks((filtered_mz, filtered_int))
    e.addSpectrum(s)

MzMLFile().store("test_filtered.mzML", e)
```

Note that in a real-world application, we would set the `mz_start` and `mz_end` parameter to an actual area of interest, for example the area between 125 and 132 which contains quantitative ions for a TMT experiment.

Similarly we could change line 13 to only report peaks above a certain intensity or to only report the top *N* peaks in a spectrum.

### 16.7.5 Memory Management

On order to save memory, we can avoid loading the whole file into memory and use the OnDiscMSExperiment for reading data.

```
from pyopenms import *
ondisc_exp = OnDiscMSExperiment()
ondisc_exp.openFile("example.mzML")

e = MSExperiment()
for k in range(ondisc_exp.getNrSpectra()):
  s = ondisc_exp.getSpectrum(k)
  if s.getNativeID().startswith("scan="):
    e.addSpectrum(s)

MzMLFile().store("test_filtered.mzML", e)
```

Note that using the approach the output data e is still completely in memory and may end up using a substantial amount of memory. We can avoid that by using the following code:

```
from pyopenms import *
ondisc_exp = OnDiscMSExperiment()
ondisc_exp.openFile("example.mzML")

consumer =
PlainMSDataWritingConsumer("test_filtered.mzML")

e = MSExperiment()
for k in range(ondisc_exp.getNrSpectra()):
  s = ondisc_exp.getSpectrum(k)
  if s.getNativeID().startswith("scan="):
    consumer.consumeSpectrum(s)

del consumer
```

Make sure you do not forget `del consumer` since otherwise the final part of the mzML may not get written to disk (and the consumer is still waiting for new data).

## 16.8 Example: Peptide Search

In MS-based proteomics, fragment ion spectra (MS2 spectra) are often interpreted by comparing them against a theoretical set of spectra generated from a FASTA database. OpenMS contains a (simple) implementation of such a "search engine" that compares experimental spectra against theoretical spectra generated from a chemical or enzymatic digest of a proteome.

In most proteomics applications, a dedicated search engine (such as Comet, Crux, Mascot, MSGFPlus, MSFragger, Myrimatch, OMSSA, SpectraST or XTandem; all of which are supported by pyOpenMS) will be used to search data. Here, we will use the internal `SimpleSearchEngineAlgorithm` from OpenMS used for teaching purposes. This makes it very easy to search an (experimental) mzML file against a fasta database of protein sequences:

```
from pyopenms import *
SimpleSearchEngineAlgorithm().search("example.mzML",
"search.fasta", protein_ids, peptide_ids)
```

This will print search engine output including the number of peptides and proteins in the database and how many spectra were matched to peptides and proteins. We can now investigate the individual peptide spectrum matches (PSM) using Python:

```
for peptide_id in peptide_ids:
  # Peptid e identification values
  print (35*"=")
  print ("Peptide ID m/z:", peptide_id.getMZ())
  print ("Peptide ID rt:", peptide_id.getRT())
  print ("Peptide scan index:",
peptide_id.getMetaValue("scan_index"))
  print ("Peptide scan name:",
peptide_id.getMetaValue("scan_index"))
  print ("Peptide ID score type:",
peptide_id.getScoreType())
  # PeptideHits
  for hit in peptide_id.getHits():
    print("  - Peptide hit rank:", hit.getRank())
    print("  - Peptide hit charge:", hit.getCharge())
    print("  - Peptide hit sequence:",
hit.getSequence())
    z = hit.getCharge()
    mz =
hit.getSequence().getMonoWeight(Residue.ResidueType.Full,
z) / z
    print("  - Peptide hit monois  otopic m/z:", mz)
    print("  - Peptide ppm error:", abs(mz    -
peptide_id.getMZ())/mz *10**6 )
    print("  - Peptide hit score:", hit.getScore())
```

We notice that the second peptide spectrum match (PSM) was found for the third spectrum in the file for a precursor at 775.38 *m*/*z* for the sequence `RPGADSDIGGFGGLFDLAQAGFR`.

```
    tsg = TheoreticalSpectrumGenerator()
    thspec = MSSpectrum()
    p = Param()
    p.setValue("add_metainfo", "true")
    tsg.setParameters(p)
    peptide =
AASequence.fromString("RPGADSDIGGFGGLFDLAQAGFR")
    tsg.getSpectrum(thspec, peptide, 1, 1)
    # Iterate over annotated ions and their masses
    for ion, peak in zip(thspec.getStringDataArrays()[0],
thspec):
        print(ion, peak.getMZ())

    e = MSExperiment()
    MzMLFile().load("searchfile.mzML", e)
    print ("Spectrum native id", e[2].getNativeID() )
    mz,i = e[2].get_peaks()
    peaks = [(mz,i) for mz,i in zip(mz,i) if i > 1500 and mz
> 300]
    for peak in peaks:
      print (peak[0], "mz", peak[1], "int")
```

Comparing the theoretical spectrum and the experimental spectrum for `RPGADSDIGGFGGLFDLAQAGFR` we can easily see that the most abundant ions in the spectrum are y8 (877.452 *m*/*z*), b10 (926.432), y9 (1024.522 *m*/*z*) and b13 (1187.544 *m*/*z*).

We can now use the OpenMS tool TOPPView for visualization of the mzML file (see Chapter 6 for more information on OpenMS and TOPPView). When loading the `searchfile.mzML` into the OpenMS visualization software TOPPView, we can convince ourselves that the observed spectrum indeed was generated by the peptide `RPGADSDIGGFGGLFDLAQAGFR` by loading the corresponding theoretical spectrum into the viewer using "Tools"->"Generate theoretical spectrum" (see Figure 16.2).

## 16.9 pyOpenMS in R

The R programming language is a powerful open-source statistical programming language that is often used in Bioinformatics. While Chapter 15 describes the available tools in R in greater detail, here we will briefly discuss how one can use an existing Python library, such as pyOpenMS, directly in R. Since there are no native wrappers for the OpenMS library in R, we will use the "reticulate" package in order to get access to the full functionality of pyOpenMS in the R programming language.

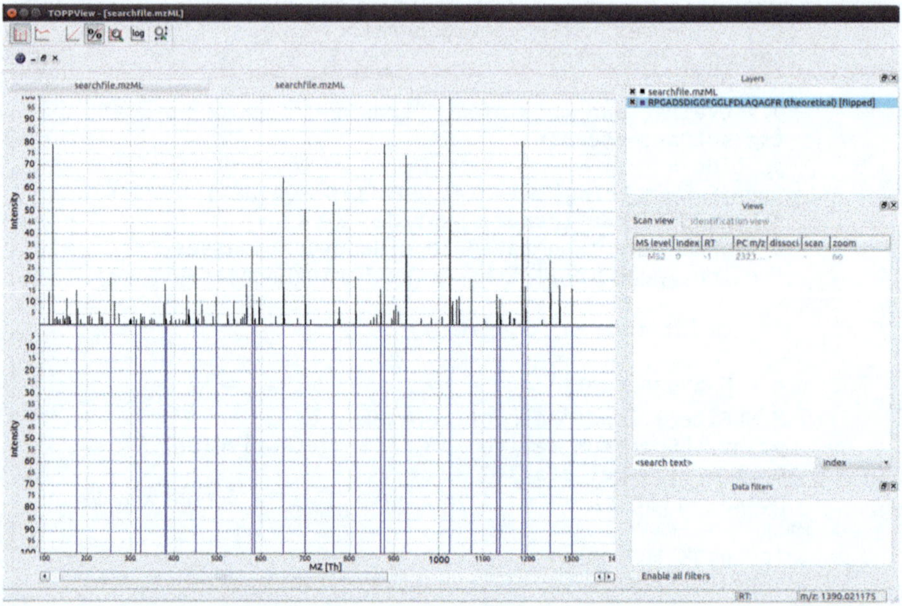


**Figure 16.2** *Peptide spectrum match, visualized with TOPPView.* A theoretical spectrum and an experimental spectrum are visualized together using the OpenMS software TOPPView.

## 16.9.1 Install the "reticulate" R Package

In order to use all pyopenms functionalities in R, we suggest you use the "reticulate" R package.

```
install.packages("reticulate")
```

Thorough documentation is available at: https://rstudio.github.io/reticulate/

Installation of pyopenms is a requirement as well and it is necessary to make sure that R is using the same Python environment.

In case R is having trouble finding the correct Python environment, you can set it by hand as in this example (using miniconda, you will have to adjust the file path to your system to make this work). You will need to do this before loading the "reticulate" library:

```
Sys.setenv(RETICULATE_PYTHON =
"/usr/local/miniconda3/envs/py37/bin/python")
```

Or after loading the "reticulate" library:

```
library("reticulate")

use_python("/usr/local/miniconda3/envs/py37/bin/python")
```

### 16.9.2 Import Pyopenms in R

After loading the "reticulate" library you should be able to import pyopenms into R.

```
library(reticulate)
ropenms=import("pyopenms", convert = FALSE)
```

This should now give you access to all of pyopenms in R. Importantly, the convert option has to be set to FALSE, since type conversions such as 64 bit integers will cause problems.

You should now be able to interact with the OpenMS library and, for example, read and write mzML files:

```
library(reticulate)
ropenms=import("pyopenms", convert = FALSE)
exp = ropenms$MSExperiment()
ropenms$MzMLFile()$store("testfile.mzML", exp)
```

which will create an empty mzML file called *testfile.mzML*.

## References

1. P. J. Cock, *et al.*, Biopython: freely available python tools for computational molecular biology and bioinformatics, *Bioinformatics*, 2009, **25**, 1422–1423.
2. S. Chaudhury, S. Lyskov and J. J. Gray, PyRosetta: a script-based interface for implementing molecular modeling algorithms using rosetta, *Bioinformatics*, 2010, **26**, 689–691.
3. E. Jones, T. Oliphant and P. Peterson, *et al.*, *SciPy: Open Source Scientific Tools for Python*, 2001.
4. T. Bald, J. Barth, A. Niehues, M. Specht, M. Hippler and C. Fufezan, pymzml—python module for high-throughput bioinformatics on mass spectrometry data, *Bioinformatics*, 2012, **28**, 1052–1053.
5. A. A. Goloborodko, L. I. Levitsky, M. V. Ivanov and M. V. Gorshkov, Pyteomics—a python framework for exploratory data analysis and rapid software prototyping in proteomics, *J. Am. Soc. Mass Spectrom.*, 2013, **24**, 301–304.

6. H. L. Röst, U. Schmitt, R. Aebersold and L. Malmström, pyOpenMS: A python-based interface to the OpenMS mass-spectrometry algorithm library, *Proteomics*, 2014, **14**, 74–77.
7. P. Kiefer, U. Schmitt and J. A. Vorholt, eMZed: an open source framework in python for rapid and interactive development of LC/MS data analysis workflows, *Bioinformatics*, 2013, **29**, 963–964.
8. W. M. Alexander, S. B. Ficarro, G. Adelmant and J. A. Marto, multiplierz v2. 0: A python-based ecosystem for shared access and analysis of native mass spectrometry data, *Proteomics*, 2017, **17**, 1700091.
9. L. I. Levitsky, J. A. Klein, M. V. Ivanov and M. V. Gorshkov, Pyteomics 4.0: five years of development of a python proteomics framework, *J. Proteome Res.*, 2018, **18**, 709–714.
10. M. Kösters, J. Leufken, S. Schulze, K. Sugimoto, J. Klein, R. Zahedi, M. Hippler, S. Leidel and C. Fufezan, pymzml v2. 0: introducing a highly compressed and seekable gzip format, *Bioinformatics*, 2018, **34**, 2513–2514.
11. A. Ressa, M. Fitzpatrick, H. Van den Toorn, A. J. Heck and M. Altelaar, Padua: A python library for high-throughput (phospho) proteomics data analysis, *J. Proteome Res.*, 2018, **18**, 576–584.
12. J. Klein and J. Zaia, psims-a declarative writer for mzml and mzidentml for python, *Mol. Cell. Proteomics*, 2019, **18**, 571–575.
13. D. H. May, K. Tamura and W. S. Noble, Detecting modifications in proteomics experiments with param-medic, *J. Proteome Res.*, 2019, **18**, 1902–1906.
14. L. P. Yunker, S. Donnecke, M. Ting, D. Yeung and J. S. McIndoe, PythoMS: A python framework to simplify and assist in the processing and interpretation of mass spectrometric data, *J. Chem. Inf. Model.*, 2019, **59**, 1295–1300.
15. M. K. łacki, M. Startek, D. Valkenborg and A. Gambin, Isospec: hyperfast fine structure calculator, *Anal. Chem.*, 2017, **89**, 3272–3277.

CHAPTER 17

# Mass Spectrometry Development Kit (MSDK): a Java Library for Mass Spectrometry Data Processing

TOMÁŠ PLUSKAL*[a], NILS HOFFMANN[b], XIUXIA DU[c] AND JING-KE WENG*[a,d]

[a]Whitehead Institute for Biomedical Research, 455 Main Street, Cambridge, MA 02142, USA; [b]Leibniz-Institut für Analytische Wissenschaften – ISAS – e.V., Otto-Hahn-Straße 6b, 44227 Dortmund, Germany; [c]University of North Carolina at Charlotte, Department of Bioinformatics and Genomics, 9331 Robert D. Snyder Rd, Charlotte, NC 28223, USA; [d]Massachusetts Institute of Technology, Department of Biology, 77 Massachusetts Ave, Cambridge, MA 02139, USA
*E-mail: pluskal@wi.mit.edu, wengj@wi.mit.edu

## 17.1 Introduction

The Mass Spectrometry Development Kit (MSDK)[1] is a Java library of algorithms for processing mass spectrometry data, with the main application being the development of data processing methods for metabolomics. The goals of the library are to provide a flexible data model with Java interfaces for mass-spectrometry-related objects (including raw mass spectra, detected features, compound identification results, spectral databases, *etc.*) and to

New Developments in Mass Spectrometry No. 8
Processing Metabolomics and Proteomics Data with Open Software: A Practical Guide
Edited by Robert Winkler

Published by the Royal Society of Chemistry, www.rsc.org

integrate existing algorithms for mass spectrometry data processing, compound identification, and database searching, into a single easy-to-deploy package. While competition among various data-processing approaches and algorithms is usually beneficial for the progress of research, the absence of a common data model and a development platform makes it difficult to compare and evaluate the results and performance of different algorithms.

The MSDK project began in 2015 as an offshoot of the graphical MZmine 2 mass spectrometry data-processing and visualization framework,[2] and a number of data-processing methods have been transferred from MZmine to MSDK. The MSDK library serves a similar purpose to the community of mass spectrometry Java developers as the OpenMS or ProteoWizard frameworks serve to the community of C/C++ developers.[3,4] The obvious advantages of Java are its multi-platform nature and robust support for automatic garbage collection, multi-threading, modularity, and encapsulation. Owing to the virtual machine abstraction of Java, a multitude of high-level, object-oriented programming languages, such as Scala, Kotlin, Clojure, Groovy, and Jython can be used to program applications based on MSDK. Although MSDK resembles the MzJava project[5] to some extent, the main difference is that MSDK is focused on data processing for metabolomics, whereas MzJava is heavily oriented towards proteomics.

## 17.2 Architecture

The MSDK library is separated into modules, based on the type or purpose of the implemented algorithms (Figure 17.1). Each module is built as a separate *.jar* package deployed to the Maven Central repository.[6] The core *msdk-datamodel* module defines the shared Java interfaces for all

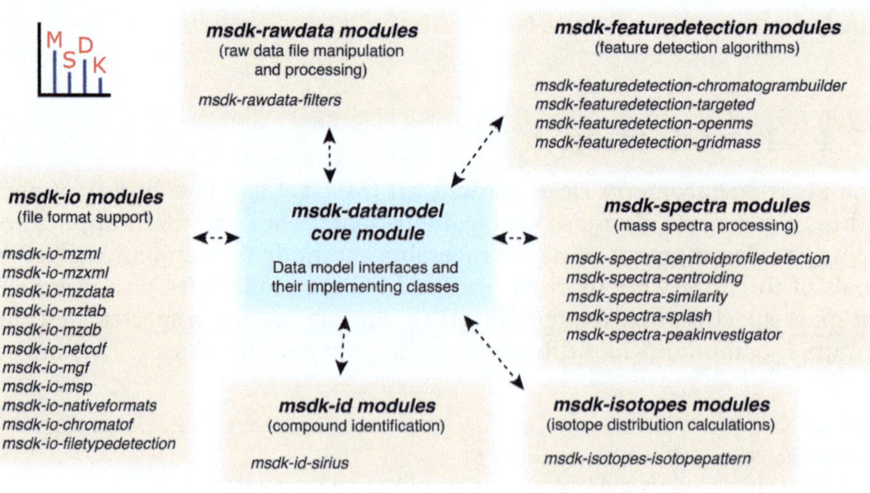


**Figure 17.1** An overview of the currently available MSDK modules. Each module is available as a separate Java *.jar* package from the Maven Central repository.[6] All modules interact through interfaces defined in the core *msdk-datamodel* package.

mass-spectrometry-related objects (*e.g.*, *RawDataFile* or *MsSpectrum*). The *msdk-datamodel* module also provides simple implementing classes for the interfaces (*e.g.*, *SimpleRawDataFile* or *SimpleMsSpectrum*). All other MSDK modules depend on the core *msdk-datamodel* module.

## 17.3 MSDK Modules

### 17.3.1 Raw Data File Format Support

The *msdk-io* modules contain support for parsing common open data formats (Table 17.1).[7] The *msdk-io-mzml* module provides support for both sequential and random access to individual spectra in the mzML format,[8] whereas other *msdk-io* modules are limited to sequential access. Convenience utility methods are available for automatic detection of file formats (in the *msdk-io-filetypedetection* module) and for heuristic classification of spectra as centroided or profile (in the *msdk-spectra-centroidprofiledetection* module).

### 17.3.2 Basic Spectra Processing Algorithms

The *msdk-rawdata-filters* module contains a selection of algorithms for raw spectra filtering, such as cropping or smoothing using rolling mean or Savitzky–Golay methods.[9] Centroiding of spectra can be performed with the *msdk-spectra-centroiding* module, which contains simple algorithms like binning or local maxima detection, as well as more sophisticated approaches like wavelet transform or the high-resolution "Exact mass" algorithm ported from MZmine 2.[2] The *msdk-spectra-splash module* provides a function to calculate the spectral hash (SPLASH) of a given mass spectrum, which can be used as a data-driven identifier of the spectrum for database submissions or publications.[10] The *msdk-spectra-similarity* module provides algorithms to calculate similarity scores of two different mass spectra, such as cosine

**Table 17.1** File formats currently supported by MSDK.

| File format | Type | Read/write support | MSDK module |
|---|---|---|---|
| mzML[8] | XML-based | Read & write | *msdk-io-mzml* |
| mzXML[27] | XML-based | Read only | *msdk-io-mzxml* |
| mzData[28] | XML-based | Read only | *msdk-io-mzdata* |
| mzTab[29] | tab-delimited text | Read & write | *msdk-io-mztab* |
| mzDB[30] | binary, SQLite-based | Read only | *msdk-io-mzdb* |
| netCDF (ANDI-MS)[31] | binary | Read & write | *msdk-io-netcdf* |
| MGF[32] | text-based | Read & write | *msdk-io-mgf* |
| MSP (NIST)[33] | text-based | Read & write | *msdk-io-msp* |
| Thermo .raw | binary, proprietary | Read only, using vendor library | *msdk-io-nativeformats* |
| Waters .raw | binary, proprietary | Read only, using vendor library | *msdk-io-nativeformats* |
| LECO ChromaTOF | comma-delimited text | Read only | *msdk-io-chromatof* |

similarity (also referred to as weighted dot product),[11] Manhattan distance, or SPLASH-based similarity measures. These algorithms are typically used for comparing experimental spectra to library records.[12]

### 17.3.3 Feature Detection

For the detection of features in chromatography-separated mass spectrometry datasets, MSDK provides the chromatogram builder algorithm contributed from MZmine 2 (the *msdk-featuredetection-chromatogrambuilder* module).[2] It also offers an interface for the *FeatureFinderMetabo* tool from the OpenMS package (the *msdk-featuredetection-openms* module).[13] Porting of the previously developed GridMass and ADAP feature detection methods is under way.[14,15] In the future, we hope that MSDK may aggregate many different feature detection algorithms under a common interface, thus allowing for easy switching and evaluation of the strengths and weaknesses of different methods.

### 17.3.4 Compound Identification

Compound identification is currently the principal bottleneck in mass-spectrometry-based metabolomics. The current trend in the field is to employ machine learning algorithms to predict chemical structures from high-resolution tandem mass spectra.[16] To this end, the *msdk-id-sirius* module provides an interface for the SIRIUS/CSI:FingerID algorithm,[17] which has been recognized as one of the most successful tools for mass spectrometry-based metabolite structure elucidation.[18]

## 17.4 Future Plans

Several open-source Java-based tools for mass spectrometry data-processing tools have been developed previously, with varying application scopes and sophistication. Many of them have since been abandoned by the original developers. Therefore, there is an opportunity to integrate the useful and promising algorithms under a common MSDK interface. This includes the GC-MS-processing methods in the tools Guineu,[19] Maltcms,[20] and Maui-VIA,[21] or CE-MS-processing methods in JDAMP.[22] In particular, Guineu and Maltcms include sophisticated algorithms for preprocessing, alignment and clustering of GCxGC-MS mass spectra, which will be included in MSDK soon. Furthermore, the algorithm-based nature and unified data model of MSDK modules allows for integration of these modules into a graphical workflow system such as KNIME, which is itself implemented in Java.[23] Development of such a KNIME-based mass spectrometry data-processing toolbox was previously attempted in the MassCascade project,[24] but the project was discontinued and does not work with recent versions of KNIME. Although MSDK already provides rich support for various file formats, the existing mzTab module will be extended to support the recently-introduced mzTab-M file format for reporting metabolomics results.[25]

## 17.5 Conclusions

The goal of the MSDK project is to provide a collection of algorithms that span the whole data processing workflow for mass spectrometry analysis of small molecules. MSDK currently includes support for many mass-spectrometry-related file formats and a selection of data processing algorithms that are actively used by other projects. Development of new MSDK modules is ongoing. Documentation for MSDK modules is available in the form of JavaDoc-generated API description at the MSDK website.[1] The source codes of the library are available on GitHub[26] either under the terms of the GNU Lesser General Public License version 2.1, or (per the licensee's choosing) under the terms of the Eclipse Public License v1.0 as published by the Eclipse Foundation. This dual licensing allows flexible integration of MSDK algorithms into both academic and commercial software packages. The binary artifacts (Java *.jar* files) of each MSDK module are deployed to the Maven Central repository.[6]

## Acknowledgements

We acknowledge the generous support of the Google Summer of Code program, which has funded the development of several MSDK modules through student projects. T.P. is a Simons Foundation Fellow of the Helen Hay Whitney Foundation. N.H. acknowledges funding and support by de.NBI, German Ministry of Education and Research, grant no. FKZ 031 L0108 A and by the Ministerium für Kultur und Wissenschaft des Landes Nordrhein-Westfalen and the Regierende Bürgermeister von Berlin - inkl. Wissenschaft und Forschung. This work is in part supported by the National Science Foundation (CHE-1709616 and MCB-1818132) and the Richard and Susan Smith Family Foundation. We thank Minura Sachinthana Dinith Kumara for donating the *msdk* username on GitHub.

## References

1. MSDK, https://msdk.github.io (accessed 15 May 2019).
2. T. Pluskal, S. Castillo, A. Villar-Briones and M. Oresic, *BMC Bioinf.*, 2010, **11**, 395.
3. H. L. Röst, T. Sachsenberg, S. Aiche, C. Bielow, H. Weisser, F. Aicheler, S. Andreotti, H.-C. Ehrlich, P. Gutenbrunner, E. Kenar, X. Liang, S.

Nahnsen, L. Nilse, J. Pfeuffer, G. Rosenberger, M. Rurik, U. Schmitt, J. Veit, M. Walzer, D. Wojnar, W. E. Wolski, O. Schilling, J. S. Choudhary, L. Malmström, R. Aebersold, K. Reinert and O. Kohlbacher, *Nat. Methods*, 2016, **13**, 741–748.
4. M. C. Chambers, B. Maclean, R. Burke, D. Amodei, D. L. Ruderman, S. Neumann, L. Gatto, B. Fischer, B. Pratt, J. Egertson, K. Hoff, D. Kessner, N. Tasman, N. Shulman, B. Frewen, T. A. Baker, M.-Y. Brusniak, C. Paulse, D. Creasy, L. Flashner, K. Kani, C. Moulding, S. L. Seymour, L. M. Nuwaysir, B. Lefebvre, F. Kuhlmann, J. Roark, P. Rainer, S. Detlev, T. Hemenway, A. Huhmer, J. Langridge, B. Connolly, T. Chadick, K. Holly, J. Eckels, E. W. Deutsch, R. L. Moritz, J. E. Katz, D. B. Agus, M. MacCoss, D. L. Tabb and P. Mallick, *Nat. Biotechnol.*, 2012, **30**, 918–920.
5. O. Horlacher, F. Nikitin, D. Alocci, J. Mariethoz, M. Müller and F. Lisacek, *J. Proteomics*, 2015, **129**, 63–70.
6. The Central Repository Search Engine, https://search.maven.org (accessed 27 May 2019).
7. E. W. Deutsch, *Mol. Cell. Proteomics*, 2012, **11**, 1612–1621.
8. L. Martens, M. Chambers, M. Sturm, D. Kessner, F. Levander, J. Shofstahl, W. H. Tang, A. Römpp, S. Neumann, A. D. Pizarro, L. Montecchi-Palazzi, N. Tasman, M. Coleman, F. Reisinger, P. Souda, H. Hermjakob, P.-A. Binz and E. W. Deutsch, *Mol. Cell. Proteomics*, 2011, **10**, R110.000133.
9. A. Savitzky and M. J. E. Golay, *Anal. Chem.*, 1964, **36**, 1627–1639.
10. G. Wohlgemuth, S. S. Mehta, R. F. Mejia, S. Neumann, D. Pedrosa, T. Pluskal, E. L. Schymanski, E. L. Willighagen, M. Wilson, D. S. Wishart, M. Arita, P. C. Dorrestein, N. Bandeira, M. Wang, T. Schulze, R. M. Salek, C. Steinbeck, V. C. Nainala, R. Mistrik, T. Nishioka and O. Fiehn, *Nat. Biotechnol.*, 2016, **34**, 1099–1101.
11. S. E. Stein and D. R. Scott, *J. Am. Soc. Mass Spectrom.*, 1994, **5**, 859–866.
12. A. Samokhin, K. Sotnezova, V. Lashin and I. Revelsky, *J. Mass Spectrom.*, 2015, **50**, 820–825.
13. E. Kenar, H. Franken, S. Forcisi, K. Wörmann, H.-U. Häring, R. Lehmann, P. Schmitt-Kopplin, A. Zell and O. Kohlbacher, *Mol. Cell. Proteomics*, 2014, **13**, 348–359.
14. V. Treviño, I.-L. Yañez-Garza, C. E. Rodriguez-López, R. Urrea-López, M.-L. Garza-Rodriguez, H.-A. Barrera-Saldaña, J. G. Tamez-Peña, R. Winkler and R.-I. Díaz de-la-Garza, *J. Mass Spectrom.*, 2015, **50**, 165–174.
15. A. Smirnov, W. Jia, D. I. Walker, D. P. Jones and X. Du, *J. Proteome Res.*, 2018, **17**, 470–478.
16. D. H. Nguyen, C. H. Nguyen and H. Mamitsuka, *Briefings Bioinf.*, 2019, **20**, 2028.
17. K. Dührkop, M. Fleischauer, M. Ludwig, A. A. Aksenov, A. V. Melnik, M. Meusel, P. C. Dorrestein, J. Rousu and S. Böcker, *Nat. Methods*, 2019, **16**, 299–302.
18. E. L. Schymanski, C. Ruttkies, M. Krauss, C. Brouard, T. Kind, K. Dührkop, F. Allen, A. Vaniya, D. Verdegem, S. Böcker, J. Rousu, H. Shen, H. Tsugawa, T. Sajed, O. Fiehn, B. Ghesquière and S. Neumann, *J. Cheminf.*, 2017, **9**, 22.

19. S. Castillo, I. Mattila, J. Miettinen, M. Orešič and T. Hyötyläinen, *Anal. Chem.*, 2011, **83**, 3058–3067.
20. N. Hoffmann, M. Keck, H. Neuweger, M. Wilhelm, P. Högy, K. Niehaus and J. Stoye, *BMC Bioinf.*, 2012, **13**, 214.
21. P. H. J. L. Kuich, N. Hoffmann and S. Kempa, *Front. Bioeng. Biotechnol.*, 2014, **2**, 84.
22. M. Sugimoto, A. Hirayama, T. Ishikawa, M. Robert, R. Baran, K. Uehara, K. Kawai, T. Soga and M. Tomita, *Metabolomics*, 2010, **6**, 27–41.
23. A. Fillbrunn, C. Dietz, J. Pfeuffer, R. Rahn, G. A. Landrum and M. R. Berthold, *J. Biotechnol.*, 2017, **261**, 149–156.
24. S. Beisken, M. Earll, D. Portwood, M. Seymour and C. Steinbeck, *Mol. Inf.*, 2014, **33**, 307–310.
25. N. Hoffmann, J. Rein, T. Sachsenberg, J. Hartler, K. Haug, G. Mayer, O. Alka, S. Dayalan, J. T. M. Pearce, P. Rocca-Serra, D. Qi, M. Eisenacher, Y. Perez-Riverol, J. A. Vizcaíno, R. M. Salek, S. Neumann and A. R. Jones, *Anal. Chem.*, 2019, **91**, 3302–3310.
26. msdk, https://github.com/msdk/msdk (accessed 15 May 2019).
27. P. G. A. Pedrioli, J. K. Eng, R. Hubley, M. Vogelzang, E. W. Deutsch, B. Raught, B. Pratt, E. Nilsson, R. H. Angeletti, R. Apweiler, K. Cheung, C. E. Costello, H. Hermjakob, S. Huang, R. K. Julian, E. Kapp, M. E. McComb, S. G. Oliver, G. Omenn, N. W. Paton, R. Simpson, R. Smith, C. F. Taylor, W. Zhu and R. Aebersold, *Nat. Biotechnol.*, 2004, **22**, 1459–1466.
28. S. Orchard, L. Montechi-Palazzi, E. W. Deutsch, P.-A. Binz, A. R. Jones, N. Paton, A. Pizarro, D. M. Creasy, J. Wojcik and H. Hermjakob, *Proteomics*, 2007, **7**, 3436–3440.
29. J. Griss, A. R. Jones, T. Sachsenberg, M. Walzer, L. Gatto, J. Hartler, G. G. Thallinger, R. M. Salek, C. Steinbeck, N. Neuhauser, J. Cox, S. Neumann, J. Fan, F. Reisinger, Q.-W. Xu, N. Del Toro, Y. Pérez-Riverol, F. Ghali, N. Bandeira, I. Xenarios, O. Kohlbacher, J. A. Vizcaíno and H. Hermjakob, *Mol. Cell. Proteomics*, 2014, **13**, 2765–2775.
30. D. Bouyssié, M. Dubois, S. Nasso, A. Gonzalez de Peredo, O. Burlet-Schiltz, R. Aebersold and B. Monsarrat, *Mol. Cell. Proteomics*, 2015, **14**, 771–781.
31. E13 Committee, *Specification for Analytical Data Interchange Protocol for Chromatographic Data*, ASTM International, West Conshohocken, PA, 2014.
32. Mascot database search | Data file format for mass spectrometry peak lists, http://www.matrixscience.com/help/data_file_help.html (accessed 15 May 2019).
33. S. E. Stein, *NIST Standard Reference Database 1A*, NIST, 2014.

CHAPTER 18

# MASSyPup64: Linux Live System for Mass Spectrometry Data Processing

ROBERT WINKLER*[a,b]

[a]Department of Biochemistry and Biotechnology, Center for Research and Advanced Studies (CINVESTAV) Irapuato, Km. 9.6 Libramiento Norte Carr. Irapuato-León, 36824 Irapuato, Gto., Mexico; [b]Mass Spectrometry Group, Max Planck Institute for Chemical Ecology, Hans-Knöll-Straße 8, 07745 Jena, Germany
*E-mail: robert.winkler@cinvestav.mx

## 18.1 Introduction

MASSyPup64 (http://www.bioprocess.org/massypup/) is a ready-to-use collection of programs and example data, built on a Linux live system, for the analysis of mass spectrometry data and data mining.[1] The initial purpose of building a Linux live system was to provide a common informatics set-up for MS data analysis courses.[2] However, the system can be run directly from a USB drive and can also be used to process MS data on local drives.

The first version of was based on 32 bit Puppy Linux (http://www.puppylinux.org/) and named 'MASSyPup'. Since most computers, now, support the 64 bit architecture, the latter versions of MASSyPup, namely 'MASSyPup64' are based on the 64 bit live Linux operating systems, such as Fatdog64 (http://distro.ibiblio.org/fatdog/web/) and antiX (https://antixlinux.com/).

New Developments in Mass Spectrometry No. 8
Processing Metabolomics and Proteomics Data with Open Software: A Practical Guide
Edited by Robert Winkler

Published by the Royal Society of Chemistry, www.rsc.org

MASSyPup64 is relatively easy to 'remaster', *i.e.*, the modified versions of the system can be compiled to a new, bootable ISO. Therefore, MASSyPup64 can be employed as an open platform for developing, testing, or distributing new MS software and/or data sets.

## 18.2 'Installation-free' MS Data Processing

The main hindrance to installing/distributing any new software is the dependency on various libraries and the operating system functions. One strategy to avoid installation problems for potential users is the development of web applications. These can be simple forms, such as ESIprot Online (http://www.bioprocess.org/esiprot/esiprot_form.php), or complex platforms, such as XCMS Online (http://xcmsonline.scripps.edu/). The downside of the web applications is the requirement to upload the data, which in the case of MS data, can be a massive load for the networks. This entails reliable internet connections. Another issue is that the source code of these web applications is usually hidden and users might not feel comfortable uploading unpublished 'private' data, more so, if the data are stored in a cloud. Finally, the risk of a possible discontinuation of the web service also has to be considered.

The distribution of precompiled binary versions of a software program, or an application that relies on 'cross-platform' programming languages, such as Java or Python, often suffers from compatibility issues, depending on the operating system or the software version. Therefore, all MASSyPup/MASSyPup64 variants were created by including the operating system, the necessary libraries and the programming languages (sometimes with different versions), so that they can run all the installed MS programs without any trouble.[2]

To provide a reproducible development/operating environment, Docker provides a standardised platform, on which the included applications run in so-called 'containers'. Such containers work on any operating system that is compatible with Docker, such as Linux, Microsoft® Windows 10® and MacOS. A variety of MS data-analysis tools are also available as 'Dockerised' versions (see Chapter 19), which facilitate the 'installation-free' use of MS data processing programs. However, Docker needs to be installed on the host operating system, whereas MASSYPup64 includes the operating system itself and can be used on any computer that allows booting from a USB drive.

Figure 18.1 depicts the conceptual difference between Docker and MASSyPup64. Docker runs separate program containers on a host operating system. In contrast, MASSyPup64 is a monolithic system that integrates the Linux operating system and any required libraries and programs, and runs directly on the host's hardware.

Thus, MASSyPup64 provides an alternative platform for computers running on operating systems that are incompatible with Docker (*e.g.*, Microsoft Windows with versions lower than 10) or where the software installation is restricted.

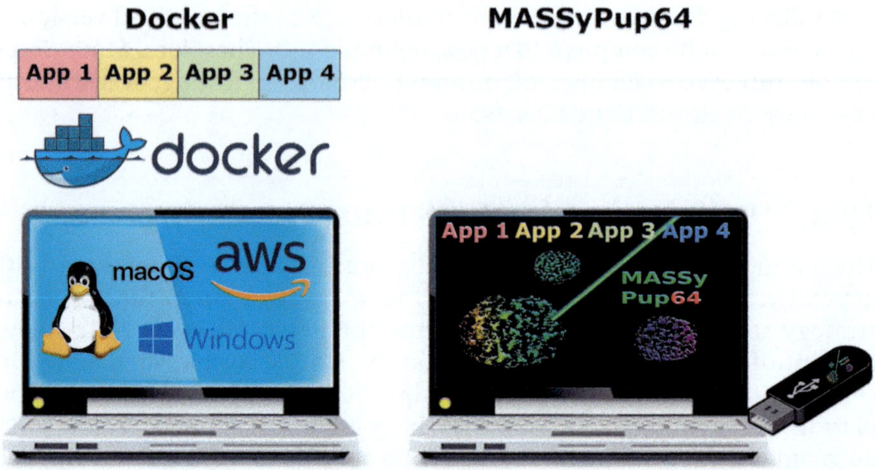


**Figure 18.1** Comparison of Docker and MASSyPup64. The latest release of MASSyPup64 is a hybrid system and provides Docker. Images in the public domain.

**Table 18.1** Programming tools included in MASSyPup64.

| Program | Description | Call | Reference |
|---|---|---|---|
| Docker | DevOps/Containerisation | `docker` | https://www.docker.com/ |
| OpenJDK | Open Java Development Kit | `java` | https://openjdk.java.net/ |
| Python 2 and 3 | Python programming language | `python/ python3` | https://www.python.org/ |

**Table 18.2** General MS data analysis tools included in MASSyPup64.

| Program | Description | Call | Reference |
|---|---|---|---|
| ESIprot | Molecular weight determination from multiply charged ions | `esiprot` | http://www.bioprocess.org/esiprot/ [3] |
| mMass | Versatile MS tool | `mmass` | http://mmass.org/ [4] |
| MZmine 2 | MS data processing (mainly LC-MS) | `mzmine` | http://mzmine.github.io/ [5] |

## 18.3 Programs Installed on MASSyPup64

The version *20190525_mp64_docker* is the first to include **Docker**. Thus, it combines the advantages of a pre-compiled system and a modular DevOps platform (see Chapter 19).

Many open source MS programs are based on **high-level programming languages**. The most popular ones are included in MASSyPup64 (see Table 18.1).

Some general MS programs have been installed for MS data visualisation and manual MS data processing (see Table 18.2).

For ESIprot, example data are installed in /usr/local/mp64.

Many MS data processing programs and workflows are based on the statistical computing and graphics environment **R** (https://www.r-project.org/). Therefore, a collection of **R packages and tools** has also been included (see Table 18.3).

Running Docker requires root access, which can be acquired by running the *sudo* command (sudo su; password 'mp64'). The Graphical User Interface (GUI) of the **Trans-Proteomic Pipeline – Petunia** (http://tools.proteomecenter.org/wiki/index.php?title=Software:TPP) can be run as a Docker image with the following command.

```
docker run -dit --user=root -p 10401:10401 -v /tmp/tppdata:/data

spctools/tpp apache2ctl -DFOREGROUND
```

After starting the container, Petunia is available from a web browser at http://localhost:10401/tpp/cgi-bin/tpp_gui.pl. To login, one can use the guest account (username: 'guest'; password: 'guest').

**Table 18.3** R packages and tools included in MASSyPup64.

| Program | Description | Call | Reference |
|---|---|---|---|
| CRAN R | Recent version of R | R | https://www.r-project.org/ |
| RStudio | (Integrated development environmet) | rstudio | https://www.rstudio.com/ |
| MALDIquantForeign | Reading/writing of MS data | R>library (MALDIquant Foreign) | http://www.strimmerlab.org/software/maldiquant/ |
| MALDIquant | Quantitative analysis of MS data | R>library (MALDIquant) | http://www.strimmerlab.org/software/maldiquant/ [6] |
| MetaboAnalystR | Metabolomics data processing | R>library (MetaboAnalystR) | https://github.com/xia-lab/MetaboAnalystR [7] |
| XCMS | Metabolomics data processing | R>library(xcms) | https://bioconductor.org/packages/release/bioc/html/xcms.html [8,9] |
| RmsiGUI | Mass spectrometry imaging platform | R>library (RmsiGUI) | https://bitbucket.org/lababi/rmsigui [10] |
| Rattle | Data Mining GUI for R | R>library (rattle) | https://rattle.togaware.com/ |

## 18.4 Preparing and Starting MASSyPup64

1 *Downloading the ISO file*

Different versions of MASSyPup(64) ISO files are deposited at Zenodo (http://doi.org/10.5281/zenodo.195529). In this tutorial, we use the most recent version, namely `20190525_mp64_docker.iso`. The link for the same is given below:

(https://zenodo.org/record/3228788/files/20190525_mp64_docker.iso?download=1)

2 *Creating a bootable USB stick*

A fast USB stick (USB 3.0) with at least 8 GB capacity is recommended for installing MASSyPup64. Please note that **all the previously existing data on the USB drive will be lost** during the installation. For creating a bootable USB stick from an ISO file, multiple tools are available for free, such as 'Rufus' for Windows (http://rufus.akeo.ie/).

However, the geek method to create a bootable USB drive (assuming a USB drive mounted on `/dev/sdb1`) is through using the Linux command line:

```
umount /dev/sdb1

mkfs.vfat /dev/sdb -I

dd if=./20190525_mp64_docker.iso of=/dev/sdb
```

3 *Booting from the MASSyPup64 USB stick*

Once the bootable USB stick is prepared, the computer can be restarted and booted from the USB drive. Depending on the BIOS configuration, it may be necessary to select the boot drive or change the boot order. The BIOS options can be entered by pressing the designated function key [F1–F12] or a special key [ESC, DEL] during the initial stage of the boot process. Some computers display a booting message about the key that needs to be pressed to enter the BIOS. Usually, the correct key can be easily found by trial and error or by a quick online search (for example, using the search string: "boot from usb lenovo thinkpad L380").

Furthermore, it may be necessary to review the boot security settings ('Booting mode UEFI/Legacy', 'allow booting from external drives', 'fast boot yes/no' *etc.*). In case the booting from the USB stick fails, a different tool for installing the ISO on the USB drive may be explored.

4 *Installing MASSyPup64 on a hard drive*

MASSyPup64 can also be installed on a hard drive (HD), for example, for quickly setting up a dedicated computer for the MS data analysis. It may also be installed alongside other operating systems, such as Microsoft Windows ('dual-boot' or 'multi-boot' systems).

All the versions of MASSyPup(64) provide a HD installer utility. For booting different operating systems, it is necessary to install an adequate **bootloader** such as Grub2 (https://www.gnu.org/software/grub/). If MASSyPup(64) is not recognised by a pre-existing boot manager, the bootloader must be installed in the MASSyPup(64) partition/volume boot record (PBR/VBR). This option is offered during the installation.MASSyPup64 requires a relatively small partition; typically, a disk space of 20 GB should be sufficient for a normal user. Please refer to the online resources about shrinking or resizing of the hard disk partitions for installing additional operating systems.

## 18.5 Example: Determination of Protein Weight with ESIprot

Once MASSyPup64 is booted, all the installed programs and data are available to the user. In addition, data on the mounted local drives can be accessed and processed, and the system can be modified. The standard user is called `'mp64'`, which is also the username, as well as the root password. Since the super user has elevated rights, the password should be changed on publicly accessible computers.

To test MASSyPup64, we first start MZmine from a console window by typing `mzmine` and then, open the data set `Cytochrome_C.mzML`, which can be found in the example data directory `/usr/local/mp64`.

Figure 18.2 shows the scan #16 of this data set, which represents the multiply charged *m/z* signals of a cythochrome C solution.

Next, ESIprot is started by typing `esiprot`. Now, the molecular weight of the major species in the protein preparation can be determined by simply entering the *m/z* values into the popup dialog presented by ESIprot and clicking the "Calculate MW" button (see Figure 18.2). The calculated value of 12 360.3 ± 0.7 Da fits the theoretical amino acid sequence with Heme-bis-L-cysteine and *N*-acetylglycine (theoretical molecular weight of 12 360.0 Da).[3]

## 18.6 Remastering the Live USB

The live system can be easily modified by the users to create their own flavours of MASSyPup64. The following procedure applies to the antiX (https://antixlinux.com/) version, MASSyPup64 release 20190525 and later.

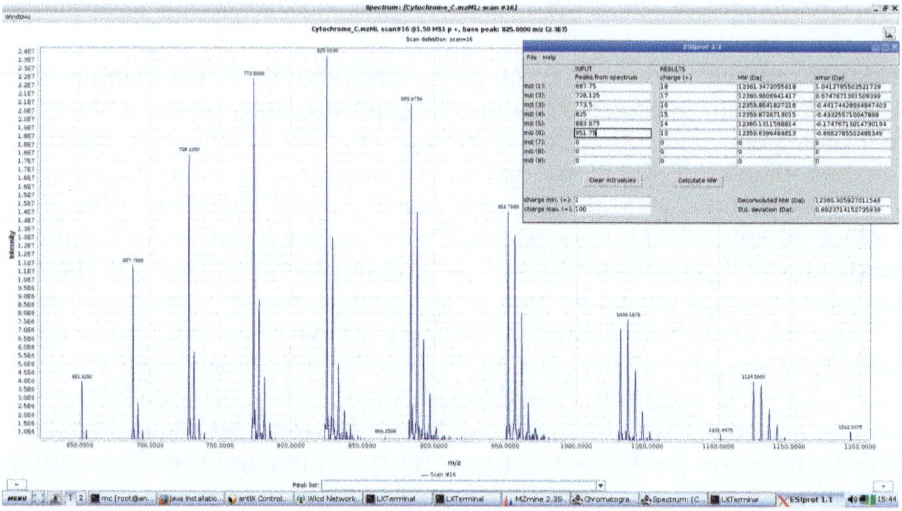


**Figure 18.2** Determination of the molecular weight of a protein from electrospray ionisation MS data, using MASSyPup64.

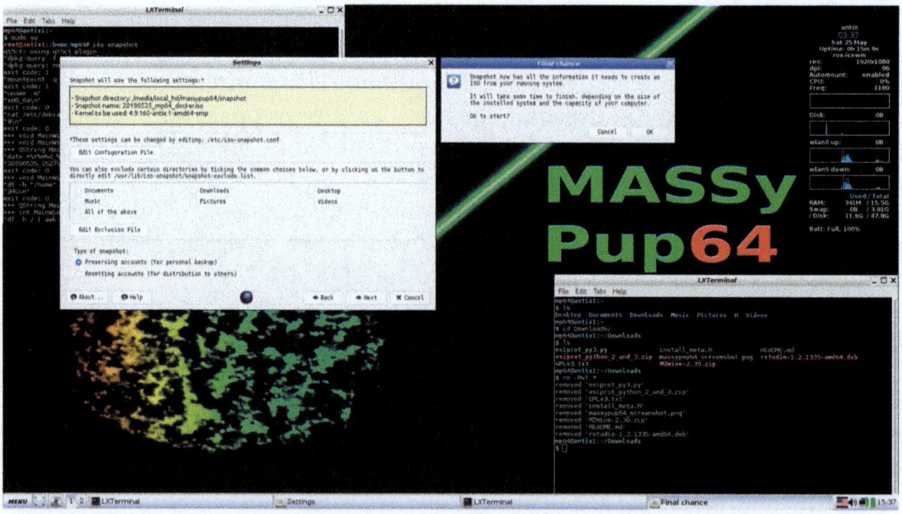


**Figure 18.3** Remastering MASSyPup64 with the `iso-snapshot` tool of antiX.

Assuming that a system is already installed on a hardware partition, additional software can be installed using the antiX tools or manually. The distribution is based on the stable version of Debian Linux, "stretch" (https://www.debian.org/). Therefore, official Debian packages can be installed with the `apt` command; for example, `apt install mmass` installs mMass (http://mmass.org/) and all the necessary dependencies.

Other programs, data sets, symbolic links *etc.* are preferably installed in the directory `/usr/local`.

After completing the modification of the system, the complete system can now be compiled to a new ISO file, by running the tool (as root): `iso-snapshot`.

Using the graphical user interface of the snapshot tool is a straightforward process (see Figure 18.3).

To keep the custom settings such as the desktop wallpaper, the account data have to be preserved. Private data, such as browser passwords *etc.*, have to be erased manually in this case.

## 18.7 Future of MASSyPup64

Providing stability and bleeding edge functions are two conflicting goals in software development. With a view to reaching a well-balanced compromise, MASSyPup64 is built on a stable Linux version, providing up-to-date MS software, programming libraries, as well as the Docker capability, for speedy adoption of the development versions. New technological advances will be integrated into MASSyPup64 in a timely manner to continuously provide a reliable and a modern platform for MS data processing.

## References

1. R. Winkler, An evolving computational platform for biological mass spectrometry: Workflows, statistics and data mining with MASSyPup64, *PeerJ*, 2015, **3**, 1–34.
2. R. Winkler, MASSyPup—an 'Out of the Box' solution for the analysis of mass spectrometry data, *J. Mass Spectrom.*, 2014, **49**, 37–42.
3. R. Winkler, ESIprot: A universal tool for charge state determination and molecular weight calculation of proteins from electrospray ionization mass spectrometry data, *Rapid Commun. Mass Spectrom.*, 2010, **24**, 285–294.
4. M. Strohalm, D. Kavan, P. Novák, M. Volný and V. Havlícek, mMass 3: A cross-platform software environment for precise analysis of mass spectrometric data, *Anal. Chem.*, 2010, **82**, 4648–4651.
5. T. Pluskal, S. Castillo, A. Villar-Briones and M. Orešič, MZmine 2: Modular framework for processing, visualizing, and analyzing mass spectrometry-based molecular profile data, *BMC Bioinf.*, 2010, **11**, 395.
6. S. Gibb and K. Strimmer, MALDIquant: A versatile R package for the analysis of mass spectrometry data, *Bioinformatics*, 2012, **28**, 2270–2271.
7. J. Chong, M. Yamamoto and J. Xia, MetaboAnalystR 2.0: From Raw Spectra to Biological Insights, *Metabolites*, 2019, **9**, 57.
8. C. A. Smith, E. J. Want, G. O'Maille, R. Abagyan and G. Siuzdak, XCMS: Processing mass spectrometry data for metabolite profiling using nonlinear peak alignment, matching, and identification, *Anal. Chem.*, 2006, **78**, 779–787.

9. R. Tautenhahn, C. Böttcher and S. Neumann, Highly sensitive feature detection for high resolution LC/MS, *BMC Bioinf.*, 2008, **9**, 504.
10. I. Rosas-Román, C. Ovando-Vázquez, A. Moreno-Pedraza, H. Guillén-Alonso and R. Winkler, Open LabBot and RmsiGUI: Community development kit for sampling automation and ambient imaging, *Michrochem. J.*, 2020, **152**, 104343.

CHAPTER 19

# *Cross-platform Software Development and Distribution with Bioconda and BioContainers*

YASSET PEREZ-RIVEROL*[a], OLIVIER SALLOU[b] AND BJÖRN A. GRÜNING[c]

[a]EMBL European Bioinformatics Institute, Cambridge, UK; [b]Institut de Recherche en Informatique et Systèmes Aléatoires (IRISA/INRIA) – GenOuest Platform, Université de Rennes, Rennes, France; [c]Bioinformatics Group, Department of Computer Science, University of Freiburg, Freiburg, 79110, Germany
*E-mail: yperez@ebi.ac.uk

## 19.1 Introduction

Modern science is built on automated data generation, software development, and large-scale data analysis. In recent years, data analysis has become more complex, moving from single applications to convoluted and integrated workflows,[1,2] creating two major challenges for software developers and the bioinformatics community: (i) software availability and (ii) reproducible experiments. The increasing use of complex workflows integrating several bioinformatics tools require substantial effort for correct

New Developments in Mass Spectrometry No. 8
Processing Metabolomics and Proteomics Data with Open Software: A Practical Guide
Edited by Robert Winkler

Published by the Royal Society of Chemistry, www.rsc.org

installation and configuration (*e.g.* conflicting system dependencies). To overcome these challenges, the bioinformatics community have created different groups, communities and platforms that define guidelines for best practices bioinformatics software development, deployment and deposition.[3]

Journal editors, funding agencies, and individual scientists have increasingly made calls for the scientific community to embrace best practices to support computational reproducibility.[4] Recently, Ioannidis *et al.* evaluated 18 published research studies that used computational methods to evaluate gene expression data, but they were able to reproduce only two of those studies.[5] Most of the failures were related with incomplete descriptions of software-based analyses and lack of description of the data analysis protocol. Recreating analyses that lack such details can require hundreds of hours of effort and may be impossible, even after consulting the original authors; which can lead to retractions.

Several groups highlighted that a good starting point for the replicability and reproducibility of the original results should be well-documented (software parameters, dependencies, *etc.*) and easily installable software.[6] Packaging and container technologies have emerged to overcome these challenges by automating the deployment of applications inside so-called software containers. Two main technologies have begun to change the field of software deployment in bioinformatics: Conda and Docker. Conda is an open-source package and environment management system for any programming language; though it is quite popular in the python community.[7] Docker containers provides an isolated environment for the installation and execution of a specific software, without affecting other parts of the system.[8,9] In 2015, two different groups explored for the first time the use of containers in bioinformatics.[9,10] However, they were limited to exploring the potential of Docker technology and not to define a complete ecosystem to tools to manage containers. In addition, these efforts were not scaled and promoted to a wide audience.

Between 2017 and 2018, the Bioconda and BioContainers communities joined efforts to create a platform that enables the generation, maintenance and deployment of bioinformatics containers into different platforms, operation systems and architectures such as Cloud, high-performance computing (HPC) or personal computers. New guidelines have been formalized to standardize the generation of containers and tools from bioinformatics software[11] to improve the quality and sustainability of research software based on software packages and containers. These guidelines encourage the adoption of packaging (*e.g.* Conda) and container (*e.g.* Docker, Singularity) technologies in bioinformatics and software development for research.[11]

In this chapter, we introduce the BioContainers (https://biocontainers.pro) and Bioconda communities to the new bioinformaticians, and data analysist arriving to the bioinformatics field. We introduce a clean and easy to understand framework that will make the published data analysis pipelines more reproducible and reusable by reviewers of journals and other bioinformaticians. Both communities have organized more than 800 contributors and collaborators that helps new bioinformaticians and developers to create their package, container and test them in different architectures. The community-driven approach of both communities guarantees the sustainability and scalability of the projects. At the time of writing, BioContainers provides more than 7900 containers that can be searched, tagged and accessed through a common web registry (https://biocontainers.pro).

## 19.2 From Tools to Bioconda Packages

The first step on the path to reproducible data analysis workflows is to make the software findable, accessible, interoperable, reusable (FAIR);[12] and in the context of bioinformatics software executable.[3] While the FAIR principles were originally designed for data, they are sufficiently general that their high-level concepts can be applied to bioinformatics software.[3]

The source of bioinformatics software should be available in public repositories such as GitHub or Bitbucket[3,13] (Figure 19.1). The use of public repositories guaranties long-term storage for the source, binary and additional files. While other alternatives such as university repositories and private FTPs tend to disappear in the long term.[3,8,13] The binaries, source code, the

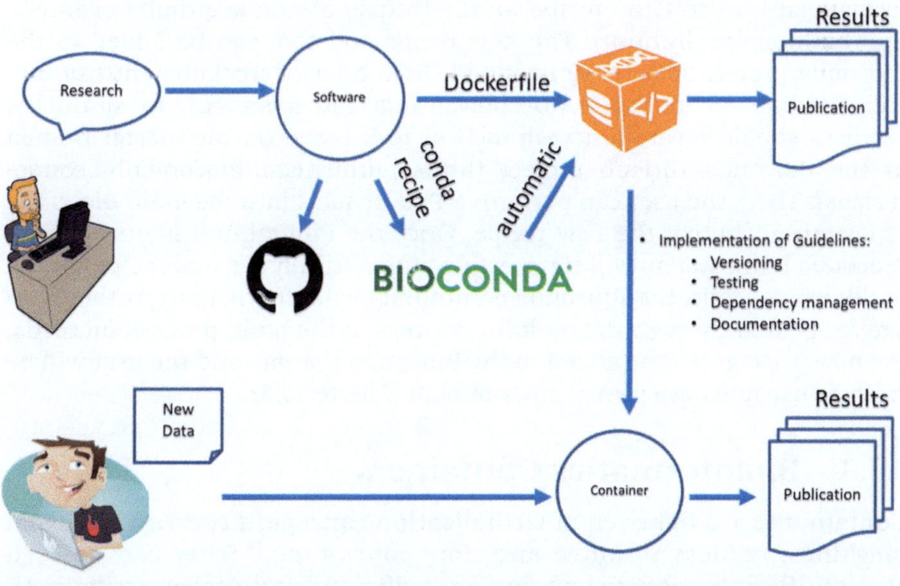


**Figure 19.1** Workflow to convert software research into reproducible pipelines.

license of the tool, and documentation pages should be provided in the software repository. In addition, it is important to make transparent how external developers can contribute to the code and project, the governance and communication processes.

We strongly recommend making the code of the bioinformatics tool open source in GitHub.[13] Open source code promotes trust among the software users, and encourages contribution from the community increasing software reuse and validation[3] (Figure 19.1).

For every bioinformatics software a Bioconda recipe can be created and the corresponding Conda package built.[7] The Conda package manager (https://conda.io) has become an increasingly popular approach to packaging software from different programming languages such as Python, Java or C++. Conda allows installations across language ecosystems by describing each software package with a recipe that defines meta-information and dependencies, as well as a build script that performs the steps necessary to build the software (Figure 19.2). Conda prepares and builds software packages within an isolated environment, transforming them into relocatable binaries. Conda are built for all three major operating systems: Linux, macOS, and Windows. Conda avoids reliance on specific systems by allowing users to generate isolated software environments, within tools that can be managed per-project, without generating conflicts or incompatibilities. These environments support reproducibility, as they can be rapidly exchanged.[7]

Before making your Conda recipe available to the community, the bioinformatician can test the recipe locally (https://bioconda.github.io/contributor/building-locally.html). The new recipe and tool can be added to the Bioconda project using a well-defined GitHub based workflow (https://bioconda.github.io/contributor/workflow.html#create-a-branch). In summary, the user should create a new branch or fork based on the master branch of the Bioconda GitHub project (https://github.com/bioconda/bioconda-recipes). Then, the user can perform a Pull Request into the main Bioconda repository including the new recipe. Once the Pull request is opened, the Bioconda build system will start testing automatically the new recipe.[7] If the build process fails, the Bioconda community will offer help to fix the package recipe and through the building process. If the build process succeeds, the new package will be added to the Bioconda channel and the user will be ready to use it using a virtual environment (Figure 19.3).

## 19.3 Bioinformatics Containers

Containers are a lightweight virtualization concept in contrast to virtual machines; *i.e.* less resource and time consuming.[14] They can be seen as more flexible tools for packaging, delivering and orchestrating both

```
{% set version = "1.0.beta" %}
{% set sha256 =
"f9b436c406f949b86c257a5efad20ff87371fbd33446533ea83afce9372ae878" %}

package:
  name: pepgenome
  version: {{ version }}

source:
  url: https://github.com/bigbio/pgatk/releases/download/PepGenome-
1.0-SNAPSHOT/PepGenome-1.0-SNAPSHOT-bin.zip
  sha256: {{ sha256 }}

build:
  noarch: generic
  number: 0

requirements:
  run:
    - openjdk  >=6
    - python

test:
  commands:
    - (pepgenome 2>&1 | grep -qF 'PepGenome' || (pepgenome && false))

about:
  home: https://github.com/bigbio/pgatk/
  license: Apache-2.0
  license_family: Apache
  summary: A java tool to map peptide and peptidoform evidences to
ENSEMBL Genome Coordinates

extra:
  notes: |-
    PepGenome is Java program that comes with a custom wrapper shell
script. For example, run it with "PepGenome -Xms512m -Xmx1g"
```

**Figure 19.2** Bioconda recipe for a proteo-genomics tool that enables mapping of peptides to genome coordinates. The recipe describes the package name, version, and requirements. In addition, it contains information about the software tool including license, short description and web page.

```
$ conda config --add channels conda-forge
$ conda config --add channels bioconda
$ conda install pepgenome
$ source activate my-env
```

**Figure 19.3** Conda installation process for a Bioconda package. The first two lines enables Conda to define the channels where the packages are stored. The third line installs the pepgenome tools from the Bioconda channel and the fourth line activates the virtual environment to be used.

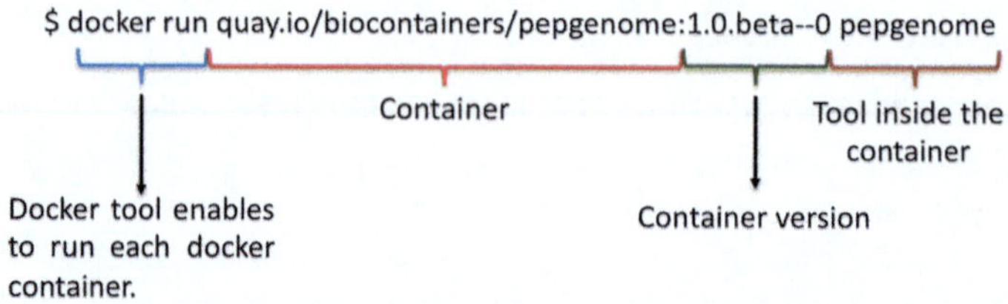


**Figure 19.4** Command-line description to run a BioContainer using docker. The BioContainer is composed of a container name and a version.

software services and applications. A container is a standard unit of software that packages up code and all its dependencies, so the application runs quickly and reliably from one computing environment to another.[14] Although lightweight operating system (OS) virtualization techniques like Solaris Zones and OpenVZ are long established, it was the release of Docker in March 2013 that led to mass adoption and even a hype around containerization. In recent years other companies such as CoreOS (https://coreos.com/rkt/) or Singularity[15] (https://sylabs.io/) have emerged to provided more secure containers and HPC solutions. Docker aims at making container technologies easy to use and among other things encourages a service-oriented architecture (SOA) and especially the microservice architecture. A principal concept of the cloud service-oriented architectures is the microservice or single tool; single container rule.[11]

The BioContainers community[8] has developed a platform to create bioinformatics containers from two main sources: (i) from Bioconda packages; (ii) from the original source code (Figure 19.1). For every Bioconda package, a container is generated automatically and push into a Biocontainers registry (https://quay.io/organization/biocontainers). Then, for the previous Bioconda package the user just need to pull and run the docker container (Figure 19.4).

A container can be also created from a Dockerfile recipe; a text document that contains all the commands a user could call on the command line to assemble an image (see example https://github.com/BioContainers/containers/blob/master/abacas/1.3.1-3-deb/Dockerfile). Using Docker build users can create an automated build that executes several command-line instructions in succession. The BioContainers community encourage creation of containers from Dockerfile only for those tools and services where it is complex to create a Bioconda package. The main reason for that decision is that the Conda package automatic build allows the community to have both the Conda package and BioContainers, while the Dockerfile only provides the container.[11]

## 19.4 Containers Deployment and Workflows

Recently, several authors have discussed the advantages of using containers in combination with workflow engines.[6,16] When running more than one container, an orchestration system is needed to coordinate and manage their execution and handle issues related to *e.g.* execution environment (HPC, Cloud); load balancing, health checks and scaling. A scientific workflow system/engine support the data analysis using several components and running a set of command-line tools in a sequential fashion, commonly on a computer cluster.[1] The main advantages of using scientific workflows engines include: (i) making automation of multi-step computations easier by combining independent tools; (ii) providing more transparency as to what the pipeline does through its more high-level workflow description and better reporting and visualization facilities, and (iii) providing robust policies to handle transient or persistent error conditions and including strategies to recover partial computations.[17]

Using containerized tools like BioContainers or Bioconda packages as the nodes in a scientific workflow has the advantage of adding isolation between processes and the complete encapsulation of tool dependencies within a container and reduces the burden to administer the system where the workflow is scheduled to run.[6,16] This means the system can be tested on a local computer and executed on a remote server and cluster without modification, using the exact same tool and environment.

Multiple workflows systems have adopted Conda and Docker as a way to handle software components deployment in different architectures. Galaxy[18] and Nextflow[19] are two excellent examples of workflow engines that natively use Conda, Docker and Singularity as the desired way to solve software tools. Figure 19.5, shows a simple peptide identification workflow using multiple search engines including MSGF+[20] and Tide.[21] Each Nextflow workflow process defines the container that will be used and the tool inside the container that will be executed. The configuration file (see example: https://github.com/bigbio/nf-workflows/blob/master/xt-msgf-nf/nextflow.config) defines how the workflow will be executed, the architecture (Cloud, HPC), queuing platform (LSF, Kubernetes). By using containers, the workflow engine doesn't need to commit efforts to resolve tool dependencies, tool version conflicts or tool incompatibilities.

Galaxy is annotating all tools with its corresponding requirements. Figure 19.6 shows a workflow where all tools are annotated with the specific version: openms (Version 2.3), xtandem (Version 15.12.15.2), fido (Version 1.0) and msgf_plus (Version 2017.07.21). These versions are then used by Galaxy to either create a Conda environment for this tool or retrieve a Container. A Galaxy admin can freely choose between Conda or Containers,

```
 process createDecoyDb {
        container 'quay.io/biocotnainers/searchgui:3.3.9--1'

        input:
        file "db.fasta" from fasta_file

        output:
        file "db_concatenated_target_decoy.fasta" into fasta_decoy_db

        script:
        """
        searchgui eu.isas.searchgui.cmd.FastaCLI -decoy -in db.fasta
        """
}

process searchMsgf {
        container 'quay.io/biocontainers/msgf_plus:2017.07.21--3'

        input:
        file "user.fasta" from fasta_decoy_db
        file msgf_fasta_index
        file mgf_file_msgf from mgf_files
        file msgf_mods

        output:
        file "*.mzid" into msgf_result, all_msgf_result

        script:
        """
        msgf_plus -d user.fasta -s ${mgf_file_msgf} -t 20ppm -ti 0,1 -thread
2 -tda 0 -inst 3 -e 1 -ntt 2 -mod ${msgf_mods} -minCharge 2 -maxCharge 4
        """
}

process searchTide {
        container 'quay.io/biocontainers/crux-toolkit:3.2--hfc679d8_0'

        input:
        file fasta_tide_index
        file mgf_file from mgf_files_2

        output:
        file "*.txt" into tide_result, all_tide_result

        script:
        """
        crux tide-search --parameter-file "${tide_config_file}"
"${mgf_file}" fasta-index
        """
}
```

**Figure 19.5** Small segment of a Nextflow workflow for peptide identification using multiple search engines (https://github.com/bigbio/nf-workflows/blob/master/xt-msgf-nf/main.nf). Each process, step defines the container that will be used, and the *script* defines the tool inside the container that will be executed.

depending on technical feasibility, *e.g.* access to Docker enabled HPC environments, or reproducibility needs.[6] Through our automatic procedures that produce and test Conda and BioContainers we guarantee that all best practice Galaxy workflows can either run in Conda or Containers.

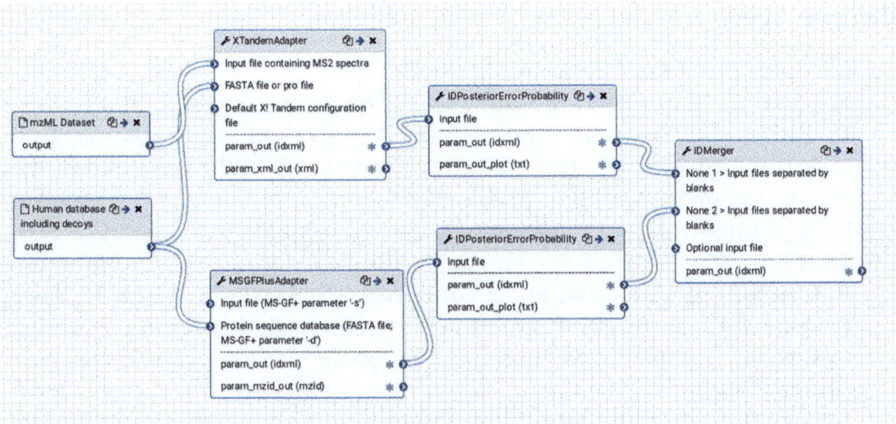


**Figure 19.6** Galaxy workflow for peptide and protein identification. The Galaxy framework provides a user interface to construct and design the workflow. A Galaxy admin can freely choose between Conda or Containers, depending on technical feasibility, *e.g.* access to Docker enabled HPC environments.

## 19.5 Conclusions

This chapter promotes and encourages the adoption of package and container technologies using the Bioconda and BioContainers communities; to improve the quality and reusability of research software. We introduced new bioinformaticians to container-based technologies by presenting simple Nextflow workflow using Biocontainers. We highlighted previously published papers including recommendations to produce containers from bioinformatics tools. The recommendations share a set of core views that are summarised: (i) for every tool in your data analysis pipeline a Conda package should be created; (ii) the encapsulated software should not be a complex environment of dependencies, tools and scripts; (iii) the developers of the software should be engaged or made aware of supporting the sustainability of the container; (iv) a tool container should be safe to reuse by any other workflow component or task through its access interface; (v) a tool container should perform a specific atomic task, so it is easier to check and use.

As with many other fields in bioinformatics, a learning curve lays ahead, but several basic yet powerful features are accessible even to the beginner and may be applied to many different use-cases. The Bioconda and BioContainers communities will help the new bioinformaticians to cross this learning curve. Both communities provided sufficiently documentations including the main help pages: Bioconda (https://bioconda.github.io/) and BioContainers (http://biocontainers-edu.biocontainers.pro/en/latest/).

## References

1. J. Leipzig, *Briefings Bioinf.*, 2017, **18**, 530–536.
2. S. Baichoo, Y. Souilmi, S. Panji, G. Botha, A. Meintjes, S. Hazelhurst, H. Bendou, E. Beste, P. T. Mpangase, O. Souiai, M. Alghali, L. Yi, B. D. O'Connor, M. Crusoe, D. Armstrong, S. Aron, F. Joubert, A. E. Ahmed, M. Mbiyavanga, P. V. Heusden, L. E. Magosi, J. Zermeno, L. S. Mainzer, F. M. Fadlelmola, C. V. Jongeneel and N. Mulder, *BMC Bioinf.*, 2018, **19**, 457.
3. R. C. Jimenez, M. Kuzak, M. Alhamdoosh, M. Barker, B. Batut, M. Borg, S. Capella-Gutierrez, N. Chue Hong, M. Cook, M. Corpas, M. Flannery, L. Garcia, J. L. Gelpi, S. Gladman, C. Goble, M. Gonzalez Ferreiro, A. Gonzalez-Beltran, P. C. Griffin, B. Gruning, J. Hagberg, P. Holub, R. Hooft, J. Ison, D. S. Katz, B. Leskosek, F. Lopez Gomez, L. J. Oliveira, D. Mellor, R. Mosbergen, N. Mulder, Y. Perez-Riverol, R. Pergl, H. Pichler, B. Pope, F. Sanz, M. V. Schneider, V. Stodden, R. Suchecki, R. Svobodova Varekova, H. A. Talvik, I. Todorov, A. Treloar, S. Tyagi, M. van Gompel, D. Vaughan, A. Via, X. Wang, N. S. Watson-Haigh and S. Crouch, *F1000Research*, 2017, **6**, 876.
4. S. R. Piccolo and M. B. Frampton, *GigaScience*, 2016, **5**, 30.
5. J. P. Ioannidis, D. B. Allison, C. A. Ball, I. Coulibaly, X. Cui, A. C. Culhane, M. Falchi, C. Furlanello, L. Game, G. Jurman, J. Mangion, T. Mehta, M. Nitzberg, G. P. Page, E. Petretto and V. van Noort, *Nat. Genet.*, 2009, **41**, 149–155.
6. B. Gruning, J. Chilton, J. Koster, R. Dale, N. Soranzo, M. van den Beek, J. Goecks, R. Backofen, A. Nekrutenko and J. Taylor, *Cell Syst.*, 2018, **6**, 631–635.
7. B. Gruning, R. Dale, A. Sjodin, B. A. Chapman, J. Rowe, C. H. Tomkins-Tinch, R. Valieris, J. Koster and T. Bioconda, *Nat. Methods*, 2018, **15**, 475–476.
8. F. da Veiga Leprevost, B. A. Gruning, S. Alves Aflitos, H. L. Rost, J. Uszkoreit, H. Barsnes, M. Vaudel, P. Moreno, L. Gatto, J. Weber, M. Bai, R. C. Jimenez, T. Sachsenberg, J. Pfeuffer, R. Vera Alvarez, J. Griss, A. I. Nesvizhskii and Y. Perez-Riverol, *Bioinformatics*, 2017, **33**, 2580–2582.
9. F. Moreews, O. Sallou, H. Menager, Y. Le Bras, C. Monjeaud, C. Blanchet and O. Collin, *F1000Research*, 2015, **4**, 1443.
10. P. Belmann, J. Droge, A. Bremges, A. C. McHardy, A. Sczyrba and M. D. Barton, *GigaScience*, 2015, **4**, 47.
11. B. Gruening, O. Sallou, P. Moreno, F. da Veiga Leprevost, H. Ménager, D. Søndergaard, H. Röst, T. Sachsenberg, B. O'Connor, F. Madeira, V. Dominguez Del Angel, M. Crusoe, S. Varma, D. Blankenberg, R. Jimenez and Y. Perez-Riverol, *F1000Research*, 2019, **7**, 742.
12. M. D. Wilkinson, M. Dumontier, I. J. Aalbersberg, G. Appleton, M. Axton, A. Baak, N. Blomberg, J. W. Boiten, L. B. da Silva Santos, P. E. Bourne, J. Bouwman, A. J. Brookes, T. Clark, M. Crosas, I. Dillo, O. Dumon, S. Edmunds, C. T. Evelo, R. Finkers, A. Gonzalez-Beltran, A. J. Gray, P. Groth,

C. Goble, J. S. Grethe, J. Heringa, P. A. t Hoen, R. Hooft, T. Kuhn, R. Kok, J. Kok, S. J. Lusher, M. E. Martone, A. Mons, A. L. Packer, B. Persson, P. Rocca-Serra, M. Roos, R. van Schaik, S. A. Sansone, E. Schultes, T. Sengstag, T. Slater, G. Strawn, M. A. Swertz, M. Thompson, J. van der Lei, E. van Mulligen, J. Velterop, A. Waagmeester, P. Wittenburg, K. Wolstencroft, J. Zhao and B. Mons, *Sci. Data*, 2016, **3**, 160018.
13. Y. Perez-Riverol, L. Gatto, R. Wang, T. Sachsenberg, J. Uszkoreit, V. Leprevost Fda, C. Fufezan, T. Ternent, S. J. Eglen, D. S. Katz, T. J. Pollard, A. Konovalov, R. M. Flight, K. Blin and J. A. Vizcaino, *PLoS Comput. Biol.*, 2016, **12**, e1004947.
14. D. Bernstein, *IEEE Cloud Comput.*, 2014, **1**, 81–84.
15. G. M. Kurtzer, V. Sochat and M. W. Bauer, *PLoS One*, 2017, **12**, e0177459.
16. Y. Perez-Riverol and P. Moreno, *Proteomics*, 2020, 1900147.
17. O. Spjuth, M. Capuccini, M. Carone, A. Larsson, W. Schaal, J. Novella, P. Di Tommaso, C. Notredame, P. Moreno and P. E. Khoonsari, *Approaches for Containerized Scientific Workflows in Cloud Environments With Applications in Life Science*, Report 2167–9843, PeerJ Preprints, 2018.
18. E. Afgan, D. Baker, B. Batut, M. van den Beek, D. Bouvier, M. Cech, J. Chilton, D. Clements, N. Coraor, B. A. Gruning, A. Guerler, J. Hillman-Jackson, S. Hiltemann, V. Jalili, H. Rasche, N. Soranzo, J. Goecks, J. Taylor, A. Nekrutenko and D. Blankenberg, *Nucleic Acids Res.*, 2018, **46**, W537–W544.
19. P. Di Tommaso, M. Chatzou, E. W. Floden, P. P. Barja, E. Palumbo and C. Notredame, *Nat. Biotechnol.*, 2017, **35**, 316–319.
20. S. Kim and P. A. Pevzner, *Nat. Commun.*, 2014, **5**, 5277.
21. S. McIlwain, K. Tamura, A. Kertesz-Farkas, C. E. Grant, B. Diament, B. Frewen, J. J. Howbert, M. R. Hoopmann, L. Kall, J. K. Eng, M. J. MacCoss and W. S. Noble, *J. Proteome Res.*, 2014, **13**, 4488–4491.

# Part C

# Conclusion

CHAPTER 20

# *Concluding Remarks and Perspectives*

ROBERT WINKLER*[a,b]

[a]Department of Biochemistry and Biotechnology, Center for Research and Advanced Studies (CINVESTAV) Irapuato, Km. 9.6 Libramiento Norte Carr. Irapuato-León, 36824 Irapuato, Gto., Mexico; [b]Mass Spectrometry Group, Max Planck Institute for Chemical Ecology, Hans-Knöll-Straße 8, 07745 Jena, Germany
*E-mail: robert.winkler@cinvestav.mx

This book provides a snapshot of currently available open-source software (OSS) for the analysis of mass spectrometry (MS) data, with a primary focus on metabolomics and proteomics. Different tools can be modified and reshuffled to customise workflows to best suit individual needs. The ongoing development and integration of analytical methods, such as combined LC-MS/NMR metabolomics, provides new challenges for data handling and understanding.[1]

The evaluation of multidimensional datasets, including (for example) ion mobility separation (IMS), different types of MS fragmentation, and NMR data, and the merging of different levels of information, such as those derived from genomics, metabolomics, proteomics, and phenomics, are becoming severe computational problems. To generate meaningful knowledge rather than just 'big data', it is therefore crucial to (a) strictly apply the criteria of *The Scientific Method*,[2] and (b) employ advanced statistics and data mining strategies. Supervised and unsupervised machine learning algorithms will

New Developments in Mass Spectrometry No. 8
Processing Metabolomics and Proteomics Data with Open Software: A Practical Guide
Edited by Robert Winkler

Published by the Royal Society of Chemistry, www.rsc.org

play key roles in dealing with complex tasks, such as predicting chemical structure from MS data and interpreting massive omics datasets.[3,4]

On the other hand, changes in MS software development and distribution paradigms will be most influential in moving forward. OSS fits well into modern strategies of integrated software development/testing/deployment; *i.e.*, the 'DevOps' concept. A large community of users and programmers is now collaborating on the construction of OSS MS data analysis tools and platforms. Fortunately, the spread of OSS has not led to a reduction in high-quality standards; it also does not oppose commercial products, since multi-licensing models offer professional services for OSS platforms, thereby combining the advantages of both worlds.

# References

1. J. L. Markley, R. Brüschweiler, A. S. Edison, H. R. Eghbalnia, R. Powers, D. Raftery and D. S. Wishart, The future of NMR-based metabolomics, *Curr. Opin. Biotechnol.*, 2017, **43**, 34–40.
2. K. R. Popper, *The Logic of Scientific Discovery*, Basic Books, Oxford, England, 1959.
3. K. Dührkop, M. Fleischauer, M. Ludwig, A. A. Aksenov, A. V. Melnik, M. Meusel, P. C. Dorrestein, J. Rousu and S. Böcker, SIRIUS 4: A rapid tool for turning tandem mass spectra into metabolite structure information, *Nat. Methods*, 2019, **16**, 299.
4. C. S. Greene, J. Tan, M. Ung, J. H. Moore and C. Cheng, Big data bioinformatics, *J. Cell. Physiol.*, 2014, **229**, 1896–1900.

# Subject Index